Dynamics of Vortex Structures in a Stratified Rotating Fluid

ATMOSPHERIC AND OCEANOGRAPHIC SCIENCES LIBRARY

VOLUME 47

For further volumes:
http://www.springer.com/series/5669

Mikhail A. Sokolovskiy • Jacques Verron

Dynamics of Vortex Structures in a Stratified Rotating Fluid

Springer

Mikhail A. Sokolovskiy
RAS, Water Problems Institute
Moscow, Russia

Jacques Verron
CNRS
Grenoble, France

This book is an extended edition of the translation of the book in Russian "Dinamika vikhrevykh structur v stratifitsirovannoy vraschayuscheysya zhidkosti" by M.A. Sokolovskiy and J. Verron. The book was published originally in Russian by Publishing House of Izhevsk Institute of Computer Science, Moscow in 2011.

Translated from Russian by Gennady N. Krichivets and Olga I. Yakovenko.

ISSN 1383-8601
ISBN 978-3-319-37564-9 ISBN 978-3-319-00789-2 (eBook)
DOI 10.1007/978-3-319-00789-2
Springer Cham Heidelberg New York Dordrecht London

Printed on acid-free paper

Springer is part of Springer Science+Business Media (www.springer.com)

Dedicated

to the blessed memory of
Professor Vadim Fedorovich Kozlov,
the Founder of Far Eastern School
of Geophysical Fluid Dynamics,
Vladivostok, Russia
(1933–2005)

and

to Emil Hopfinger,
Directeur de Recherche CNRS,
Ex-Director of Laboratoire des Ecoulements
Géophysiques et Industriels,
Grenoble, France

Foreword

This monograph is an expanded English translation of the Russian book published in 2011 by the Izhevsk Institute of Computer Science Publishing House under the same name and reviewed by Dr. G.M. Reznik.

The objective of this book is studying, in the framework of quasi-geostrophy, the role of stratification in the synoptic-scale vortex dynamics problems for atmosphere and ocean, and the effect of a bottom topography on large/meso-scale currents and vortices. The book summarizes the long-term joint studies of the authors in vortex dynamics of stratified rotating fluid.

The book is intended for experts in physical oceanography and meteorology, hydrodynamics, dynamical systems, and teachers, post-graduate students, and students in those fields.

Mikhail Sokolovskiy
Water Problems Institute
Russian Academy of Science
3, Gubkina Str., 119333
Moscow, Russia
sokol@aqua.laser.ru

Jacques Verron
Laboratoire des Ecoulements
Géophysiques et Industriels
UMR 5519, CNRS, BP53 X
38041, Grenoble Cedex, France
jacques.verron@legi.grenoble-inp.fr

Moscow-Grenoble
March 2013

Preface

This book is a summary of our 18-year period of joint work in the field of vortex dynamics in a stratified rotating fluid under INTAS 94-3614 (1995–1997), INTAS/AIRBUS 04-80-7297 (2009–2011), PICS 5805 (2011–2013) as well as within the framework of Groupement de Recherche Europeen (GDRE) "Regular and chaotic hydrodynamics" (2006–2009) and during mutual scientific visits in Grenoble and Moscow (1996–2012).

This monograph is an expanded English translation of the Russian book [858] published in 2011 by the Izhevsk Institute of Computer Science Publishing House under the same name and reviewed by Dr. G. M. Reznik. The translation was made by G. N. Krichivets and O. I. Yakovenko.

We have made the following changes as compared with the Russian edition of the book: Sects. 3.2.2 and 3.5.2, Chap. 4, Appendices A and B, and Index are added; Introduction, Sects. 2.2.2.1 and 3.5.3, and the list of references are expanded, most pictures are now in color, and found misprints are corrected. Appendix A is written by Emil Hopfinger. It covers the description of laboratory experiments on heton interaction which partially fit our theoretical results. In Appendix B, Mikhail Sokolovskiy tells a short biography of his teacher, Professor Vadim F. Kozlov.

Part of the results has been obtained in collaboration with V. F. Kozlov[††], V. M. Gryanik, K. V. Koshel, V. G. Makarov, Z. Kizner, V. N. Zyryanov, X. Carton, P. A. Davies, E. Hopfinger, S. Valcke, B. N. Filyushkin, I. M. Vagina, and N. G. Kozhelupova, whose contributions to the work and kind permission to include their results in the book are much appreciated.

We are also grateful to A. V. Aksenov, M. S. Apfel'baum, H. Aref[††], E. N. Benilov, V. L. Berdichevsky, A. V. Borisov, H. Borth, V. I. Byshev, Yu. D. Chashechkin, V. A. Cherkashin[††], S. M. Corréard, V. B. Darnitskiy, E. N. Dolgopolova, F. V. Dolzhanski[††], T. N. Doronina, E. V. Ermanyuk, A. I. Fel'zenbaum[††], J.-B. Flór, Y. Fukumoto, R. F. Ganiev, A. I. Ginzburg, L. V. Gogish, A. N. Golubyatnikov, A. V. Gotovtsev, S. K. Gulev,

[††]Deceased.

A. Yu. Gurulev, M. A. Guzev, L. Kh. Ingel', V. M. Kaistrenko, M. V. Kalashnik, M. G. Khublaryan[††], R. Khvoles, T. R. Kil'matov, V. I. Klyatskin, G. K. Korotaev, M. N. Koshlyakov, A. G. Kostyanoy, V. V. Kozlov, E. A. Kulikov, L. G. Kurakin, N. P. Kuzmina, S. S. Lappo[††], L. Ya. Lyubavin, I. S. Mamaev, V. V. Meleshko[††], A. P. Mirabel, T. Miyazaki, S. V. Muzylev, P. K. Newton, R. I. Nigmatulin, V. V. Novotryasov, V. L. Okulov, N. Paldor, G. N. Panin, E. G. Pavia, E. N. Pelinovsky, M. S. Permyakov, X. Perrot, S. V. Prants, V. V. Pukhnachov, A. B. Rabinovich, J. N. Reinaud, G. M. Reznik, K. A. Rogachev, P. B. Rutkevich, E. A. Ryzhov, E. A. Sagomonyan, G. I. Shapiro, A. I. Shavlyugin, G. V. Shevchenko, J. Sommeria, D. V. Stepanov, G. G. Sutyrin, T. G. Talipova, M. V. Tevs, D. V. Treschev, O. O. Trusenkova, A. N. Vul'fson, O. I. Yakovenko, V. I. Yudovich[††], A. G. Zatsepin, P. O. Zavialov, V. V. Zhmur and D. V. Zyryanov for useful discussions of some problems considered in this book.

We express our gratitude to V. M. Gryanik, Z. Kizner, K. V. Koshel and O. I. Yakovenko for helpful comments on the Russian version of this book.

We are extremely grateful to Emil Hopfinger who kindly agreed to write Appendix A.

The authors hope that the presented results will be of interest to experts in geophysical fluid dynamics and physical oceanography. We will appreciate critical analysis of the results of our work.

Moscow-Grenoble
March 2013

Mikhail Sokolovskiy
Jacques Verron

Contents

Chapter 1
The Introductory Chapter

Abstract In Sect. 1.1 of this chapter we explain the subject of Geophysical Fluid Dynamics, and give the description of main vortex structures that have become objects of the present book: (a) *Heton* – a two-layer vortex with opposite rotations in different layers, and (b) *Intrathermocline lens*, which is studied in this work as a vortex patch in the intermediate layer of a three-layer ocean model. In this section, we propose also a short review of works in the related topics. Section 1.2 is a mathematical introduction. It contains all main formulae which are used further in Chaps. 2 and 3.

1.1 Introduction

The theory of vortex motion of an inviscid incompressible fluid [11, 12, 66, 82, 87, 169, 285, 308, 438, 485, 490, 504, 510, 511, 541, 548, 565, 593, 595, 610, 625, 673, 713–716, 744, 745, 788, 789, 808, 890, 911, 950, 987], going back to the classical works of Helmholtz, Gröbli, Kirchhoff, Rankin, Greenhill, Taylor, Poincaré [see 37, 102, 640, 702], has developed mainly due to the need to understand the properties of atmospheric cyclones and anticyclones. Indeed, simple two-dimensional hydrodynamic models of discrete vortices provide some insight on the type of interaction occurring between elementary vortices and into the structure of the velocity field they induce.

However, many effects intrinsic to motion (especially, vortex motion) in the atmosphere or ocean cannot be explained without allowance being made for the rotation of the medium as a whole and the heterogeneity (stratification) of the density field that forms under the effect of gravity. The solution of some important hydrodynamic problems of a planetary nature has become possible with the development of *Geophysical Fluid Dynamics* (GFD) [40, 82, 196, 242, 295, 304, 403, 404, 449, 583, 584, 599, 624, 626, 689, 690, 743, 759, 798, 822, 858, 896, 922, 969, 977, 988] dealing with this class of problem (an independent branch of hydromechanics which has developed over the last three decades), the setting of domains of

M.A. Sokolovskiy and J. Verron, *Dynamics of Vortex Structures in a Stratified Rotating Fluid*, Atmospheric and Oceanographic Sciences Library 47,
DOI 10.1007/978-3-319-00789-2_1,

flow parameters dominated by vortex motions [143, 169, 194, 239, 318, 357, 798], and the generalization of methods of Hamiltonian dynamics to the description of geophysical processes [102, 304, 404, 822]. Thus, within the context of the quasi-geostrophic approximation, which is valid for fast rotating stably stratified fluids [404, 689], Gryanik and co-authors successively created the theories of discrete vortices for a two-layer fluid [315], a stratified medium comprising an arbitrary number of uniform-density layers [327–329], and a continuously stratified fluid [316, 317]. The contour dynamics method (CDM) has been developed by Kozlov et al. [477] for the description of finite-size vortices in a two-layer fluid and generalized later for three layers [844–846] and for the case of continuous stratification [457]. These works formed the basis for new studies, the results of which are partially reflected in this publication.

This book focuses substantially on the analysis of the dynamics of both discrete and distributed baroclinic vortices with zero total intensity in a two-layer medium (with constant density within each layer), since the two-layer model has the main features of the large-scale (mesoscale) dynamics of the atmosphere and ocean [404, 689]. Some results deal with hetons in a three-layer fluid. Unlike classical (barotropic) vortices in an ideal fluid, the baroclinic vortices possess a reserve of not only kinetic but also available potential (thermal) energy. As shown in [315], the baroclinic nature of vortices radically changes both the structure of the velocity fields they induce and the character of vortex interactions. Structures comprising two vortices with zero total intensity feature an important *self-motion* property (a two-layer pair moving without changing in shape or intensity [315]). In particular, in the case of two point vortices concentrated in different layers of a two-layer fluid and having equal intensities with opposite signs, each induces a rectilinear and uniform motion in the other.

The notion of *heton* was introduced by Hogg and Stommel [350] with the aim of emphasizing the ability of a baroclinic vortex pair to transfer heat. The word "heton" is derived from "heat". Indeed, when the geostrophic and hydrostatic approximations [404, 574, 689], conventional for GFD, are valid, any vortex of the top (bottom) layer, which has negative (positive) intensity, induces a downward local distortion of the interface between the layers. Such situation is referred to as a "warm heton". When the vorticity sign of each vortex changes, the sign of the curvature of the interface also changes, the situation turning into a "cold heton".

Since under conditions of stable stratification, the bottom layer should be denser and colder, the integral amount of heat in the domain containing vortices of this type will clearly be anomalous to that in any domain with the same volume. It is therefore obvious that the motion of hetons consisting of combinations of oppositely rotating vortices has a greater effect on the redistribution of heat (heat and salt, in the case of the ocean) than the motion of any other vortex structures. The notion of "heton" is also used to refer to vortices with finite horizontal dimensions (vertical patches, i.e., domains with constant values of potential vorticity Π_1 and Π_2 in the top and bottom layer, respectively). When the centers of the vortex patches in different layers are vertically aligned, the heton is said to have a "vertical" axis, otherwise the axis is said to be "tilted" (Fig. 1.1). The possible formation mechanisms of a warm (cold)

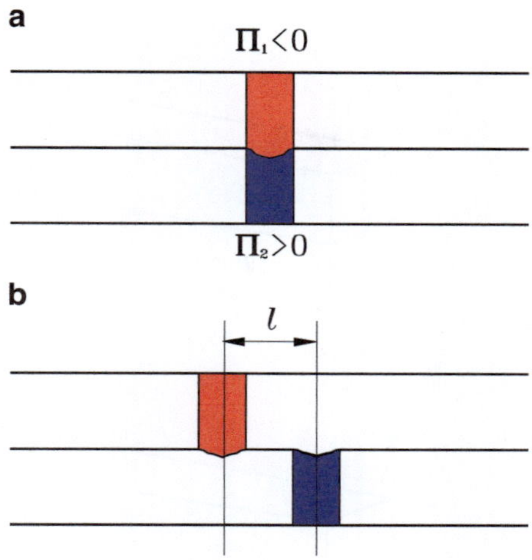

Fig. 1.1 Schematic representation of a two-layer distributed heton with a vertical (**a**) and tilted axis (**b**). Π_1 and Π_2 are the potential vorticities for the top (*red*) and bottom (*blue*) layers, respectively

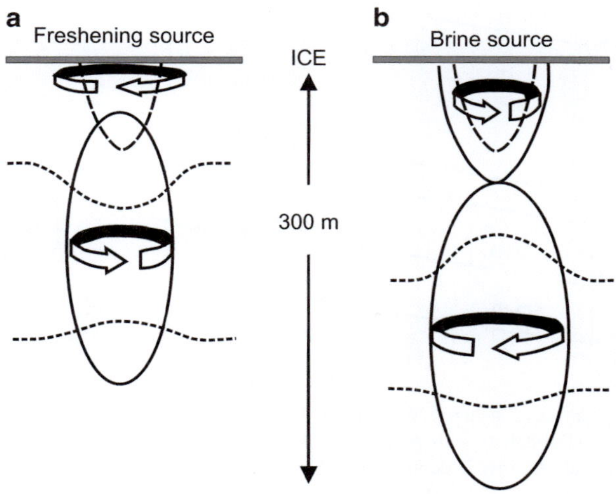

Fig. 1.2 Visualization of the formation mechanism of a warm (**a**) and cold (**b**) hetons according to [162] (Published with permission of American Geophysical Union)

heton with a vertical axis under ice in the ocean due to ice melting are given in Fig. 1.2. The formation scheme of a heton with a tilted axis due to the mechanism of baroclinic instability of the coastal current' front of potential vorticity anomaly (PVA[1] — in the terminology of the authors of [633]) is given in Fig. 1.3 (see also

[1]PVA concept is in wide use in geophysical applications, in particular, in the analysis of drift inertial recirculations in the ocean [160], in the studies of evolution of tropic cyclones [623], long-living mesoscale convective systems [729].

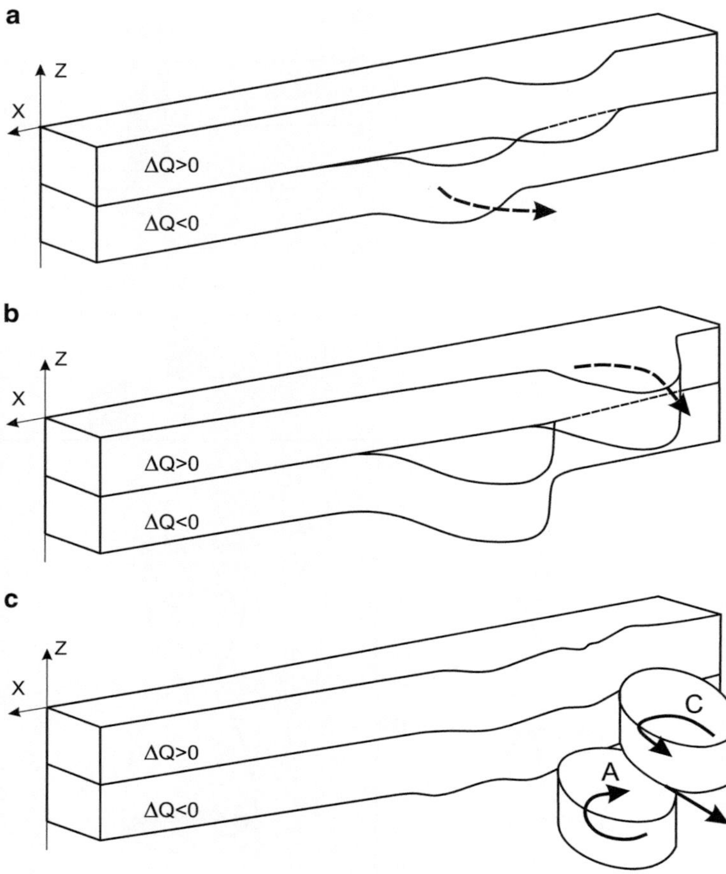

Fig. 1.3 The formation of a heton with a tilted axis due to baroclinic instability mechanism according to [633] (Published with permission of American Meteorological Society). Here the anomality of potential vorticity is denoted by ΔQ

[178]). Hetons can form also from unstable jets in the deep ocean [275,587]. Indeed, they have been observed south of the Gulf Stream; they are composed of a Gulf Stream cold-core ring associated with a Sargasso Sea water anticyclone.

The vortex structure, which has been derived from *in situ* observations south of Australia along 132° E in the Antarctic circumpolar current in January–March 1977 onboard "Professor Zubov" research vessel and which represents a heton with a tilted axis, is shown in Fig. 1.4. Naturally hetonic structures described also in [57] in relation to Agulhas eddies near the continental slope of southern Africa and in [763] in relation to tidal vortices in the neighborhood of Academy Bay in the Sea of Okhotsk. In summary, many observations of oceanic and atmospheric vortices confirm the existence of hetons.

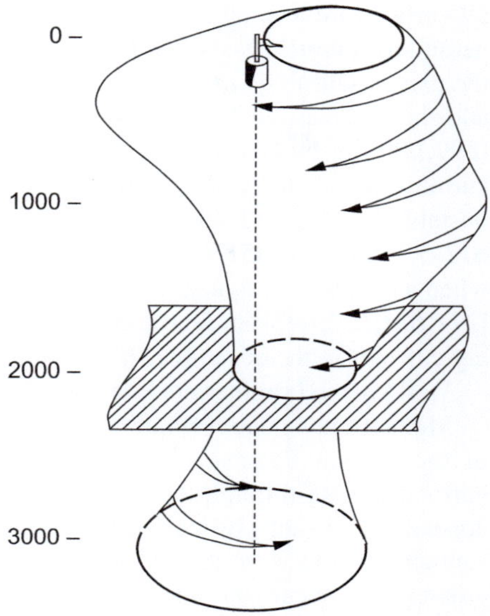

Fig. 1.4 The spatial structure of a heton with a tilted axis derived from direct instrumental observations (current meter, hydrographic and XBT data) in the Antarctic Circumpolar Current according to [803] (Published with permission of American Meteorological Society)

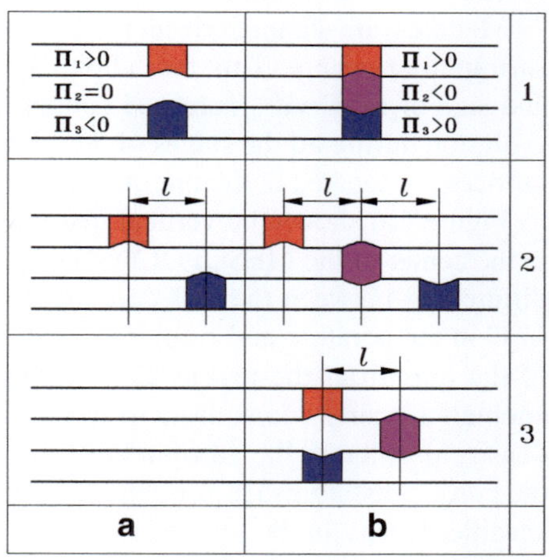

Fig. 1.5 Schematic representation of a three-layer distributed heton with equal-thickness layers (**1**) with a *vertical* axis, (**2**) with *tilted* axis and (**3**) with *arched* axis in the case of a three-layer fluid; (**a**) 1-modal heton; (**b**) 2-modal heton or S-vortex (in terminology of Morel and McWilliams [632]). *Red*, *magenta* and *blue* colors correspond to the upper, middle and lower layers, respectively

A hierarchy of hetons exists in the multilayer (in particular, three-layer) model [325–327, 329]. The integral vorticity of each heton is zero. The vertical distribution of vorticity follows the structure of baroclinic modes of potential-vortex operator.

Therefore, it is reasonable to refer to such baroclinic structures as *m-modal* hetons[2] ($m = 1, 2, \ldots, M-1$). A single baroclinic mode, i. e., just a heton, exists in the case of $M = 2$, corresponding to a two-layer model. When $M > 2$, only the first among the m-*modal* hetons is most close to the two-layer one (Fig. 1.5). When

[2]In our opinion, the term *modon*, which was used for such vortices in [327, 844], would be better. However, this term has been assigned to distributed dipole vortex structures [872].

all vortices are located along the same vertical, we will still say that such *m-modal* heton has a *vertical* axis, while when the vortices are spaced some distance apart, the axis is said to be *arc-wise*. Hetons can be generated in a laboratory by sources and sinks of mass [309, 310], by mechanical and locally spinning the top-layer fluid [899, 900], or by heat sources [342]. Such vortices naturally form when baroclinic currents, associated with the phenomenon of deep convection in the ocean, become unstable [59, 119, 132, 199, 200, 220–222, 224, 229, 321, 324, 524–528, 540, 571, 576, 612, 668, 669, 864, 951, 952, 978, 983, 984, 1017]. The heton idealization is also used when analyzing the dynamics of tropical cyclones and hurricanes in the atmosphere [271, 621, 622, 897], surface temperature anomalies [162–165], instability of boundary currents in the ocean [823], and intrathermocline vortices [352, 408].

Well known intrathermocline vortices are Mediterranean lenses (meddies).

Mediterranean anticyclonic and cyclonic lenses form over the continental slope of the Iberian Peninsula. Dynamic instability of a Mediterranean near-bottom current flowing through the Strait of Gibraltar into the Atlantic Ocean causes the appearance of such vortices. A Mediterranean Water (MW) jet is directed along the southern and western slopes of the peninsula [60, 108, 393, 556]. When the flow crosses bottom canyons and hills, there occurs both downwelling along the canyons [13, 20], and the separation of the stream in the seaward direction [262].

Meddies are characterized by high temperature and salinity with respect to surrounding waters. This fact allows us to determine their position in the ocean and size of MW with increased content of heat and salt, as well as to study their evolution during all the stages of their lifetime up to their decay. As a rule, all these vortices are located in the intermediate ocean layer.

Figure 1.6 shows the vertical distributions of temperature, salinity, and density in the center of the MESOPOLYGON lens and in the background region [379]. The differences between the lens core and the surrounding waters can vary from 1° to 4 °C in temperature and from 0.3 to 1.0 PSU in salinity depending on the distance of the lens from the region of its formation [258]. The temperature and salinity anomalies nearly compensate to form a homogeneous density. Indeed, the vertical density profiles in the surrounding waters and in the lens center are close, while the core is distinguished for density homogeneity. However, the absolute values of specific density in the lens cores vary in the range from 27.5 to 28.0, and the lens cores are located at depths from 800 to 1,400 m [261].

Figure 1.7 shows a T/S-diagram for the maximum values of temperature and salinity in the meddy cores based on the data catalogue for the period 1968–2007 [261]. The variability range for these characteristics is related to the different regions and mechanisms of the meddy formation. The long-term measurements show that the main regions of lens formation are the canyons of the Gulf of Cadiz [13, 46, 176, 393, 883], the region of Cape Saint Vincent [752, 806, 813], canyons of the continental slope and western coast of the Iberian Peninsula [107, 111] as well as the region near Cape Ortega [681, 682].

According to long-term observations, it appears that the Cadiz Bay canyons, the region of the Saint-Vincent Cape [13, 19, 46, 108, 813] as well as canyons along the western coast of Portugal [408, 681] are the main areas of meddies' formation.

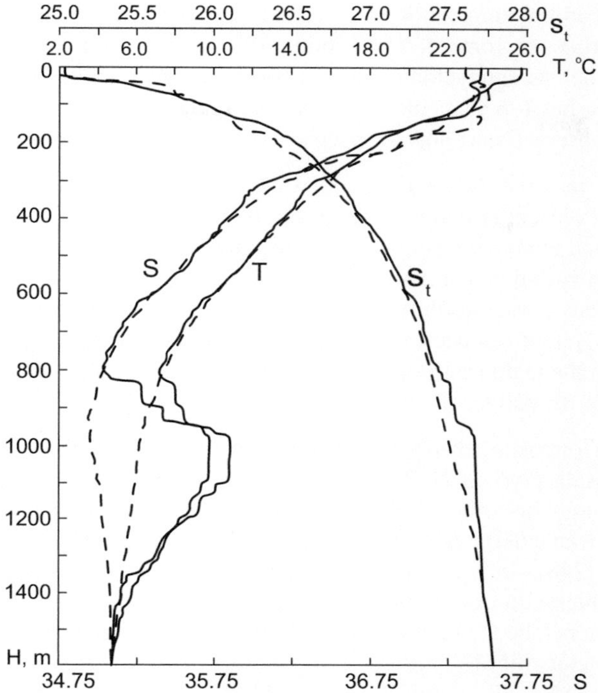

Fig. 1.6 Vertical distributions of temperature, salinity, and density in the center of the MESOPOLYGON lens (*solid lines*) and background region (*dashed lines*) [379]

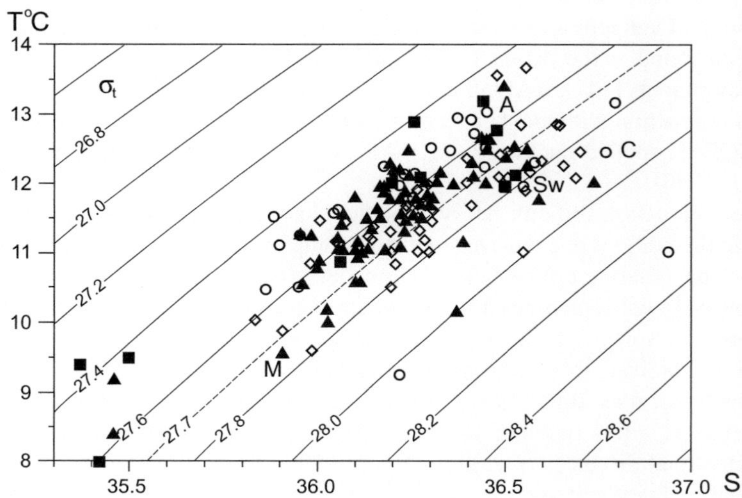

Fig. 1.7 T/S diagram for maximum values temperature and salinity in the lens cores. Depth of the cores: ■ < 800 m; ○ 800–1,000 m; ▲ 1,000–1,200 m; ◊ > 1,200 m; **Sw** Cadiz Gulf [883]; **M** Tropical Atlantic [1001]; **C** Cadiz Gulf [13]; **A** Azores Frontal Zone [13]

Usually, in the initial moment of water volume separation from the MW jet a dipole system of vortices is formed. The lifetime of cyclonic vortices does not exceed 0.5–1 year, whereas anticyclonic vortices (they are called meddies) live 4–5 years on the average, but 4–6 years old meddies were found in some cases [1001].

Available observations allow to specify two stages of anticyclonic lens evolution:

- A "young" lens represents a dipole system. An anticyclonic lens, as a rule, has a two-core vertical structure in the stage of formation. The water volume of a cyclonic lens is considerably smaller, and the core features are less marked in comparison with the meddies.
- An "old" lens possesses only one anticyclonic core, and the cyclonic vortex of the dipole system has decayed [257]. These vortices in the intermediate depths accomplish the main transport of MW heat and salt over long distances from the area of their formation.

Vortices of opposite rotations exist in the region of formation adjacent to the Iberian Peninsula [145, 262]. The vortex concentration in this region is high. In the period from November 1993 up to April 1994, 19 lenses were found and tracked in the frame of the AMUSE project (**A** **M**editerranean **U**ndercurrent **S**eeding **E**xperiment) [748]. In the study of meddies it appears necessary to take into account the interaction of vortices between themselves and with the vortices of cyclonic rotation. The mean currents and bottom topography of this region are the background factors [266].

In the SEMANE 99 (**S**uivi des **E**aux **M**éditerranéennes en **A**tlantique **N**ord–**E**st) experiments in the southern part of Gulf of Cadiz, similar situations were observed: two closely located meddies of different sizes interacted together and with an cyclonic vortex of smaller size also consisting of warm and saline MW [145]. These meddies had a marked two-core vertical structure according to the distribution of temperature and salinity. The maximum values of their characteristics have been found at the depths of 750–800 and 1,100–1,300 m. In the SEMANE 2000 experiments [147], RAFOS floats allowed the tracking of a meddy during and after its interaction with another meddy, and a priori with cyclones (RAFOS is the word SOFAR spelled backwards. SOFAR is an acronym for **SO**und **F**ixing **A**nd **R**anging) [770].

Observations are difficult because of the chaotic nature of motion: the floats, placed in the body of one lens, can perform complex motion, when it interacts with another lens. Satellite altimetry may allow detecting meddies at the ocean surface [263,869,914], but sometimes it is impossible to evaluate the character of interaction and motion of vortices with respect to time and space, which would coincide with the results obtained from data of hydrological observations [147]. As a rule, at distances larger than 200–400 km, it is possible to meet only anticyclonic vortices which migrate with a speed of 1–3 cm/s at distances up to 4,000 km. The meddies decay mostly at distances of 2,000–2,500 km from the their origin. It is important to emphasize that these vortices act as a "container" in carrying anomalous values of temperature, salt, oxygen, and other properties at long-distances.

Some properties of meddies will be investigated later in the Chap. 3.

1.2 The Mathematical Introduction

This section gives a general formulation of the problem for the two-layer quasi-geostrophic model and describes an original algorithm of two-layer version of contour dynamic method, which is widely used in Chap. 3 in solving different problems of vortex dynamics. A scheme of linear analysis of the stability of a distributed heton is given. The structure of simplest types of external field satisfying the conditions of identically zero distribution of potential vorticity is given. Equations of motion are derived for the limiting case of discrete vortices. The applicability conditions of trilinear coordinates are specified for the qualitative analysis of motions of three- and four-vortex systems. Dynamics of discrete vortices is investigated in Chap. 2. Next, a concise formulation of the vortex problems (finite-core and discrete) in the framework of the three-layer quasi-geostrophic model is given.

1.2.1 The Derivation of Potential Vortex Conservation Equations

The majority of the results presented in this book have been obtained in the two-layer quasi-geostrophic model; so, we will deal with the approximation of two-layer, ideal, incompressible fluid with constant densities ρ_1, ρ_2 ($\rho_1 \leq \rho_2$) and undisturbed depths h_1 and h_2 for the top and bottom layers, respectively, rotating with an angular velocity of Ω. We also use the Boussinesq and hydrostatic assumptions. Now the equations of motion can be written as

$$u_{jt} + u_j u_{jx} + v_j u_{jy} - f v_j = -\frac{p_{jx}}{\rho_0}, \tag{1.1}$$

$$v_{jt} + u_j v_{jx} + v_j v_{jy} + f u_j = -\frac{p_{jy}}{\rho_0}, \tag{1.2}$$

$$p_{jz} = \rho_j g, \tag{1.3}$$

$$u_{jx} + v_{jy} + w_{jz} = 0, \tag{1.4}$$

where u, v, w are velocity vector components along the axes x, y, z, respectively, of a rectangular coordinate system with z axis directed downward from the surface; p is pressure, $f = 2\Omega$ is the Coriolis parameter; ρ_0 is the mean density value; g is the acceleration due to gravity; the subscripts t, x, y, z denote partial derivatives with respect to time and appropriate spatial coordinates, and $j = 1, 2$ specifies the layer index.

Let us formulate the boundary conditions on a vertical.

The "rigid lid" condition on the top boundary of the top layer is

$$z=0: \qquad w_1=0; \tag{1.5}$$

the kinematic conditions and pressure continuity condition on the interface between layers give

$$z=h_1-\eta: \qquad w_1=-\frac{D_1\eta}{Dt}, \qquad w_2=-\frac{D_2\eta}{Dt}, \qquad p_1=p_2; \tag{1.6}$$

the flow without separation condition at the bottom is

$$z=h_1+h_2-\zeta: \qquad w_2=-\frac{D_2\zeta}{Dt}. \tag{1.7}$$

Here $\eta=\eta(x,y,t)$ and $\zeta=\zeta(x,y)$ are perturbations of the interface between layers and the bottom relief, respectively (the former function is unknown, while the latter is assumed specified); $D_j/Dt \equiv \partial/\partial t+u_j\partial/\partial x+v_j\partial/\partial y$ is a two-dimensional operator of total time derivative $(j=1,2)$.

The initial conditions and boundary conditions for horizontal variables will be formulated below depending on the problem considered.

Introducing $\omega_j=v_{jx}-u_{jy}$ (denotation for the vertical component of the relative vortex in the jth layer) and eliminating the pressure from (1.1) to (1.2), we obtain the equation

$$\frac{D_j(\omega_j+f)}{Dt}-\left(\omega_j+f\right)w_{jz}=0, \qquad j=1,2. \tag{1.8}$$

Integrating (1.8) with respect to z successively within the top and bottom layers and using the boundary conditions (1.5), (1.6) and (1.7), we obtain

$$\frac{D_1}{Dt}\left(\frac{\omega_1+f}{h_1-\eta}\right)=0, \tag{1.9}$$

$$\frac{D_2}{Dt}\left(\frac{\omega_2+f}{h_2-\zeta+\eta}\right)=0. \tag{1.10}$$

Assuming $|\eta| \ll h_1$ and $|\zeta| \ll h_2$, we obtain the approximate estimate

$$\frac{\omega_1+f}{h_1-\eta} \approx \frac{f}{h_1}+\frac{1}{h_1}\left(\omega_1+\frac{f}{h_1}\eta\right), \qquad \frac{\omega_2+f}{h_2-\zeta+\eta} \approx \frac{f}{h_2}+\frac{1}{h_2}\left[\omega_2+\frac{f}{h_2}(\zeta-\eta)\right].$$

Now (1.9) and (1.10) become

$$\frac{D_j\Pi_j}{Dt}=0, \qquad j=1,2, \tag{1.11}$$

where

$$\Pi_1 = \omega_1 + \frac{f}{h_1}\eta \quad \text{and} \quad \Pi_2 = \omega_2 + \frac{f}{h_2}(\zeta - \eta) \tag{1.12}$$

are potential vorticity in the layers.

Integrating hydrostatic equations (1.3) with respect to z, and taking into account (1.5), (1.6) and (1.7), we obtain

$$p_1 = \rho_1 g z + p_0, \qquad 0 \le z < h_1 - \eta,$$

$$p_2 = \rho_1 g(h_1 - \eta) + \rho_2 g(z - h_1 + \eta) + p_0, \qquad h_1 - \eta \le z \le h_1 + h_2 - \eta,$$

where $p_0(x, y, t)$ is surface pressure. For layer pressure anomalies in the layers with respect to their values under hydrostatic equilibrium, we have

$$\tilde{p}_1 = p_1 - \rho_1 g z = p_0,$$

$$\tilde{p}_2 = p_2 - \rho_1 g h_1 - \rho_2 g(z - h_1) = p_0 + g\Delta\rho\eta,$$

where $\Delta\rho = \rho_2 - \rho_1$. From here we have

$$\tilde{p}_2 - \tilde{p}_1 = g\Delta\rho\eta. \tag{1.13}$$

For further analysis, it is common to write geostrophic relationships that can be derived from (1.1) to (1.2), assuming balance between the Coriolis forces and pressure gradients:

$$u_j = -\frac{p_{jy}}{\rho_0 f} = -\frac{\tilde{p}_{jy}}{\rho_0 f}, \quad v_j = \frac{p_{jx}}{\rho_0 f} = \frac{\tilde{p}_{jx}}{\rho_0 f}, \quad j = 1, 2. \tag{1.14}$$

Equations (1.1), (1.2), (1.3), (1.4), (1.5), (1.6), (1.7), (1.8), (1.9), (1.10), (1.11), (1.12), (1.13) and (1.14) are written in dimensional form. A change to dimensionless variables can be performed by the following scaling:

$$(x', y') = D(x, y), \quad h'_j = Hh_j, \quad (\eta', \zeta') = RoH(\eta, \zeta),$$

$$(u'_j, v'_j) = U(u_j, v_j), \quad t' = (D/U)t,$$

$$\Pi'_j = (U/D)\Pi_j, \quad \tilde{p}'_j = \rho_0 f U D p'_1,$$

where capital letters denote the scales of appropriate variables, and strokes refer to dimensional variables, $Ro = U/Df = O(10^{-2})$[3] is Rossby number [851].

---

[3]This estimate is valid for the values of U, D typical of the atmosphere and ocean and for mid latitudes.

In this case, the dimensionless potential vortices

$$\Pi_1 = \nabla^2 p_1 + \frac{\eta}{h_1}, \quad \Pi_2 = \nabla^2 p_2 + \frac{\zeta - \eta}{h_2} \tag{1.15}$$

still satisfy equations of the type of (1.11), and, instead of (1.13) and (1.14) we obtain

$$\eta = F\left(p_2 - p_1\right), \tag{1.16}$$

$$u_j = -p_{jy}, \quad v_j = p_{jx}, \quad j = 1,2. \tag{1.17}$$

Here $\nabla^2 = \partial^2/\partial x^2 + \partial^2/\partial y^2$ is two-dimensional Laplace operator, $F = f^2D^2/g'(h_1+h_2)$ is Froude number, $g' = g\Delta\rho/\rho_0$ is reduced acceleration due to gravity. Equations (1.17) show that, in the geostrophic approximation, the pressure anomaly essentially serves as the stream function of quasi-two-dimensional motion.

The nine equations (1.11), (1.15), (1.16) and (1.17) form a closed system of equations for determining nine unknown functions η and p_j, u_j, v_j, Π_j $(j = 1,2)$.

1.2.2 Formal Solution. Integral Invariants

To separate part of variables in (1.15) and (1.16), we introduce the barotropic stream function

$$p = h_1 p_1 + h_2 p_2, \tag{1.18}$$

allowing us to derive from (1.1), considering (1.16), independent equations for p and η:

$$\nabla^2 p = h_1 \Pi_1 + h_2 \Pi_2 - \zeta, \tag{1.19}$$

$$\nabla^2\left(\frac{\eta}{F}\right) - \gamma^2\left(\frac{\eta}{F}\right) = \Pi_1 - \Pi_2 - \frac{\zeta}{h_2}. \tag{1.20}$$

Here we use the denotation

$$\gamma = \sqrt{\frac{F}{h_1 h_2}} = \frac{D}{\lambda}, \tag{1.21}$$

where $\lambda = \sqrt{g'h_1h_2/(h_1+h_2)}/f$ is the internal Rossby deformation radius [404, 689].

In virtue of (1.16) and (1.18), we obtain an expression for pressure anomalies in the layers

$$p_1 = p - h_2\left(\frac{\eta}{F}\right), \tag{1.22}$$

$$p_2 = p + h_1\left(\frac{\eta}{F}\right), \tag{1.23}$$

whence it can be seen that function η/F determines the baroclinic part of the flow.

Suppose that potential vorticity in the layers can be represented as

$$\Pi_j = \Pi_{j0} + \sum_{\alpha=1}^{A_j} \Pi_j^\alpha(x, y, t), \quad j = 1, 2, \tag{1.24}$$

where Π_{j0} are some constants, and functions Π_j^α are finite functions satisfying Eq. (1.11). Hereafter, we use Latin letters to denote layers and Greek letters for vorticities within layers.

Suppose that, according to (1.19) and (1.20), functions p_0 and η_0 determine some external field, such that

$$\nabla^2 p_0 = h_1\Pi_{10} + h_2\Pi_{20} - \zeta, \tag{1.25}$$

$$\nabla^2\left(\frac{\eta_0}{F}\right) - \gamma^2\left(\frac{\eta_0}{F}\right) = \Pi_{10} - \Pi_{20} - \frac{\zeta}{h_2}, \tag{1.26}$$

and, according to (1.22) and (1.23),

$$p_{10} = p_0 - h_2\left(\frac{\eta_0}{F}\right), \tag{1.27}$$

$$p_{20} = p_0 + h_1\left(\frac{\eta_0}{F}\right). \tag{1.28}$$

The solution of Eqs. (1.19) and (1.20) can be written as

$$p(x, y, t) = p_0 + h_1 \sum_{\alpha=1}^{A_1} \iint_{S_1^\alpha} \Pi_1^\alpha(x', y', t) G(r) dx' dy' + h_2 \sum_{\alpha=1}^{A_2} \iint_{S_2^\alpha} \Pi_2^\alpha(x', y', t) G(r) dx' dy', \tag{1.29}$$

$$\frac{\eta(x,y,t)}{F} = \frac{\eta_0}{F} + \sum_{\alpha=1}^{A_2} \iint\limits_{S_2^\alpha} \Pi_2^\alpha(x',y',t) G_0(r) dx'dy'$$
$$- \sum_{\alpha=1}^{A_1} \iint\limits_{S_1^\alpha} \Pi_1^\alpha(x',y',t) G_0(r) dx'dy', \tag{1.30}$$

where

$$G(r) = (1/2\pi)\ln\gamma^* r, \quad G_0(r) = -(1/2\pi)\mathrm{K}_0(\gamma r) \tag{1.31}$$

are Green functions of the operators of the left parts of Eqs. (1.19) and (1.20): ∇^2 and $\nabla^2 - \gamma^2$ respectively; $r = \sqrt{(x-x')^2 + (y-y')^2}$; $\mathrm{K}_0(z)$ is the modified zero-order Bessel function [725].[4] Here, $\gamma_* = D/R_{reg}$ and R_{reg} is a regularizing scale of length equal, for example, to the Oboukhov–Rossby scale ([315]) or to radius of sphere in problems of vortex dynamics on sphere ([90, 91]).

Substituting (1.29) and (1.30) in (1.22) and (1.23), considering (1.27) and (1.28), we obtain

$$p_j(x,y,t) = p_{j0} + h_j \sum_{\alpha=1}^{A_j} \iint\limits_{S_j^\alpha} \Pi_j^\alpha \left[G(r) + \frac{h_{3-j}}{h_j} G_0(r)\right] dx'dy'$$
$$+ h_{3-j} \sum_{\alpha=1}^{A_{3-j}} \iint\limits_{S_{3-j}^\alpha} \Pi_{3-j}^\alpha \left[G(r) - G_0(r)\right] dx'dy', \quad j = 1,2. \tag{1.32}$$

It can be readily seen that there exist integral invariants: total potential vorticity

$$Q = \sum_{j=1}^{2} h_j \sum_{\alpha=1}^{A_j} \iint\limits_{S_j^\alpha} \Pi_j^\alpha dx'dy',$$

momentum components

$$P_x = \sum_{j=1}^{2} h_j \sum_{\alpha=1}^{A_j} \iint_{S_j^\alpha} \Pi_j^\alpha x' dx'dy', \quad P_y = \sum_{j=1}^{2} h_j \sum_{\alpha=1}^{A_j} \iint\limits_{S_j^\alpha} \Pi_j^\alpha y' dx'dy', \tag{1.33}$$

---

[4] Hereafter, we will use, without special mention, analogous denotations $\mathrm{K}_n(z)$ for modified Bessel function of the nth order.

and moment of impulse of the system

$$M = \sum_{j=1}^{2} h_j \sum_{\alpha=1}^{A_j} \iint_{S_j^\alpha} \Pi_j^\alpha \left[(x')^2 + (y')^2 \right] dx' dy'.$$

1.2.3 Contour Dynamics Method

Expressions (1.32) are just a formal solution for pressure anomalies, since the expressions under the integral sign in the right-hand side are unknown functions. Now suppose that all Π_j^α $(j = 1, 2; \alpha = 1, 2, \ldots, A_j)$ are constants, i. e., the potential vorticity has a piece-wise constant distribution. In such case, the appropriate constants can be factorized outside the integral, and the surface integrals can be transformed into contour integrals with the use of Stokes theorem:

$$\begin{aligned} p_j = p_{j0} + h_j \sum_{\alpha=1}^{A_j} \Pi_j^\alpha \oint_{C_j^\alpha} N \left[M(r) + \frac{h_{3-j}}{h_j} M_1(r) \right] dv \\ + h_{3-j} \sum_{\alpha=1}^{A_{3-j}} \Pi_{3-j}^\alpha \oint_{C_{3-j}^\alpha} N \left[M(r) - M_1(r) \right] dv, \quad j = 1, 2. \end{aligned} \tag{1.34}$$

Here, C_j^α are the contours of domains S_j^α, and notations

$$N = \frac{(x'-x)(\partial y'/\partial v) - (y'-y)(\partial x'/\partial v)}{r^2},$$

$$M = \frac{r^2}{4\pi} \left(\ln \gamma^* r - \frac{1}{2} \right), \quad M_1 = \frac{1}{2\pi\gamma^2} \left[\gamma r \mathrm{K}_1(\gamma r) - 1 \right]$$

are introduced; v is a linear characteristic measured in the positive direction (counterclockwise) along each contour C_j^α.

Expressions (1.17) with the help of contour integrals, can be used to write formulas for velocities. If constants Π_j^α are known, all unknown functions are determined only by the configuration of contours C_j^α of evolving domains S_j^α. This fact forms the basis of an effective and vivid Lagrangian approach — the so-called contour dynamic method (CDM) [191, 303, 304, 726, 1006], developed for the case of two-layer fluid in [477].[5] Thus, formulas (1.34) are the theoretical basis for the two-layer CDM version.

[5] Similar results in the generalization of CDM for two-layer case were later and independently obtained in [343, 710].

In the realization of CDM, a set of marker points (nodes) ($x_j^{\alpha k}$, $y_j^{\alpha k}$), numbered successively in the positive direction ($k{=}0,1,\ldots,K_j^{\alpha}$), is specified, where $(x_j^{\alpha 0},y_j^{\alpha 0}){=}(x_j^{\alpha K_j^{\alpha}},y_j^{\alpha K_j^{\alpha}})$. Without loss of generality, we can suppose that the parameter ν, which changes continuously and monotonically along the contour, takes integer values k in the nodes.

The equations of motion of the nodes are given by

$$\frac{d}{dt}x_j^{\alpha k}=u_j^{\alpha k},\quad \frac{d}{dt}y_j^{\alpha k}=v_j^{\alpha k},\tag{1.35}$$

$$x_j^{\alpha k}\Big|_{t=0}=x_{j0}^{\alpha k},\quad y_j^{\alpha k}\Big|_{t=0}=y_{j0}^{\alpha k},$$

$$j=1,2,\quad \alpha=1,2,\ldots,A_j,\quad k=0,1,2,\ldots,K_j^{\alpha}.$$

The right-hand parts of Eq. (1.35) for any such point can be calculated with the help of (1.17) and (1.34).

Equation (1.35) will be solved by using fourth-order Runge-Kutta procedure with a step $\Delta t = 0.01$ and with "optimal" Gill formulas [336]. The number of nodes in numerical calculations was commonly taken equal to 120 per a unit-radius circumference. As might be required during the calculation process, it was increased in proportion to the increasing length of the contour. Functions $x_j^{\alpha k}(\nu), y_j^{\alpha k}(\nu)$, which parametrically determine the contours C_j^{α}, were integrated and interpolated with the help of periodic cubic splines [455]. The major elements of the CDM-program were Makarov's program complex [557] and his "contour surgery" procedure [558], see also [233]. The latter allows one to cut off long and thin vortex lines,[6] and remove overlapping boundary segments when vortex patches with the same potential vorticities approach one another (and even merge). The same procedure admits the mechanism of artificial "dissipation", ignoring the vortex structures whose contours contain nodes whose number is less than some specified value $K_{\min}$. With the condition that the area loss of the vortex patch will not exceed half percent of its total area, we obtain that at 120 nodes corresponding to a unit-radius circumference, $K_{\min} = 6$ [851].

The implementation scheme of "contour surgery" by algorithm [558] is given in Fig. 1.8, showing how appropriate segments of vortex patch contours are replaced by linear segments in the case of intersection (or self-intersection) of auxiliary 2ε-contours.

Thus, the application of CDM procedure enables one to calculate all contour integrals in the right-hand part of the Eq. (1.34). Finally, the fields of pressure

[6]The problem of vortex line formation (filamentation), which is of indisputable interest in itself [234, 388, 589, 590, 707, 708], is not discussed in this study.

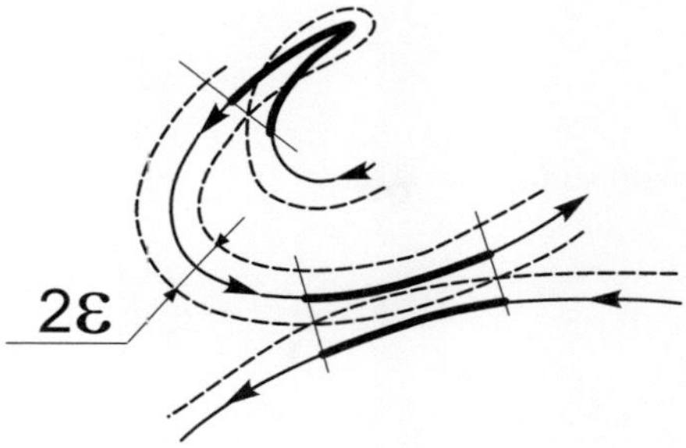

Fig. 1.8 The scheme of "contour surgery" algorithm (According to Makarov [558], see also [851]. Published with permission of Cambridge University Press)

anomalies in the layers, and hence the components of velocity vector, can be evaluated, after the substitution of stationary terms p_{j0} ($j = 1,2$) from (1.27) to (1.28), where p_0 and η_0 are solutions of Eqs. (1.25) and (1.26), respectively.

1.2.4 Stationary Axisymmetric Solution

Suppose that $\zeta = \Pi_{10} = \Pi_{20} = 0$ and a set of concentric circular vortex patches with radii R_j^α ($j = 1, 2$; $\alpha = 1, 2, \ldots, A_j$) is specified in each layer. In this case, $p_0 \equiv 0$ and $\eta_0 \equiv 0$ can be taken as the solutions of (1.25) and (1.26), implying the absence of external velocity field.

With the condition of axial symmetry, Eqs. (1.19) and (1.20) become

$$\frac{1}{r}\frac{d}{dr}\left(r\frac{dp}{dr}\right) = \sum_{j=1}^{2} h_j \sum_{\alpha=1}^{A_j} \Pi_j^\alpha, \tag{1.36}$$

$$\frac{1}{r}\frac{d}{dr}\left[r\frac{d}{dr}\left(\frac{\eta}{F}\right)\right] - \gamma^2\frac{\eta}{F} = \sum_{j=1}^{2}(-1)^j \sum_{\alpha=1}^{A_j} \Pi_j^\alpha \tag{1.37}$$

and they should be integrated under conditions

$$p|_{R_j^\alpha - 0} = p|_{R_j^\alpha + 0}, \quad \eta|_{R_j^\alpha - 0} = \eta|_{R_j^\alpha + 0}, \quad j = 1,2, \ \alpha = 1,2,\ldots,A_j, \tag{1.38}$$

$$\left.\frac{dp}{dr}\right|_\infty \to 0, \quad \left.\frac{d\eta}{dr}\right|_\infty \to 0. \tag{1.39}$$

For the construction of solutions (1.36), (1.37), (1.38) and (1.39) it is enough to consider two auxiliary problems

$$\frac{1}{r}\frac{d}{dr}\left(r\frac{dp}{dr}\right) = \begin{cases} 1, & r \le \bar{R}, \\ 0, & r > \bar{R}, \end{cases} \tag{1.40}$$

$$\frac{1}{r}\frac{d}{dr}\left(r\frac{d\eta}{dr}\right)-\gamma^2\eta=\begin{cases}F, & r\leq\bar{R},\\ 0, & r>\bar{R},\end{cases} \tag{1.41}$$

with solutions

$$p^{(\bar{R})}=\frac{\bar{R}^2}{4}\begin{cases}\left(r/\bar{R}\right)^2, & r\leq\bar{R},\\ 1+2\ln\left(r/\bar{R}\right), & r>\bar{R},\end{cases} \tag{1.42}$$

$$\eta^{(\bar{R})}=\frac{F}{\gamma^2}\begin{cases}\gamma\bar{R}\mathrm{K}_1\left(\gamma\bar{R}\right)\mathrm{I}_0\left(\gamma r\right)-1, & r\leq\bar{R},\\ -\gamma\bar{R}\mathrm{I}_1\left(\gamma\bar{R}\right)\mathrm{K}_0\left(\gamma r\right), & r>\bar{R}.\end{cases} \tag{1.43}$$

Here I_n is nth-order Bessel function of imaginary argument [725].

It follows

$$p=\sum_{j=1}^{2}h_j\sum_{\alpha=1}^{A_j}\Pi_j^{\alpha}p^{(R_j^{\alpha})}, \tag{1.44}$$

$$\eta=\sum_{j=1}^{2}(-1)^j\sum_{\alpha=1}^{A_j}\Pi_j^{\alpha}\eta^{(R_j^{\alpha})}. \tag{1.45}$$

Then, (1.44) and (1.45) specify a class of axisymmetric steady states, satisfying the system of Eqs. (1.36), (1.37), (1.38) and (1.39). Their stability will be studied in Chap. 3.

Using successively (1.22)–(1.23) and (1.17), we can derive from (1.44) to (1.45) expressions for the radial distributions of azimuthal velocities in both layers:

$$V_j(r)=\sum_{k=1}^{2}h_k\sum_{\alpha=1}^{A_k}\Pi_k^{\alpha}\frac{dp^{(R_k^{\alpha})}}{dr}+h_{3-j}\sum_{k=1}^{2}(-1)^{j-k}\sum_{\alpha=1}^{A_k}\Pi_k^{\alpha}\frac{d\eta^{(R_k^{\alpha})}}{dr}. \tag{1.46}$$

In the specific case of a single two-layer vortex of cylindrical shape $\mathrm{A}_1=\mathrm{A}_2=1$ and $R_1^1=R_2^1\equiv R$, those expressions take the simplest form

$$V_j(r)=V_p^{(R)}\sum_{k=1}^{2}h_k\Pi_k-h_{3-j}V_\eta^{(R)}\sum_{k=1}^{2}(-1)^{j-k}\Pi_k, \tag{1.47}$$

where

$$V_p^{(R)}=\frac{dp^{(R)}}{dr}=\frac{R}{2}\begin{cases}r/R, & r\leq R,\\ R/r, & r>R,\end{cases} \tag{1.48}$$

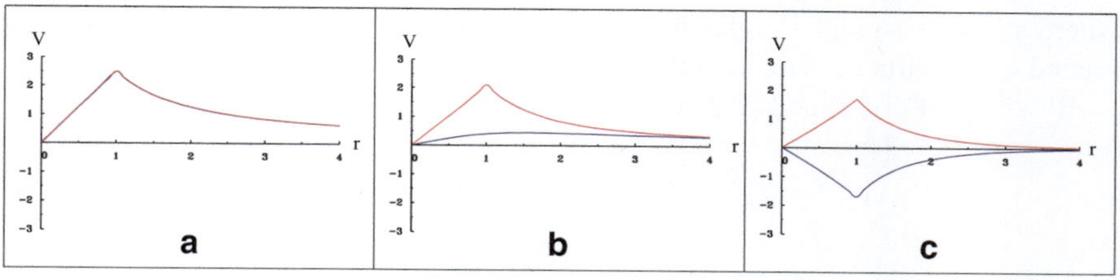

Fig. 1.9 The radial distributions of azimuthal velocities for the top (*red*) and bottom (*blue*) layers at $\gamma = 1$ and (**a**) $\Pi_1 = \Pi_2 = 5$, (**b**) $\Pi_1 = 5, \Pi_2 = 0$, (**c**) $\Pi_1 = -\Pi_2 = 5$

$$V_\eta^{(R)} = \frac{d\eta^{(R)}}{dr} = R \begin{cases} \mathrm{K}_1(\gamma R)\mathrm{I}_1(\gamma r), & r \le R, \\ \mathrm{I}_1(\gamma R)\mathrm{K}_1(\gamma r), & r > R. \end{cases} \tag{1.49}$$

Examples of radial distributions of azimuthal velocities (1.47), (1.47) and (1.49) are given in Fig. 1.9.

1.2.5 An Approach to Studying the Stability of a Axisymmetric Two-Layer Vortex

Suppose that for the stationary solution (1.47) in the case of development of its instability, the contour lines of constant potential vorticity in layers can be described by the parametric equations

$$r_j = f_j(\theta, t; r_0), \quad r_0 = f_j(\theta, 0; r_0), \quad j = 1, 2. \tag{1.50}$$

Since those lines coincide with fluid contours, their motion equations can be written as

$$f_j f_{jt} + V_j^{(\theta)} f_{j\theta} - V_j^{(r)} f_j = 0, \quad j = 1, 2, \tag{1.51}$$

where subscripts t and θ imply partial differentiation with respect to appropriate variables, and $V^{(r)}$ and $V^{(\theta)}$ are the radial and transverse projections of velocity vector. We use the method of small perturbations, assuming that

$$f_j(\theta, t; r_0) = r_0 + \varepsilon_j(\theta, t; r_0), \quad |\varepsilon_j| \ll r_0. \tag{1.52}$$

Linearizing relationships (1.51) yields

$$\varepsilon_{jt} + \frac{1}{\rho}\bar{V}_j^{(\theta)}\varepsilon_{j\theta} - \tilde{V}_j^{(r)} = 0, \quad j = 1, 2, \tag{1.53}$$

where symbols $\bar{()}$ and $\tilde{()}$ refer to the undisturbed ($\varepsilon_j \equiv 0$) and disturbed (linearly related to ε_j) parts of velocity field.

By virtue of (1.22) and (1.23), we have

$$\bar{V}_1^{(\theta)} = \bar{V}_p^{(\theta)} - h_2 \bar{V}_\eta^{(\theta)}, \quad \bar{V}_2^{(\theta)} = \bar{V}_p^{(\theta)} + h_1 \bar{V}_\eta^{(\theta)}, \tag{1.54}$$

$$\tilde{V}_1^{(r)} = \tilde{V}_p^{(r)} - h_2 \tilde{V}_\eta^{(r)}, \quad \tilde{V}_2^{(r)} = \tilde{V}_p^{(r)} + h_1 \tilde{V}_\eta^{(r)}, \tag{1.55}$$

where

$$\bar{V}_p^{(\theta)} = \frac{d\bar{p}}{dr}, \quad \bar{V}_\eta^{(\theta)} = \frac{d}{dr}\left(\frac{\bar{\eta}}{F}\right), \tag{1.56}$$

$$\tilde{V}_p^{(r)} = -\frac{1}{r}\frac{\partial \tilde{p}}{\partial \theta}, \quad \tilde{V}_\eta^{(r)} = -\frac{1}{r}\frac{\partial}{\partial \theta}\left(\frac{\tilde{\eta}}{F}\right). \tag{1.57}$$

With those assumptions, we have from (1.29) to (1.30):

$$p = \bar{p} + \tilde{p} = \iint_{-\infty}^{\infty} \left[h_1 \Pi_1(x_1, y_1, t) + h_2 \Pi_2(x_1, y_1, t)\right] G(R) dx_1 dy_1, \tag{1.58}$$

$$\frac{\eta}{F} = \frac{\bar{\eta}}{F} + \frac{\tilde{\eta}}{F} = \iint_{-\infty}^{\infty} \Big[\Pi_2(x_1, y_1, t) - \Pi_1(x_1, y_1, t)\Big] G_0(R) dx_1 dy_1, \tag{1.59}$$

where $G(R)$ and $G_0(R)$ are Green functions (1.31); $R = \sqrt{(x - x_1)^2 + (y - y_1)^2}$ is the distance between observation and integration points.

Let us change to integration variables (λ, ρ) in formulas (1.58) and (1.59) by formulas

$$x_1 = f_j(\lambda, t; \rho)\cos\lambda, \quad y_1 = f_j(\lambda, t; \rho)\sin\lambda, \quad j = 1, 2.$$

Since

$$dx_1 dy_1 = f_j(\lambda, t; \rho) f_{j\rho}(\lambda, t; \rho) d\lambda d\rho,$$

we obtain with first order accuracy with respect to ε_j

$$G(R) f_j(\lambda, t; \rho) f_{j\rho}(\lambda, t; \rho) = \rho G(R) + \frac{\partial}{\partial \rho}\Big[G(R)\rho\, \varepsilon_j\Big]. \tag{1.60}$$

Let us denote

$$J^{(n)} = \int_0^{2\pi} G(R) e^{in\phi} d\phi, \quad J_0^{(n)} = \int_0^{2\pi} G_0(R) e^{in\phi} d\phi,$$

where $R = \sqrt{z^2 + w^2 - 2zwcos\phi}$.

According to [725], we have

$$J^{(0)}(z, w) = \begin{cases} \ln w, & z < w, \\ \ln z, & z > w, \end{cases} \tag{1.61}$$

$$J^{(n)}(z, w) = -\frac{1}{2m} \begin{cases} (z/w)^n, & z < w, \\ (w/z)^n, & z > w, \end{cases} \quad n \geq 1, \tag{1.62}$$

$$J_0^{(n)}(z, w) = -\begin{cases} \mathrm{I}_n(\gamma z)\mathrm{K}_n(\gamma w), & z < w, \\ \mathrm{I}_n(\gamma w)\mathrm{K}_n(\gamma z), & z > w, \end{cases} \quad n \geq 0. \tag{1.63}$$

Considering (1.60), we carry out integration by parts and separate the unperturbed and perturbed parts in (1.58) and (1.59) to obtain

$$\bar{p} = \int_0^\infty \Big[h_1 \Pi_1(\rho) + h_2 \Pi_2(\rho)\Big] \rho\, J^{(0)}(r, \rho) d\rho, \tag{1.64}$$

$$\tilde{p} = -\int_0^{2\pi}\!\!\int_0^\infty \Big[h_1 \Pi_1'(\rho)\varepsilon_1(\lambda, t; \rho) + h_2 \Pi_2'(\rho)\varepsilon_2(\lambda, t; \rho)\Big] \rho\, G(R) d\lambda\, d\rho, \tag{1.65}$$

$$\frac{\bar{p}}{F} = \int_0^\infty \Big[\Pi_2(\rho) - \Pi_1(\rho)\Big] \rho\, J_0^{(0)}(r, \rho) d\rho, \tag{1.66}$$

$$\frac{\tilde{p}}{F} = -\int_0^{2\pi}\!\!\int_0^\infty \Big[\Pi_2'(\rho)\varepsilon_2(\lambda, t; \rho) - \Pi_1'(\rho)\varepsilon_1(\lambda, t; \rho)\Big] \rho\, G_0(R) d\lambda\, d\rho, \tag{1.67}$$

where now $R = \sqrt{r^2 + \rho^2 - 2r\rho\cos(\lambda - \theta)}$.

Assuming for individual modes of perturbations

$$\varepsilon_j(\theta, t; \rho) = \varepsilon_j(\rho) e^{im(\theta - \delta t)}, \quad m \geq 1, \tag{1.68}$$

and substituting (1.68) in (1.65) and (1.67) and using (1.57), we come to

$$\bar{V}_p^{(\theta)} = \int_0^\infty \Big[h_1 \Pi_1(\rho) + h_2 \Pi_2(\rho)\Big] \rho \frac{\partial J^{(0)}(r, \rho)}{\partial r} d\rho,$$

$$\bar{V}_\eta^{(\theta)} = \int_0^\infty \Big[\Pi_2(\rho) - h_1 \Pi_1(\rho)\Big] \rho \frac{\partial J_0^{(0)}(r, \rho)}{\partial r} d\rho,$$

$$\tilde{V}_p^{(r)} = e^{im(\theta-\delta t)} \frac{im}{r} \int_0^\infty \Big[h_1 \Pi_1'(\rho)\varepsilon_1(\rho) + h_2 \Pi_2'(\rho)\varepsilon_2(\rho) \Big] \rho J^{(m)}(r,\rho) d\rho,$$

$$\tilde{V}_\eta^{(r)} = e^{im(\theta-\delta t)} \frac{im}{r} \int_0^\infty \Big[\Pi_2'(\rho)\varepsilon_2(\rho) - \Pi_1'(\rho)\varepsilon_1(\rho) \Big] \rho J_0^{(m)}(r,\rho) d\rho.$$

Now Eqs. (1.52), considering (1.53), (1.54), (1.55), (1.56) and (1.57), become

$$\Big[-\delta + P(\rho) - h_2 Q(\rho) \Big] \varepsilon_1(\rho) - M_m(\rho) + h_2 N_m(\rho) = 0, \tag{1.69}$$

$$\Big[-\delta + P(\rho) + h_1 Q(\rho) \Big] \varepsilon_2(\rho) - M_m(\rho) - h_1 N_m(\rho) = 0, \tag{1.70}$$

where

$$P(\rho) = \frac{1}{\rho} \int_0^\rho \Big[h_1 \Pi_1(\rho_1) + h_2 \Pi_2(\rho_1) \Big] \rho_1 d\rho_1,$$

$$Q(\rho) = \frac{\gamma}{\rho} \Bigg\{ \mathrm{K}_1(\gamma\rho) \int_0^\rho \Big[\Pi_2(\rho_1) - \Pi_1(\rho_1) \Big] \rho_1 \mathrm{I}_0(\gamma\rho_1) d\rho_1 - \mathrm{I}_1(\gamma\rho) \int_\rho^\infty \Big[\Pi_2(\rho_1) - \Pi_1(\rho_1) \Big] \rho_1 \mathrm{K}_0(\gamma\rho_1) d\rho_1 \Bigg\},$$

$$M_m(\rho) = \frac{1}{\rho} \int_0^\infty \Big[h_1 \Pi_1'(\rho_1)\varepsilon_1(\rho_1) + h_2 \Pi_2'(\rho_1)\varepsilon_2(\rho_1) \Big] \rho_1 J^{(m)}(\rho,\rho_1) d\rho_1,$$

$$N_m(\rho) = \frac{1}{\rho} \int_0^\infty \Big[\Pi_2'(\rho_1)\varepsilon_2(\rho_1) - \Pi_1'(\rho_1)\varepsilon_1(\rho_1) \Big] \rho_1 J_0^{(m)}(\rho,\rho_1) d\rho_1.$$

The spectral problem (1.69) and (1.70) determines the stability properties of each mode (clearly, the condition $Im\delta > 0$ must hold for unstable modes), and it can be used for arbitrary functions $\Pi_1(\rho)$ and $\Pi_2(\rho)$. In Chap. 3 this algorithm will be applied to actual problems for studying the stability of two-layer vortices.

1.2.6 The Structure of Simplest Types of External Field

Let us consider some examples of external fields where Eqs. (1.25) and (1.26) have an analytical solution.

- Suppose that bottom topography perturbations have a "horizontal-cylindrical" distribution (i. e., depend only on one horizontal coordinate, e. g., y): $\zeta = \zeta(y)$. Now we have

$$p_0 = -\int_0^y \zeta(y_1)(y-y_1)dy_1 - U(t)y,$$

$$\eta_0 = \frac{F}{2\gamma h_2}\left[\int_{-\infty}^{y} \zeta(y_1)e^{-\gamma(y-y_1)}dy_1 + \int_{y}^{\infty} \zeta(y_1)e^{\gamma(y-y_1)}dy_1\right] + c_1(t)e^{-\gamma y} + c_2(t)e^{\gamma y} \qquad (1.71)$$

with arbitrary functions $U(t)$, $c_1(t)$ and $c_2(t)$. In the specific case $\Pi_{10} = \Pi_{20}$ and $c_1(t) = c_2(t) \equiv 0$, we obtain a barotrophic current along the x with a velocity of $U(t)$ over ridges and troughs parallel to it.
- Suppose that $h_1\Pi_{10} + h_2\Pi_{20} = 0$ and

$$\zeta(x,y) = \begin{cases} \sigma, & \sqrt{x^2+y^2} \le R, \\ 0, & \sqrt{x^2+y^2} > R. \end{cases}$$

In that case, we have an axisymmetric solution (see (1.42) and (1.43))

$$p_0 = -\sigma p^{(R)},$$
$$\eta_0 = -\left(\Pi_{20} - \Pi_{10}\right)\frac{F}{\gamma^2} - \frac{\sigma}{h_2}\eta^{(R)}, \qquad (1.72)$$

Corresponding to a trapped topographic vortex over an elevation (when $\sigma > 0$) or a depression (when $\sigma < 0$) in the form of a vertical cylinder of revolution.
- When $\Pi_{10} = \Pi_{20} = 0$, we can assume

$$p_0 = -\sigma p^{(R)} - U(t)y,$$
$$\eta_0 = -\frac{\sigma}{h_2}\eta^{(R)}, \qquad (1.73)$$

which corresponds to a total field of a topographic vortex and an ongoing flow.

- In the case of horizontal bed ($\zeta \equiv 0$), the external field can be represented as

$$p_0 = \left(h_1 \Pi_{10} + h_2 \Pi_{20}\right) \frac{y^2}{2} - U(t) y,$$
$$\eta_0 = -\left(\Pi_{20} - \Pi_{10}\right) \frac{F}{\gamma^2} + c_1(t) e^{-\gamma y} + c_2(t) e^{\gamma y}. \tag{1.74}$$

The expressions (1.71) and (1.74) exemplify flows that can be used as a background for studying, with the use of CDM, the motion of vortex patches within a wide range of hydrodynamic and geophysical problems.

1.2.7 A Limiting Case of Discrete Vortices

The case of discrete vortices with intensities κ_j^α is obtained as the limit

$$\lim_{\substack{S_j^\alpha \to 0 \\ \Pi_j^\alpha \to \infty}} \Pi_j^\alpha S_j^\alpha = \kappa_j^\alpha,$$

where S_j^α is an undisturbed circular domain. Now

$$\Pi_j^\alpha = \kappa_j^\alpha \delta\left(x - x_j^\alpha\right) \delta\left(y - y_j^\alpha\right), \tag{1.75}$$

where δ is Dirac delta function, and (1.34) transforms into

$$p_j(x,y) = \frac{h_j}{2\pi} \left\{ \sum_{\alpha=1}^{A_j} \kappa_j^\alpha \left[\ln \gamma^* r - \frac{h_{3-j}}{h_j} \mathrm{K}_0(\gamma r)\right] + \sum_{\alpha=1}^{A_{3-j}} \frac{h_{3-j}}{h_j} \kappa_{3-j}^\alpha \left[\ln \gamma^* r + \mathrm{K}_0(\gamma r)\right] \right\}, \quad j = 1, 2, \tag{1.76}$$

where $r = \sqrt{\left(x - x_j^\alpha\right)^2 + \left(y - y_j^\alpha\right)^2}$. Considering that here we have a single linear scale (Rossby radius), we have to assume $D = 1$ in (1.21). From (1.17) and (1.76), we can obtain the equations of motion of point vortices:

$$\dot{x}_j^\alpha = -\frac{h_j}{2\pi} \left\{ \sum_{\substack{\beta=1 \\ \beta \neq \alpha}}^{A_j} \kappa_j^\beta \frac{y_j^\alpha - y_j^\beta}{(r_{jj}^{\alpha\beta})^2} \left[1 + \frac{h_{3-j}}{h_j} \gamma r_{jj}^{\alpha\beta} \mathrm{K}_1\left(\gamma r_{jj}^{\alpha\beta}\right)\right] + \sum_{\beta=1}^{A_{3-j}} \kappa_{3-j}^\beta \frac{h_{3-j}}{h_j} \frac{y_j^\alpha - y_{3-j}^\beta}{(r_{j(3-j)}^{\alpha\beta})^2} \left[1 - \gamma r_{j(3-j)}^{\alpha\beta} \mathrm{K}_1\left(\gamma r_{j(3-j)}^{\alpha\beta}\right)\right] \right\}, \tag{1.77}$$

$$\dot{y}_j^\alpha = \frac{h_j}{2\pi}\left\{\sum_{\substack{\beta=1\\ \beta\neq\alpha}}^{A_j} \kappa_j^\beta \frac{x_j^\alpha - x_j^\beta}{(r_{jj}^{\alpha\beta})^2}\left[1+\frac{h_{3-j}}{h_j}\gamma r_{jj}^{\alpha\beta}\mathrm{K}_1\left(\gamma r_{jj}^{\alpha\beta}\right)\right]\right.$$

$$\left. + \sum_{\beta=1}^{A_{3-j}} \kappa_{3-j}^\beta \frac{h_{3-j}}{h_j}\frac{x_j^\alpha - x_{3-j}^\beta}{(r_{j(3-j)}^{\alpha\beta})^2}\left[1-\gamma r_{j(3-j)}^{\alpha\beta}\mathrm{K}_1\left(\gamma r_{j(3-j)}^{\alpha\beta}\right)\right]\right\}. \quad (1.78)$$

Here $r_{ij}^{\alpha\beta} = \sqrt{(x_i^\alpha - x_j^\beta)^2 + (y_i^\alpha - y_j^\beta)^2}$, $\alpha = 1, 2, \ldots, A^j$, $\quad j = 1, 2.$

Expressions for integral invariants (1.32) become

$$Q = \sum_{j=1}^{2} h_j \sum_{\alpha=1}^{A_j} \kappa_j^\alpha, \quad P_x = \sum_{j=1}^{2} h_j \sum_{\alpha=1}^{A_j} x_j^\alpha \kappa_j^\alpha,$$

$$P_y = \sum_{j=1}^{2} h_j \sum_{\alpha=1}^{A_j} y_j^\alpha \kappa_j^\alpha, \quad M = \sum_{j=1}^{2} h_j \sum_{\alpha=1}^{A_j} \left[(x_j^\alpha)^2 + (y_j^\alpha)^2\right]\kappa_j^\alpha. \quad (1.79)$$

The system (1.77) and (1.78) can be rewritten in the Hamiltonian form

$$\dot{q}_j^\alpha = \frac{\partial \mathscr{H}}{\partial p_j^\alpha}, \qquad \dot{p}_j^\alpha = -\frac{\partial \mathscr{H}}{\partial q_j^\alpha},$$
$$\alpha = 1, 2, \ldots, A^j, \quad j = 1, 2, \quad (1.80)$$

where $q_j^\alpha = x_j^\alpha$ are generalized coordinates, $p_j^\alpha = y_j^\alpha \kappa_j^\alpha/2$ are generalized momenta, and

$$\mathscr{H} = -\frac{1}{4\pi}\sum_{j=1}^{2} h_j \left\{\sum_{\substack{\alpha,\beta=1\\ \alpha\neq\beta}}^{A_j} \kappa_j^\alpha \kappa_j^\beta \left[\ln\gamma^* r_{jj}^{\alpha\beta} - \frac{h_{3-j}}{h_j}\mathrm{K}_0\left(\gamma r_{jj}^{\alpha\beta}\right)\right]\right.$$

$$\left. + \sum_{\alpha,\beta=1}^{A_j, A_{3-j}} \frac{h_{3-j}}{h_j}\kappa_j^\alpha \kappa_{3-j}^\beta \left[\ln\gamma^* r_{j(3-j)}^{\alpha\beta} + \mathrm{K}_0\left(\gamma r_{j(3-j)}^{\alpha\beta}\right)\right]\right\} \quad (1.81)$$

is a Hamiltonian, coinciding with the energy of interaction between vortices.

It can be readily seen, that integral invariants M, $\mathscr{H}$ and the combination $(P_x)^2 + (P_y)^2$ are mutually involutive (i. e., the corresponding Poisson brackets are zero), and, according to Liouville theorem [485,486], in the case of three vortices in a two-layer fluid[7], as well as in a homogeneous medium [25,27,29,31,32,34,35,89,97–99,

[7]This is also true for an n-layer fluid with arbitrary n.

102, 105, 113, 330–333, 485, 489, 766, 772, 773, 809, 876, 886], the problem always has a regular solution. In a barotropic fluid, in the general case [36, 101, 384, 485, 595, 654–656, 1030], the problem of four vortices is nonintegrable. In [38, 102, 243, 384] it is shown that at

$$Q = P_x = P_y = 0 \tag{1.82}$$

reduction to three-vortex problems is possible; thus, under the conditions (1.82), the four-vortex problem also becomes integrable.

This result does not depends on the specific form of the Hamiltonian, and hence holds valid for the model of a two-layer fluid. Unlike the case of homogeneous fluid, where many cases of integrable vortex problem at $A_1 + A_2 > 3$ [102, 485, 595, 597] are known, the number of such results for stratified medium is much less [324, 325, 350, 351, 851, 852, 1003].

1.2.8 Phase Portraits. Choreographies

Studying the topological properties of a Hamiltonian, depending only on the distance between vortices, is the basis of the qualitative analysis of relative motion. Clearly, the phase portrait of dynamic system in the plane of generalized coordinates and momentums or (equivalently) physical coordinates of vortices can be obtained only in a two-vortex problem or when the vortex configuration has A axes of symmetry.

When the number of symmetry axes is not large enough, it is advantageous in the qualitative analysis of a dynamic system to use special coordinates allowing phase portraits to be drawn on the plane. The so-called *trilinear coordinates* (t_1, t_2, t_3) are a good example. A procedure for their construction was proposed in [25] and tested in [29, 772, 773] as applied to the analysis of the motion of three vortices.

The key idea of the scheme, which we will formulate here for the case of $A_1 = 1$, $A_2 = 2$ and $h_1 = h_2 = 1/2$, is the use of a combination of integral invariants (1.79) in the form

$$L = 4\left[QM - (P_x)^2 - (P_y)^2\right]. \tag{1.83}$$

As can be readily seen, the variable L can be expressed in terms of quadratic distances between the vortices, thus enabling the relative motion to be analyzed.

The coordinates (t_1, t_2, t_3), satisfying at $L \neq 0$ the obvious identity

$$t_1 + t_2 + t_3 = 3, \tag{1.84}$$

are introduced by the formulas

$$t_1 = \frac{3\kappa_2^1 \kappa_2^2 \left(r_{22}^{12}\right)^2}{L}, \quad t_2 = \frac{3\kappa_1^1 \kappa_2^2 \left(r_{12}^{12}\right)^2}{L}, \quad t_3 = \frac{3\kappa_1^1 \kappa_2^1 \left(r_{12}^{11}\right)^2}{L} \tag{1.85}$$

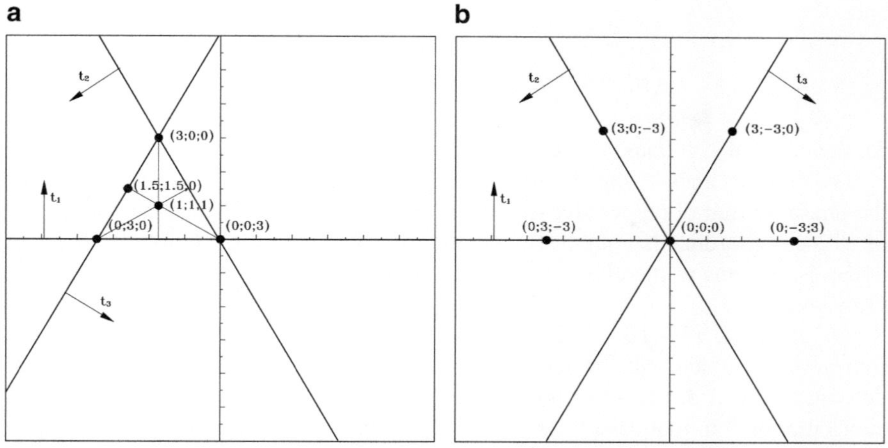

Fig. 1.10 Scheme of the trilinear coordinates with specified orientation of axes at (**a**) $L \neq \neq 0$ and (**b**) $L = 0$

and characterize the distances from any point of the plane to the coordinate axes constructed on the sides of an equiangular triangle with a height of 3 (see Fig. 1.10a).

If $L = 0$, we have

$$t_1 + t_2 + t_3 = 0 \tag{1.86}$$

and

$$t_1 = 3\kappa_2^1\kappa_2^2\left(r_{22}^{12}\right)^2, \quad t_2 = 3\kappa_1^1\kappa_2^2\left(r_{12}^{12}\right)^2, \quad t_3 = 3\kappa_1^1\kappa_2^1\left(r_{12}^{11}\right)^2. \tag{1.87}$$

A geometric interpretation of trilinear coordinates for this case is given in Fig. 1.10b.

In a specific problem (when the initial coordinates of vortices and the intensities κ_j^α are specified), it is necessary to identify on the plane of variables (t_1, t_2, t_3) a "physical domain" (PD), in which the triangle inequality holds true for the mutual distances between any three points. In terms of trilinear coordinates, this inequality takes the form

$$(\kappa_1^1 t_1)^2 + (\kappa_2^1 t_2)^2 + (\kappa_2^2 t_3)^2 \leq 2(\kappa_1^1\kappa_2^1 t_1 t_2 + \kappa_1^1\kappa_2^2 t_1 t_3 + \kappa_2^1\kappa_2^2 t_2 t_3). \tag{1.88}$$

Clearly, the points of PD boundary (where (1.88) turns into equality) always correspond to a continuum of collinear states of the three vortices.

Eliminating t_1 from (1.84) to (1.86), we obtain more compact expressions instead of (1.88)

$$\left[\left(\kappa_1^1 + \kappa_2^1\right) t_2 + \left(\kappa_1^1 + \kappa_2^2\right) t_3 - 3\kappa_1^1\right]^2 \leq 4\kappa_2^1\kappa_2^2 t_2 t_3 \tag{1.89}$$

and

$$\left[\left(\kappa_1^1+\kappa_2^1\right)t_2+\left(\kappa_1^1+\kappa_2^2\right)t_3\right]^2\leq 4\kappa_2^1\kappa_2^2 t_2 t_3 \tag{1.90}$$

for the cases of $L \neq 0$ and $L = 0$, respectively.

The contour lines of the Hamiltonian (1.81) in coordinates (t_1, t_2, t_3) determine the phase portrait of the relative motion of vortices in the PD. Thus, the application of trilinear coordinates notably extends the potentialities of qualitative analysis of dynamic systems to the class of arbitrary motions of vortex structures comprising three vortices.

As noted in [38, 102, 243, 244], when conditions (1.82) are satisfied, the four-vortex problem can be reduced to three-vortex one, because the symmetry determined by (1.82) makes it possible to formally exclude one of the vortices from consideration.[8] In that case at $M \neq 0$, the condition (1.84) is true, and the trilinear coordinates become

$$t_1=\frac{3\kappa_2^1\kappa_2^2\left(r_{22}^{12}\right)^2}{2M(\kappa_1^1+\kappa_2^1+\kappa_2^2)},\quad t_2=\frac{3\kappa_1^1\kappa_2^2\left(r_{12}^{12}\right)^2}{2M(\kappa_1^1+\kappa_2^1+\kappa_2^2)},\quad t_3=\frac{3\kappa_1^1\kappa_2^1\left(r_{12}^{11}\right)^2}{2M(\kappa_1^1+\kappa_2^1+\kappa_2^2)}. \tag{1.91}$$

However, if $M = 0$, then (1.86) is true and

$$t_1=\frac{3\kappa_2^1\kappa_2^2\left(r_{22}^{12}\right)^2}{\kappa_1^1+\kappa_2^1+\kappa_2^2},\quad t_2=\frac{3\kappa_1^1\kappa_2^2\left(r_{12}^{12}\right)^2}{\kappa_1^1+\kappa_2^1+\kappa_2^2},\quad t_3=\frac{3\kappa_1^1\kappa_2^1\left(r_{12}^{11}\right)^2}{\kappa_1^1+\kappa_2^1+\kappa_2^2}. \tag{1.92}$$

The graphic representation of trilinear coordinates in this case at $M \neq 0$ and $M = 0$ is similar to that given in Fig. 1.10a, b, respectively.

Of great importance in celestial mechanics and vortex theory are closed trajectories, which correspond to periodic solutions of the problem. Looking ahead, we note that, in the interpretation of the relative motion of vortices with the use of trilinear coordinates, periodic motions will be identified automatically, since the appropriate phase curves begin and end on PD boundary. The trajectories of such vortices (or material bodies) are referred to as *choreographies* [103, 172, 827, 828].

Though the solution of three-body problem in the form of a closed circular trajectory was first obtained by Lagrange as long ago as 1772 [172], and next, starting from works of Hénon [344] and Moore [627], the class of known solutions (for celestial mechanics problems with homogenous Newtonian potential) has extended considerably (see also [205, 407]), the notion of *c*horeography was introduced by Simó [827] recently (2002).

---

[8] Below (in problem with $A_1 = A_2 = 2$) this procedure will be applied to the vortex $\binom{2}{1}$. Hereafter, $\binom{\alpha}{j}$ denotes the αth vortex in the jth layer.

The choreographies can be

- *absolute*, if the trajectories are closed in a fixed coordinate system;
- *relative*, if the trajectories become closed in a coordinate system, which steadily rotates about a center of rotation (or moves with a constant translational velocity, if the center of rotation is at infinity);
- *simple*, if all bodies (or vortices) move along the same trajectory;
- *complex*, if at least one body moves along a separate trajectory.

Borisov, Mamaev, and Kilin in [101,103,104] were the first to apply the notion of choreography to vortex dynamics in a homogeneous fluid on a plane and a sphere, see also [406,912].

Because of the two-layer character of vortex structures considered in this work, the problem of finding choreographies gains even more in importance. As shown below, a possible situation in a two-layer medium is when the vortices have common trajectories but belong to different layers.

Not setting ourselves the task to systematically study choreographies, we, nonetheless, will be repeatedly mention the cases of existence of this specific class of vortex motion.

1.2.9 Three-Layer Model Equations

The above reasoning was carried out under a two-layer quasi-geostrophic model. Here we will give the main equations applicable to the approximation of a three-layer ocean with uniform-density layers and arbitrary nonzero thicknesses. In [832], studying the stability of a quasi-geostrophic current in a zonal channel, it is noted that, despite the lack of complete analogy between the models of continuously stratified and three-layer flows, the latter can be used for qualitative description of the first three normal modes of the continuous system.

Under the quasi-geostrophic approximation without external forcing, the conservation laws hold true within each layer

$$\frac{D_j \Pi_j}{Dt} = 0, \qquad j = 1,2,3, \tag{1.93}$$

(compare with (1.11)) for the components of three-dimensional potential vortex Π, which is related with the appropriate vortex $\mathbf{p}$ of hydrodynamic pressure perturbation (relative to the hydrostatic-equilibrium pressure) by the linear differential operator

$$\Pi = \nabla^2 \mathbf{p} + T\mathbf{p}, \tag{1.94}$$

$$\Pi = \begin{pmatrix} \Pi_1 \\ \Pi_2 \\ \Pi_3 \end{pmatrix}, \quad \mathbf{p} = \begin{pmatrix} p_1 \\ p_2 \\ p_3 \end{pmatrix}, \quad T = \begin{pmatrix} -\frac{F_1}{h_1} & \frac{F_1}{h_1} & 0 \\ \frac{F_1}{h_2} & -\frac{F_1+F_2}{h_2} & \frac{F_2}{h_2} \\ 0 & \frac{F_2}{h_3} & -\frac{F_2}{h_3} \end{pmatrix},$$

where $F_n = \rho_0(fD)^2/(g\Delta\rho_n H)$, $\Delta\rho_n = \rho_{n+1} - \rho_n$, $n = 1,2$. Here ρ_j $(j = 1,2,3)$ are fluid densities in the layers, $H = h(h_1 + h_2 + h_3)$ is the total depth, and h and D are the characteristic vertical and horizontal scales, such that $h_1 + h_2 + h_3 = 1$, other denotations are the same as in Sect. 1.2.1.

For the interfaces between layers, we have $\eta_j(x,y,t) = F_j(p_{j+1} - p_j)$, $j = 1,2$. Diagonalization by the method described in detail in [405] transforms (1.94) into

$$\mathbf{W} = \nabla^2 \mathbf{w} + U\mathbf{w}, \tag{1.95}$$

where $\mathbf{p} = Q\mathbf{w}$, $\mathbf{W} = S\Pi$, $U = \lambda E$,

$$Q = (\mathbf{q}^{(1)}, \mathbf{q}^{(2)}, \mathbf{q}^{(3)})$$
$$= \begin{pmatrix} 1 & \frac{h_3\lambda_3}{\lambda_3-\lambda_2} & -\frac{F_1}{h_1\lambda_3} \\ 1 & \frac{1}{\lambda_3-\lambda_2}\left(h_3\lambda_3 + \frac{F_2}{h_2}\right) & -\left(\frac{F_1}{h_1\lambda_3} + 1\right) \\ 1 & \frac{1}{\lambda_3-\lambda_2}\left(h_3\lambda_3 + \frac{F_2}{h_2} + \lambda_2 + \frac{F_1(h_1+h_2)}{h_1h_2}\right) & -\left(\frac{F_1}{h_1\lambda_3} + 1 + \frac{h_2}{F_2}\left(\lambda_3 + \frac{F_1(h_1+h_2)}{h_1h_2}\right)\right) \end{pmatrix},$$

$$S = Q^{-1}$$
$$= \begin{pmatrix} h_1 & h_2 & h_3 \\ -\frac{h_2}{F_2}\left(\lambda_3 + \frac{F_1(h_1+h_2)}{h_1h_2}\right) & \frac{h_2}{F_2}\left(\lambda_3 + \frac{F_1(h_1+h_2)+F_2h_1}{h_1h_2}\right) & -1 \\ -\frac{1}{\lambda_3-\lambda_2}\left(\lambda_2 + \frac{F_1(h_1+h_2)}{h_1h_2}\right) & \frac{1}{\lambda_3-\lambda_2}\left(\lambda_2 + \frac{F_1(h_1+h_2)+F_2h_1}{h_1h_2}\right) & -\frac{1}{\lambda_3-\lambda_2}\frac{F_2}{h_2} \end{pmatrix},$$

$$\lambda_1 = 0,\ \lambda_{2,3} = -\frac{1}{2}\left[\frac{F_1}{h_1} + \frac{F_1+F_2}{h_2} + \frac{F_2}{h_3} \mp \sqrt{\left(\frac{F_1}{h_1} + \frac{F_1+F_2}{h_2} + \frac{F_2}{h_3}\right)^2 - 4\frac{F_1F_2}{h_1h_2h_3}}\right]. \tag{1.96}$$

Here λ_j are the eigenvalues of the spectral problem $T\mathbf{q} + \lambda\mathbf{q} = 0$; $\mathbf{q}^{(1)}$, $\mathbf{q}^{(2)}$, $\mathbf{q}^{(3)}$ are its eigenvectors, E is an identity matrix.

In the simplest case, where $\Delta\rho_1 = \Delta\rho_2$ and $h_1 = h_3$, we have

$$F_1 = F_2 = F, \quad \lambda_2 = -F/h_1, \quad \lambda_3 = -F/h_1h_2,$$
$$Q = \begin{pmatrix} 1 & 1/2 & h_2 \\ 1 & 0 & -2h_1 \\ 1 & -1/2 & h_2 \end{pmatrix}, \qquad S = \begin{pmatrix} h_1 & h_2 & h_1 \\ 1 & 0 & -1 \\ 1/2 & -1 & 1/2 \end{pmatrix}.$$

In any point (x,y), the components w_j of the auxiliary vector $\mathbf{w}$ from the right-hand part of (1.95) can be determined with the use of Green function G_j:

$$w_j(x,y) = \iint_{-\infty}^{+\infty} W_j G_j(r)dx'dy', \quad j = 1,2,3,$$

$$G_j = \frac{1}{2\pi}\begin{cases} \ln \gamma^* r, & j = 1, \\ -\mathrm{K}_0(\gamma_1 r), & j = 2, \\ -\mathrm{K}_0(\gamma_2 r), & j = 3, \end{cases}$$

where $r = \sqrt{(x-x')^2 + (y-y')^2}$ and $\gamma_{1,2} = \sqrt{-\lambda_{2,3}}$. Let us suppose, as in Sect. 1.2.3, that the potential vorticities Π_j are piecewise-constant functions of the type $\Pi_j = \sum_{\alpha=1}^{A_j} \Pi_j^\alpha$ (A_j is the number of vortex patches in the jth layer); now the expressions for pressure perturbations in layers can be expressed in terms of combination of contour integrals

$$p_j = \sum_{\alpha=1}^{A_j} \Pi_j^\alpha \oint_{C_j^\alpha} NM(r)dv + \sum_{m=2}^{3} q_{jm} \sum_{n=1}^{3} s_{mn} \sum_{\alpha=1}^{A_n} \Pi_n^\alpha \oint_{C_n^\alpha} NM_{m-1}(r)dv,$$
$$j = 1,2,3. \tag{1.97}$$

Here q_{jn}, s_{jn} are elements of the matrices Q and S introduced above, $M_{1,2} = \frac{1}{2\pi\gamma_{1,2}^2}\left[\gamma_{1,2} r \mathrm{K}_1(\gamma_{1,2} r) - 1\right]$, and other denotations have the same meaning as in (1.32).

The behavior of interfaces η_1 (between the top and middle layers) and η_2 (between the middle and bottom layers) is determined by the equations

$$\eta_j = F_j(p_{j+1} - p_j), \qquad j = 1,2. \tag{1.98}$$

The system of equations describing the motion of fluid particles at the boundaries of vortex patches C_j^α can be solved numerically with the help of CDM procedures [844].

On studying the stability of three-layer vortex. Obviously, in the three-layer model, any axisymmetric (circular) distribution of potential vorticity (in particular, at $A_1 = A_2 = A_3 = 1$) represents a steady state and induces a velocity field of the type

$$V_j(r) = \sum_{n=1}^{3} q_{jn} \sum_{i=1}^{3} s_{ni} \Pi_m^1 H_{n-1}^{(i)}, \quad j = 1,2,3, \tag{1.99}$$

where

$$H_0^{(i)} = \frac{1}{2}\begin{cases} r/\rho_i, \ r \le \rho_i, \\ \rho_i/r, \ r > \rho_i, \end{cases}$$

$$H_l^{(i)} = \begin{cases} \mathrm{K}_1(\gamma_l\rho_i)\mathrm{I}_1(\gamma_l r), \ r \le \rho_i, \\ \mathrm{I}_1(\gamma_l\rho_i)\mathrm{K}_1(\gamma_l r), \ r > \rho_i, \end{cases} \qquad l = 1,2\,,$$

and ρ_i is the radius of the vortex patch of the ith layer.

The general scheme of linear analysis of the stability of such three-layer vortex is the same as that described in Sect. 1.2.5. The expressions (1.52) and (1.68) Still have the form

$$f_j(\theta, t; \rho_j) = \rho_j + \varepsilon_j(\theta, t; \rho_j), \quad |\varepsilon_j| \ll \rho_j,$$

$$\varepsilon_j(\theta, t; \rho_j) = \varepsilon_j(\rho_j)e^{im(\theta-\delta t)}, \quad m \ge 1,$$

where m is the number of the azimuthal mode, but now $j = 1, 2, 3$.

The method of perturbation in the powers of the small parameter ε_j, enabling the linearization of the equations of motion written for f_j

$$f_j f_{jt} + V_j^{(\theta)} f_{j\theta} - V_j^{(r)} f_j = 0, \quad j = 1, 2, 3,$$

allows the problem of stability of a three-layer vortex to be reduced to the solution of the spectral problem that has a system of characteristic equations

$$|B - \delta E| = 0$$

with a real matrix B of the type

$$b_{nj} = \sum_{k=1}^{3} s_{kn}\left(q_{nk}T_{nk}^{(k-1)} - \delta_{nj}\frac{m}{\rho_n}U_{k-1}(\rho_n)\right), \quad n, j = 1,2,3.$$

Here δ_{nj} is Kronecker delta, $T_{nn}^{(0)} = 1/2$, $T_{nn}^{(1,2)} = m\mathrm{I}_n(\gamma_{1,2}\rho_n)\mathrm{K}_n(\gamma_{1,2}\rho_n)$ $(n = 1,2,3)$. The variables U_j and $T_{nk}^{(j)}$ at $n \ne k$ depend on the vertical structure of the three-layer vortex. A version of expressions for them is given below.

The following six combinations are possible:

$$\begin{aligned} &1)\ \rho_1 \ge \rho_2 \ge \rho_3, \quad 2)\ \rho_1 \ge \rho_3 \ge \rho_2, \\ &3)\ \rho_2 \ge \rho_1 \ge \rho_3, \quad 4)\ \rho_2 \ge \rho_3 \ge \rho_1, \\ &5)\ \rho_3 \ge \rho_1 \ge \rho_2, \quad 6)\ \rho_3 \ge \rho_2 \ge \rho_1. \end{aligned} \tag{1.100}$$

The particular case of a cylindrical vortex with $\rho_1 = \rho_2 = \rho_3$, is contained in any inequality out of (1.100).

Without going into details of simple but cumbersome calculations, we give expressions for $T_{nk}^{(j)}$ and U_j only for the first out of the six variants (1.100):

$$T_{12}^{(0)} = \frac{1}{2}\left(\frac{\rho_2}{\rho_1}\right)^{(m-1)}, \quad T_{13}^{(0)} = \frac{1}{2}\left(\frac{\rho_3}{\rho_1}\right)^{(m-1)}, \quad T_{21}^{(0)} = \frac{1}{2}\left(\frac{\rho_2}{\rho_1}\right)^{(m+1)},$$

$$T_{23}^{(0)} = \frac{1}{2}\left(\frac{\rho_3}{\rho_2}\right)^{(m-1)}, \quad T_{31}^{(0)} = \frac{1}{2}\left(\frac{\rho_3}{\rho_1}\right)^{(m+1)}, \quad T_{32}^{(0)} = \frac{1}{2}\left(\frac{\rho_3}{\rho_2}\right)^{(m+1)},$$

$$T_{12}^{(j)} = \frac{\rho_1}{\rho_2} m \mathrm{I}_m(\gamma_j\rho_2)\mathrm{K}_m(\gamma_j\rho_1), \quad T_{13}^{(j)} = \frac{\rho_1}{\rho_3} m \mathrm{I}_m(\gamma_j\rho_3)\mathrm{K}_m(\gamma_j\rho_1),$$

$$T_{21}^{(j)} = \frac{\rho_2}{\rho_1} m \mathrm{I}_m(\gamma_j\rho_2)\mathrm{K}_m(\gamma_j\rho_1), \quad T_{23}^{(j)} = \frac{\rho_2}{\rho_3} m \mathrm{I}_m(\gamma_j\rho_3)\mathrm{K}_m(\gamma_j\rho_2),$$

$$T_{31}^{(j)} = \frac{\rho_3}{\rho_1} m \mathrm{I}_m(\gamma_j\rho_3)\mathrm{K}_m(\gamma_j\rho_1), \quad T_{32}^{(j)} = \frac{\rho_3}{\rho_2} m \mathrm{I}_m(\gamma_j\rho_3)\mathrm{K}_m(\gamma_j\rho_2),$$

$$U_0(\rho_1) = \frac{1}{2}\left(s_{11}\Pi_1\rho_1^2 + s_{12}\Pi_2\rho_2^2 + s_{13}\Pi_3\rho_3^2\right)/\rho_1,$$

$$U_0(\rho_2) = \frac{1}{2}\left((s_{11}\Pi_1 + s_{12}\Pi_2)\rho_2 + s_{13}\Pi_3\rho_3^2/\rho_2\right),$$

$$U_0(\rho_3) = \frac{1}{2}\left(s_{11}\Pi_1 + s_{12}\Pi_2 + s_{13}\Pi_3\right)\rho_3,$$

$$U_j(\rho_1) = \mathrm{K}_1(\gamma_j\rho_1)\Big(s_{j+1,1}\Pi_1\rho_1\mathrm{I}_1(\gamma_j\rho_1) + s_{j+1,2}\Pi_2\rho_2\mathrm{I}_1(\gamma_j\rho_2)$$
$$+ s_{j+1,3}\Pi_3\rho_3\mathrm{I}_1(\gamma_j\rho_3)\Big),$$

$$U_j(\rho_2) = s_{j+1,1}\Pi_1\mathrm{I}_1(\gamma_j\rho_2)\left(2\rho_2\mathrm{K}_1(\gamma_j\rho_2) - \rho_1\mathrm{K}_1(\gamma_j\rho_1)\right)$$
$$+ s_{j+1,2}\Pi_2\rho_2\mathrm{I}_1(\gamma_j\rho_2)\mathrm{K}_1(\gamma_j\rho_2) + s_{j+1,3}\Pi_3\rho_3\mathrm{I}_1(\gamma_j\rho_3)\mathrm{K}_1(\gamma_j\rho_2),$$

$$U_j(\rho_3) = \mathrm{I}_1(\gamma_j\rho_3)\Big[s_{j+1,1}\Pi_1\Big(2\rho_3\mathrm{K}_1(\gamma_j\rho_3) - \rho_1\mathrm{K}_1(\gamma_j\rho_1)\Big)$$
$$+ s_{j+1,2}\Pi_2\Big(2\rho_3\mathrm{K}_1(\gamma_j\rho_2) - \rho_2\mathrm{K}_1(\gamma_j\rho_2)\Big)$$
$$+ s_{j+1,3}\Pi_3\rho_3\mathrm{K}_1(\gamma_j\rho_3)\Big].$$

Here $j = 1,2$ and $m \geq 1$. The structure of those terms for other five variants in (1.100) is analogous.

The formation of a vortex structure over an isolated submarine obstacle in a three-layer fluid. Let us suppose that, similar to the case considered in paragraph 1.6, we have a vertically homogeneous flow $U(t)$, incoming onto a small-height elevation (at least, not exceeding the thickness of the bottom layer) with the shape of a circular cylinder with unit radius. Now the external field will be determined by anomalies of the hydrodynamic pressure described by

$$p_{0j} = -U(t)y - \sigma c_j(\gamma_1, \gamma_2, r), \quad j = 1,2,3, \tag{1.101}$$

$$c_j = \begin{cases} \dfrac{r^2}{4} - \dfrac{q_{j2}s_{23}}{s_{13}(\gamma_1)^2}\Big[1-\gamma_1 \mathrm{K}_1(\gamma_1)\mathrm{I}_0(\gamma_1 r)\Big] - \dfrac{q_{j3}s_{33}}{s_{13}(\gamma_2)^2}\Big[1-\gamma_1 \mathrm{K}_1(\gamma_2)\mathrm{I}_0(\gamma_2 r)\Big], & r \le 1, \\ \dfrac{1+\ln r^2}{4} - \dfrac{q_{j2}s_{23}}{s_{13}(\gamma_1)^2}\Big[1-\gamma_1 \mathrm{I}_1(\gamma_1)\mathrm{K}_0(\gamma_1 r)\Big] - \dfrac{q_{j3}s_{33}}{s_{13}(\gamma_2)^2}\Big[1-\gamma_1 \mathrm{I}_1(\gamma_2)\mathrm{K}_0(\gamma_2 r)\Big], & r \ge 1. \end{cases} \tag{1.102}$$

The expressions for velocities in the layers can be written as

$$\begin{aligned} u_{0j} &= U(t) + \frac{\sigma y}{r} d_j(\gamma_1, \gamma_2, r), \\ v_{0j} &= -\frac{\sigma x}{r} d_j(\gamma_1, \gamma_2, r), \quad j = 1,2,3, \end{aligned} \tag{1.103}$$

where

$$d_j = \begin{cases} \dfrac{r}{2} + \dfrac{q_{j2}s_{23}}{s_{13}(\gamma_1)^2}\mathrm{K}_1(\gamma_1)\mathrm{I}_1(\gamma_1 r) + \dfrac{q_{j3}s_{33}}{s_{13}(\gamma_2)^2}\mathrm{K}_1(\gamma_2)\mathrm{I}_1(\gamma_2 r), & r \le 1, \\ \dfrac{1}{2r} + \dfrac{q_{j2}s_{23}}{s_{13}(\gamma_1)^2}\mathrm{I}_1(\gamma_1)\mathrm{K}_1(\gamma_1 r) + \dfrac{q_{j3}s_{33}}{s_{13}(\gamma_2)^2}\mathrm{I}_1(\gamma_2)\mathrm{K}_1(\gamma_2 r), & r \ge 1. \end{cases} \tag{1.104}$$

Using formulas (1.103) and (1.104), we can obtain the critical values of parameter $\sigma = \sigma_j^*$, at which a first stagnant point appears in each layer (it is clear that at $U > 0$, such point should be at the boundary of the elevation at $y = -1$). Introducing the notation $\theta_j = U/\sigma_j^*$, we obtain

$$\theta_j = \frac{1}{2} + \frac{q_{j2}s_{23}}{s_{13}} L_1(\gamma_1) + \frac{q_{j3}s_{33}}{s_{13}} L_1(\gamma_2), \quad j = 1,2,3, \tag{1.105}$$

where $L_1(\beta) = \mathrm{K}_1(\beta)\mathrm{I}_1(\beta)$ are monotonically decreasing functions with limiting values of 1/2 and 0 at $\beta \to 0$ and $\beta \to \infty$, respectively. Since $\gamma_2 > \gamma_1$, we have

$$0 < \frac{1}{2} - L_1(\gamma_2) < \theta_1 = \frac{1}{2} - \frac{\lambda_3 L_1(\gamma_1) - \lambda_2 L_1(\gamma_2)}{\lambda_3 - \lambda_2} < \frac{1}{2} - L_1(\gamma_1) < \frac{1}{2},$$

$$\theta_1 < \theta_1 + \frac{F_2\Big(L_1(\gamma_1) - L_1(\gamma_2)\Big)}{h_2 h_3\Big((\gamma_2)^2 - (\gamma_1)^2\Big)} < \theta_2 = \frac{1}{2}$$

$$-\frac{\Big(\lambda_3 + F_2/(h_2h_3)\Big)L_1(\gamma_1) - \Big(\lambda_2 + F_2/(h_2h_3)\Big)L_1(\gamma_2)}{\lambda_3 - \lambda_2} < \frac{1}{2} - L_1(\gamma_1) < \frac{1}{2},$$

$$\frac{1}{2} < \frac{1}{2} + L_1(\gamma_2)\left(\frac{1}{h_3} - 1\right) < \theta_3 = \frac{1}{2}$$
$$- \frac{1}{h_3(\lambda_3 - \lambda_2)} \left[\left(h_3\lambda_3 + \frac{F_2}{h_2} + \lambda_2 + \frac{F_1(h_1+h_2)}{h_1 h_2} \right) L_1(\gamma_1) \right.$$
$$\left. - \left(h_3\lambda_3 + \frac{F_2}{h_2} + \lambda_3 + \frac{F_1(h_1+h_2)}{h_1 h_2} \right) L_1(\gamma_2) \right]$$
$$< \frac{1}{2} + L_1(\gamma_1)\left(\frac{1}{h_3} - 1\right) < \frac{1}{2h_3},$$

whence

$$0 < \theta_1 < \theta_2 < \frac{1}{2} < \theta_3 < \frac{1}{2h_3}, \tag{1.106}$$

where $\theta_{1,2} \to 1/2 - 0$, $\theta_3 \to 1/2 + 0$ at $\gamma_1, \gamma_2 \gg 1$ and $\theta_{1,2} \to 0$, $\theta_3 \to 1/(2h_3)$ at $\gamma_1, \gamma_2 \ll 1$.

The value of $\gamma_{1,2}$ being small requires that the solid-cap conditions be satisfied at the interfaces between layers for protecting the top and middle layers from the effect of finite perturbations of bed relief. Therefore, at small $\gamma_{1,2}$, we have $\sigma_{1,2} \to \infty$. Essentially, both limits $\gamma_{1,2} \to \infty$ and $\gamma_{1,2} \to 0$ correspond to barotropy conditions: the first case corresponds to a homogeneous ocean, and the second case, to a three-layer ocean with interacting layers and theoretically infinite density jumps at interfaces between them. In both cases, the critical values of the topographic parameter show the relationship $\sigma^* = 2a$, where $\sigma^* = \sigma_1^* = \sigma_2^* = \sigma_3^*$, $a = U$ for the first limiting case, and $\sigma^* = \sigma_3^*$, $a = Uh_3$ for the second case, the fact that is in agreement with the well-known result in [375].

From (1.106), in the general case, we obtain

$$\sigma_1^* > \sigma_2^* > 2U > \sigma_3^* > 2Uh_3 > 0, \tag{1.107}$$

and thus, the flow is the most (least) subject to the impact of bottom topography in the lower (top) layer.

The limiting case of discrete vortices. The dynamics of three- and N-layer vortices was studied in detail by Gryanik and Tevs [327–329, 897]; here, we will show only the main equations.

As was the case with two-layer fluid, the equations of motion of vortices can be written in the Hamiltonian form

$$\dot{q}_j^\alpha = \frac{\partial \mathcal{H}}{\partial p_j^\alpha}, \qquad \dot{p}_j^\alpha = -\frac{\partial \mathcal{H}}{\partial q_j^\alpha},$$
$$\alpha = 1, 2, \ldots, A^j; \quad j = 1, 2, 3, \tag{1.108}$$

where $q_j^\alpha = x_j^\alpha$ are generalized coordinates, $p_j^\alpha = h_j y_j^\alpha \kappa_j^\alpha$ are generalized momentums, and the Hamiltonian takes the form

$$\dot{q}_j^\alpha = \frac{\partial \mathcal{H}}{\partial p_j^\alpha}, \qquad \dot{p}_j^\alpha = -\frac{\partial \mathcal{H}}{\partial q_j^\alpha},$$

$$\alpha = 1, 2, \ldots, A^j; \quad j = 1, 2, 3, \tag{1.109}$$

where $q_j^\alpha = x_j^\alpha$ are generalized coordinates, $p_j^\alpha = h_j y_j^\alpha \kappa_j^\alpha$ are generalized momentums, and the Hamiltonian takes the form

$$\mathcal{H} = -\frac{1}{4\pi} \sum_{j,m=1}^{3} \sum_{\substack{\alpha,\beta=1 \\ \alpha\neq\beta}}^{A_j,A_m} \kappa_j^\alpha \kappa_m^\beta \left[q_{j1} s_{1m} \ln \gamma^* r_{jm}^{\alpha\beta} - \sum_{k=1}^{2} q_{jk} s_{km} \mathrm{K}_0 \left(\gamma_k r_{jm}^{\alpha\beta} \right) \right].$$

Chapter 2
Dynamics of Discrete Vortices

Abstract This chapter deals mostly with the dynamics of discrete vortices in a two-layer fluid and has the following structure. First, two vortices (a simplest heton) are used to demonstrate the characteristic features of a system of baroclinic vortices, after which, the results of analytical and numerical studying of vortex ensembles are studied, starting from relatively complex (systems of A and $A+1$ vortices) with arbitrary A to simpler ones with $A=2$. A class of motions of axisymmetric vortex structures in an external deformation field. The possibility of formation of chaotic regimes is examined. The last subsection of the chapter gives formulas describing the motion of simplest stationary vortex structures in a three-layer fluid.

The structure of the chapter is as follows. First, we consider the simplest case of two vortices in a two-layer fluid, reflecting the major characteristic features of vortex interaction in a layered stratified fluid (Sect. 2.1). Next (now, in the opposite direction—from the general to the special), we consider the properties of two-layer vortex structures (hetons) of the form $\kappa_1^\alpha = -\kappa_2^\alpha = \kappa$, $\alpha = 1,2,\ldots,A$, where $A = A_1 = A_2$ (Sect. 2.2) with arbitrary A (Sect. 2.2.1) and, separately, in the particular case of $A = 2$ (Sect. 2.2.2), as well as structures of the type $\kappa_1^1 \equiv \kappa_1$, $\kappa_2^1 = \kappa_2^2 = \ldots = \kappa_2^\alpha \equiv \kappa_2$, $\alpha = 1,2,\ldots,A$, $A = A_2, A_1 = 1$ (Sect. 2.3) for arbitrary A (Sect. 2.3.1) with a particular case $\kappa_1 = -2\kappa_2$, $A = 2$ (Sect. 2.3.2). In the same section, the asymmetric (in terms of intensity distribution) case $\kappa_1^1 = \kappa_2^2 = -\kappa$, $\kappa_2^1 = 2\kappa$ is considered. Section 2.4 gives formulas for the kinematic characteristics of stationary vortex structures in a three-layer fluid.

This chapter is based on the authors' papers [325, 847, 849, 851–858].

In Sects. 2.1, 2.2 and 2.3, the top and bottom layers are assumed to have equal thicknesses $h_1 = h_2 = 1/2$.

M.A. Sokolovskiy and J. Verron, *Dynamics of Vortex Structures in a Stratified Rotating Fluid*, Atmospheric and Oceanographic Sciences Library 47,
DOI 10.1007/978-3-319-00789-2_2, © Springer International Publishing Switzerland 2014

2.1 Two Vortices in a Two-Layer Fluid

The specificity of vortex interactions in a two-layer fluid can be seen even in the simplest case $A_1 + A_2 = 2$, so we begin the preliminary consideration from this case. This class of motions has been studied also in [315, 350, 1003].

The possible variants include (a) $A_1 = 2, A_2 = 0$ or $A_1 = 0, A_2 = 2$[1] and (b) $A_1 = A_2 = 1$. Equations (1.77) and (1.78) for those variants yield,

(a)

$$u_1^1 = -\frac{\kappa_1^2(y_1^1 - y_1^2)}{4\pi\left(r_{11}^{12}\right)^2}\left[1 + \gamma r_{11}^{12}\mathrm{K}_1\left(\gamma r_{11}^{12}\right)\right],$$

$$u_1^2 = -\frac{\kappa_1^1(y_1^2 - y_1^1)}{4\pi\left(r_{11}^{12}\right)^2}\left[1 + \gamma r_{11}^{12}\mathrm{K}_1\left(\gamma r_{11}^{12}\right)\right];$$

$$v_1^1 = \frac{\kappa_1^2(x_1^1 - x_1^2)}{4\pi\left(r_{11}^{12}\right)^2}\left[1 + \gamma r_{11}^{12}K_1\left(\gamma r_{11}^{12}\right)\right],$$

$$v_1^2 = \frac{\kappa_1^1(x_1^2 - x_1^1)}{\left(4\pi r_{11}^{12}\right)^2}\left[1 + \gamma r_{11}^{12}K_1\left(\gamma r_{11}^{12}\right)\right]; \tag{2.1}$$

(b)

$$u_1^1 = -\frac{\kappa_2^1(y_1^1 - y_2^1)}{4\pi\left(r_{12}^{11}\right)^2}\left[1 - \gamma r_{12}^{11}\mathrm{K}_1\left(\gamma r_{12}^{11}\right)\right],$$

$$u_2^1 = -\frac{\kappa_1^1(y_2^1 - y_1^1)}{4\pi\left(r_{12}^{11}\right)^2}\left[1 - \gamma r_{12}^{11}\mathrm{K}_1\left(\gamma r_{12}^{11}\right)\right];$$

$$v_1^1 = \frac{\kappa_2^1(x_1^1 - x_2^1)}{4\pi\left(r_{12}^{11}\right)^2}\left[1 - \gamma r_{12}^{11}\mathrm{K}_1\left(\gamma r_{12}^{11}\right)\right],$$

$$v_2^1 = \frac{\kappa_1^1(x_2^1 - x_1^1)}{4\pi\left(r_{12}^{11}\right)^2}\left[1 - \gamma r_{12}^{11}\mathrm{K}_1\left(\gamma r_{12}^{11}\right)\right], \tag{2.2}$$

respectively. In the general case, the vortices are involved in circular motions with the angular velocities

[1] Formulas (1.77) and (1.78) show that, with layer thicknesses assumed equal, those variants yield equivalent results; therefore, we may consider only one of them, e.g., the first.

(a)

$$\omega = \frac{\kappa_1^1 + \kappa_1^2}{4\pi\left(r_{11}^{12}\right)^2}\left[1 + \gamma r_{11}^{12} \mathrm{K}_1\left(\gamma r_{11}^{12}\right)\right], \tag{2.3}$$

(b)

$$\omega = \frac{\kappa_1^1 + \kappa_2^1}{4\pi\left(r_{12}^{11}\right)^2}\left[1 - \gamma r_{12}^{11} \mathrm{K}_1\left(\gamma r_{12}^{11}\right)\right] \tag{2.4}$$

about vorticity centers with coordinates

(a)

$$\left(x_c, y_c\right) = \frac{\kappa_1^1\left(x_1^1, y_1^1\right) + \kappa_1^2\left(x_1^2, y_1^2\right)}{\kappa_1^1 + \kappa_1^2}, \tag{2.5}$$

(b)

$$\left(x_c, y_c\right) = \frac{\kappa_1^1\left(x_1^1, y_1^1\right) + \kappa_2^1\left(x_2^1, y_2^1\right)}{\kappa_1^1 + \kappa_2^1}. \tag{2.6}$$

A system of two vortices (belonging to different layers) rotating at $\kappa_1^1 + \kappa_2^1 \neq 0$ in accordance with (2.4) about the center (2.6) will be referred to as a *two-tier top*.

Formulas (2.1), (2.2), (2.3), (2.4), (2.5) and (2.6) indicate to a specific feature of the two-layer model [315]:

(!) The interactions between vortices belonging to the same layer and vortices belonging to different layers follow different laws.[2]

This can be most clearly seen in the examples of one-layer and two-layer vortex pairs, where we have $\kappa_1^1 = -\kappa_1^2 \equiv \kappa$ in the case (a) and $\kappa_1^1 = -\kappa_2^1 \equiv \kappa$ in the case (b), and, clearly, the vorticity center lies at infinity, and the angular velocities (2.3) and (2.4) vanish.

Indeed, suppose that, at the initial moment, both vortices lie on the x-axis spaced l apart. Now formulas (2.1) and (2.2) yield the following expressions for the velocities of vortex pairs (*heton with tilted axis*):

(a)

$$\begin{aligned} u_1^1 &= u_1^2 = 0, \\ v_1^1 &= v_1^2 = \pm\frac{\gamma\kappa}{4\pi L}\left[1 + L\mathrm{K}_1(L)\right] \equiv \pm\frac{\gamma\kappa}{4\pi} f_1(L), \end{aligned} \tag{2.7}$$

[2]Below, we will refer to this statement more than once, using the sign (**!**).

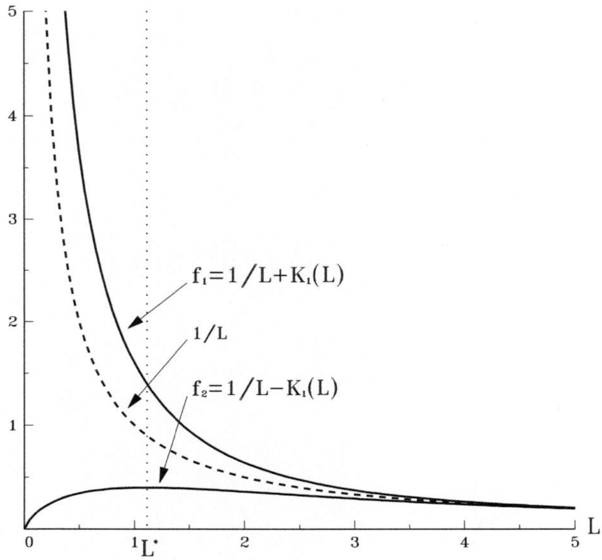

Fig. 2.1 Relationships (2.7) and (2.8) for $f_1(L)$ and $f_2(L)$. The *vertical dashed line* shows the position of the maximum of $f_2(L)$

(b)

$$u_1^1 = u_2^1 = 0,$$
$$v_1^1 = v_2^1 = \pm\frac{\gamma\kappa}{4\pi L}\left[1 - L\mathrm{K}_1(L)\right] \equiv \pm\frac{\gamma\kappa}{4\pi} f_2(L), \tag{2.8}$$

where $L = \gamma l$. Here at $\kappa > 0$, the top (bottom) sign corresponds to the situation where the vortex with positive intensity occupies the left (right) position; when $\kappa < 0$, the orientation of vortex pair motion is inverse. Formulas (2.7) and (2.8) and Fig. 2.1, where the behavior of functions $f_1(L)$ and $f_2(L)$ is shown (compared with function $f(L) = 1/L$), demonstrate their significant difference. It should be mentioned that *heton* velocity (2.8) is a *nonmonotonic* function of the distance between vortices:

$$f_2(L) \sim -L\ln L \to 0 \quad \text{at} \quad L \ll 1, \qquad f_2(L) \sim 1/L \to 0 \quad \text{at} \quad L \gg 1,$$

and $f_2(L^*) = \max f_2(L)$, where the value $L^* = 1.114$ is the solution of the equation

$$f_2'(L) = \mathrm{K}_0(L) + \frac{1}{L}\mathrm{K}_1(L) - \frac{1}{L^2} = 0.$$

Thus, a heton with a *vertical* axis is always immobile, while the motion of a heton with a *tilted* axis is translational, its velocity attaining its maximum at $l = L^*/\gamma$. This fact will be of key importance below in the analysis of motion of vortex structures at $A_1 + A_2 > 2$.

2.2 2A Vortices in a Two-Layer Fluid

A system of vortices with $2A \geq 4$ can be integrated when it has additional symmetries [26, 68, 439, 595, 635].

Suppose that $A_1 = A_2 = A$, and vortices with intensities $\kappa_1^1 = \kappa_1^2 = \ldots = \kappa_1^A \equiv \kappa$, $\kappa_2^1 = \kappa_2^2 = \ldots = \kappa_2^A \equiv -\kappa$ (here we assume $\kappa < 0$) in the top and bottom layers lie in the vertices of a regular A-vortex polygon.

2.2.1 *The Case of Arbitrary A*

Considering the assumed symmetry conditions, the relative motion equations can be reduced to equations describing the evolution of an equivalent heton. The method of such reduction of Hamiltonian systems is described in [26, 595] in the context of the model of two-dimensional vortices.[3]

Following [26], we search for solution of (1.77) and (1.78) in the form

$$x_j^\alpha = r(t)\cos\left[\varphi_j(t) + \frac{2\pi(\alpha-1)}{A}\right],$$
$$y_j^\alpha = r(t)\sin\left[\varphi_j(t) + \frac{2\pi(\alpha-1)}{A}\right], \tag{2.9}$$

where (r, φ_j) are polar coordinates, and, as before, $j = 1, 2$ is layer number, $\alpha = 1, 2, \ldots, A$ is heton number.

Now, from (1.77) and (1.78), we obtain the reduced equations

$$\dot{r} = -\frac{\kappa}{4\pi r}\left\{\frac{A}{2}s(0, A\varphi) - \gamma r\left[\sum_{\alpha=1}^{A-1} s(\alpha, 0)S(\alpha, 0) + \sum_{\alpha=0}^{A-1} s(\alpha, \varphi)S(\alpha, \varphi)\right]\right\}, \tag{2.10}$$

[3] In works [26, 595], sets of polygonal vortex configurations lied on circumferences of different radiuses. The two-layer model, containing no singularity in the case of coincidence of coordinates of vortices in different layers, allows the method to be extended to the interesting case of equal radiuses as considered below.

$$\dot{\varphi} = \frac{\kappa}{4\pi r^2}\left\{-\frac{1}{2}+\gamma r\left[\sum_{\alpha=1}^{A-1} S(\alpha,0)+\sum_{\alpha=0}^{A-1} S(\alpha,\varphi)\right]\right\} \tag{2.11}$$

for the equivalent heton. Here, $\varphi = \varphi_{3-j} - \varphi_j$ and

$$s(c,d) = \frac{\sin\left(\frac{2\pi c}{A}+d\right)}{1-\cos\left(\frac{2\pi c}{A}+d\right)},$$

$$S(c,d) = \left|\sin\left(\frac{\pi c}{A}+\frac{d}{2}\right)\right| \mathrm{K}_1\left[2\gamma r\left|\sin\left(\frac{\pi c}{A}+\frac{d}{2}\right)\right|\right].$$

The invariance condition for the Hamiltonian (1.81), considering the formulas for finite products in expressions under logarithm [724], becomes

$$\begin{aligned}\Phi(R,\varphi,A) &= \ln R\left|\sin\frac{A\varphi}{2}\right| + \mathrm{K}_0\left(2R\sin\frac{\varphi}{2}\right)\\ &+\sum_{\alpha=1}^{A-1}\left\{\mathrm{K}_0\left(2R\sin\frac{\pi\alpha}{A}\right)+\mathrm{K}_0\left[2R\left|\sin\left(\frac{\pi\alpha}{A}+\frac{\varphi}{2}\right)\right|\right]\right\} = \mathrm{const},\end{aligned} \tag{2.12}$$

with denotation $R = \gamma r$.

Figure 2.2 gives phase portraits—contour lines of function (2.12)—for the absolute motion of vortices at $A = 3$ and $A = 6$, when we have three and six radial symmetry axes, respectively. The phase curves are constructed in (x, y) axes at different values of γ (see explanations for the figure), and with passage to coordinates $(\gamma x, \gamma y)$, the linear scale in Fig. 2.2b increases twofold. The domains $\{1\}$, $\{2\}$, and $\{3\}$ correspond to different types of motion.

The motions of the type $\{1\}$ are either (a) counter motions of two-layer dipole structures with subsequent exchange of partners,[4] if at the initial moment, we have an A-symmetric construction with $2A$ vertices, unequal adjacent faces, and alternating vortices of the top and bottom layers in its vertices, or (b) pairwise coupling of vortices from different layers into A receding vortices with tilted axes, i. e. the second part of the previous scenario, if the configuration is a regular polygon with $2A$ vertices. In such case, the phase curves corresponding to the trajectories of a single two-layer pair lie in two subdomains $\{1\}$, separated by a subdomain $\{2\}$.

[4] At $A = 2$, we have an analogue of Greenhill problem [510] with the only difference that in this case the vortices involved in the pair motion belong to different layers.

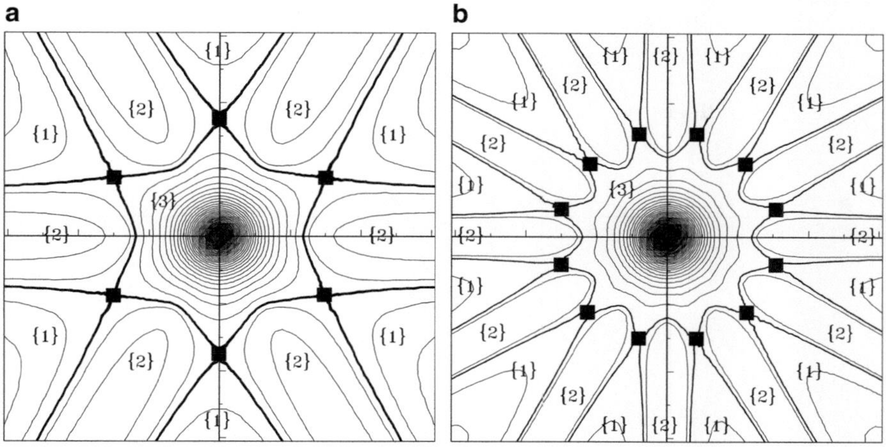

Fig. 2.2 Phase portraits—isolines of function (2.12)—in (x, y) axes of a system of A vortices, uniformly distributed on circumferences with the same radius in each layer: $A = 3,\ \gamma = 1$ (**a**); $A = 6,\ \gamma = 2$ (**b**). The denotations $\{1\}$, $\{2\}$, and $\{3\}$ refer to different types of motion. *Square markers* show the intersection points of separatrices (*thick lines*)

The solutions of type $\{2\}$ correspond to the following motions:

- if at the initial moment, all hetons axes have a tilt, which ensures their motion toward the center of symmetry, we have successively:
  - counter motion of A two-layer pairs,
  - a decrease in the tilt of their axes down to zero (with simultaneous cessation of translational motion—"remote impact"),
  - change in the signs of axis tilts, accompanied by the passage to reverse motion of all hetons and their moving away in A directions;
- if the axis of each heton initially has the opposite slope, all A two-layer pairs will move away from the center along A radial axes of symmetry;
- if heton axes are vertical at the initial moment, then, owing to the interaction between layers, the tilts will immediately become nonzero, allowing the hetons to move away following the scenario described in the previous item.

In the phase portrait, this type of motion corresponds to curves which now coincide with superimposed trajectories of vortices in the top and bottom layers, and those trajectories lie not in every A subdomains of type $\{2\}$, but in the next nearest ones.

Examples of motion of types $\{1\}$ and $\{2\}$ for $A = 6$ are given in Fig. 2.3. The time intervals for calculations are chosen so that in Fig. 2.3b, the final coordinates of each vortex coincide with the initial position of its partner, and those in Fig. 2.3a are mirror images of the initial coordinates about an axis of symmetry, individual for each heton. Here and below, red color corresponds to the top layer and blue color to bottom one. Note that in the terminology of [595], the motions the types of $\{1\}$ and $\{2\}$ give illustrations of the *direct* and *exchange* collisions, respectively. It is

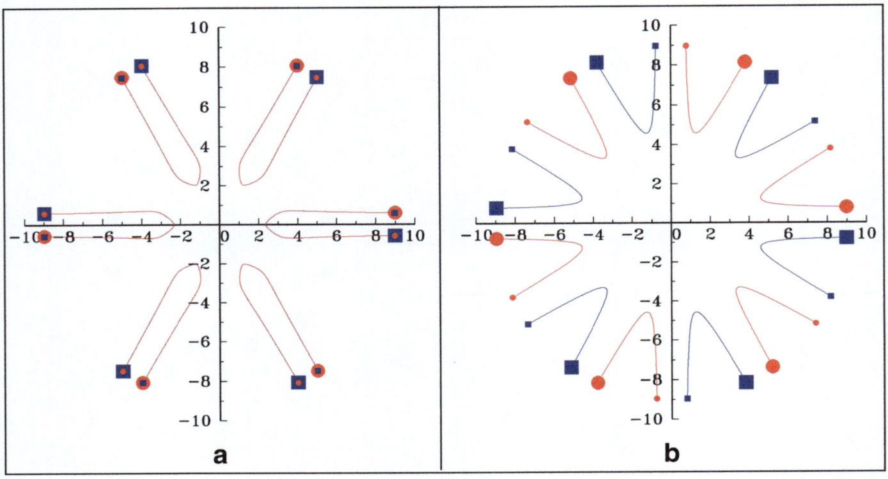

Fig. 2.3 Trajectories of motion of types $\{1\}$ (**a**) and $\{2\}$ (**b**) for a system of two-layer vortices with $\kappa > 0$, $A = 6$, $r = 9$, $\gamma = 2$, and the initial angles $\varphi = \pi/24$ (**a**) and $\varphi = \pi/18$ (**b**). The initial positions of vortices are given by large markers (*circles* for the top layer and *squares* for the bottom layer), and their final positions are given by small markers

worth noting that the motions of the type $\{2\}$ form through baroclinic mechanisms and have no barotropic analogue.

The central part of the phase portrait corresponds to quasi-barotropic motions of type $\{3\}$, in which vortex rotations in the layers have opposite directions (*mutual capture* [595]). During their motion, the vortex structures of both layers always lie in the vertices of regular A-gons with periodically varying side lengths. The side lengths are minimal in the moments when the appropriate A-gons superimpose on one another (and we have A hetons with vertical axes) and maximal when the vortices form a regular $2A$-gonal structure (i.e. one of the A-gons is turned through $\pi/2A$ relative to the other). Conversely, the modulus of angular rotation velocity of each A-gon is maximal in the former case and minimal (in particular, zero—in separatrix intersection points) in the latter case. The limiting configurations of trajectories are A-gonal stars (near the separatrix) and circumferences in the limit with polygon side tending to zero. It is important to note that *intralayer* interactions are predominant for this type of motion. The influence of one layer onto another can be seen only in that vortex trajectories are not circular, and the side length of the appropriate A-gon and the angular velocity of its rotation are not constants but periodic time functions.

However, purely periodic solutions of this type exist only for $A = 2$; while when $A > 2$, the motion is only quasi-periodic. Below, the properties of motions of type $\{3\}$ will be considered in more detail; in particular, it will be shown that they can be unstable as well.

Table 2.1 The values of R_{max} and R_{min} for some values of A

A	2	3	4	5	6	7	8	9	10	11
R^S_{max}	1.128	1.767	2.278	2.779	3.274	3.770	4.269	4.773	5.280	5.790
R^S_{min}	0.717	1.176	1.573	1.943	2.302	2.656	3.009	3.363	3.719	4.076

Unstable stationary saddle states of vortex system in Fig. 2.2 correspond to points marked by dark squares. We find the polar coordinates of saddles by equating the right-hand parts of Eqs. (2.10) and (2.11) to zero. From (2.10), we obtain the identity

$$\varphi = \pi/A,$$

and (2.11) transforms in a transcendent relationship

$$\frac{1}{2R} = S\left(0, \frac{\pi}{A}\right) + \sum_{\alpha=1}^{A-1} \left[S(\alpha, 0) + S\left(\alpha, \frac{\pi}{A}\right)\right], \tag{2.13}$$

where, as well as in (2.12), $R = \gamma r$.

Let us denote by R^S_{max} the solution of (2.13) for any fixed A, then the equation for R^S_{min} (the distance from the center to the nearest coordinate of the separatix) can be derived from the condition that the appropriate image points belong to the same phase curve, i.e., by equating the values of function Φ from (2.10) at $R = R^S_{max}$ and $R = R^S_{min}$:

$$\begin{aligned} &\ln \frac{R_{max}}{R_{min}} + \mathrm{K}_0\left(2R_{max} \sin \frac{\pi}{2A}\right) + \sum_{\alpha=1}^{A-1} \left\{ \mathrm{K}_0\left(2R_{max} \sin \frac{\pi\alpha}{A}\right) \right. \\ &\left. -2\mathrm{K}_0\left(2R_{min} \sin \frac{\pi\alpha}{A}\right) + \mathrm{K}_0\left[2R_{max} \sin \frac{\pi(2\alpha+1)}{2A}\right] \right\} \\ &-\left(\ln \frac{\gamma}{2} + \mathrm{Eu}\right) = 0, \end{aligned} \tag{2.14}$$

where Eu ≈ 0.577 is Euler number.

A set of numerical solutions of Eqs. (2.13) and (2.14) is given in Table 2.1, whence it can be seen that both the values R^S_{max} and R^S_{min} and the amplitudes $R^S_{max} - R^S_{min}$ increase with A.

In Fig. 2.4, summarizing the results of a large series of numerical calculations, the values of $R^S_{max}(A)$ and $R^S_{min}(A)$ are given by circular and triangular hollow markers. It is interesting to note that both R^S_{max} and R^S_{min} are practically linear functions of A. The hollow circles and triangles show the positions of $R^N_{min}(A)$ and $R^N_{max}(A)$, i.e., the maximal and minimal distances from the center to the boundaries of the

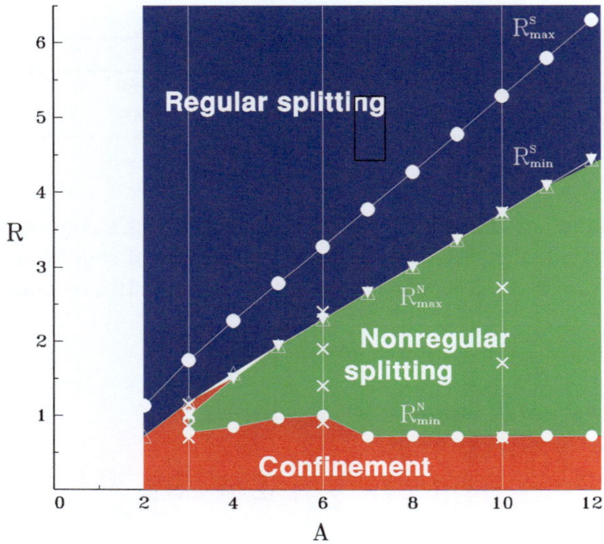

Fig. 2.4 The state diagram of an A-symmetrical configuration of hetons with originally vertical axes for $A = 2, 3, \ldots, 12$. *Dashed* and *dotted lines* show the continuation to the origin of coordinates of the imaginary curves passing via $R^S_{min}(A)$ and $R^S_{max}(A)$. Crosses correspond to the calculation parameters given in Fig. 2.6. Other explanations are given in the text

instability domain.[5] Obviously, the domain "regular scatter" in this figure refers to solutions of type $\{2\}$ (solutions of type $\{1\}$ for hetons with originally vertical axes cannot appear), and the "confinement" domain refers to finite quasi-stable solutions of type $\{3\}$.

Examples of "confined" trajectories are given in Fig. 2.5. In all cases shown here, the initial radial coordinates of vortices satisfy the conditions $R = R^N_{min} - 0.001$ (or $R = R^N_{max} + 0.001$ for external trajectories in Fig. (2.5b, c)), i.e., near the interface between different regimes. Figure 2.5a confirms the above-mentioned periodicity of both relative and absolute motions of vortices at $A = 2$. Figure 2.5b, c demonstrates an interesting feature of cases $A = 3$ and $A = 4$ (it can also be seen at $A = 5$, though within a very narrow interval), consisting in that the "confinement" domain (Fig. 2.4) is doubly connected for them: a part of interval $[R^N_{min}; R^S_{min}]$ belongs to "irregular scatter" domain. The realization of such scatter is illustrated by Fig. 2.6.2a, 3a. In addition to Fig. 2.5, examples of finite captured trajectories are also given in Fig. 2.6.1, 4a.

Dynamics of heton populations has a complex character, implying the coexistence of regular and chaotic regimes [199, 200, 229, 324, 351, 526, 737, 738]. Their

[5]Markers of the same type (in this case, triangular) are used for the minimal values of one function and the maximal values of the other one with the aim to more clearly demonstrate the closeness of the curves passing via them, as mentioned below, at $A \geq 6$.

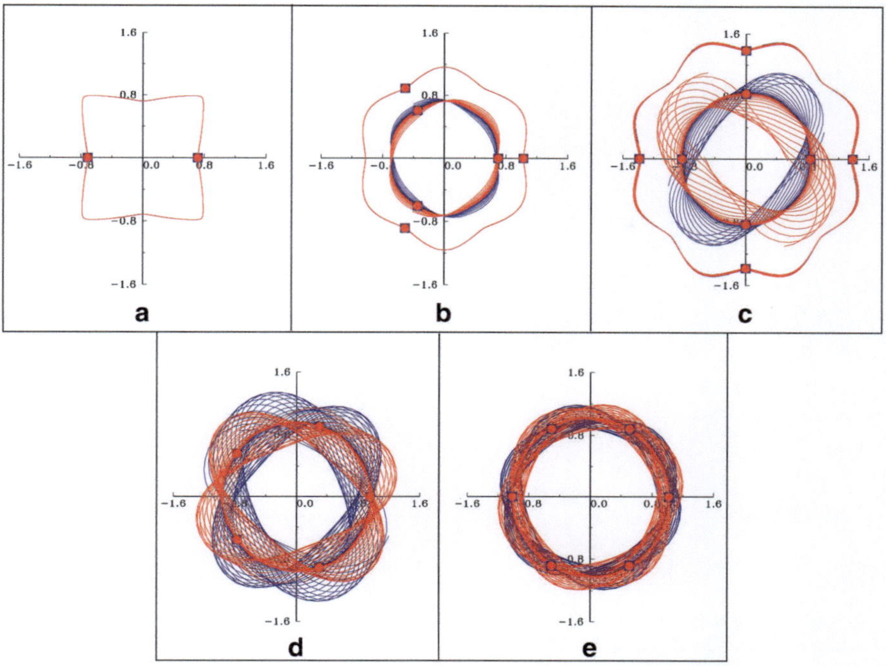

Fig. 2.5 Trajectories of hetons, whose axes were originally vertical and which are localized near vortices of regular A-gons lying within "confinement" zone within time intervals equal to periods of relative motion: (**a**) $A = 2$, $R = 0.715$; (**b**) $A = 3$, $R = 0.769$ and $R = 1.025$; (**c**) $A = 4$, $R = 0.832$ and $R = 1.386$; (**d**) $A = 5$, $R = 0.955$; (**e**) $A = 6$, $R = 1.017$. The *circular* and *square* (coinciding) markers show the initial positions of the top and bottom layers, respectively

realization depends on problem parameters and initial conditions. Simplest examples of nonintegrable situations appear even at the random perturbations of heton initial positions in the case of A-symmetrical heton configurations considered here.

Indeed, numerical experiments show that small random (white noise) perturbations ε^{α} ($\max \varepsilon^{\alpha} \leq 0.01$) in the radial coordinates of vortices

$$R_j^{\alpha}\big|_{t=0} = \bar{R} + \varepsilon^{\alpha} \tag{2.15}$$

make the symmetric solution (2.10) unstable in the interval $\bar{R} \in \left[R_{min}^N, R_{max}^N\right]$, where $R_{max}^N \leq R_{min}^S$ (see the domain "irregular scatter" in Fig. 2.4).

Analysis of numerical experiments shows that two types of instability may exist:

- When $A \in [3; 5]$, the instability is due to the baroclinic interaction between vortices from *different* layers (the equivalent-barotropic subsystem of vortices in each layer is stable [318, 325, 495, 495, 496, 595]). In this case, the instability zones on the phase plane lie in domain $\{3\}$—their outer boundaries lie at a final

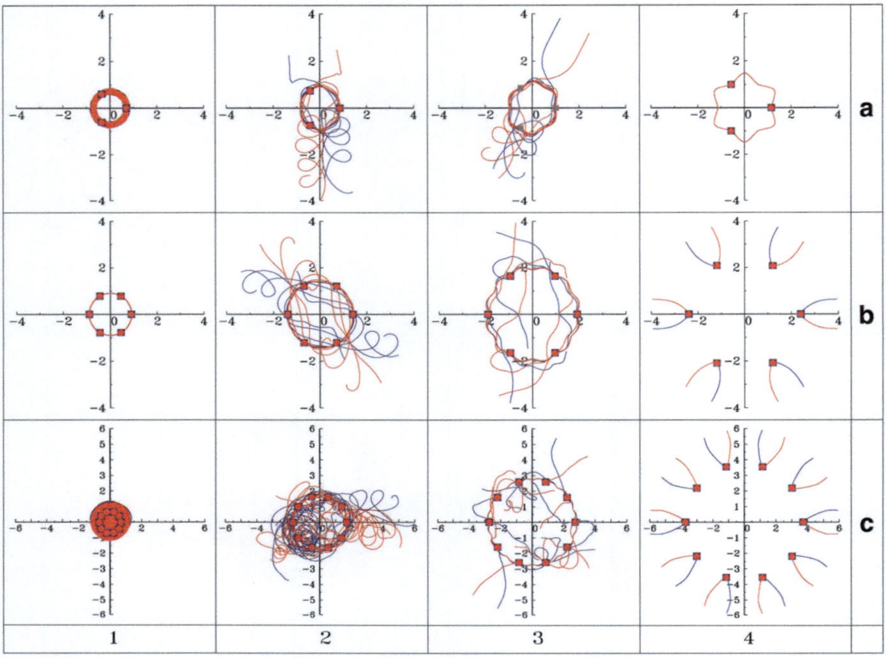

Fig. 2.6 Trajectories of motion of hetons with originally vertical axes localized near the vortices of regular A-gons with their radial coordinates perturbed in accordance with formulas (2.15), where their mean distance from the center is given by the equality $\bar{R} = r + \Delta r(i-1)$, where i is the number of the column in the figure: (**a**) $A = 3$, $r = 0.7$, $\Delta r = 0.15$, (**b**) $A = 6$, $r = 0.9$, $\Delta r = 0.5$, (**c**) $A = 10$, $r = 0.7$, $\Delta r = 1.01$ (these parameters correspond to the positions of crosses in Fig. 2.4)

distance from separatrices, and hence the stochasticity layer, which forms near separatrix intersections, has practically no effect on the behavior of the system of vortices.

- With increasing A (at $A \geq 6$), in addition to the interaction between layers, the barotropic instability, associated with the interaction between vortices within the same layer, is gaining in importance. Figure 2.4 shows that, in such cases, the outer boundaries of instability domains lie on a circumference approaching the separatrix points at minimal distance R^S_{min} from the center. As for the internal boundary of the irregular scatter domain, this boundary at $A \geq 7$ has an asymptote determined by the value of R^S_{min} for $A = 2$.

Note that the coexistence of chaotic and quasi-ordered motions under analogous formulation of the problem for the case of $A = 6$ has been mentioned in [351]. The boundaries of instability domains, established in [351], coincide to within ε with those given in Fig. 2.4.

Examples of "scatter" trajectories are given in Fig. 2.6. These are regular scatter (2.6.4b), (2.6.4c) and irregular scatter (2.6.2, 2.6.3); note that the initial stage of

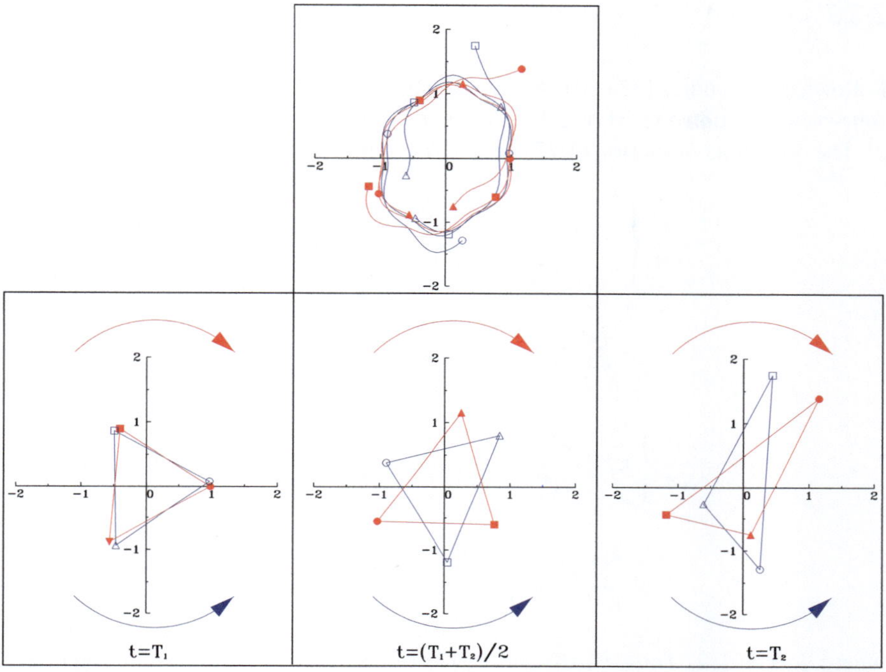

Fig. 2.7 *Top*: detailed description of a three-heton system reaching the infinite regime according to the scenario in Fig. 2.6.3a. *Bottom*: a set of synchronous triangular configurations constructed on vortices of the top (*full line*) and bottom (*dashed line*) layers. The *arc arrows* show the appropriate rotations at the stage preceding the scatter of vortices

scatter in Fig. 2.6.2a, 3a is quasi-ordered. The top part of Fig. 2.7 gives a segment of vortex trajectories (for the latter case) at the short stage of passage to the scatter regime after all vortices have performed two full turns along closed star-like trajectories. The initial and final moments of this time interval are denoted by T_1 and T_2, respectively. The bottom part of the figure shows in more detail the configuration of vortex triads in both layers at the initial, intermediate, and final moments. Later, the motion of the two-layer vortex pair will be practically rectilinear, while the two two-layer pairs in the lower part will shift in the opposite direction, taking part in vortex "leapfrog".

Overall, studying the behavior of the originally A-symmetrical heton configuration in the instability domain is a problem still to be solved.[6]

[6]In the limit at $\gamma \to 0$, i.e., $\Delta\rho \to \infty$, vortices lying in different layers almost do not interact, and we have the problem [68, 94, 318, 438, 496, 510, 595, 635] of independent evolution of A-gonal vortex structures—anticyclonic in the top layer and cyclonic in the bottom layer.

2.2.2 *Case* $A = 2$

Following the studies [324, 350, 851, 852, 1003], we will give the main properties of four-vortex motions, satisfying the conditions (1.82).

The equations of motion (1.77) and (1.78) in this case become

$$u_j^\alpha = -\frac{1}{4\pi}\left\{\kappa_j^{3-\alpha}\frac{y_j^\alpha - y_j^{3-\alpha}}{\left(r_{jj}^{\alpha(3-\alpha)}\right)^2}\left[1+\gamma r_{jj}^{\alpha(3-\alpha)}\mathrm{K}_1\left(\gamma r_{jj}^{\alpha(3-\alpha)}\right)\right]\right.$$
$$\left.+\sum_{\beta=1}^{2}\kappa_{3-j}^{\beta}\frac{y_j^\alpha - y_{3-j}^\beta}{\left(r_{j(3-j)}^{\alpha\beta}\right)^2}\left[1-\gamma r_{j(3-j)}^{\alpha\beta}\mathrm{K}_1\left(\gamma r_{j(3-j)}^{\alpha\beta}\right)\right]\right\}, \quad (2.16)$$

$$v_j^\alpha = \frac{1}{4\pi}\left\{\kappa_j^{3-\alpha}\frac{x_j^\alpha - x_j^{3-\alpha}}{\left(r_{jj}^{\alpha(3-\alpha)}\right)^2}\left[1+\gamma r_{jj}^{\alpha(3-\alpha)}\mathrm{K}_1\left(\gamma r_{jj}^{\alpha(3-\alpha)}\right)\right]\right.$$
$$\left.+\sum_{\beta=1}^{2}\kappa_{3-j}^{\beta}\frac{x_j^\alpha - x_{3-j}^\beta}{\left(r_{j(3-j)}^{\alpha\beta}\right)^2}\left[1-\gamma r_{j(3-j)}^{\alpha\beta}\mathrm{K}_1\left(\gamma r_{j(3-j)}^{\alpha\beta}\right)\right]\right\}, \quad (2.17)$$

where $j, \alpha = 1,2$. Thus, (2.16) and (2.17) represent a system of eight equations. Let us suppose that the conditions $\kappa_1^1 = \kappa_1^2 = -\kappa_2^1 = -\kappa_2^2 \equiv \kappa < 0$, hold, i.e., the intensity of all vortices, anticyclonic in the top layer and cyclonic in the bottom layer, are the same in absolute value, and thus the condition of zero total intensity ($Q = 0$) is satisfied.

Let us consider in detail the three variants: (a) $P_x + P_y = 0, M \neq 0$ (a centrally symmetric configuration of four vortices), (b) $P_x + P_y \neq 0$, $M = 0$ (reflectional symmetry about the y axis), (c) $P_x + P_y = M = 0$ (a particular case of the problem 2.2.1 at $A = 2$).

2.2.2.1 Two Hetons with Zero Total Linear Momentum and Nonzero Angular Momentum

For the problem under consideration, the trilinear coordinates (1.91) and the conditions (1.89) become

$$t_1 = \frac{3\kappa\left(r_{22}^{12}\right)^2}{M}, \qquad t_2 = -\frac{3\kappa\left(r_{12}^{12}\right)^2}{M}, \qquad t_3 = -\frac{3\kappa\left(r_{12}^{11}\right)^2}{M}, \quad (2.18)$$

$$t_2 t_3 \geq \frac{9}{4} \quad \text{and} \quad \begin{cases} t_1 \geq 6, & M > 0, \\ t_1 \leq 0, & M < 0. \end{cases} \quad (2.19)$$

The phase portraits of relative motion of the four-vortex system—the contour lines of the Hamiltonian (1.81), written in coordinates (2.18)—are given for our case in Fig. 2.8. The denotations $\{1\}$, $\{2\}$, and $\{3\}$ here have the same meaning as in Fig. 2.2. The "nonphysical" domains of the phase plane are blacked.

Before passing to the analysis of different types of motion, we consider their main properties. The solutions for $\{1\}$ and $\{2\}$, as was the case with $A > 2$, give examples of exchangeless and exchange interaction between two two-layer pairs, respectively, while the solutions for $\{3\}$ exemplify mutual confinement.

When interpreting the behavior of phase curves, one should take into account that their reaching PD boundary corresponds to the collinear location of vortices. If the curve begins and ends on domain boundary, the image point in the phase space is involved in periodic oscillations between those points, and the relative motions of the vortex system are also periodic time functions. As can be seen from Fig. 2.8, the phase curves of motions of the types $\{2\}$ and $\{3\}$ reach the boundary of PD, while those of type $\{1\}$ do not reach it. Thus, it is clear that it will be enough to consider only the set of collinear states as the initial conditions for motions for the types of $\{2\}$ and $\{3\}$. For motions of type $\{1\}$, it is enough to place, at the initial moment, vortices with alternating strength signs in the vertices of rhombs containing some *base rhomb* (its meaning will be elucidated below).

When $M \neq 0$, the hyperbolic singular point of the phase portrait corresponds to a rhomb rotating with constant angular velocity

$$\omega = -\frac{\dot{x}_1^1}{y_1^1} = \frac{\dot{y}_2^1}{x_2^1}. \tag{2.20}$$

From (2.16) to (2.17) and (2.20), we have the equation

$$2\left[1 - \frac{R}{\cos\psi}\mathrm{K}_1\left(\frac{R}{\cos\psi}\right)\right]\sin^2 2\psi = 1 + 2R\tan\psi\mathrm{K}_1\left(2R\tan\psi\right), \tag{2.21}$$

representing a dispersion relationship between variables $R = \gamma r$ and ψ, whose meaning is clear from Fig. 2.9a. From (2.20) we obtain an expression for angular velocity

$$\omega = -\frac{\kappa\gamma^2}{4\pi R^2\tan^2\psi}\left\{1 + 2R\tan\psi\mathrm{K}_1\left(2R\tan\psi\right)\right.$$
$$\left. -4\sin^2\psi\left[1 - \frac{R}{\cos\psi}\mathrm{K}_1\left(\frac{R}{\cos\psi}\right)\right]\right\}. \tag{2.22}$$

Figure 2.9b gives functions $\psi_S(R)$ and $\omega(R)$ as solutions of Eqs. (2.21) and (2.22), respectively. At $\psi = \pi/4$, when the rhomb degenerates into a square, R attains its minimal value, equal to 1.128 (the vertical dashed line in the figure) and coinciding with the appropriate value $R_{max}^{(S)}$ from Table 2.1.

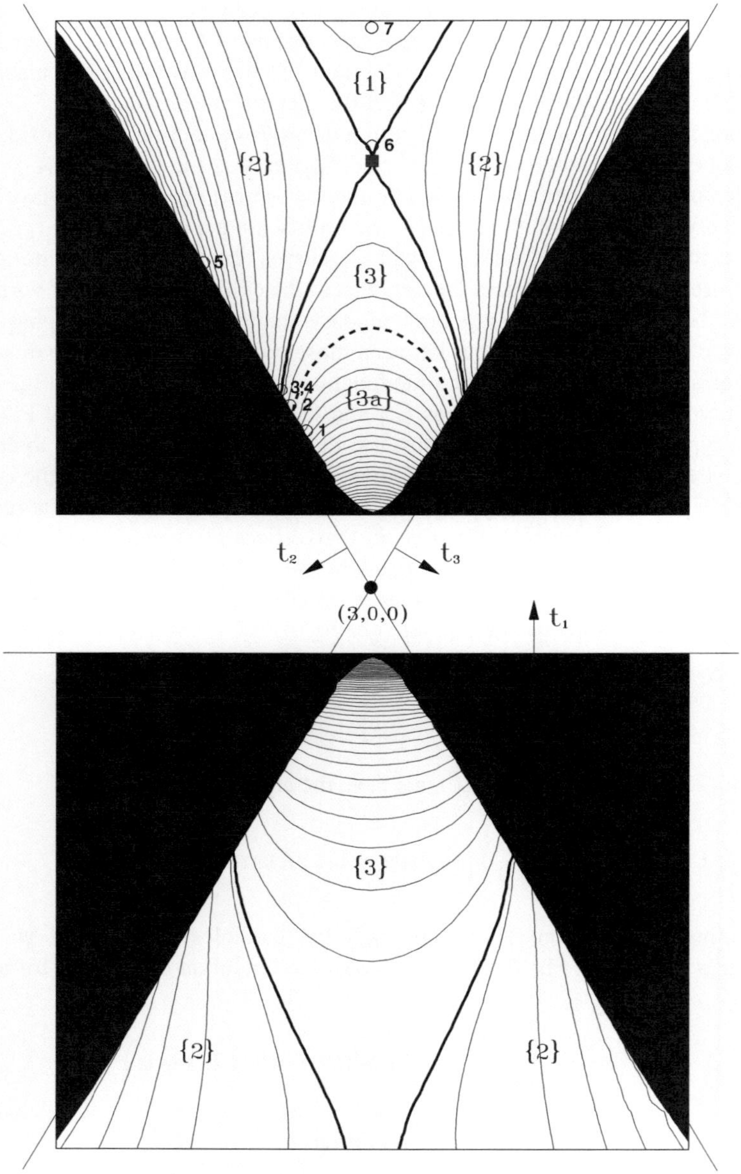

Fig. 2.8 Phase portraits in trilinear coordinates of the problem with $A = 2$, $Q = P_x = P_y = 0$ and $\gamma^2 M = 0.8$ (*top*), $\gamma^2 M = -0.8$ (*bottom*). The separatrices between different types of motion are shown by *thick lines*. The orientation of axes and the coordinates of the marked point are given. The positions of the *numbered circular markers* correspond to the trilinear initial coordinates of the image point in the phase domain for numerical experiments given by Figs. 2.10, 2.11, 2.12, 2.13 and 2.14

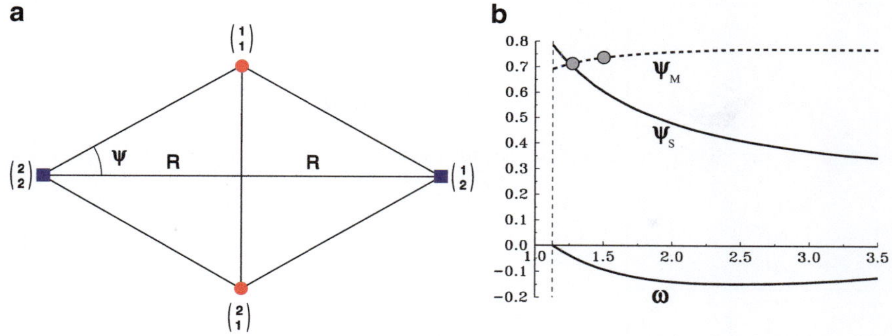

Fig. 2.9 Layout of vortices in the corners of a diamond (**a**); curves ω, ψ_S and ψ_M are traced as functions of R (**b**). Coordinates of *circle markers* on the curve $\psi_M(R)$ correspond to the initial conditions of the experiments illustrated by Fig. 2.14

From (2.21) it follows that in this case,

$$\frac{1}{2R} = \sqrt{2}\mathrm{K}_1\left(\sqrt{2}R\right) + \mathrm{K}_1\left(2R\right), \tag{2.23}$$

and (2.22) yields $\omega = 0$. In the same figure, the curve $\psi_M(R)$ specifies the parameters of the rhomb satisfying the conservation condition of momentum M—the second formula in (2.24), when $\gamma^2 M = 0.8$. Clearly, the parameters of the *basic rhomb*, specified above, are determined by the coordinates of the intersection point of curves ψ_S and ψ_M.

For the analysis of individual types of motion of a system of vortices corresponding to the same phase portrait, it is necessary to chose a fixed value of the total momentum M. The coordinates of vortices at the initial time moments (at $\kappa = 1$) for collinear and mutually orthogonal configurations must satisfy the relationships

$$X_2^1 = \sqrt{\gamma^2 M/2 + \left(X_1^1\right)^2} \quad \text{and} \quad X_2^1 = \sqrt{\gamma^2 M/2 + \left(Y_1^1\right)^2}, \tag{2.24}$$

respectively (hereafter, we use the notations $(X_j^\alpha, Y_j^\alpha) \equiv \gamma(x_j^\alpha, y_j^\alpha)$). The coordinates of other vortices can be found from the conditions of axial symmetry.

We will start with discussing the properties of finite trapped motions of types $\{3a\}$ and $\{3\}$, which is the most substantial in the problem under consideration. Figure 2.10[7] gives characteristic examples of trajectories and the corresponding choreographies. The initial conditions for X_2^1 are specified by the first formula (2.24).

[7]Hereafter, all figures containing trajectories of vortices use the denotation of the form $t = (t_1, t_2, t_3)$ for trilinear coordinates corresponding to the initial vortex configuration.

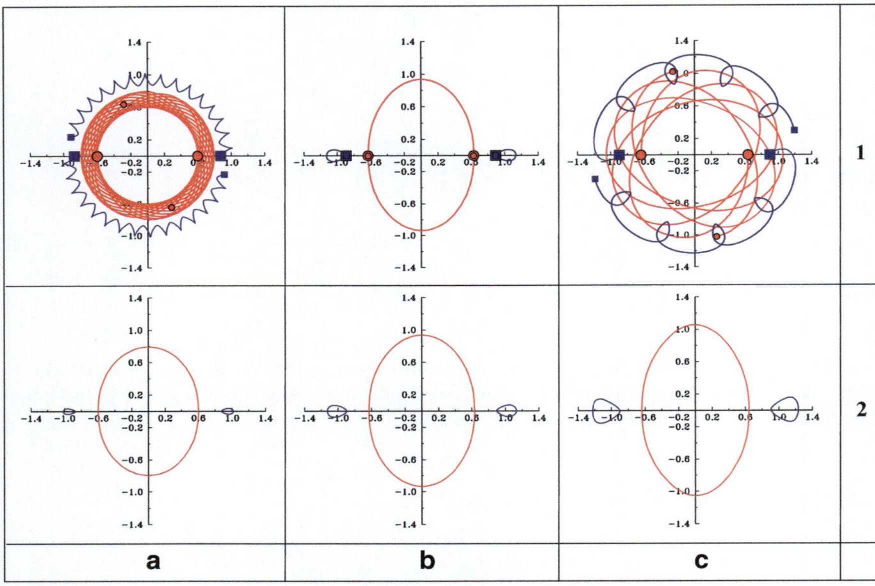

Fig. 2.10 Trajectories of absolute motion (**1**) and the appropriate choreographies (**2**) for motions of the types {3a} and {3} at $\gamma^2 M = 0.8$ and the following initial conditions: (**a**) $X_1^1 = 0.6$, $t = (11.4000, -0.2770, -8.1230)$, (**b**) $X_1^1 = 0.6319$, $t = (11.9895, -0.2577, -8.718)$, (**c**) $X_1^1 = 0.6404$, $t = (12.1517, -0.2528, -8.8988)$. The initial conditions of experiments (a), (b), (c) are the points in the phase plane marked by numbers 1, 2, 3 in Fig. 2.8, respectively. Additional details see in the text

Figure 2.10.1a gives a typical trajectory of the type {3a}, reflecting anticyclonic rotations of vortices in the top layer and cyclonic rotations in the bottom layer. In the coordinate system, uniformly rotating counterclockwise along with bottom-layer vortices, we obtain *tripole-shaped complex relative choreographies* (Fig. 2.10.2a), involving vortices of the given system. Clearly, when $X_1^1 \to 0$, the top-layer vortices will rotate along nearly circular trajectories with theoretically infinitesimal radius and infinitely large angular velocity. The limiting choreographies will degenerate into three stationary points.

The second column shows trajectories and *tripole-shaped absolutely complex choreographies*, which coincide with them and have a trajectory on the phase plane which is the boundary between domains {3a} and {3}. It is new state of the vortex structure—*double capture*, which is of purely baroclinic nature.

In the third column, the bottom-layer vortices, which are spaced further apart, lie in the zone of predominant influence of top-layer vortices and become involved in the anticyclonic rotation induced by them (type {3}). Clearly, the relative choreographies in Fig. 2.10.2c (unlike Fig. 2.10.2a) have been obtained in a coordinate system now rotating clockwise.

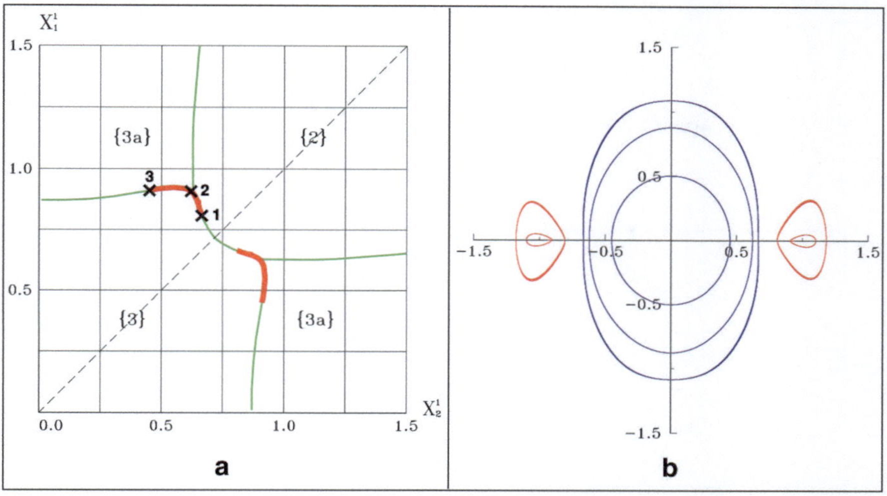

Fig. 2.11 (**a**) Vortex-system state diagram at $A = 2$, $P = M = 0$ with their collinear initial position in the parameter plane (X_2^1, X_1^1). Numbered crosses show values of geometrical parameters for existence of the choreographies, presented in the panel (**b**). Explanations are the text

To demonstrate more vividly the direction of rotation of bottom-layer vortices, the calculations given in Fig. 2.10.1a, 1c were interrupted before the beginning of the self-overlapping stage of their trajectories. The final (calculated) coordinates are shown by smaller markers than the initial ones. The final position of vortices is not shown in Fig. 2.10.1b as unnecessary.

In Fig. 2.10.1b, stationary vortex structure, we named *double capture*, corresponds to the point of the phase portrait in Fig. 2.8 (upper panel), where the dashed line (a boundary between the regions {3a} and {3}) approaches the boundary of PD. Hereby the successive change of regimes in panels 2.10.1a – 2.10.1c occurs when the distance X_1^1 increases having a constant value of the total angular momentum M. This takes place when distances between vortices in the initial collinear configuration are set by the first formula (2.24).

Curved boundary sections between areas with different regimes (red solid lines in Fig. 2.11a) correspond to the total set of possible values of distances between vortices in a collinear construction which can initialize such stationary regimes. This figure was drawn in the plane of parameters X_2^1, X_1^1, that are now independent. In particular, it shows that absolute choreographies can take place not only on the boundary between the areas of regimes {3} and 3a, but also during the transition from motion of type {3} to that one of type {2}. Numbered crosses in Fig. 2.11a mark points with pair values X_2^1, X_1^1, that are related with three choreographies in Fig. 2.11b. External, intermediate and internal curves correspond to points 1, 2 and 3. Boundary trajectories (external and internal) being limiting for this class of stationary solutions, as outside these parts of curves in Fig. 2.11a no absolute choreographies exist. In case 3 the common trajectory of upper-layer vortices

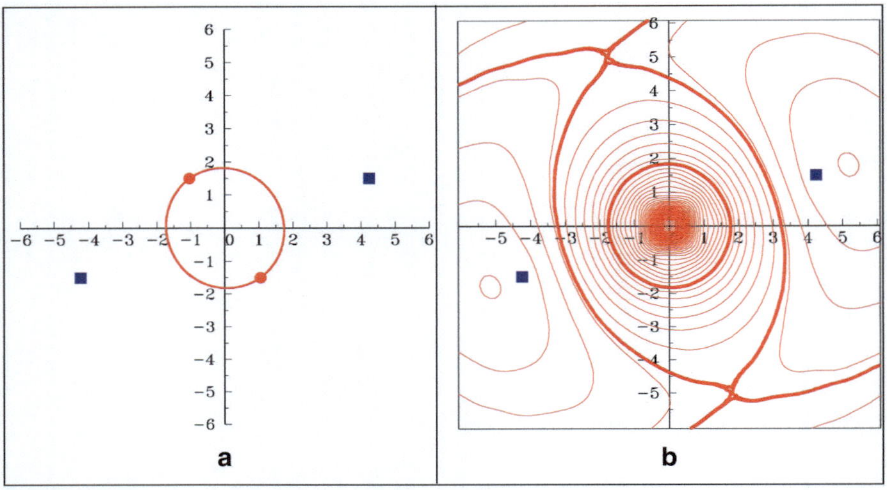

Fig. 2.12 (**a**) Absolute choreography in the case of a diamond initial state for the limit case with practically motionless vortices of the lower layer. *Markers* indicate the initial positions of the vortices. (**b**) Stream-function isolines in the upper layer, when vortices in this layer are free and the lower-layer vortices are fixed. *Thick lines* depict a zero stream-function isoline and a separatrix

becomes close to the circular one, and vortices in the bottom layer remain practically motionless, that's why in Fig. 2.11b trajectories of bottom-layer vortices for this variant degenerate into points.

It is clear that such solutions can be obtained not only for collinear initial states, but also, for example, for diamond-shape construction. In Fig. 2.12a, where markers show the initial vortex positions in the angles of a diamond, there is show a choreography for the case similar to the third variant in the previous figure.

Let's discuss now a following auxiliary problem. In the upper layer we set two free anticyclonic vortices, and in the bottom layer we place two fixed (motionless) cyclonic vortices; the coordinates of all vortices matching the initial positions of vortices in Fig. 2.12a. Panel 2.12b demonstrates stream-function isolines for the upper layer of such problem. In the upper layer, such vortex structure generates in top layer advection of liquid particles of three types: (a) isolated closed cyclonic circulations whose centers are a little displaced with respect to the motionless lower vortices, (b) particles' anticyclonic rotation along quasi-elliptic orbits in the central part of the area, (c) particles' rotation along closed trajectories outside the separatrix.

We can see, that the stream-function isoline, where the upper-layer vortices were located in the initial moment (a thick line), substantially coincides with the corresponding trajectory of the upper-layer vortices in the previous problem. As the coincidence of the trajectory and the stream-function isoline is possible only in the stationary case. We can consider the results of this numerical experiment as an indirect evidence of the existence of a stationary solution in the form of two equal motionless vortices of one layer, located symmetrically at the periphery of a

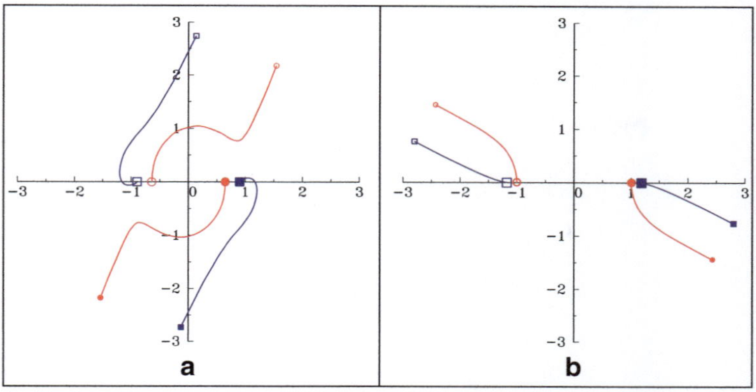

Fig. 2.13 Trajectories of absolute motion of type $\{2\}$ for two hetons at $\gamma^2 M = 0.8$ and the initial conditions with $X_1^1 = 0.6405$, $t = (12.1536, -0.2528, -8.9008)$ (**a**), $X_1^1 = 0.75$, $t = (14.4375, -0.2002, -11.2373)$ (**b**). The points 4 and 5 in the phase plane (Fig. 2.8) correspond to the initial conditions of experiments (a) and (b)

quasi-circular choreography surrounded by two equal second-layer vortices of the opposite sign.

So, the structure of the obtained choreographies allows us to describe the general features of motions associated with them:

- The top-layer vortices move along quasi-elliptic orbits in anticyclonic direction and always remain at the opposite ends of the same diameter of the figure.
- The two bottom-layer vortices are involved in synchronous rotations about two peripheral centrally symmetrical stationary points.
- Every quarter of the period, the vortices alternatively form either a collinear or a diamond-shape configuration (the latter forms in the moments when the appropriate image point in the phase plane crosses PD symmetry axis).
- The rotation periods of vortices in the top and bottom layers are in the ratio of $1:2$.
- The choreographies characterize the stationary solutions of the system of Eqs. (2.16) and (2.17).
- The properties mentioned above refer to *relative* motions in cases *(a)* and *(c)* and *absolute* motions in case (2) in the Fig. 2.10.
- A change of sign of κ clearly changes the cyclonicity of each vortex, though the qualitative properties of the motion remain the same.

Two examples of motion of type $\{2\}$ for two hetons are given in Fig. 2.13. It is important to note that the initial positions of the vortices in Fig. 2.10.3a and in Fig. 2.13a differ only slightly, though the motions refer to different types, and the trajectories have essentially different topological properties, since the appropriate phase curves lie on different sizes of the separatrix in its neighborhood. In this figure, large-size markers, as before, show the initial positions of vortices, while

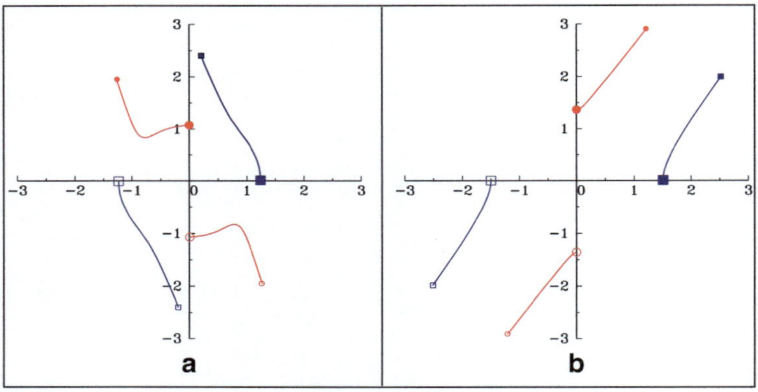

Fig. 2.14 Trajectories of absolute motion of type $\{1\}$ for two hetons at $\gamma^2 M = 0.8$ and the initial conditions corresponding to vortices lying in the vertices of the rhomb with $R = X_2^1 = 1.241$, $t = (29.1012, -13.0506, -13.0506)$ (**a**), $R = X_2^1 = 1.5$, $t = (39.7500, -18.3750, -18.3750)$ (**b**). In the phase plane (Fig. 2.8), points 6 and 7 correspond to the initial conditions of experiments (a) and (b)

small-size markers show their intermediate positions at equal time intervals. At the initial stage of motion, the vortices can be seen to "try" to form a uniformly rotating rhomb-like configuration (at this stage, the image point comes most closely to the singularity represented by the intersection of separatrices), but next they pass into infinite regime.

In this case, the configuration of vortices was collinear at the initial moment. The process of exchangeless nonfrontal interaction can be seen more vividly in an experiment with opposite motion at the initial stage. So, the second stage of interaction (after the maximal approach of the two pairs) will be defined as a mirror transformation of all trajectories about the straight line where the vortices were located originally.

Characteristic trajectories of motions of type $\{1\}$ are given in Fig. 2.14, where Fig. 2.14a gives an example of a nearly critical case. Thus, Figs. 2.10.3a, 2.13a and 2.14a give an insight into motions of types $\{3\}$, $\{2\}$, and $\{1\}$ when their phase curves lie near the separatrices. All three variants demonstrate (contraction of marker nodes placed at regular time intervals) the tendency of vortices to localize in the vertices of the base rhomb and to move along circular orbits with the subsequent passage to scenarios characteristic of each specified type. In Fig. 2.14b, where the initial rhomb is already much greater than the base one, the scatter of vortices begins immediately. The full picture of exchange interaction for this type of motions, as well as in Fig. 2.13, can be obtained from symmetry considerations.

Note that detailed analysis of interaction between two pairs of discrete vortices in a homogeneous fluid with different types of collisions has been carried out by Meleshko and Konstantinov in [595].

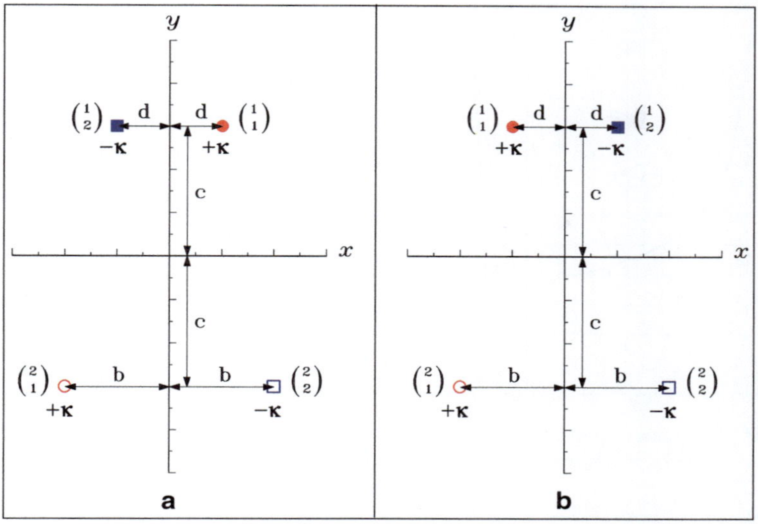

Fig. 2.15 Initial layout of vortices at nonzero total linear momentum: (**a**) the case of motion in opposite directions, (**b**) the case of motion in the same direction. *Circles* are used for the top-layer vortices and *squares*, for the bottom-layer vortices

2.2.2.2 Two Hetons with Nonzero Total Linear Momentum and Zero Angular Momentum

Now let us consider the case of nonzero total linear momentum, though with the assumption that the layout of vortices is symmetrical about the y axis, as shown in 2.15. Although the sufficient conditions of integrability [102, 103, 485, 1030] are not met in this case, the presence of symmetry allows us to state that all solutions are regular [595].

The initial configuration in Fig. 2.15a serves as a starting position for the counter motion of hetons, while that in Fig. 2.15b serves as that for their motion in the same direction. Clearly, the cases of $d = 0$, $b \neq 0$, and $d \neq 0$, $b = 0$ are equivalent and limit cases and belong to both those types of interaction. When $d = b = 0$, the total momentum vanishes; this situation will be considered in the following subsection.

Case $d = 0$. Suppose that the axis of heton with number 1 is vertical at the initial time. For this situation, Fig. 2.16 gives diagrams of possible states of the vortex system on parameter plane (γ, b) for three fixed values of c. Examples of motion of types $\{2a\}$, $\{2b\}$ and $\{3\}$ at $c = 2.5$ are given in Fig. 2.17. To simplify the identification of individual configurations, the positions of vortices in appropriate times are connected by segments and the vortices themselves are shown by the same markers as in schemes in Fig. 2.15. The characteristic properties of different types of interaction are as follows

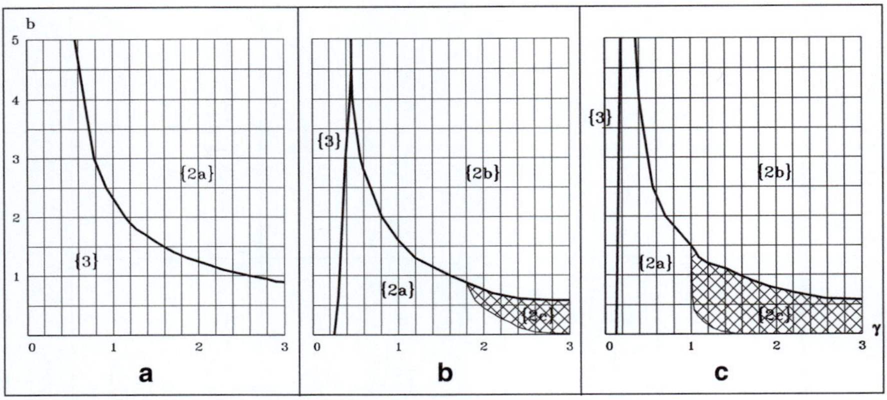

Fig. 2.16 Diagrams of possible types of motions of a four-vortex system on (γ, b) plane at $d = 0$ and (**a**) $c = 0$, (**b**) $c = 2.5$, (**c**) $c = 5$. The denotations are explained in the text

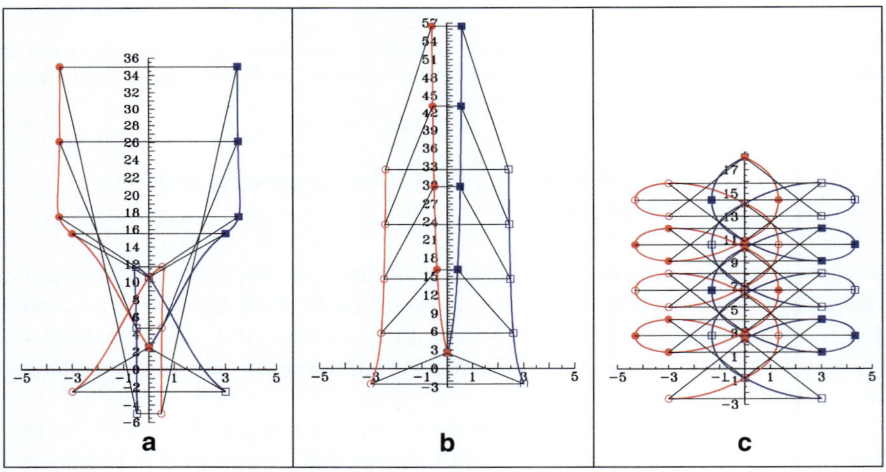

Fig. 2.17 Examples of trajectories of different types of motion at $d = 0$, $b = 3$, $c = 2.5$ (the appropriate diagram is given in Fig. 2.16.b): (**a**) $\gamma = 0.45$—type $\{2a\}$, (**b**) $\gamma = 0.7$—type $\{2b\}$ and (**c**) $\gamma = 0.2$—type $\{3\}$

- The motions of type $\{2a\}$ can be described by the following scheme. At the first stage, the second pair $\binom{2}{1}\binom{2}{2}$ starts moving in the positive direction of the axis y, decreasing the slope of its axis. At the same time, the first pair $\binom{1}{1}\binom{1}{2}$ acquires the slope of its axis, which initiates its motion in the same direction. The distance between vortices of the first pair increases at this stage. After a time, the axis of the second pair becomes vertical, next it changes the sign of its slope, and this heton starts linear and reverse motion. The divergence of the vortices of the first pair ceases, and we have two hetons moving in opposite directions but, clearly, with different velocities. In Fig. 2.17a, the following configurations can be identified:

(a) two triangular configurations, where one heton has a vertical axis,
(b) a trapezoidal configuration (a limiting one, at which the velocities of first-heton vortices change their signs),
(c) two trapezoidal figures, demonstrating uniform divergence of hetons.

- At the interaction of type $\{2b\}$, the heton $\binom{1}{1}\binom{1}{2}$, after forming a nonzero tilt of its axis, gets a momentum in the positive direction of the y-axis and moves farther from the heton $\binom{2}{1}\binom{2}{2}$, which moves in the same direction. The configurations shown in Fig. 2.17b at regular time intervals, demonstrate a steady increase in the height of trapezoids in whose vertices the vortices are situated.
- A notable phenomena are the motions of type $\{2c\}$, which correspond to the hatched domains in Fig. 2.16, and the lines separating the domains $\{2a\}$ and $\{2b\}$. In this case, we see the states when, similar to the situation in "klapstos" strike in billiard, pairs "exchange" their momentums: the incoming heton stops, while the one originally motionless acquires a translational velocity and moves away.
- The motions of type $\{3\}$ are a baroclinic analogue of the classical "vortex leapfrog" [931], when pairs steadily and periodically increase or decrease their size, outstripping one another. However, this analogy is only visual. First, unlike the barotropic case, where the velocity of the internal (slipping) pair increases due to a decrease in its arm, in this case, the sign of the tilt changes in the internal pair. Thus, its own motion could be expected to change its direction to the opposite; however, this does not take place because of the "pushing" effect of the external pair. Second, in the baroclinic variant, the small (in this case, even zero) tilt of one pair does not prevent the appearance of the "leapfrog",[8] but rather facilitates it. Highlighted in Fig. 2.17c, where the evolution of vortices is shown within an interval of two periods, are collinear structures and triangular structures for which one heton has an instantanelus vertical axis.

  Relative choreographies for this class of motions are shown in Fig. 2.18, where the role of parameter γ, which determines the mutual effect of vortices in different layers, can be vividly seen. Each closed curve, representing deformed circumferences, whose deformation degree increases with increasing γ, is a common relative trajectory of vortices from the same layer.

As can be seen from Fig. 2.19a, b, for which the difference between γ values is 10^{-4}, the interface between domains $\{2a\}$ and $\{3\}$ corresponds to a set of unstable states. Figure 2.19a shows that, when the vortices of the inner heton are in their highest point, the corresponding trapezoidal configuration is close to the limiting in

[8]The well-known Love condition for the appearance of "leapfrog" regime [545] in our notations takes the form $d/b > 3 - 2\sqrt{2}$.

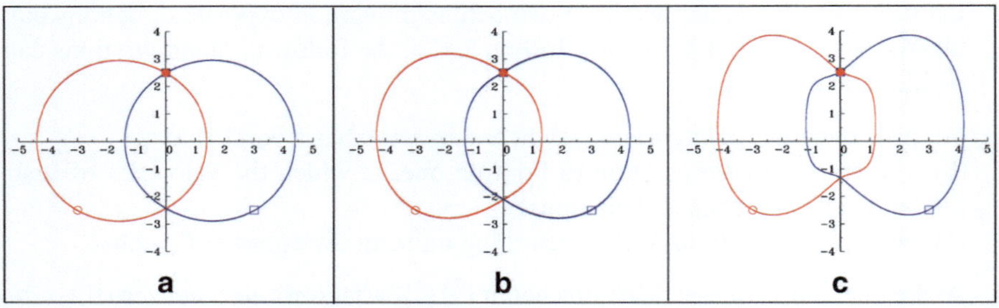

Fig. 2.18 Relative choreographies in a translationally moving coordinate system for motions of type $\{3\}$ at $d=0$, $b=3$, $c=2.5$: (**a**) $\gamma=0.1$, (**b**) $\gamma=0.2$, (**c**) $\gamma=0.3$. *Markers* show the initial positions of vortices

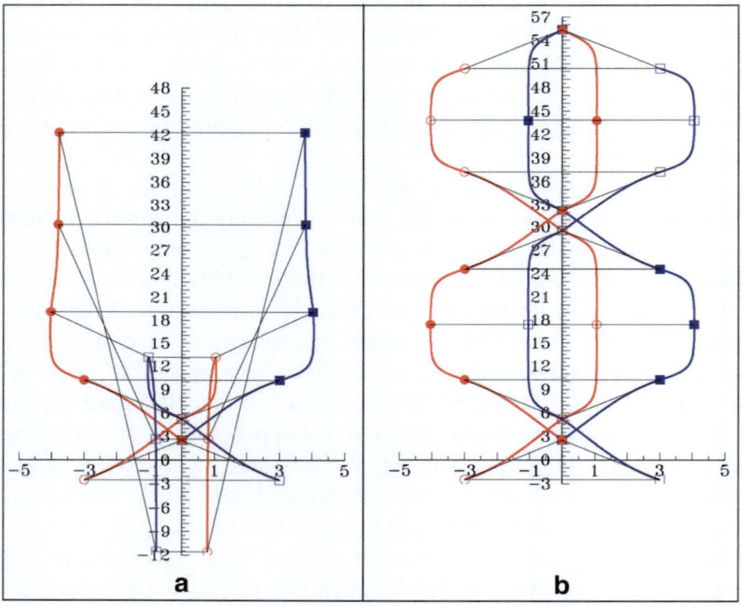

Fig. 2.19 Trajectories of vortices at $d=0$, $b=3$, $c=2.5$: (**a**) $\gamma=0.3854$—type $\{2a\}$, (**b**) $\gamma=0.3853$—type $\{3\}$

the sense that the balance between the momenta of the inner and outer pairs still shows a shift toward the former. An infinitesimal increase in stratification leads to the opposite relationship between the momenta, and we observe a passage from the motions of type $\{2a\}$ to motions of type $\{3\}$, for which trajectories are represented in an interval one period long. Obviously, the initial stages of motions are practically the same in both figures.

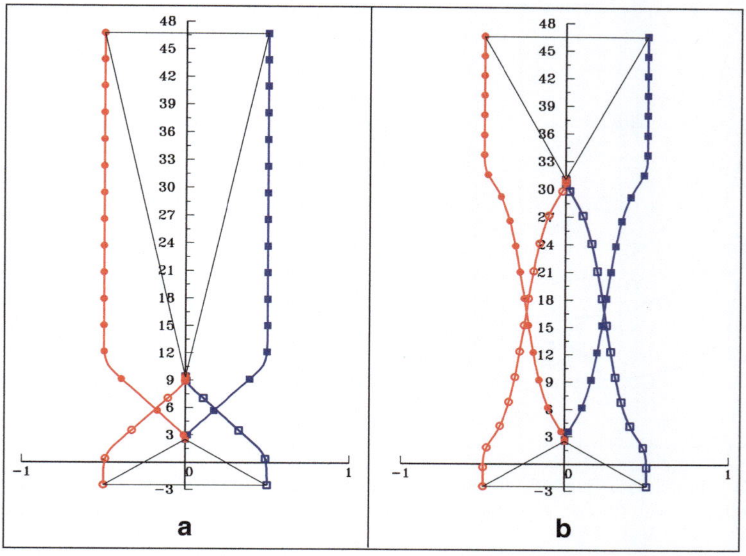

Fig. 2.20 Trajectories of vortices of type $\{2c\}$ at $d=0$, $b=0.5$, $c=2.5$: (**a**) $\gamma=2.0$, (**b**) $\gamma=2.8$

Trajectories of vortices at the interaction of type $\{2c\}$ are given in Fig. 2.20, where the horizontal axis is strongly stretched to make the picture more vivid. Only the initial and final triangular configurations are given here. This type of motion (klapstos—inelastic collision at a distance) is an illustration of the baroclinic mechanism in the most effective form.

A common property of interactions of types $\{2a\}$, $\{2b\}$ and $\{2c\}$ is, as before, their exchangeless character.

Note that all examples of behavior of trajectories are given here for the case $c=2.5$. The effect of parameter c can be assessed from the comparison of Fig. 2.16a–c. Thus, at $c=0$ (Fig. 2.16a, where a neighborhood of the horizontal axis should not be taken into consideration because of singularity), the factor of within-layer interactions of type $\{3\}$ gains much in significance , while motions $\{2b\}$ and $\{2c\}$ are completely absent in the considered interval $\gamma \subset [0,3]$. As the initial distance between hetons increases, the role of quasi-barotropic motions decreases, but the domains of existence for the motions of hetons with weak interaction between them (type $\{2a\}$ and, especially, type $\{2c\}$) become wider.

Counter collision of hetons. Now let us suppose that $d \neq 0$, and the initial layout of vortices is the one shown in Fig. 2.15a. A summary of the results of studying possible states of vortex structures for this case is given in Fig. 2.21. In addition to diagrams with parameters corresponding to motions involving heton collisions (Fig. 2.21b, c), we also give a diagram for the limiting case of originally collinear states ($c=0$, Fig. 2.21a).

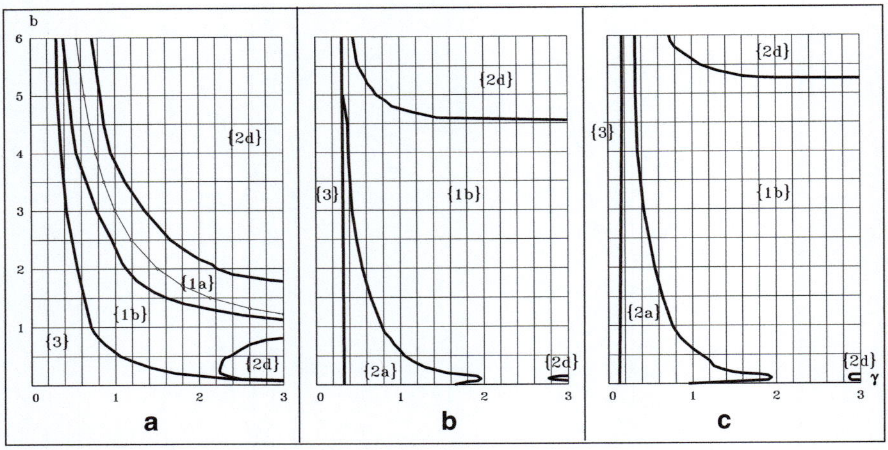

Fig. 2.21 Diagrams of possible types of motion of a four-vortex system on the plane (γ, b) at $d = 1$ and (a) $c = 0$, (b) $c = 2.5$, (c) $c = 5$. Explanation of denotations are given in the text

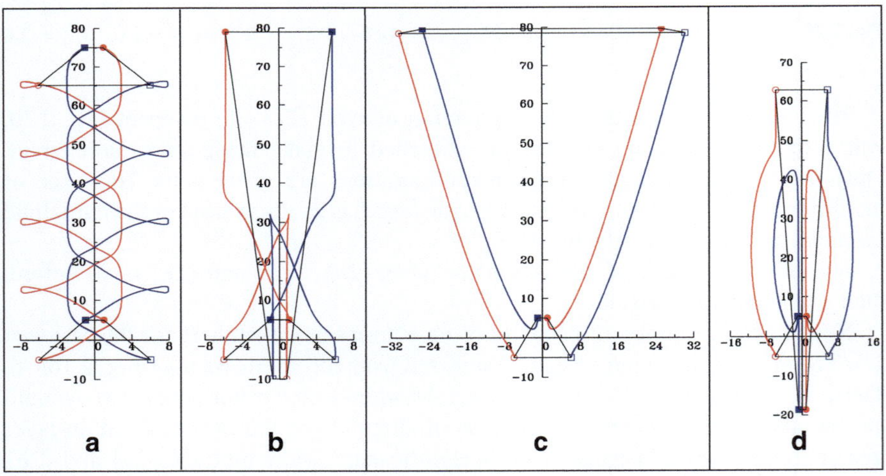

Fig. 2.22 Trajectories of vortices at $d = 1$, $b = 6$, $c = 5$ (the appropriate diagram is given in Fig. 2.21): (a) $\gamma = 0.15$—type $\{3\}$, (b) $\gamma = 0.30$—type $\{2a\}$, (c) $\gamma = 0.60$—type $\{1b\}$, (d) $\gamma = 0.90$—type $\{2d\}$

First, we consider examples of different types of motions with counter collision of two-layer pairs where $c \neq 0$. A series of four possible scenarios with $c = 5$ and successively growing values of γ is given in Fig. 2.22, where the positions of vortices only in the initial and final moments are connected by segments of straight lines.

- Type $\{3\}$. Vortex leapfrog (Fig. 2.22a), which appears at small γ, shows the same properties as in the previous section (Figs. 2.17c and 2.19b), the only difference being that the starting configuration is trapezoidal with four vortices rather than triangular. Here the trajectories are shown for a time interval with a length of two periods.

- Type $\{2a\}$. With increasing γ and the same set of geometric parameters, we pass into the domain $\{2a\}$ (Fig. 2.21c). It corresponds to motions in which one of the two-layer pairs changes its motion to the opposite. The topological properties of trajectories in Figs. 2.22b, 2.16a and 2.19a are similar except that now the heton with number 1 has a nonzero tilt at the initial moment. This fact ensures the initial countermotion of hetons. However, a continuous motion of the first heton is impossible because of the counteraction of the "stronger" second heton. Therefore, vortices $\binom{1}{1}$ and $\binom{1}{2}$ have to interchange places to start moving in opposite directions. Further, the roles of hetons change, and now the "weaker" second heton starts reverse motion. As the result, we have two pairs moving away from one another. Using once again the "billiard" analogy, we can identify the motion of type $\{2a\}$ with elastic collision.
- The subsequent growth in parameter γ contributes to an increase in the intensity of within-layer interaction. After the initial stage of counter motion and the corresponding approach of hetons, each vortex of the first pair is captured by the nearest vortex of the second pair. The newly formed two-layer pairs move away from one another in different directions along straight lines, forming some angle, and the distance between those pairs steadily increases (Fig. 2.22c). Thus, regrouping of the type

$$\binom{1}{1}\binom{1}{2}+\binom{2}{1}\binom{2}{2}\to\binom{1}{1}\binom{2}{2}+\binom{1}{2}\binom{2}{1}.$$

takes place. This type of exchange interaction will be denoted as $\{1b\}$.
- Type $\{2d\}$. As can be seen from Fig. 2.22d, the farther increase in the strength of layer-to-layer contact allows the first heton to slip between the vortices of the heton with larger arm and to continue its original motion in the negative direction along the y-axis. This example is of interest because temporal exchange of partners takes place after all, as in the previous case, though later an inverse transfer will be seen: the vortices of the first heton, after moving along loops, will again form a two-layer pair. Obviously, the subsequent increase in γ will lead to a decrease in the size of the loops. When $\gamma \gg 1$, the slip will be such that the pairs will drift apart exerting almost no effect on one another.

  Thus, class $\{2\}$ of exchangeless interactions $\big(\{2a\}$—an elastic frontal collision with scatter in opposite directions, $\{2b\}$—concurrent scattering, and $\{2c\}$—inelastic collision "klapstos"$\big)$ now includes one more type $\{2d\}$—slipping.

Now let us consider the particular case $c = 0$, which is of interest in itself, because the full set of collinear configurations cannot realize at $c \neq 0$. A diagram of possible states of a four-vortex system at $d = 1$ is given in Fig. 2.21a.

Note the appearance of a new type of interaction $\{1a\}$. The transition via the boundary between domains $\{1b\}$ and $\{1a\}$ in Fig. 2.19a can be seen in a numerical experiment illustrated by Fig. 2.23b, c. Figure 2.23a, b give examples of motions

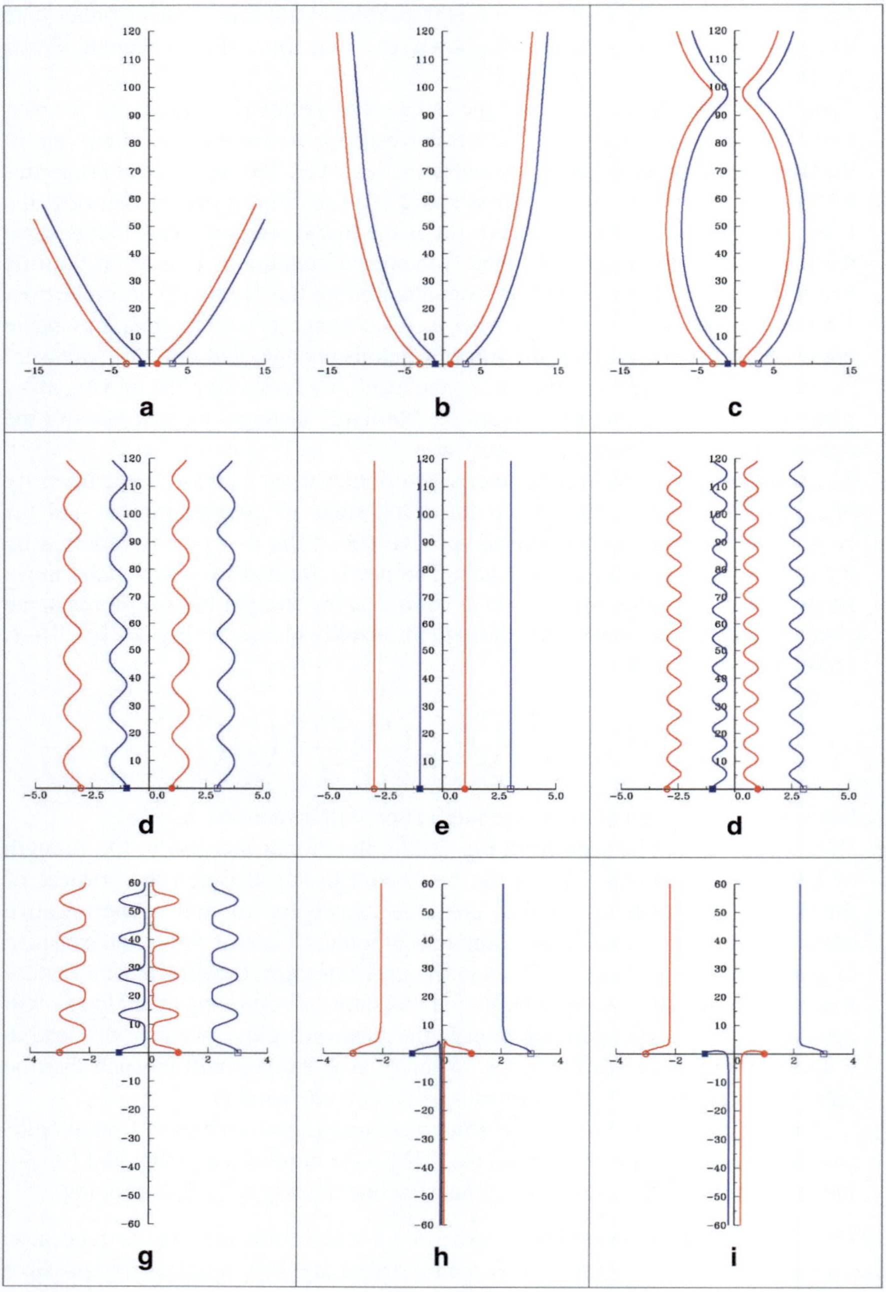

Fig. 2.23 Trajectories of vortices at $d = 1$, $b = 3$, $c = 0$ (the appropriate diagram is given in Fig. 2.21a): **(a)** $\gamma = 0.74$—type $\{1b\}$, **(b)** $\gamma = 0.76$—type $\{1b\}$, **(c)** $\gamma = 0.77$—type $\{1a\}$, **(d)** $\gamma = 0.90$—type $\{1a\}$, **(e)** $\gamma = 0.9985$—type $\{1a\}$, **(f)** $\gamma = 1.18$—type $\{1a\}$, **(g)** $\gamma = 1.33$—type $\{1a\}$, **(h)** $\gamma = 1.34$—type $\{2d\}$, **(i)** $\gamma = 2.00$—type $\{2d\}$

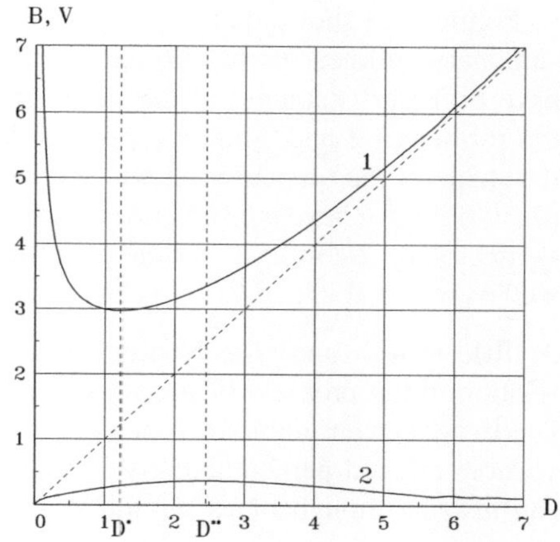

Fig. 2.24 Curve 1 is a dispersion curve $B(D)$, i.e. a solution of Eq. (2.26), and curve 2 is a profile of the translational velocity (2.25) of a stationary construction of two hetons V. The *vertical dashed lines* $D = D^*$ and $D = D''$ correspond to extremums of functions $B(D)$ and $V(D)$, and the *tilted dashed line* is the asymptote $B \sim D$ at $D \gg 1$

of type $\{1b\}$ which are drifting away of two-layer vortex pairs $\binom{1}{1}\binom{2}{2}$ and $\binom{1}{2}\binom{2}{1}$. Since the initial configurations in this case are collinear, the stage of counter motion, available in Fig. 2.22c, is not present here. The angle of scatter can be seen to depend on parameter γ, though in any case, the interaction between the pairs will weaken, as is typical of such motions.

A characteristic feature of the new type of motions $\{1a\}$ (Fig. 2.23c–g) is the intense and permanent interaction between the two hetons. In the general case, the hetons are involved in translational motion with periodic oscillations, alternatively approaching one another or scattering. A notable feature of this type of motion is the existence of an intermediate stationary regime (it corresponds to the thin line in diagram 2.21a), at which the collinear vortex structure with alternating configuration of vortices in the top and bottom layers moves strictly translationally with a constant velocity (Fig. 2.23e)

$$V = \frac{\kappa}{4\pi}\left[\frac{D^2+4DB-B^2}{D(B^2-D^2)} + \mathrm{K}_1(D+B) - \mathrm{K}_1(B-D) + \mathrm{K}_1(2D)\right]. \qquad (2.25)$$

Here, $D = \gamma d$ and $B = \gamma b$ satisfies the dispersion equation

$$\frac{(D-B)^2}{2DB(D+B)} = 2\mathrm{K}_1(B+D) + \mathrm{K}_1(2D) + \mathrm{K}_1(2B). \qquad (2.26)$$

A graphical representation of the solution of Eq. (2.26) and the velocity (2.25) is given in Fig. 2.24, where B and V are functions of parameter D.

Those collinear structures are a particular case of the so-called *Kizner' heton quartets*, obtained in [424], while (2.26) and (2.25)—are particular cases of Eqs. (2.13) and (2.14) from [424] with variables (a, b, h, l) denoted as $(\kappa, \kappa, D, -B)$, respectively. Since in such solid-body vortex structures the vortices in the upper and lower layers are alternating, we will call them *zigzag–shaped translating heton quartets*.

Figure 2.23 shows that at $\gamma < \gamma_s$ or $\gamma > \gamma_s$ (where $\gamma_s = 0.9985$ corresponds to a stationary collinear vortex structure at $d = 1$), quasi-periodic forms of trajectories have different character: at the initial stage, the interacting hetons move apart in the former case and come close in the latter case. The qualitative dependence of the shape of the trajectory on γ is shown in Fig. 2.23c–g. At some value $\gamma = \gamma_{cr} \in [0.33; 0.34]$ we have a bifurcation to solution of type $\{2d\}$ which is sliding apart of vortices (Fig. 2.23h–i). It is clear, that the stage of counter motion is absent here, as well as in Fig. 2.23a, b.

A little more about the Kizner' quartets. The concept of a hetonic quartet in relation to the problem of transitions in baroclinic modons [427]. A heton quartet is a two-layer ensemble of four synchronously translating quasi-geostrophic point vortices aligned perpendicularly to the axis of their translation. A hetonic quartet can also be though of as an anti-symmetric pair of hetons with specially fitted parameters, the strengths and separations of the hetons. Kizner [424] provided, a complete analysis of the nonlinear stability of hetonic quartets to antisymmetric perturbations had been provided. Of special interest are the quartets made up of overlapping hetons that are, generally, stable, but are located relatively close to the border of the stability region in the parameter space. An overlapping quartet represents a discrete model of a baroclinic modon with an overlap of the upper and lower vorticity lumps [418, 423]. A hetonic quartets of this type, being subjected to prolonged action of small perturbations, may make a transition to a pair of hetons travelling in opposite directions. After getting rid of the overlap, the stronger heton accelerates and increases its separation. In terms of modons, this corresponds to the transition to a non-overlapping quasi-elliptical modon state [427, 428]. Kizner [424] discovered some similarity between his hetonic quartets and their transitions, on the one hand, and the finite-core hetons and their break-down (this will be considered in Chap. 3 of this book), on the other. At small separations, a finite-core heton splits up into two smaller baroclinic pairs, whereas at a greater separation, the finite-core heton evolves towards a quasi-stable state. These correspond well with Kizner's results. In fact, small separation between the vorticity patches means their considerable overlap, while a large separation makes the patches less overlapping, i.e. more similar to point-vortex hetons.

2.2.2.3 Two Hetons with Zero Total Linear and Angular Momenta

In the case of double symmetry, qualitative analysis of the system can be carried out not only in trilinear, but also in rectangular coordinates, as it was made in the previous section for arbitrary A and as shown in Fig. 2.25a for $A = 2$.[9] The critical distances R^S_{max} and R^S_{min} for the case under consideration are given in Table 2.1, with the Eq. (2.21) for determining R^S_{max} now becoming (2.23).

[9]Note that Hogg and Stommel were first to obtain analogous phase curves in [350].

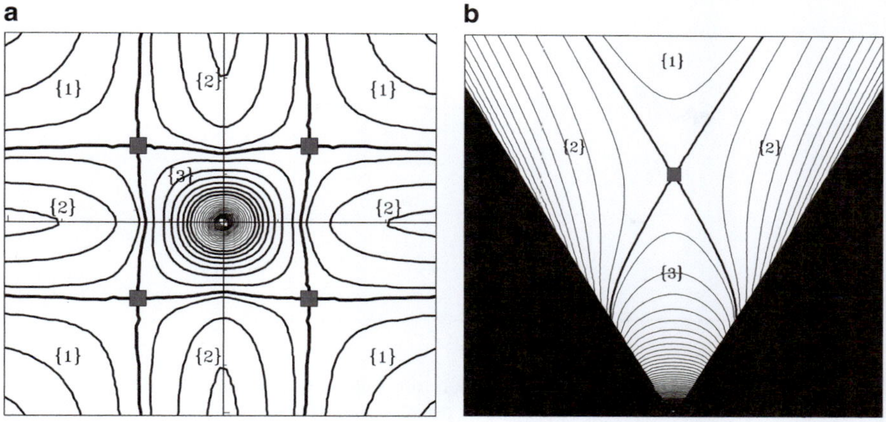

Fig. 2.25 Phase portrait (isolines of the Hamiltonian (1.81)) in coordinates (X,Y) (**a**) and an equivalent portrait in trilinear coordinates (**b**) for a system of two hetons with zero total linear and angular momenta. The denotations are the same as in Figs. 2.2 and 2.8

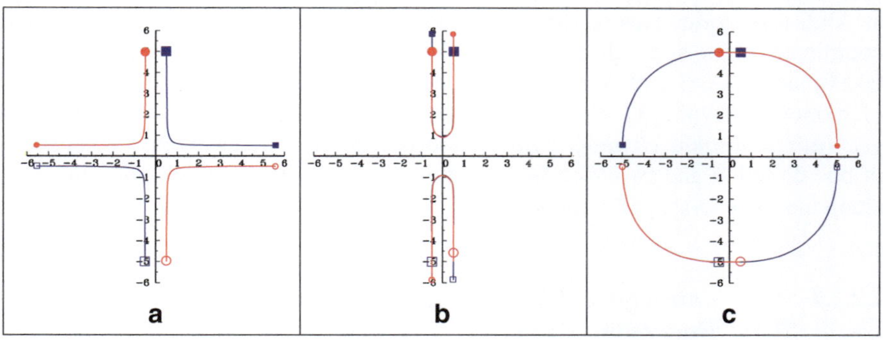

Fig. 2.26 Trajectories of motion in coordinate system (x, y) for four vortices, which compose two hetons at $b = d = 0.5$, $c = 5$ and (**a**) type $\{1\}$, $\gamma = 2$, (**b**) type $\{2\}$, $\gamma = 1$, (**c**) type $\{3\}$, $\gamma = 0.1$.

Examples of trajectories of different types are given in Fig. 2.26. Here the initial coordinates of vortices are the same for all three cases; the passage from one type to another was carried out by varying parameter γ (the values of γ are given in figure caption). As before, the initial positions of vortices are shown by larger markers.

The motions of types $\{1\}$ and $\{2\}$ exemplify exchange and exchangeless collisions of two two-layer pairs. Clearly, Fig. 2.26a, b are analogues of Fig. 2.3a, b, respectively. As can be seen from Fig. 2.26b, the baroclinic pairs had passed through the segments of counter motion after which, changing the signs of the tilts of their axes to the opposite, they moved apart in the opposite directions in such a way that the trajectories of vortices of the top and bottom layers were imposed on one another.

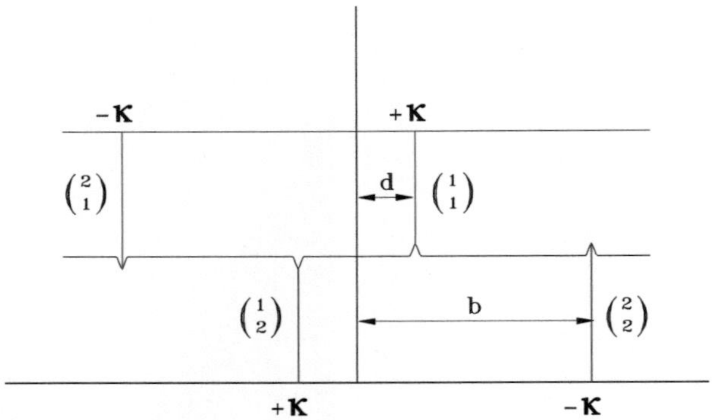

Fig. 2.27 Initial layout (*vertical section*) of four-vortex collinear system with conditions (2.27)–(2.28) satisfied. Adapted from [847] (Fig. 1) by permission of IOP Publishing

In Fig. 2.26c, to avoid overcharging the figure, the time interval was chosen such that the stage of analogous mutual superimposition has not began by the end of calculation. Such O-shaped[10] trajectories are the family of closed lines. In this case, at A=2, trajectories have a shape close to the circular one in the vicinity of the coordinate origin, and a shape of A-gonal star in the vicinity of the separatrix (as it takes place at A=3 and A=6; see Sect. 2.2.1, Figs. 2.2 and 2.5). The trajectories of motions of type $\{3\}$ represent a new class of choreographies which we call *generalized absolute choreographies* as following: two vortices of upper layer move in one direction and two vortices of bottom layer move in the opposite direction along the same O-shaped trajectories.

2.2.2.4 Vortex Structures: Warm Heton–Cold Heton, Two Antihetons, Two "Horizontal" Pairs

In Sect. 2.2.2, vortices with the same signs were assumed to be placed in each layer. Here, within the framework of the four-vortex problem, we consider the case

$$\kappa_1^1 = \kappa_2^1 = -\kappa_1^2 = -\kappa_2^2 \equiv \kappa, \tag{2.27}$$

where each layer contains a cyclone–anticyclone pair. In this case, by virtue of (**!**), the interaction between point vortices has qualitatively different properties. An example of the initial layout of vortices for this case is given in Fig. 2.27. In virtue of the conservation of linear and angular momenta of the vortex system, the properties of reflection symmetry, specified by the initial conditions, will remain true whatever the motions of vortices, i.e.

$$x_1^1 = -x_2^1, \quad y_1^1 = -y_2^1, \quad x_1^2 = -x_2^2, \quad y_1^2 = -y_2^2,$$

[10]Terminology of Hogg & Stommel [350].

and at $t = 0$, it will be enough to assume

$$x_1^1 = d, \quad x_2^2 = b, \quad y_1^1 = y_1^2 = 0. \tag{2.28}$$

Indeed, varying the parameters d and b, we can simulate

- the interaction of two hetons—warm and cold, if the predominant mutual influence is that of the vortices composing two-layer pairs $\binom{1}{1}$ and $\binom{2}{2}$, $\binom{2}{1}$ and $\binom{1}{2}$,
- the interaction of two antihetons,[11] if the predominant mutual influence is that of vortices with the same sign belonging to different layers $\binom{1}{1}$ and $\binom{1}{2}$, $\binom{2}{1}$ and $\binom{2}{2}$; obviously, one antiheton will be cyclonic and the other will be anticyclonic,
- the interaction of "horizontal" pairs, if the predominant influence is that between vortices of the same layer $\binom{1}{1}$ and $\binom{2}{1}$, $\binom{1}{2}$ and $\binom{2}{2}$.

Before passing to studying the general properties of motions of this vortex system, we will note some particular cases, following directly from Eqs. (2.16) to (2.17).

(a) When $d = b$, we have the case of two hetons—a warm one (left) and a cold one (right)—with vertical axes. This type of interaction (exchange of partners between hetons) was studied by Hogg and Stommel [351].
(b) If we rotate both vortices in one layer through 180° about vertical axis, we will obtain a quasi-barotropic pair, moving along the y-axis with the velocity

$$\dot{y}_1^2 = \dot{y}_2^2 = \dot{y}_1^1 = \dot{y}_2^2 = -\frac{\kappa\gamma}{2\pi}\mathrm{K}_1(2\gamma d).$$

(c) If the dispersion relationship

$$\frac{B^2 + D^2}{2DB} = (B+D)\mathrm{K}_1(B+D) + (B-D)\mathrm{K}_1(B-D) \\ + D\mathrm{K}_1(2B) + B\mathrm{K}_1(2D) \tag{2.29}$$

holds, we have a collinear configuration of 4 vortices, which rotates as a solid structure about the center $(0, 0)$ with constant angular velocity

$$\omega = \frac{\kappa\gamma^2}{4\pi D}\left[\frac{B^2 + 3D^2}{2D(B^2 - D^2)} - \mathrm{K}_1(B+D) - \mathrm{K}_1(2D) - \mathrm{K}_1(B-D)\right], \tag{2.30}$$

where $D = \gamma d$ and $B = \gamma b$. Further we will call them *zigzag-shaped rotating heton quartets*.

The relationship (2.29) between D and B, as well as the distribution of angular velocity (2.30) along the dispersion curve are given in Fig. 2.28. Note that stable

[11] According to the definition given by Hogg and Stommel [350], antiheton is a combination of two vortices with the same sign, belonging to different layers.

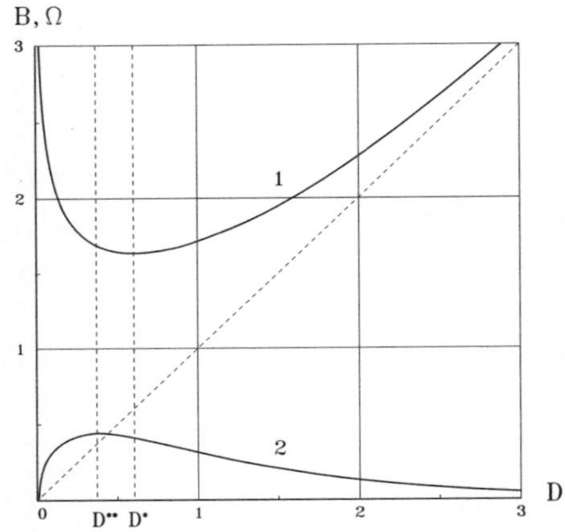

Fig. 2.28 **1** Graphical solution of Eq. (2.29), **2** the angular velocity of the collinear construction (2.30); Here $\Omega = 4\pi\omega$. The *vertical dashed lines* $D = D^*$ and $D = D^{**}$ correspond to extremums of the first and second curves, respectively, and the tilted line determines the asymptote $B \sim D$ at $D \gg 1$. Adapted from [847] (Fig. 2) by permission of IOP Publishing

solutions for $B(D)$ and $\Omega(D)$ exist only at $D < D^{**}$. Let's note also the qualitative similarity of Figs. 2.24 for translation and 2.28 for rotation zigzag-shaped quartets. Dispersion curves that relate the geometrical parameters D and B have identical asymptotics at $D \ll 1$ and $D \gg 1$. Minimum values of B are reached at $D^* = 1.1228$ and $D^* = 0.6105$; and maximum values of the translation and angular velocities have place at $D^{**} = 2.4518$ and $D^{**} = 0.3827$, correspondingly.

The asymptotic regimes at strong and weak stratification are as follows.

(d) If $\gamma \ll 1$ and $d, b = O(1)$, the Eqs. (2.16) and (2.17) yield

$$\dot{x}_j^\alpha \sim \frac{\kappa(-1)^{3-\alpha}}{2\pi} \times \frac{y_j^\alpha - y_j^{3-\alpha}}{\left(r_{jj}^{\alpha(3-\alpha)}\right)^2}, \quad \dot{y}_j^\alpha \sim \frac{\kappa(-1)^{3-\alpha}}{2\pi} \times \frac{x_j^\alpha - x_j^{3-\alpha}}{\left(r_{jj}^{\alpha(3-\alpha)}\right)^2}.$$

Obviously, only intralayer interaction takes place in this case. With such choice of initial conditions (Fig. 2.27), the vortices in the layers move as two "horizontal" pairs parallel to the y-axis in opposite directions

$$\dot{y}_1^1 = \dot{y}_1^2 = -\dot{y}_2^1 = -\dot{y}_2^2 = -\frac{\kappa}{2\pi(d+b)}.$$

(e) If $\gamma \gg 1$ and $d, b = O(1)$, we have

$$\dot{x}_j^\alpha \sim -\frac{\kappa}{4\pi}\left[(-1)^{3-\alpha}\frac{y_j^\alpha - y_j^{3-\alpha}}{\left(r_{jj}^{\alpha(3-\alpha)}\right)^2} + \sum_{\beta=1}^{2}(-1)^{3-\beta}\frac{y_j^\alpha - y_{3-j}^\beta}{\left(r_{j(3-j)}^{\alpha\beta}\right)^2}\right],$$

$$\dot{y}_j^\alpha \sim \frac{\kappa}{4\pi}\left[(-1)^{3-\alpha}\frac{x_j^\alpha - x_j^{3-\alpha}}{\left(r_{jj}^{\alpha(3-\alpha)}\right)^2} + \sum_{\beta=1}^{2}(-1)^{3-\beta}\frac{x_j^\alpha - x_{3-j}^\beta}{\left(r_{j(3-j)}^{\alpha\beta}\right)^2}\right],$$

and clearly, the interaction of vortices from opposite layers is of "barotropic" character.

Since in the case under consideration, the total intensity and total momentum of the system of vortices are zero, i.e., the conditions (1.82) hold, we can, as in Sect. 2.2.2.1, use trilinear coordinates, excluding, for example, the vortex $\binom{2}{2}$. In that case, we have

$$t_1 = \frac{3\kappa}{M}(r_{12}^{11})^2, \quad t_2 = -\frac{3\kappa}{M}(r_{12}^{21})^2, \quad t_3 = -\frac{3\kappa}{M}(r_{11}^{12})^2,$$

and the PD is defined by the conditions

$$t_2 t_3 \geq \frac{9}{4} \quad \text{and} \quad \begin{cases} t_1 \geq 6, & \text{if} \quad M > 0, \\ t_1 \leq 0, & \text{if} \quad M < 0. \end{cases}$$

Further we assume that always $b > a$, and thus $M = \kappa(d^2 - b^2) < 0$. A set of typical phase portraits is given in Fig. 2.29.

When stratification is strong ($\gamma = 0.1$), the interaction between the layers is very weak, and the vortices tend to move along lines $t_3 = const$ (Fig. 2.29a), and thus, two "horizontal" dipoles, moving apart, will form: $\binom{1}{1}\binom{2}{1}$ and $\binom{1}{2}\binom{2}{2}$, as noted in item (d). This type of motion is denoted as $\{1\}$ in Fig. 2.29.

When stratification is moderate ($\gamma = 0.5$ in Fig. 2.29b) for relatively large values of coordinate t_3 (i.e., the distances between vortices in the same layer) we have type $\{4\}$—a type of interaction at which "horizontal" dipoles, when close enough to one another, exchange partners and become "tilted" dipoles (hetons):

$$\begin{pmatrix}1\\1\end{pmatrix}\begin{pmatrix}2\\1\end{pmatrix} + \begin{pmatrix}1\\2\end{pmatrix}\begin{pmatrix}2\\2\end{pmatrix} \longrightarrow \begin{pmatrix}1\\1\end{pmatrix}\begin{pmatrix}2\\2\end{pmatrix} + \begin{pmatrix}2\\1\end{pmatrix}\begin{pmatrix}1\\2\end{pmatrix}.$$

An example of this type of motion is given in Fig. 2.30. Since the collinear configuration is not reached at this type of interaction, the initial conditions (2.28) are inapplicable here; however, in that case they are chosen so that we still have $M = -4$.

If the interaction between layers is even stronger ($\gamma = 1.0$ and $\gamma = 4.0$ in Fig. 2.29), two more regimes appear:

- that of type $\{2\}$ (in Fig. 2.29c, d), when initially, the vertically related vortices with opposite signs (hetons) remain themselves during the motion, i.e., we see configurations of the type $\binom{1}{1}\binom{2}{2}$ or $\binom{2}{1}\binom{1}{2}$;
- that of type $\{3\}$ (in the same figures), when, unlike previous regimes, the trajectories of all vortices remain finite, and the degree of symmetry of trajectories about the straight line $t_2 = t_3$ increases with γ. In the general case, this regime involves the motion of antihetons $\binom{1}{1}\binom{1}{2}$ and $\binom{2}{1}\binom{2}{2}$ along trajectories within a limited domain with different angular velocities.

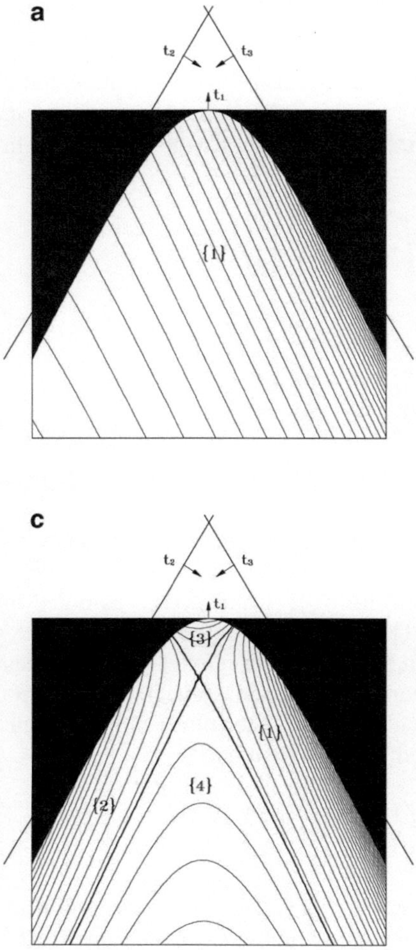

Fig. 2.29 Phase portraits of a four-vortex system in trilinear coordinates (t_1, t_2, t_3) for $M = -4$ and different values of stratification parameter: (**a**) $\gamma = 0.1$, (**b**) $\gamma = 0.5$, (**c**) $\gamma = 1.0$ and (**d**) $\gamma = 4.0$. The *thick lines* are separatrices, separating the parts of PD corresponding to different types of interactions between vortices $\{1\}$–$\{4\}$. Their characteristic properties are discussed in the text. Adapted from [847] (Fig. 3) by permission of IOP Publishing

The stationary elliptic singular point of the phase portrait, which lies on PD boundary, corresponds to steady "solid-body" rotation of a collinear vortex structure with constant angular velocity (2.30), i.e. zigzag–shaped rotating heton quartets.

Figure 2.31 gives examples of trajectories of types $\{1\}$ and $\{3\}$ at $d = 0, b = 2$ and successively increasing values of γ. The figure shows that, at fixed initial conditions, there exists a critical value of stratification parameter (here, $\gamma = 0.95659$), at which the type of motion changes and a transition from open (infinite)

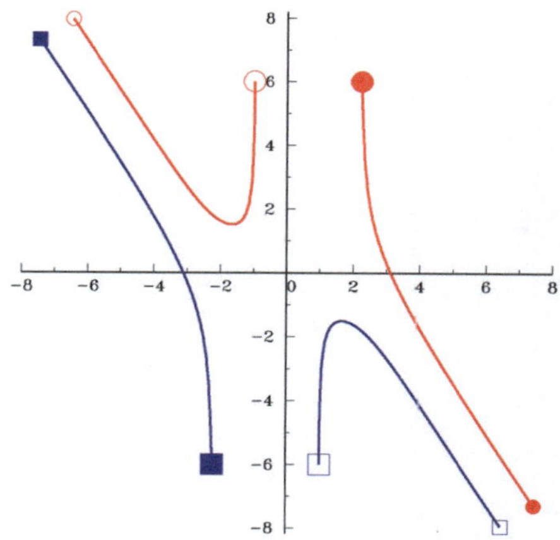

Fig. 2.30 Trajectories of vortices for motions of type $\{4\}$ at $\gamma = 0.5, M = -4$. The initial positions of vortices are given by *circular markers*. Adapted from [847] (Fig. 4) by permission of IOP Publishing

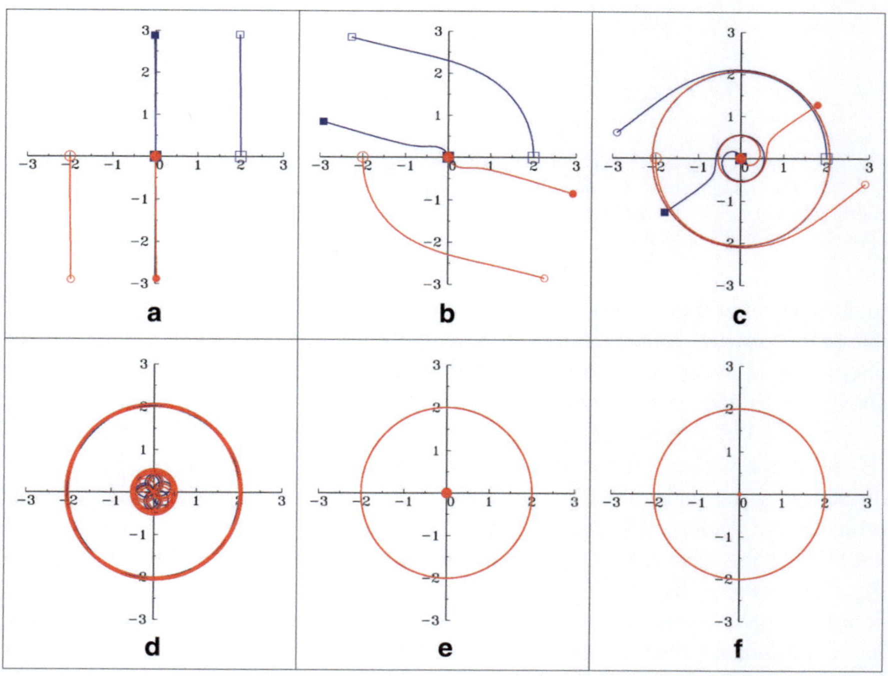

Fig. 2.31 Examples of vortex trajectories for motions of type $\{1\}$—panels (**a–c**)—and type $\{3\}$—panels (**d–f**). Here $d = 0, b = 2, M = -4$ and (**a**) $\gamma = 0.1$, (**b**) $\gamma = 0.9$, (**c**) $\gamma = 0.95659$, (**d**) $\gamma = 0.95660$, (**e**) $\gamma = 1.2$, (**f**) $\gamma = 1.5$. The full (*dashed*) lines correspond to the trajectories of vortices of the top (*bottom*) layer. Adapted from [847] (Fig. 5) by permission of IOP Publishing

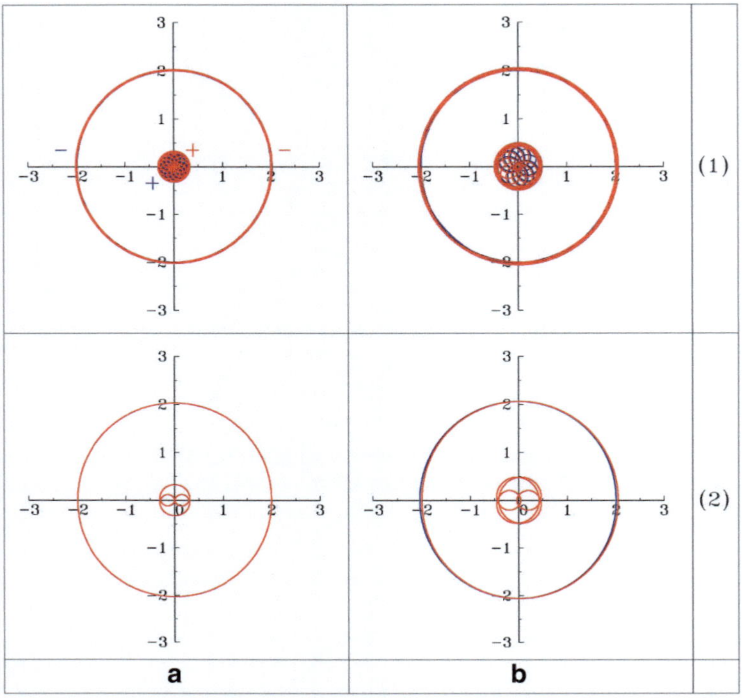

Fig. 2.32 Vortex trajectories (**1**) and relative choreographies (**2**) for motions of type {3} at $\gamma = 1.0, M = -4$ and (**a**) $d = 0$ and (**b**) $d = -0.04$. The angular velocities of rotation of moving coordinate axes $\omega = -0.00001640$ (**a**) and $\omega = -0.00001665$ (**b**). Adapted from [847] (Fig. 6) by permission of IOP Publishing

to closed (finite) trajectories, i.e., from regime {1} to regime {3} takes place. This is due to the growing contribution of the effect of joint rotation because of the stronger interaction between layers. On the other hand, this series of figures confirms the above-mentioned effect of symmetrization of the vortex structure with increasing γ.

Figure 2.31d–f show that, when the conditions for realization of motions of type {3} hold, the vortices of the outer anticyclonic antiheton, which are subject to the influence of internal vortices of positive cyclonicity, rotate along nearly circular orbits in the cyclonic direction. As for the trajectories of vortices of the internal antiheton, they have a more complex structure. The situation can be clarified by Fig. 2.32, where, in addition to vortex trajectories, choreographies in a rotating coordinate system are also given. We see that, in their motions, the vortices of the internal antiheton successively and periodically pass via halves of the internal and external closed figures. In a fixed coordinate system, those figures do not loop and the trajectories gradually fill the appropriate domain.

Note that such representation of choreographies is just one possible out of many others, and Fig. 2.33 illustrates the possibility to choose an optimal value of the angular velocity ω of coordinate system to obtain the simplest choreography. Only

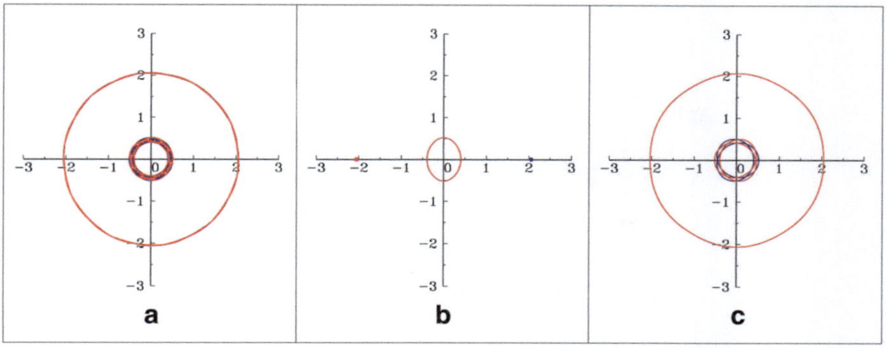

Fig. 2.33 Vortex trajectories (**a**) and relative choreographies (**b**), (**c**) for motions of type $\{3\}$ at $\gamma = 1.0, M = -4, d = 0.4$. The angular velocities of rotation of moving coordinate axes are $\omega = 0.00017341$ in panel (**b**) and $\omega = -0.0001141$ in panel (**c**)

two possible variants are given there, and in the fragments 2.33b, the value of ω coincides with the rotation velocity of vortices of the external antiheton. Therefore, vortices $\binom{2}{1}$ and $\binom{2}{2}$ are given by fixed points, while the vortices of the internal antiheton $\binom{1}{1}$ and $\binom{1}{2}$ move along a closed quasi-elliptic figure.

Clearly, the vortex structures shown in Figs. 2.31d–f, 2.32 and 2.33 are Z-shape tripoles.

An exceptional place in this class of configurations belongs to a stationary construction, whose parameters satisfy the Eq. (2.29). As mentioned above, such vortex system rotates as a solid body; thus, the vortices of the internal and external antihetons rotate with constant angular velocity (2.30) along circular orbits. Figure 2.34 shows the types of motions that are possible in the neighborhood of this steady state. All figures show only the initial stage, while the trajectories do not overlap. The positions of adjacent vortices at the initial and final moments are connected by line segments. In the case 2.34a, we have undisturbed stationary solution, where at $\gamma = 1$, the distances between vortices are determined by parameters $d^* = 0.346$, $b^* = 1.700$, satisfying (2.29). Obviously, in panel 2.34a, the segments align along straight lines. Continuation of calculations bring to a superposition of trajectories of the vortices from different layers, i.e. as in Fig. 2.25a, we observe *generalized absolute Choreography*.

In the other four cases, the dispersion equation is perturbed adding $\Delta d = 0.023$, the total momentum M of the vortex system remaining constant. The choice of such value of Δd is based on the following reasons:

- at $d = d^* + \Delta d$ (Fig. 2.34b) and at $d = d^* - \Delta d$ (Fig. 2.34d), the phase curves in Fig. 2.29 belong to domain $\{3\}$ of finite motions in the form of Z-shaped tripolar structures,
- however, at $d = d^* + 2\Delta d$ (Fig. 2.34c) and at $d = d^* - 2\Delta d$ (Fig. 2.34e), phase trajectories lie in domain $\{2\}$ in the former case and domain $\{1\}$ in the latter case.

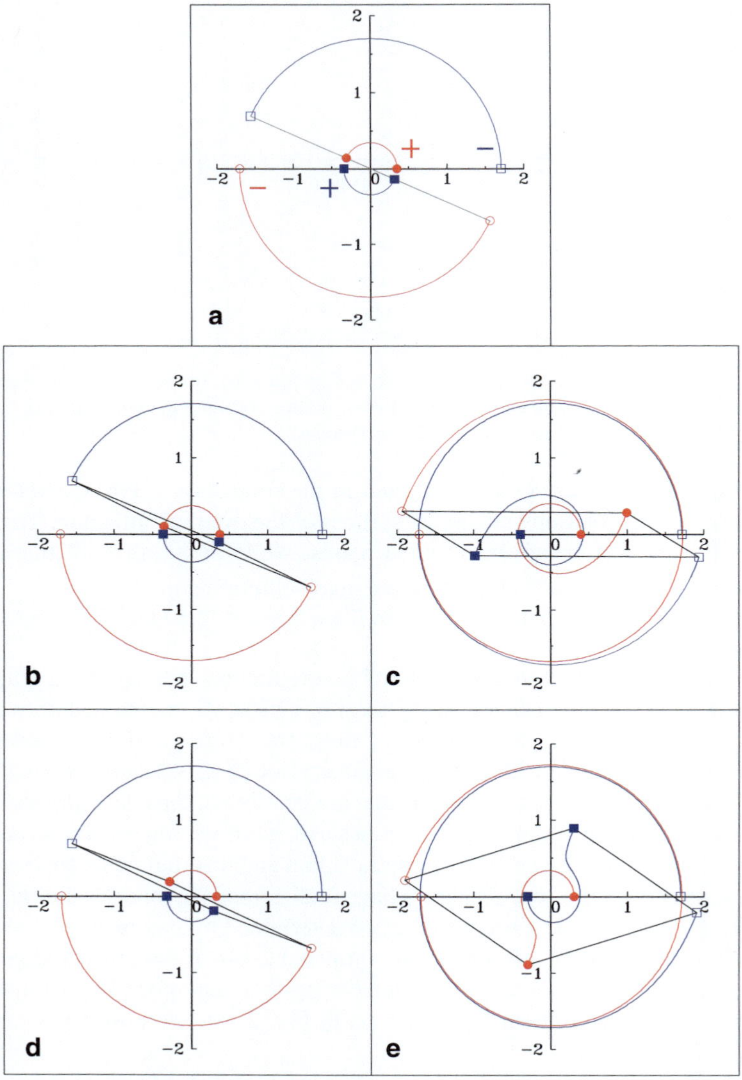

Fig. 2.34 Structural transitions in the neighborhood of a steady collinear vortex structure *rotating heton quartet* at $\gamma = 1, M = -2.7703$ and (**a**) $d = 0.346$, (**b**) $d = 0.369$, (**c**) $d = 0.392$, (**d**) $d = 0.323$, (**e**) $d = 0.300$. Adapted from [847] (Fig. 7) by permission of IOP Publishing

It can be readily seen that, with increasing d, the collinearity is disturbed because the vortices of the internal antiheton at the initial stage of motion start somewhat falling behind (Fig. 2.34b). This effect becomes more evident with the subsequent increase in d (Fig. 2.34c), resulting in the formation of two hetons, which later drift apart. Conversely, a decrease in the distance d leads to the acceleration of the motion of internal vortices (Fig. 2.34d) and the further transformation of the vortex structure into a configuration of two "horizontal" pairs drifting apart (Fig. 2.34e).

2.3 $A+1$ Vortices in a Two-Layer Fluid

In this section, we consider possible motions of a system of discrete vortices in a two-layer fluid, when A vortices lie in the bottom layer and only one vortex lies in the top layer (the opposite situation is not considered because of symmetry consideration). Let us consider successively the results of studying vortex structures at $\kappa_1^1 = \mu\kappa,\ \kappa_2^\alpha \equiv \kappa,\ \alpha = 1, 2, \ldots, A$ for two cases

(I) $A \geq 2, \quad P_x = P_y = 0,$
(II) $A = 2, \quad P_x, P_y \neq 0.$

2.3.1 *Vortex Structures with Zero Total Momentum at A ≥ 2 (Free Motion)*

The condition of zero momentum in the case (I) is always satisfied if the vortex system at the initial time moment has A axes of symmetry. Let us suppose that μ is an arbitrary value, and at the initial moment, the bottom-layer vortices lie in the vertices of a regular A-gone, while the only bottom-layer vortex occupies the central position in the coordinate origin.[12]

Because of the invariance of total components of the linear and angular momenta, we will search for the solution of (1.77)–(1.78) in the form

$$x_1^1 = y_1^1 = 0, \tag{2.31}$$

$$\left(x_2^\alpha; y_2^\alpha\right) = r\left\{\cos\left[\varphi + \frac{2\pi(\alpha-1)}{A}\right]; \sin\left[\varphi + \frac{2\pi(\alpha-1)}{A}\right]\right\},$$
$$\alpha = 1, 2, \ldots, A, \tag{2.32}$$

where r and φ are time functions.

[12]Thus, some results of [133, 430–432, 495, 595, 634, 635, 907, 908] are generalized to the case of a two-layer rotating fluid.

The substitution of (2.31)–(2.32) into (1.77)–(1.78) yields

$$\dot{r} = -\frac{\kappa\gamma}{4\pi}\sum_{\beta=1}^{A-1}\cos\frac{\pi\beta}{A}\mathrm{K}_1\left(2\gamma r\sin\frac{\pi\beta}{A}\right) \equiv 0, \tag{2.33}$$

$$\begin{aligned}\dot{\varphi} &= \frac{\kappa}{4\pi r^2}\left\{\frac{A-1}{2} + \mu\left[1-\gamma r\mathrm{K}_1(\gamma r)\right] + \gamma r\sum_{\beta=1}^{A-1}\sin\frac{\pi\beta}{A}\mathrm{K}_1\left(2\gamma r\sin\frac{\pi\beta}{A}\right)\right\} \\ &\equiv \frac{\kappa\gamma^2}{4\pi R^2}\Omega_0 \equiv \frac{\kappa\gamma^2}{4\pi}F\left(A,\,\mu;R\right),\end{aligned} \tag{2.34}$$

where, as before, $R=\gamma r$, and Ω_0 denotes the expression in curly brackets.

Function $F\left(A,\,\mu;R\right)$ has the following properties (part of them is obvious, while others will be demonstrated graphically):

(a) $F \sim \left(A-1\right)/R^2$ at $R\to 0$;
(b) $F \sim \left[\left(A-1\right)/2+\mu\right]/R^2$ at $R\to\infty$,
$F \sim -\left(A+1\right)/2R^2$ at $R\to\infty$ in the particular case $\mu=-A$;
(c) F is a monotonically decreasing positive function at

$$\mu \geq \left(1-A\right)/2 \geq -1/2; \tag{2.35}$$

(d) F changes its sign in some point R_0 and becomes nonmonotonic, acquiring its minimal value in point $R=R_{min}$, if

$$\mu < \left(1-A\right)/2 < -1/2. \tag{2.36}$$

Value R_{min} can be found as the solution of the transcendent equation

$$\begin{aligned}\Omega_{min} &= 2\left\{\frac{A-1}{2} + \mu\left[1-R\mathrm{K}_1(R)\right] + R\sum_{\beta=1}^{A-1}\sin\frac{\pi\beta}{A}\mathrm{K}_1\left(2R\sin\frac{\pi\beta}{A}\right)\right\} \\ &+ R^2\left[-\mu\mathrm{K}_0(R) + 2\sum_{\beta=1}^{A-1}\sin^2\frac{\pi\beta}{A}\mathrm{K}_0\left(2R\sin\frac{\pi\beta}{A}\right)\right] = 0,\end{aligned} \tag{2.37}$$

derived from the condition $F'_R=0$.

Now let us represent equations $\Omega_0=0$ and $\Omega_{min}=0$ (see (2.34) and (2.37)) as

$$\Omega_0^A = \Omega_0^\mu \quad \text{and} \quad \Omega_{min}^A = \Omega_{min}^\mu, \tag{2.38}$$

gathering in the left and right parts of equations expressions depending only on $(A;R)$ and only on $(\mu;R)$, respectively. The points of intersection of curves with markers A and μ in Fig. 2.35 determine the parameters of graphic solution of equations (2.38).

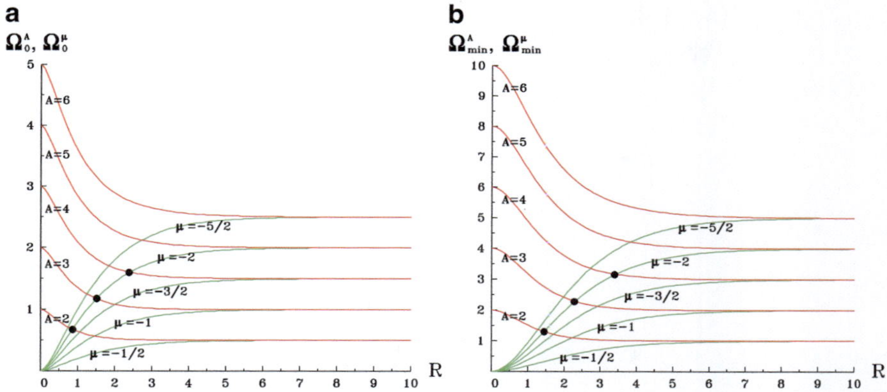

Fig. 2.35 Graphical solution of Eq. (2.37). The *brown lines* correspond to the plots of functions with superscript A, and the *green lines*, to those with superscript μ. The abscissas of the intersection points of curves determine the unknown values of R_0 (**a**) and R_{min} (**b**). The *circular black markers* show the parameters of solutions for $\mu = -2$

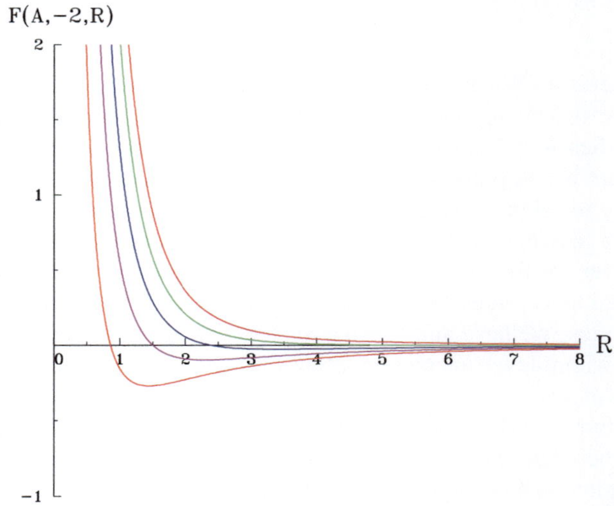

Fig. 2.36 Curves $F(A, \mu; R)$ as functions of R at $\mu = -2$ and $A = 2$ (*brown*), 3 (*magenta*), 4 (*blue*), 5 (*green*), 6 (*red*)

Figures 2.35, 2.36 and 2.37 demonstrate properties (a)–(d) for the angular velocity of rotation of the bottom-layer vortex structure.

Thus, when $\mu = 0$, the A-gonal configuration of bottom-layer vortices with equal intensities κ, if it is stable, always rotates as a solid configuration in the direction determined by the sign of κ. When $\mu \neq 0$, under the influence of the top-layer vortex, the angular velocity of the vortex polygon either increases, when, $\mu > 0$, or

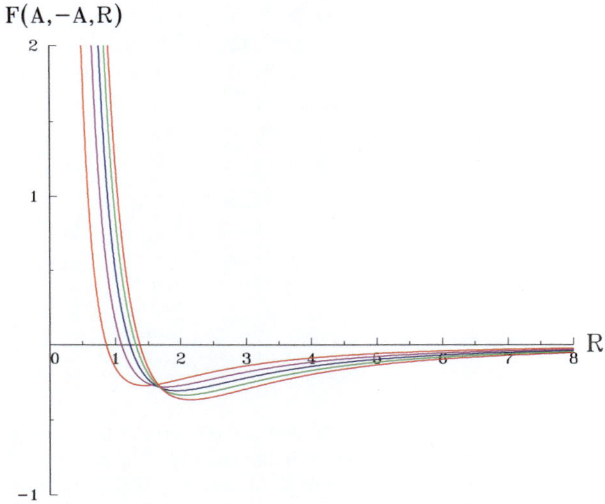

Fig. 2.37 Curves $F(A, \mu; R)$ as functions of R at $\mu = -A$ and $A = 2$ (*brown*), 3 (*magenta*), 4 (*blue*), 5 (*green*), 6 (*red*)

decreases otherwise. When the first condition (2.38) is satisfied, which is possible only when $\mu < 0$, it changes its sign in point $R = R_0$. The effect of the central vortex is maximal when $R = R_{min} > R_0$. At $R > R_{min}$, its influence on the bottom-layer vortex structure is still predominant and the latter rotates in the direction, which is determined by the sign of the intensity of the top-layer vortex. An A-gonal vortex structure lying in a layer and rotating as a solid construction about a central vortex in another layer (in this case, these are the bottom and top layers, respectively), will be referred to as *roundabout*. If the polygon rotates in the direction induced by the central vortex, we have an *ordinary roundabout*. If this vortex, because of the predominant within-layer interaction, rotates in the opposite direction, we have an *inverse roundabout*.[13]

The equation $F = 0$ at fixed μ is a dispersion relationship between A and R, determining the *static state* of the vortex system [595]. This situation radically differs from that in a homogeneous fluid. Indeed, in two-dimensional hydrodynamics, the existence conditions for analogous static configurations for any pair A, μ are uniform with respect to R. For example, in a medium with uniform density for $A = 2, 3, 4$ at any R,[14] we readily obtain $\mu = -0.5, -1, -2$, respectively.

---

[13]We introduced this terminology in [851, 853] for the case $A = 2$; a generalization for an arbitrary A is given here.

[14]The case of $A = 3$, for the first time mentioned in [907], has been studied in sufficient detail in [595].

Table 2.2 The values of R_0, R_{min} and $F^{(m)}$ for the given A

A	2	3	4	5	6	7	8	9	10	11
R_0	0.860	1.090	1.221	1.307	1.369	1.416	1.452	1.482	1.506	1.526
R_{min}	1.440	1.759	1.940	2.060	2.146	2.211	2.261	2.301	2.334	2.361
$-F^{(m)}$	0.271	0.289	0.305	0.334	0.365	0.396	0.428	0.460	0.493	0.525

In the case of vortex structure with zero total circulation ($\mu = -A$), the condition (d) for the change of sign of F is satisfied at any A. The Table 2.2, where $F^{(m)} = F(A, -A, R_{min}) < 0$, gives the main parameters of such structures.

Note that the growth in A is accompanied by an increase in both R_0 and R_{min} and the minimal values of $|F|$. This table is supplemented by Fig. 2.37, showing the character of behavior of function F for this case.

All properties of axisymmetric configurations of $A+1$ vortices can be immediately derived from the equations of motion (2.33)–(2.34). Of doubtless interest it is to study the stability of both the configurations themselves at different A and, in particular, their static states (if the first condition (2.38) is satisfied). Note, the problem of the stability of A-symmetric two-layer vortex structures in the general case is far from being solved. However, in the particular case $A = -\mu$ and $A = 2, 3$, it was solved recently by Kizner [426]. The studying of the stability of those vortex structures goes beyond the scope of present work.

Let us consider in more detail the particular case of three vortices with zero total intensity ($A = 2$, $\mu = -A$). In this case, the formula (2.34) gives us the expression for the angular velocity of vortes structure:

$$\omega = \frac{\kappa\gamma^2}{4\pi R^2}\left[2R\mathrm{K}_1(R) + R\mathrm{K}_1(2R) - \frac{3}{2}\right]. \tag{2.39}$$

Such symmetrical vortex structures in a homogeneous fluid or plasma are widely known as *tripolar vortices* (or *tripoles*) [58,71,100,136,143,146,149,150,181,185, 278–280,347,430,435,436,461,462,542,559,597,631,667,675,705,771,932,935, 936,941,959–961], rotating with constant angular velocity about the central vortex in the direction determined by the sign of its vorticity.

In a two-layer fluid, this is a particular case of a *roundabout*, which is *ordinary* at $R > R_0$ and *inverse* at $R < R_0$; the angular rotation velocity is determined by (2.34). The appropriate values of R_0 and R_{min} are given in Table 2.2. The initial segments of circular trajectories for ordinary and inverse roundabouts are given in Fig. 2.38b in the left and right parts, respectively. Given in the central part is the *static state* of roundabout at $R = R_0$. In this and the subsequent figures in this and the following section, the position of the top-layer vortex is shown by a triangle, and those of bottom-layer vortices are shown by a circle and a square. The size of the marker and the thickness of the tail of the angular arrow is proportional to the modulus of vortex intensity. If in the vortex system the central vortex is located in the upper layer, it is natural to call this system Λ-*shaped tripole*; if it is located in the bottom layer

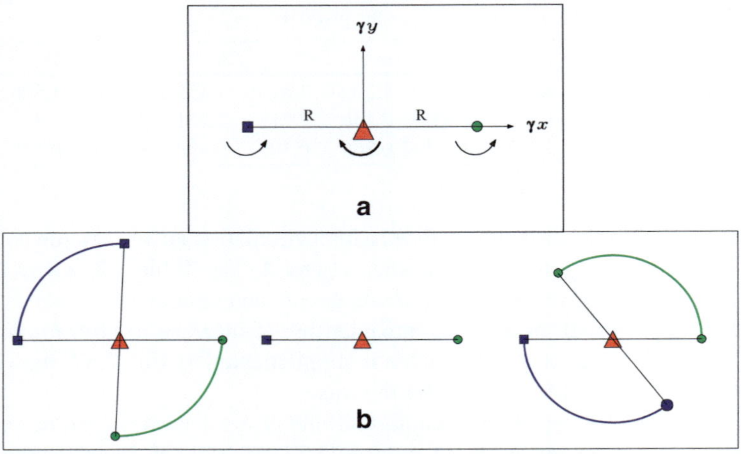

Fig. 2.38 (**a**) Initial distribution of vortices at $A = 2, \kappa = 1, \mu = -2$ with directions of their rotation. (**b**) Initial segments of vortex trajectories at $R > R_0$, $R = R_0$ and $R < R_0$ (*left to right*). Line segments connect the positions of vortices for the initial and final (calculation) moments

the system should be called V*-shaped tripole*. Note, that these tripolar structures are complex absolute choreographies.

An exceptional property of the case $A = 2$ is that if we change places of a bottom-layer and a top-layer vortices, then at $Q = P_x = P_y = 0$, the configuration of three vortices will remain solid-body whatever the value of the total momentum of the system.

Indeed, suppose that $\kappa_1^1 = \kappa_2^2 = -\kappa$; $\kappa_2^1 = 2\kappa$. Now the "strong" central vortex and one of the peripheral ones lie in the same layer (in this case, in the bottom layer), while the second vortex lies in the other layer. In virtue of (**!**), the impact of the central vortex on vortex-satellites will be different. The rotation center in this case does not coincide with the central vortex but lies on the segment connecting it and the top-layer vortex, and its distance from the latter vortex is

$$R_c = R - \frac{2R^2 \mathrm{K}_1(R)}{R\mathrm{K}_1(2R) + 3/2}. \tag{2.40}$$

Such solid-body vortex structure will be referred to as *symmetric eccentric round-about*. The angular velocity of its rotation is

$$\omega = \frac{\kappa \gamma^2}{4\pi R^2}\left[R\mathrm{K}_1(2R) + \frac{3}{2}\right], \tag{2.41}$$

where the sign of ω always coincides with the intensity sign of the stronger vortex (in this case, κ_2^1).

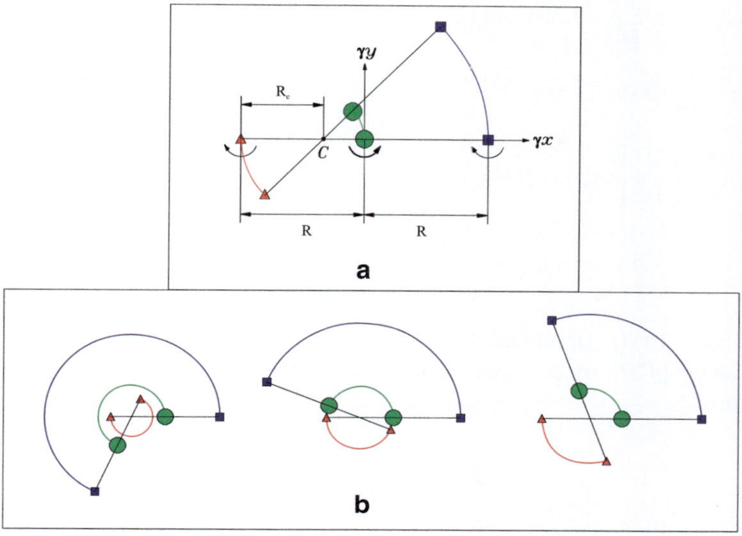

Fig. 2.39 (**a**) Initial distribution of vortices with directions of their rotation and an illustration of the character of motion of the configuration at $\kappa_1^1 = \kappa_2^2 = -1$; $\kappa_2^1 = 2$. (**b**) Initial segments of vortex trajectories at $R_c < R/2$, $R_c = R/2$ and $R_c > R/2$ (from *left to right*)

Figure 2.39a demonstrates a scheme of the initial positions of vortices for this case and elucidates the character of motion of peripheral vortices, while Fig. 2.39b gives the trajectories of motion of vortices within a short time interval for the cases when the value of R_c given by (2.40) satisfies the conditions $R_c < R/2$, $R_c = R/2$, and $R_c > R/2$.

Obviously, vortex trajectories in Fig. 2.38b are *absolute choreographies*, but for an *eccentric roundabout* generally are different. However, in the particular case of $R_c = R/2$ (central part of Fig. 2.39b), trajectories of different layer vortices $\binom{1}{1}$ and $\binom{2}{1}$ coincide, and this gives us one more example *generalized absolute choreography*.

2.3.2 *Vortex Structures with Zero Total Momentum at $A \geq 2$ (Motion in an External Field)*

Let us suppose that the motion of an A-symmetric structure of $A+1$ vortices takes place in the presence of an external field determined by a streamfunction in the form [319, 692]

$$\Psi = -\frac{1}{2}\Big[S(t)\left(x^2 - y^2\right) - \Omega(t)\left(x^2 + y^2\right)\Big]. \tag{2.42}$$

In this case, the Eqs. (2.33) and (2.34) become

$$\dot{\rho} = -\frac{\kappa}{4\pi} s\rho \sin\left[2\left(\varphi + \frac{2\pi(\alpha-1)}{A}\right)\right], \tag{2.43}$$

$$\begin{aligned}\dot{\varphi} = \frac{\kappa}{4\pi}\Bigg\{&\frac{1}{\rho^2}\left[\frac{A-1}{2} + \mu\left[1-\gamma\rho \mathrm{K}_1(\gamma\rho)\right] + \gamma\rho\sum_{\beta=1}^{A-1}\sin\frac{\pi\beta}{A}\mathrm{K}_1\left(2\gamma\rho\sin\frac{\pi\beta}{A}\right)\right]\\ &+\omega - s\cos\left[2\left(\varphi + \frac{2\pi(\alpha-1)}{A}\right)\right]\Bigg\},\end{aligned} \tag{2.44}$$

where $s(t) = 4\pi S(t)/\kappa$ and $\omega(t) = 4\pi\Omega(t)/\kappa$.

Following [410], we consider the Jacobian of the system of Eqs. (2.43) and (2.44) in the form

$$J = \frac{\kappa}{4\pi}\begin{pmatrix} J_{11} & J_{12} \\ J_{21} & J_{22}\end{pmatrix}, \tag{2.45}$$

where

$$\begin{aligned}
J_{11} &= -s\sin\left[2\left(\varphi + \frac{2\pi(\alpha-1)}{A}\right)\right],\\
J_{12} &= 2s\rho\cos\left[2\left(\varphi + \frac{2\pi(\alpha-1)}{A}\right)\right],\\
J_{21} &= -\frac{2}{\rho^3}\left[\frac{A-1}{2} + \mu\left[1-\gamma\rho \mathrm{K}_1(\gamma\rho)\right] + \gamma\rho\sum_{\beta=1}^{A-1}\sin\frac{\pi\beta}{A}\mathrm{K}_1\left(2\gamma\rho\sin\frac{\pi\beta}{A}\right)\right]\\
&\quad+\frac{\gamma^2}{\rho}\left[\mu \mathrm{K}_0(\gamma\rho) - 2\sum_{\beta=1}^{A-1}\sin^2\frac{\pi\beta}{A}\mathrm{K}_0\left(2\gamma\rho\sin\frac{\pi\beta}{A}\right)\right],\\
J_{22} &= 2s\sin\left[2\left(\varphi + \frac{2\pi(\alpha-1)}{A}\right)\right].
\end{aligned}$$

2.3.2.1 Analysis of Static States

Under the conditions of static equilibrium, where $\dot{\rho} = \dot{\varphi} = 0$ and $\omega = \omega_0$, $s = s_0$, stationary points are determined by polar angles

$$\varphi = \varphi_0(A,\alpha,n) = \pi\left[\frac{n-1}{2} - \frac{2(\alpha-1)}{A}\right] \tag{2.46}$$

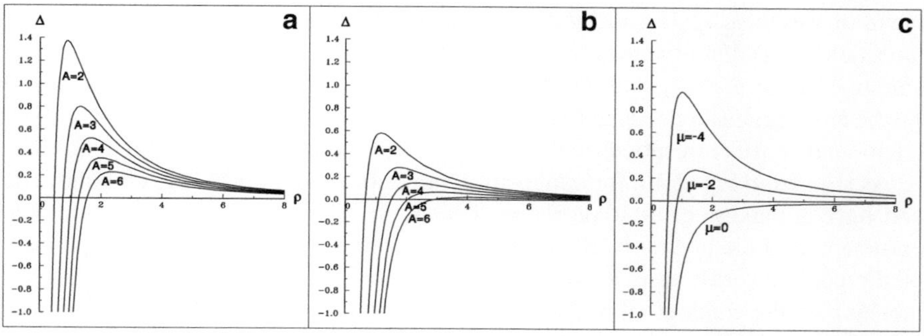

Fig. 2.40 The curves $\Delta(\rho)$, on which the static-state condition (2.46) is satisfied for multifields: (**a**) $\mu = -5$, (**b**) $\mu = -3$ for the given values of A and (**c**) $A = 2$ for the given values of μ

and radii $\rho = \rho_0(A, n, \mu, \omega_0, s_0)$, such that

$$\frac{1}{\rho_0^2}\left[\frac{A-1}{2} + \mu\left[1 - \gamma\rho_0 K_1(\gamma\rho_0)\right] + \gamma\rho_0 \sum_{\beta=1}^{A-1} \sin\frac{\pi\beta}{A} K_1\left(2\gamma\rho_0 \sin\frac{\pi\beta}{A}\right)\right]$$
$$+\omega_0 + (-1)^n s_0 = 0, \tag{2.47}$$

where $n = 1, 2$ and $\alpha = 1, 2, \ldots, A$.

In this case, the Jacobian (2.45) takes a simpler form

$$J^{(0)} = \frac{\kappa}{4\pi}\begin{pmatrix} J_{11}^{(0)} & J_{12}^{(0)} \\ J_{21}^{(0)} & J_{22}^{(0)} \end{pmatrix}, \tag{2.48}$$

where now

$$J_{11}^{(0)} = J_{22}^{(0)} = 0, \quad J_{12}^{(0)} = 2 s_0 \rho_0 (-1)^{n-1},$$
$$J_{21}^{(0)} = \frac{1}{\rho_0}\left\{2\left(\omega_0 + s_0(-1)^n\right) + \gamma^2\left[\mu K_0(\gamma\rho_0)\right.\right.$$
$$\left.\left. -2\sum_{\beta=1}^{A-1} \sin^2\frac{\pi\beta}{A} K_0\left(2\gamma\rho_0 \sin\frac{\pi\beta}{A}\right)\right]\right\} \tag{2.49}$$

and, as before, $n = 1, 2$.

If it exists, the solution of Eqs. (2.46) and (2.47), given the external parameters, determines the static polar coordinates (ρ_0, φ_0) of the bottom-layer vortex. The particular case ($\omega_0 = s_0 = 0$) was considered in the previous section.

The introduction of a negative rotation parameter ω_0 reduces the value of ρ_0. Taking into account the deformation factor (parameter s_0) and the positive values of ω_0 changes the situation qualitatively, as can be seen, in particular, from Fig. 2.40, where the solutions of Eq. (2.47) are given graphically. On the curves, shown in the plane of variables (ρ, Δ), the static-state conditions are satisfied, i.e., $\rho = \rho_0$, for multifields at different values of A and μ. Here, the variable Δ denotes one of the variables $\Delta_{\mp} = \omega_0 \mp s_0$, and the distance ρ is measured along the axis determined by the appropriate angle φ_0 in (2.46). Obviously, the particular case of $\omega_0 = s_0 = 0$ mentioned above corresponds to the intersection points of the curves with the ρ-axis.

As seen from Fig. 2.40, for nonpositive values of μ, all curves are nonmonotonic and have a maximum at some value $\Delta = \Delta_{max}$. Since one of the causes of the existence of static positions of bottom-layer vortices may be the opposite polarity of the central vortex in the top layer, it is evident that at a fixed value of μ, an increase in the number of vortices in the bottom layer A causes a decrease in Δ_{max}. Indeed, with an addition of vortex satellites, the cyclonic component of their collective rotational motion increases, thus preventing the possible cessation of structure rotation because of anticyclonic spin from the side of the core. A similar tendency can be seen with a decrease in core intensity (Fig. 1.9a, b).

2.3.2.2 Stationary Solutions at $A = 2$

Let us consider the particular case of a tripole $A = 2$. In this case (2.46), (2.47) and the expression for $J_{21}^{(0)}$ in (2.49) become simpler, yielding

$$\varphi = \varphi_0(\alpha, n) = \pi\left(\frac{n+1}{2} - \alpha\right), \qquad n, \alpha = 1, 2, \tag{2.50}$$

$$\frac{1}{\rho_0^2}\left[\frac{1}{2} + \mu\left[1 - \gamma\rho_0 \mathrm{K}_1(\gamma\rho_0)\right] + \gamma\rho_0 \mathrm{K}_1(2\gamma\rho_0)\right] + \omega_0 + (-1)^n s_0 = 0, \tag{2.51}$$

$$J_{21}^{(0)} = \frac{1}{\rho_0}\left[2\left(\omega_0 + s_0(-1)^n\right) + \gamma^2\left[\mu \mathrm{K}_0(\gamma\rho_0) - 2\mathrm{K}_0(2\gamma\rho_0)\right]\right]. \tag{2.52}$$

The eigenvalues of the matrix (2.48) are specified by expressions

$$\lambda_{1,2} = \pm\sqrt{J_{12}^{(0)} J_{21}^{(0)}}, \tag{2.53}$$

whence it follows that, if Eq. (2.51) has solutions for both $n = 1$ and $n = 2$, then, in virtue of sign alternation of $J_{12}^{(0)}$ (see (2.49)), we always have two stable and two unstable solutions.

Returning to Fig. 2.40, we note that, if both values Δ_- and Δ_+ are greater than Δ_{max}, we have no static positions. However, if one of those values is less than Δ_{max}, then, at $A = 2$, variants with one, two, three, or four pairs of static points are possible. In the latter case, the condition

$$0 < \Delta_-, \Delta_+ \leq \Delta_{max}. \tag{2.54}$$

should be valid. Below, we consider an even narrower class of vortex structures with zero total intensity: $A = 2$ and $\mu = -2$.

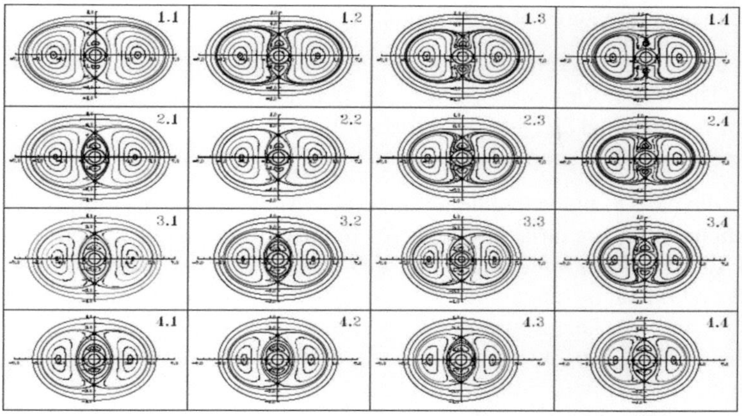

Fig. 2.41 Gallery of trajectories of bottom-layer vortices corresponding to 16 small circular markers in Fig. 2.42. The values of external parameters for the extreme (angular) panels are as follows: **(1.1)**: $\omega_0 = 0.1574$, $s_0 = 0.0569$; $\rho_{1,h} = (\pm 0.9377;0)$, $\rho_{1,e} = (\pm 3.6893;0)$, $\rho_{2,e} = (0;\pm 1.0989)$, $\rho_{2,h} = (0;\pm 2.1412)$; **(1.4)**: $\omega_0 = 0.2024$, $s_0 = 0.0569$; $\rho_{1,h} = (\pm 0.9862;0)$, $\rho_{1,e} = (\pm 2.9151;0)$, $\rho_{2,e} = (0;\pm 1.2612)$, $\rho_{2,h} = (0;\pm 1.6849)$; **(4.1)**: $\omega_0 = 0.1574$, $s_0 = 0.0284$; $\rho_{1,h} = (\pm 0.9664;0)$, $\rho_{1,e} = (\pm 3.1595;0)$, $\rho_{2,e} = (0;\pm 1.0436)$, $\rho_{2,h} = (0;\pm 2.4296)$; **(4.4)**: $\omega_0 = 0.2024$, $s_0 = 0.0284$; $\rho_{1,h} = (\pm 1.0249;0)$, $\rho_{1,e} = (\pm 2.5598;0)$, $\rho_{2,e} = (0;\pm 1.1413)$, $\rho_{2,h} = (0;\pm 1.9821)$. Here, the values $n = 1,2$ correspond to the first subscripts of ρ, while denotations "e" and "h" refer to the coordinates of elliptic and hyperbolic points, respectively. The *thick lines* are separatrices separating the domains with different types of motion

In Fig. 2.40, this case corresponds to the median curve in panel (c), where $\Delta_{max} \equiv \Delta_{max}\Big|_{\mu=-2}^{A=2} = 0.27094$. The solution (2.50)–(2.51) shows that stationary points always lie on the x and y axes. The value of Δ_- is responsible for stationary points on the x-axis with the singular points on the left branch of the curve being always hyperbolic, and those on its right branch always elliptic. At the same time, Δ_+ characterizes the singular points along the y-axis, in which case elliptic points lie on the left branch and hyperbolic points lie on its right branch.

Figure 2.41 shows a series of 16 variants of bottom-layer vortex trajectories (the top-layer vortex is fixed and lies in the origin of coordinates) for different combinations of external parameters, at which 4 pairs of singular points are always present. Those examples demonstrate the role of rotation and deformation components of the external field (2.42) on the dynamics of bottom-layer vortices. A common feature of those figures is the presence of two separatrices—an internal and an external—both having a similar character and being either heteroclinic or homoclinic. Their topological properties depend on to which class the vortex structure belongs. First, we note that a common feature of this case and the free-tripole problem is the presence of a central domain *inverse roundabouts*, where two vortices-satellites rotate in the cyclonic direction along a common trajectory. In the external domain (beyond both separatrices) the rotation is also cyclonic in all cases, because the parameter $\omega_0 > 0$ plays the determining role in this case.

Note the main features of those motions:

1. Two classes of different motions can be identified—S (similar) and D (dissimilar). The trajectories corresponding to them are given in the panels above and below the diagonal (1.1)–(4.4), respectively.
2. The comparison of graphs for those two cases shows that

   - In the case S, the external and internal separatrices are topologically similar (though they are turned through 90° relative to one another), hence the name S for this class of motions. In this case, the separatrices consist of two closed domains, tangent in two hyperbolic points, i.e., these are heteroclinic structures. In lateral domains with elliptic fixed points on the x-axis (within the external separatrix), each vortex moves along its own trajectory in the cyclonic direction. Within the internal separatrix, the vortices are involved in anticyclonic rotations within hole-like domains (along individual trajectories) and within the domain between separatrices (along a common trajectory), while in the central domain, their motions are cyclonic—this is an inverse roundabout.
   - In the case D, there exist two types of homoclinic separatrices, one of which (external) passes through two hyperbolic points on the y-axis with a closed loop corresponding to each such point, and the whiskers, running out of the saddle points, merge and envelope four vortex domains—two domains with centers on the y-axis, belonging to the same homoclinic structure (with anticyclonic vortex motion), and two domains with centers on the x-axis (with cyclonic vortex motion), belonging to the internal homoclinic structure. The second (internal) separatrix passes via two hyperbolic singular points on the x-axis and consists of two homoclinic loops, and the whiskers of saddle points envelope the central domain, which can be regarded as a heteroclinic structure along with the lateral loops of the internal separatrix; in this case, in the narrow domain between the separatrices, also forms an intermediate flow-through domain, in which bottom-layer vortices move along a common trajectory in the anticyclonic direction. The topologic difference between the separatrices is that the homoclinic loops lie within separatrix whiskers for the external domain and beyond the zone enveloped by the whiskers for the internal domain.
   - Thus, the main distinctions of those cases are as follows:

     (a) the internal separatrix in the case S envelopes both the central domain with cyclonic rotation and the vertically situated lunar domains with anticyclonic motion of vortices, thus forming a heteroclinic structure (hyperbolic points have no whiskers); in the case D, the central domain is enveloped by whiskers coming from saddle points of the internal separatrix, and they, along with the lateral loops of the homoclinic structure (the right and the left) envelope the trajectories of vortices rotating cyclonically;

(b) the external separatrix in the case S consists of two heteroclinic loops, lying on the x-axis, while in the case D, the whiskers of the saddle points of the external separatrix envelope the entire central part of the phase portrait, including the homoclinic loops of this separatrix, which now lie on the y-axis;

(c) the anticyclonic flow-through motion envelopes the central, upper, and lower domains in the case S and the central and lateral domains in the case D.

3. When the conditions $\Delta_- < \Delta_{max}$ and $\Delta_+ > \Delta_{max}$ are satisfied simultaneously, there exist pairs of parameters, such that the external and internal separatrices merge with the formation of the so-called reconnection of separtrices, and thus, the flow-through domains disappear (see panels (1.1), (2.2), (3.3), and (4.4). The conditions for reconnection to appear are as follows

$$F(a)+a^2\Delta_- = F(b)+b^2\Delta_+, \tag{2.55}$$

where $F(z) = 2\mu F_1(z) + F_2(2z)$ and $F_1(z) = \ln z + \mathrm{K}_0(z)$, $F_2(z) = \ln z - \mathrm{K}_0(z)$, while a and b are solutions of Eqs. (2.50) and (2.51) for $n=1$ and for $n=2$, respectively. The Eq. (2.55) is equivalent to the requirement that the values of the Hamiltonian coincide on the two separatrices passing simultaneously through hyperbolic points on the x and y axes. In the case of reconnection, the motions are cyclonic in the central, lateral, and external domains and anticyclonic in the lunar domains.

4. If a vortex structure belongs to class S, an increase in both parameters ω_0 and s_0 is accompanied by an increase in the vertical size of lunar domains up to the limiting one (when its point maximally remote from the center and lying on the y-axis merges with its hyperbolic singular point, i.e., a reconnection takes place); next, the separatrices surrounding it split on the x-axis, and those domains become loop-shaped—i.e., passage to class D takes place.

Figure 2.42 gives a diagram of the state of a tripole on the plane of variables (ω_0, s_0). The reconnection condition (2.55) is satisfied on full lines. The dashed lines intersecting on the horizontal axis satisfy the equations $\Delta_- = \omega_0 - s_0 = \Delta_{max}$ and $\Delta_+ = \omega_0 + s_0 = \Delta_{max}$. The domain marked by dots contains the points of the plane (ω_0, s_0) in which the condition (2.55) is satisfied when 4 pairs of singular points exist. In this case, $\Delta_{max} \equiv \Delta_{max}\Big|_{\mu=-2}^{A=2} = 0.2709$. The position of markers correspond to parameters (ω_0, s_0), for which numerical experiments, shown in Figs. 2.41, 2.43, 2.44, 2.45 and 2.46 have been carried out. In particular, one can see that diagonal small-size circular markers lie on the reconnection line.

Some peculiarities of the behavior of separatrices during reconnection can be seen in the diagonal panels (1.1)–(4.4) in Fig. 2.41. Indeed, with the downward motion along the top branch of reconnection line in Fig. 2.42, i.e., when both ω_0 and s_0 decrease, the elliptic and hyperbolic stationary points approach along both x and y axes.

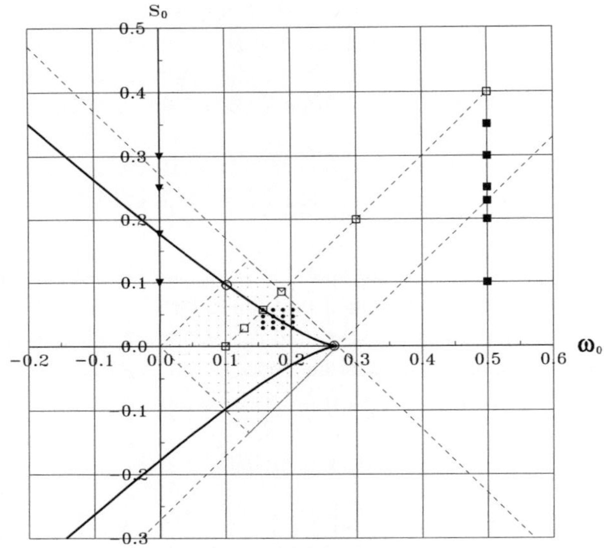

Fig. 2.42 Diagram of different states of tripole in an external stationary field on the plane of parameters (ω_0, s_0) at $\mu = -2$. The *dashed lines* intersecting on the horizontal axis correspond to the conditions $\Delta_{\mp} = \Delta_{max}$; the domain marked by *dots* contains the points of the plane where the condition (2.54) is satisfied. The reconnection condition holds on *full lines*, allowing the merging of separatrices. The *tilted dashed segment* passes through markers corresponding to experimental parameters in Fig. 2.44

In addition, Fig. 2.43 shows the peculiarities of separatrices near the limiting points in the marked domain in Fig. 2.42. Thus, Fig. 2.43a gives the case $\omega_0 = s_0$, where the straight line $\Delta_- = 0$ intersects the curve of static states in Fig. 2.40 in a single point; hence, we have on the x-axis only a pair of hyperbolic stationary points, whereas the coordinates of elliptic lateral points tend to their infinite limits. In the diagram 2.42, this experiment corresponds to the circle at the intersection of the reconnection line and the rectilinear boundary of the marked domain. At a small positive value of Δ_- (equal to 0.00615 in Fig. 2.43b), the lateral domains of the closed trajectories with centers in elliptic points with coordinates $x_{1e} = \pm 15.6173$ have a finite size. In the neighborhood of the opposite limit on the reconnection line ($\Delta_+ \sim \Delta_- \sim \Delta_{max}$ at $s_0 \to 0$), all hyperbolic and elliptic points approach pairwise (Fig. 2.43c) and, in the limit $s_0 = 0$, they merge and degenerate (Fig. 2.43d). In this case, clearly the circumference with a radius of $\rho^* = |x_{1e}| = |x_{1h}| = |y_{1e}| = |y_{1h}|$ is an isoline of zero angular velocity of roundabout rotation. Thus, both within and beyond this circumference, we have an inverse roundabout. Note that in diagram 2.42, the appropriate markers coincide visually, since the points of plane (ω_0, s_0) representing the experiments 2.43a and 2.43b, as well as 2.43c and 2.43d are very close to one another.

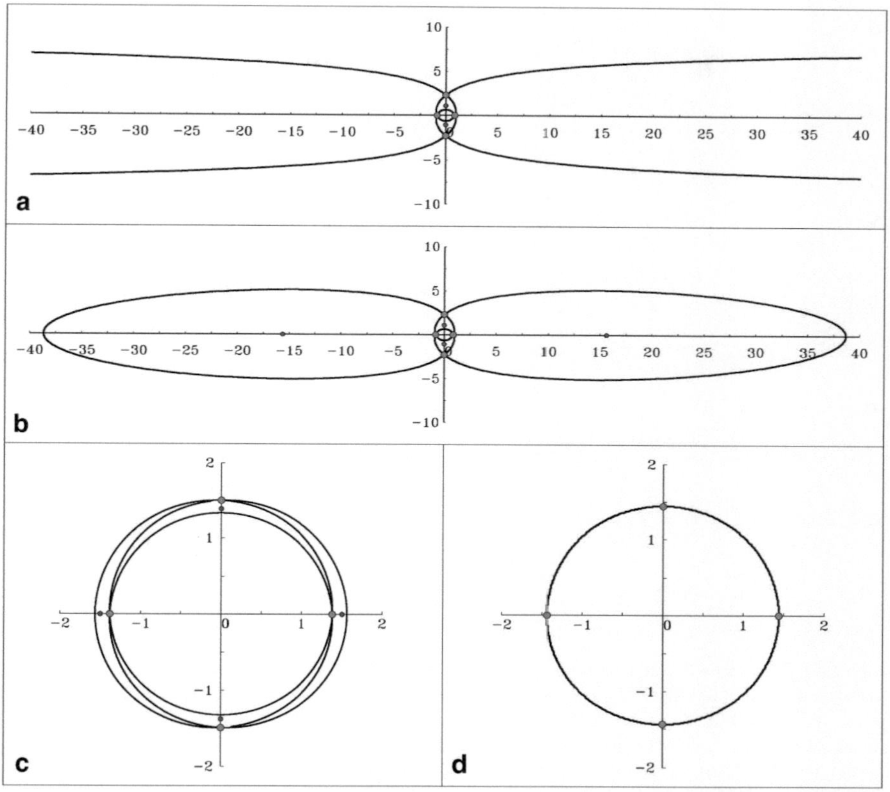

Fig. 2.43 The set of separatrices in the presence of reconnection in limiting points of the domain marked in Fig. 2.42 and their neighborhoods: (**a**) $\omega_0 = s_0 = 0.09889$ ($\Delta_+ = 0.19778, \Delta_- = 0$); $\rho_{1h} = (\pm 0.86015;0)$, $\rho_{1e} = (\pm\infty;0)$, $\rho_{2e} = (0;\pm 1.06489)$, $\rho_{2h} = (0;\pm 2.30437)$, (**b**) $\omega_0 = 0.10235$, $s_0 = 0.0962$ ($\Delta_+ = 0.19855, \Delta_- = 0.00615$), $\rho_{1h} = (\pm 0.86413;0)$, $\rho_{1e} = (\pm 15.61728;0)$, $\rho_{2e} = (0;\pm 1.06624)$, $\rho_{2h} = (0;\pm 2.29656)$, (**c**) $\omega_0 = 0.2660$, $s_0 = 0.0006$ ($\Delta_+ = 0.2666, \Delta_- = 0.2654$), $\rho_{1h} = (\pm 1.38299;0)$, $\rho_{1e} = (\pm 1.50328;0)$, $\rho_{2e} = (0;\pm 1.38591)$, $\rho_{2h} = (0;\pm 1.49980)$, (**d**) $\omega_0 = 0.27094$, $s_0 = 0$ ($\Delta_+ = \Delta_- = 0.27094$); $\rho^* = 1.4407$. Coordinates of elliptic points are marked by *small-size circles* and those of hyperbolic points are marked by *large-size markers*. The appropriate markers in Fig. 2.41 are also shown by *circles*

Thus, the passage via a reconnection changes the type of separatrix from heteroclinic (S) to homoclinic (D). For structures of class D, reverse transformation takes place with a decrease in ω_0 and s_0.

Now let us examine the topological properties of separatrices at variations in the external field that conserve the value Δ_-; a series of separatrix configurations for that case is given in Fig. 2.44. In the diagram 2.42, the appropriate markers (hollow squares) lie in a tilted dashed line determined by the equation $\Delta_- = 0.10045$. Under such conditions, clearly, the coordinates of stationary points on the x-axis also remain unchanged: $\rho_{1h} = (\pm 0.93770;0)$, $\rho_{1e} = (\pm 3.68926;0)$. In Fig. 2.44a, b, where $\omega_0 = 0.5$ and $\omega_0 = 0.3$, respectively, the value Δ_+ is greater than Δ_{max}, hence no stationary points on the y-axis.

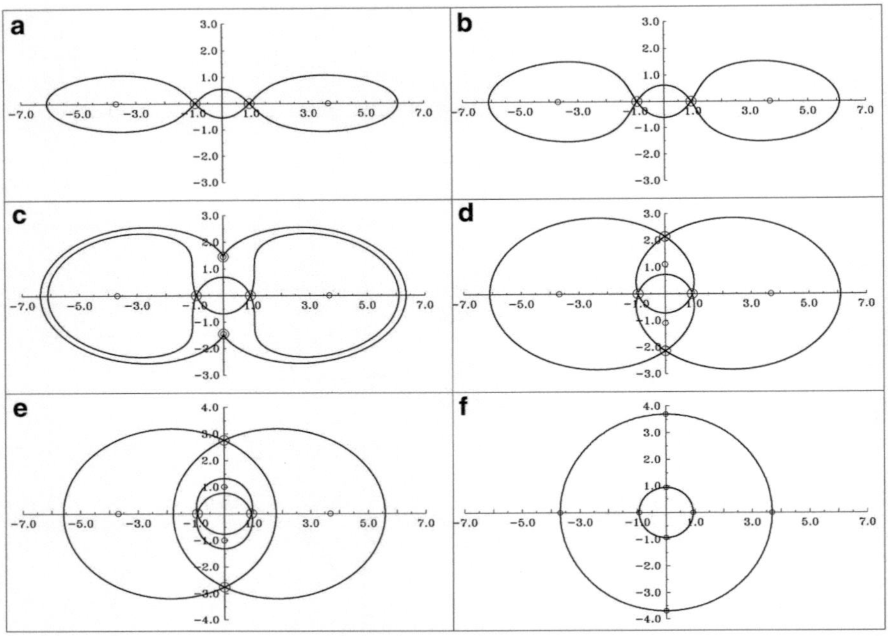

Fig. 2.44 The set of separatrices for the motion of bottom-layer vortices at fixed $\Delta_- = 0.10045$: **(a)** $\omega_0 = 0.5$, $s_0 = 0.39955$; **(b)** $\omega_0 = 0.3$, $s_0 = 0.19955$; **(c)** $\omega_0 = 0.185695$, $s_0 = 0.085240$, $\rho_{2e} = \rho_{2h} = (0;\pm 1.44087)$, **(d)** $\omega_0 = 0.15735$, $s_0 = 0.05690$, $\rho_{2e} = (0;\pm 1.09887)$, $\rho_{2h} = (0;\pm 2.14116)$, **(e)** $\omega_0 = 0.12885$, $s_0 = 0.02840$, $\rho_{2e} = (0;\pm 1.00123)$, $\rho_{2h} = (0;\pm 2.75941)$, **(f)** $\omega_0 = 0.10045$, $s_0 = 0$, $r_1 = |\rho_{1h}| = 0.93770$, $r_2 = |\rho_{2e}| = 3.68926$ (markers in the diagram 2.42 are *hollow squares*)

As can be seen from the figures, with decreasing ω_0, all circulation cells along the y-axis increase in size. In Fig. 2.44c, which exemplifies the case of $\Delta_+ = \Delta_{max}$, we see the appearance of a pair of degenerated stationary points on the y-axis and an external separatrix associated with them. Next, Fig. 2.44d, gives the separatrix for the case of reconnection—(the case (1.1), Fig. 2.41), Fig. 2.44e gives that for motions of class S (here, the values of s_0 are the same as in the bottom line of Fig. 2.41), and Fig. 2.44f gives the separatrix for the case $s_0 = 0$.

The latter example is of interest because, in the figures, the circular trajectories with radii of r_1 and r_2, which now are not separatrices, show a change in the cyclonic motion from an inverse roundabout with dominating within-layer interaction to an ordinary roundabout, where the bottom-layer vortices rotate in the direction induced by the top-layer anticyclone, and next changes again to an inverse roundabout (now due to the predominant effect of the cyclonic component of the external field). Note that in Fig. 2.43d, where the radii of the two circumferences coincide because of the reconnection, the ring domain for the ordinary roundabout degenerate into a circumference of static positions of bottom-layer vortices. This is confirmed by the median curve in Fig. 2.40c, which can be interpreted as the dependence $\omega_0(\rho_0)$ at $s_0 = 0$. According to the figure, at $-\infty < \omega_0 \leq 0$, we always have one static state

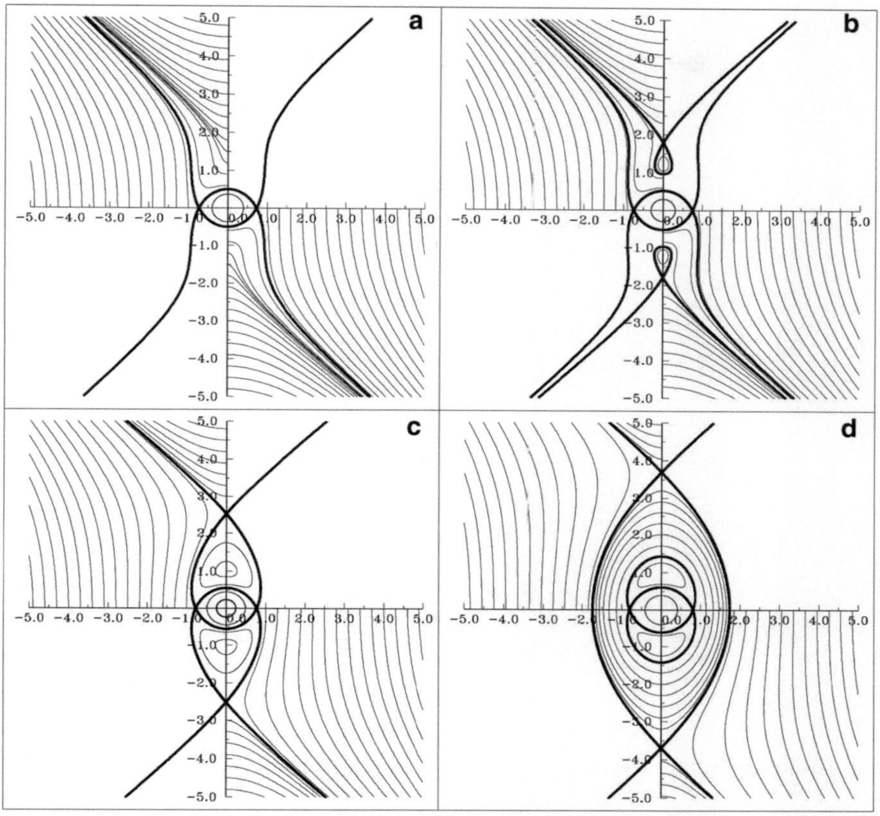

Fig. 2.45 Trajectories of bottom-layer vortices starting from points on the x and y axes at $\omega_0 = 0$: (**a**) $s_0 = 0.30$, $\rho_{1h} = (\pm 0.72845; 0)$, (**b**) $s_0 = 0.25$, $\rho_{1h} = (\pm 0.74490; 0)$, $\rho_{2e} = (0; \pm 1.21069)$, $\rho_{2h} = (0; \pm 1.78998)$; (**c**) $s_0 = 0.17759$, $\rho_{1h} = (\pm 0.77168; 0)$, $\rho_{2e} = (0; \pm 1.03044)$, $\rho_{2h} = (0; \pm 2.51862)$, (**d**) $s_0 = 0.10$, $\rho_{1h} = (\pm 0.80539; 0)$, $\rho_{2e} = (0; \pm 0.93727)$, $\rho_{2h} = (0; \pm 3.69914)$ (*triangular markers* in diagram 2.42). The thick trajectories coincide with separatrices

of the roundabout on a circumference with radius ρ_0, while at $0 < \omega_0 < \Delta_{max} = 0.27094$, we have two static states on circumferences with radii $\rho_{01} \le \rho_{02}$, where $\rho_{02} \to \infty$ when $\omega_0 \to 0$ and $\rho_{01} = \rho_{02} = \rho^* = 1.4407$ when $\omega_0 = \omega_0^* = 0.27094$. When $\omega_0 > \omega_0^*$, the angular velocity of the roundabout is positive everywhere.

The results of studying the influence of parameter s_0 on the dynamics of bottom-layer vortices at zero rotational component of the external field are given in Fig. 2.45. In addition to the separatrices, this figure also gives the trajectories of vortices that initially lie on the x and y axes. When the shear component of the external field is strong enough (Fig. 2.45a, where $\Delta_- < 0$, and $\Delta_+ > \Delta_{max}$), we have a single pair of hyperbolic stationary points on the x-axis. In this case, closed trajectories exist only within a small central domain, while beyond it, the vortices are carried away by external flow.

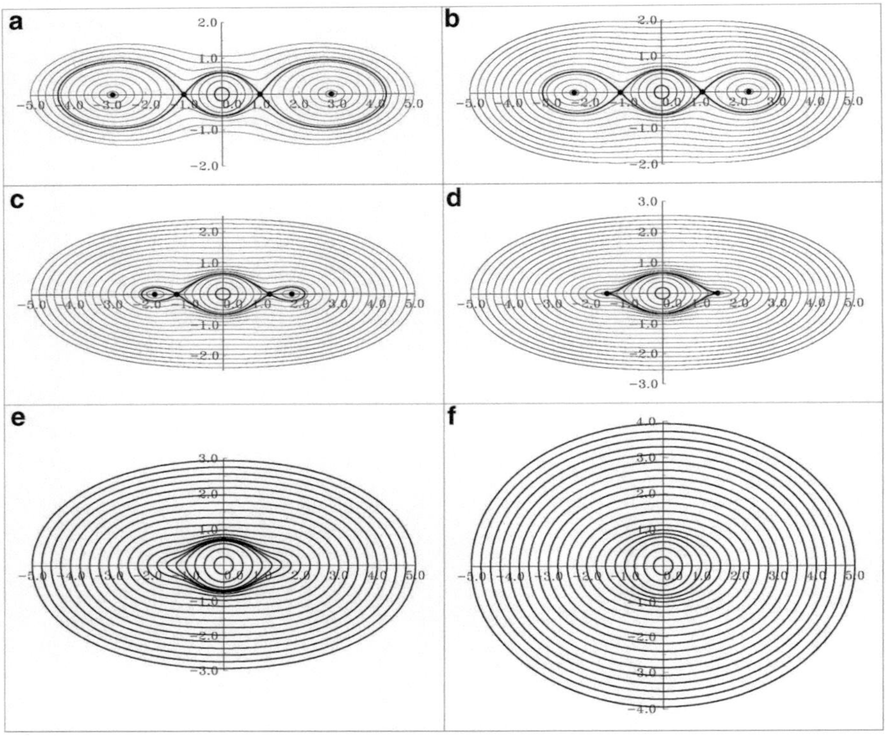

Fig. 2.46 Trajectories of bottom-layer vortices starting from points in the x and y axes at $\omega_0 = 0.5$: (**a**) $s_0 = 0.35$, $\rho_{1h} = (\pm 0.99183;0)$, $\rho_{1e} = (\pm 2.85344;0)$, (**b**) $s_0 = 0.30$, $\rho_{1h} = (\pm 1.06911;0)$, $\rho_{1e} = (\pm 2.28190;0)$, (**c**) $s_0 = 0.25$, $\rho_{1h} = (\pm 1.21069;0)$, $\rho_{1e} = (\pm 1.78998;0)$, (**d**) $s_0 = 0.22906$, $\rho_{1h} = \rho_{1e} = (\pm 1.44087;0)$; (**e**) $s_0 = 0.20$; (**f**) $s_0 = 0.10$ (*black square* markers in diagram 2.42)

In Fig. 2.45b, conversely, $\Delta_+ < \Delta_{max}$, so two pair of singular points—elliptic and hyperbolic—appear on the y-axis. An interesting fact is the formation, in a purely translational flow, of top and bottom closed domains that form because of the capture of bottom-layer vortices by the central anticyclonic core of the top layer, with the captured domains separated from the central domain by a narrow zone of flow-through infinite motions of vortices. A tendency toward the capture becomes more significant at a successive decrease in s_0: Fig. 2.45c gives the case of a reconnection, and Fig. 2.45d shows an analogue of motions of class S with a closed internal separatrix and a domain (external for it) of finite anticyclonic motions of vortices along a common trajectory.

A series of illustrations of the results of numerical experiments for studying the effect of s_0, but now with a considerable rotational component of the external field ($\omega_0 = 0.5$), is shown in Fig. 2.46. Here, all trajectories of bottom-layer vortices are closed. When $\Delta_- < \Delta_{max}$ (panels a, b, c), we have two pairs of singular points on the x axis, and the set of trajectories is represented by three cells of closed cyclonic circulations in a cyclonic external flow. When $\Delta_- = \Delta_{max}$ (panel d), the singular points merge and we have two fixed points. Note that in that case, $\rho_e = \rho_h = 1.44087$, which obviously coincides with the value of ρ^* in Fig. 2.43d, since the appropriate points in the diagram (Fig. 2.42) lie on the same straight line $\Delta_- = \Delta_{max}$. The further decrease in parameter s_0 (i.e., an increase in Δ_-) leads to a system of centered embedded trajectories with cyclonic rotation of bottom-layer vortices. Now, if we consider the fragments of this figure in the inverse order, we will see that, with increasing deformation parameter s_0, the picture of trajectories of bottom-layer vortices at $\Delta_- = \Delta_{max}$ first shows the appearance of a curve with zero rotation frequency of bottom-layer vortices; next this curve transforms into a heteroclinic separatrix (which also can be regarded as two homoclinic separatrices with whiskers connecting around the central domain), which envelopes the three-cell structure of closed circulations.

In the following section, we will analyze the rotation frequencies of bottom-layer vortices as a function of their distances from the central bottom-layer vortices. This study directly relates to the problem of chaotic regime formation in the behavior of vortices with nonstationary perturbations of external field taken into account.

2.3.2.3 Analysis of Optimal Perturbation Frequencies at $A = 2$

The chaotic regimes in dynamics systems are known to be governed by the geometric characteristics of the domains of the so-called nonlinear resonances of the main series [441, 442, 1007]. To determine them, one needs to know the rotation frequencies of vortices about the closed trajectories. This allows one to determine the parameters of a nonstationary perturbation optimal for the formation of chaotic regimes [380, 440] or for the manifestation of some specific effects of chaotization [443].

Figure 2.47 gives the dependences of rotation frequencies ν for the bottom-layer vortices on their initial position x_0 or y_0 on the x or y axis for some trajectory sets in Fig. 2.41. Clearly, zero values of $\nu(x_0)$ and $\nu(y_0)$ are attained on separatrices.

It is often convenient to pass to the action–angle variables, in which case each trajectory is uniquely determined by the value of the action variable. Since such passage is difficult in the problem in question, we use the initial position x_0 or y_0 of a trajectory to identify it. Those initial coordinates uniquely determine the trajectory, i.e., there exists a one-to-one correspondence between x_0 or y_0 and the action. Note that by rotation frequency in point x_0, we mean the value $\nu(x_0) = \mathrm{Sign}(\varpi)2\pi/T(x_0)$, where T is the period and ϖ is the angular velocity, which, unlike $\nu(x_0)$, is not constant along the trajectory.

Following [93, 1007], we say that the nonlinear resonance that corresponds to a trajectory with rotation frequency $\nu(x_0) = (m/n)\nu^*$ has the *rotation frequency* m/n. The integer n, equal to the number of islands (elliptic and hyperbolic points) on the phase portrait of Poincaré mappings, will be called *resonance multiplicity*, while m, i.e. the number of periods required for a particle to pass around all islands, will be called *resonance order*. Here ν^* is used to denote the frequency of nonstationary perturbation, which will be chosen based on the analysis of frequency relationships (as it is done, for example, in Figs. 2.47 and 2.48).

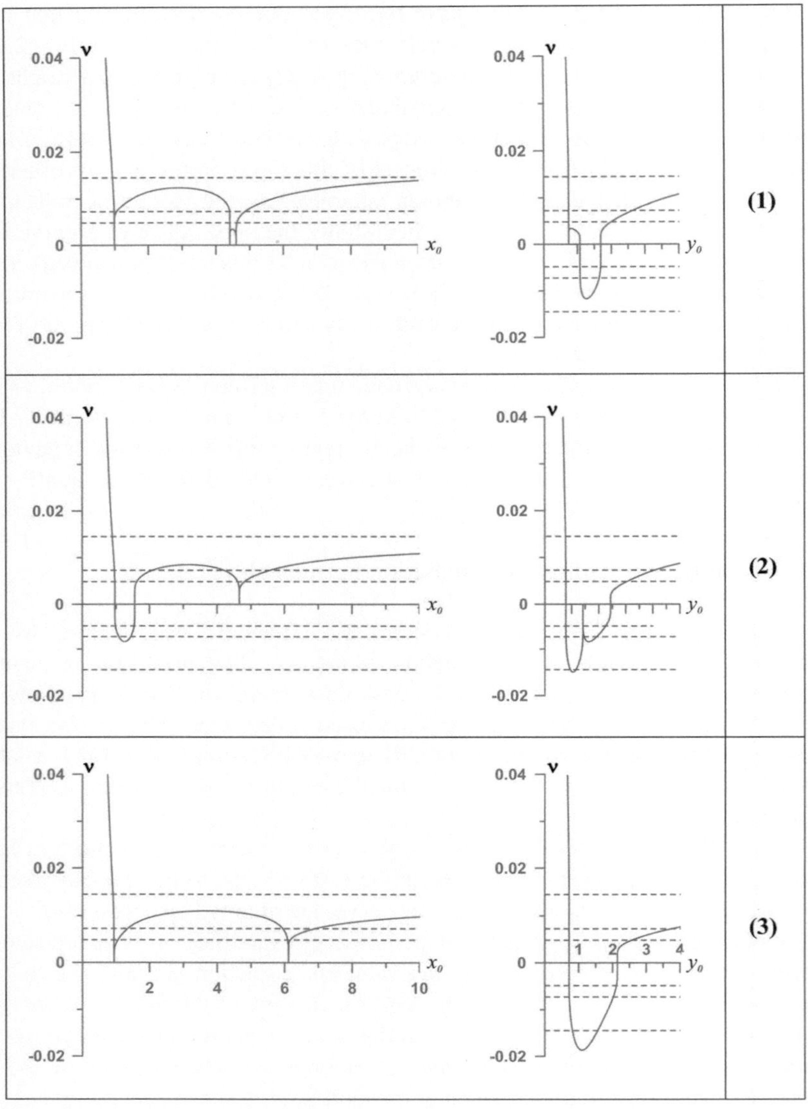

Fig. 2.47 Rotation frequency of bottom-layer vortices vs. the initial position on the x_0 (*left*) and y_0 (*right*) axis. The calculations are given for characteristic versions of Fig. 2.41: **(1)** (1.4); **(2)** (4.1); **(3)** (1.1). The *dashed lines* correspond to the levels of $|\nu| = 0.0145$, $\nu/2$, $\nu/3$

Let us evaluate the parameters at which the appearance of chaotic regimes and a chaotization domain is likely [380, 440, 443]. Note that the modulus of the maximal rotation frequency in the closed central, peripheral, and flow-through domains is of the order of $\nu \sim 0.004 \div 0.020$. Since the domains of nonlinear resonance

1:1 are most often the largest in the system, so, to generate the effective enough chaotization, the value of nonstationary perturbation frequency is commonly taken somewhat lesser than the maximal rotation frequency.

In the examples considered, the domain between separatrices is narrow, as well as the closed-circulation domains, situated along the y-axis, which are also relatively small. At the same time, the large 1:1 resonance strongly deforms the trajectories in those domains. Therefore, in numerical experiments, we choose the perturbation frequency such that its half be smaller than the maximal rotation frequency. Such a choice allows us to identify more vividly the effects of chaotization, which are determined by the positions of resonance domains with rotation number 1:2. The latter, the perturbation amplitude being the same, have much lesser frequency width than that at a resonance of the type of 1:1.

It is worth mentioning that in the domains containing an elliptic point, each rotation frequency corresponds to a unique trajectory, and the maximal frequency is attained in the elliptic point. In the closed domains lying between different separatrices, the maximal frequency corresponds to a single trajectory, while each frequency less than the maximal one, corresponds to two trajectories.

Let us consider a perturbation that leads to chaotization in the neighborhood of separatrices, but does not destroy the regular barrier in the closed domain between separatrices. This allows us to illustrate the appearance of a transport corridor between the inner and outer domains of the phase portrait. To implement such a scenario, we assume $\nu^* = 0.01450$, for which $\nu^*/2$ will be smaller than the maximal frequency in all closed domains but greater than the maximal frequency in the domains between separatrices (in the case (1.4), even $\nu^*/3$ will be greater than the latter). This is illustrated by Fig. 2.47, where the levels ν^*, $\nu^*/2$, and $\nu^*/3$ are shown by dashed lines. The intersections of such lines with the rotation frequency curve gives the x- or y-coordinates of the intersection points of the resonance trajectory and the appropriate coordinate axis, i.e., the required values of x_0 or y_0. It can be seen that, with the perturbation parameters thus chosen, the trajectories corresponding to 1:1 resonance, exist only in closed domains near the y-axis. They are localized near elliptic points and lie far enough from the trajectories of 1:2 resonances. Therefore, the overlapping of only the domains of 1:2 and 1:3 or (when the perturbation amplitude is small enough) 1:3 and 1:4 resonances can be expected [380, 440]. In this case, there exists a probability that a narrow enough stochastic layer will form in the neighborhood of separatrices, while the closed domains and, more significantly, the domain between the separatrices will mostly remain regular [443]. In such situation, a regular barrier should persist in this domain. However, in the case of reconnection of separatrices, stochastic layers merge and a chaotic transport corridor forms from the inner into the outer domain.

The regular barrier described above can be destroyed not only due to reconnection of separatrices, but also due to an increase in perturbation amplitude or due to the favorable choice of perturbation frequency, which corresponds to reconnection of nonlinear resonances of the same order in the neighborhood of a trajectory with maximal rotation frequency, the perturbation amplitude being small enough.

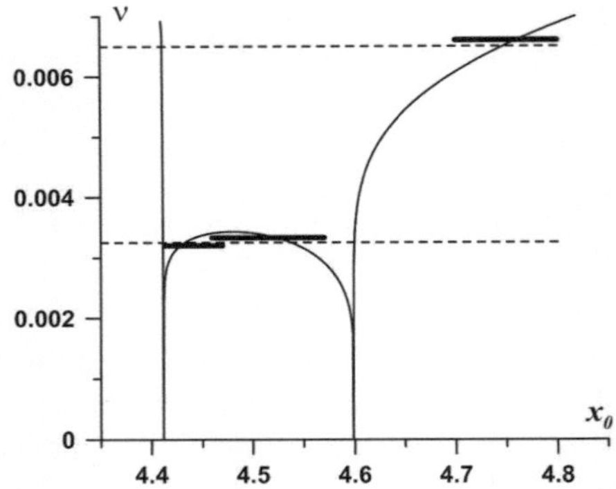

Fig. 2.48 Zoomed left fragment of Fig. 2.47c. The *dashed lines* show the levels $\nu^* = 0.065$ and $\nu^*/2$. Thick segments mark the domains of x-coordinates of the trajectories captured by nonlinear 1:1 resonance (in the outer domain) and two nonlinear 1:2 resonances (in the closed flow-through domain)

The situation is illustrated in Fig. 2.48, where thick lines show the initial positions of the trajectories captured by 1:2 resonances in the closed domain between the separatrices and 1:1 resonance in the outer domain at perturbation frequency of $\nu^* \sim 0.0065$. It can be seen that at such perturbation frequency, the two resonance domains of 1:2 type lie close to the trajectory with maximal rotation frequency and that reconnection of their separatrices can take place at some perturbation amplitude [443]. The result is that a chaotic transport corridor should form across the zone between the separatrices. The 1:2 resonance domains lie close enough to both separatrices, but the trajectory with the maximal rotation frequency lies closer to the inner separatrix and the resonance domains 1:2 and 1:3 should overlap completely. On the side of the outer separatrix, the distance between the domains with the same resonances is larger, so the overlapping will be only partial. Here, the domain of inner 1:2 resonance will be destroyed completely because of the reconnection with the inner 1:3 resonance, and the domain of outer 1:2 resonance will be destroyed only partially. However, this is enough for the transport corridor to form.

2.3.2.4 Origination of Chaos at $A = 2$

The analysis of rotation frequencies along trajectories allowed us to predict some effects of chaotization for a non-stationary perturbation of the external field. In this section, we will verify those considerations by direct calculations. To identify the domains of chaotization and the major effects it causes, it is convenient to represent the calculation results as Poincaré sections. Because of the symmetry, in most cases, we will show only the position of one bottom-layer vortex once in each perturbation period, and only if need be, will we show analogous illustrations for the second vortex.

Figure 2.49 gives the situation with a barrier in the domain between separatrices. This figure gives the results of calculations with parameters (1.4), (4.1) in Fig. 2.41, i.e., before and after separatrix reconnection, and with parameters (1.1) at a reconnection. As noted above, the perturbation frequency ($\nu^* = 0.0145$) was chosen based on the condition that the barrier between the separatrices would not be destroyed.

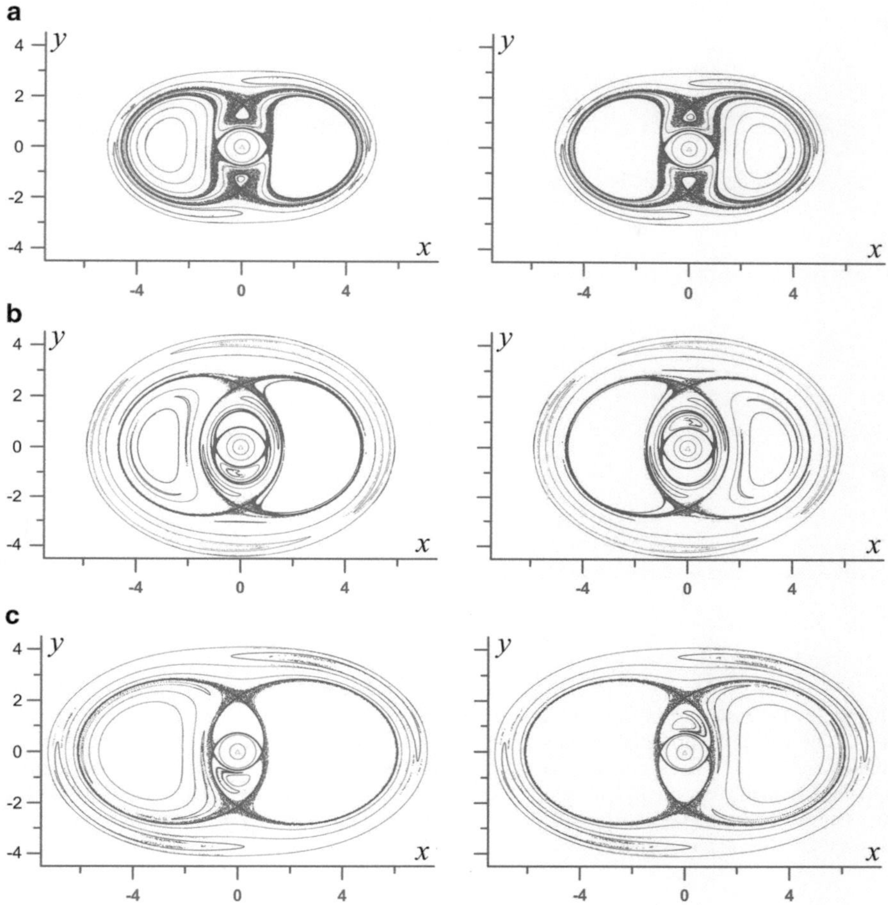

Fig. 2.49 Left and right columns represent Poincaré sections for the left and right (in terms of the initial position) bottom-layer vortices, respectively, for the same variants as shown in Fig. 2.47. In all cases, the perturbation frequency $\nu^* = 0.0145$, and the amplitude $\varepsilon = 0.01$. Nonperturbed separatrices from Fig. 2.47 are given: (**a**) (1.4); (**b**) (4.1); (**c**) (1.1). The symbols for chaotic domains and nonlinear resonances are slightly larger than those for quasi-regular trajectories

Suppose that

$$s(t) = s_0(1+\varepsilon \sin \nu t), \quad \omega(t) = \omega_0(1+\varepsilon \sin \nu t), \tag{2.56}$$

where $\varepsilon = 0.01$.

In all cases, we see irregular behavior of vortices near separtrices with occasional flip-over[15] between the motion regime in the domain between separatrices and the

[15]The notion of "flip-over" between different motion regimes was first introduced by Glukhovskii and Klyatskin [299], see also [437].

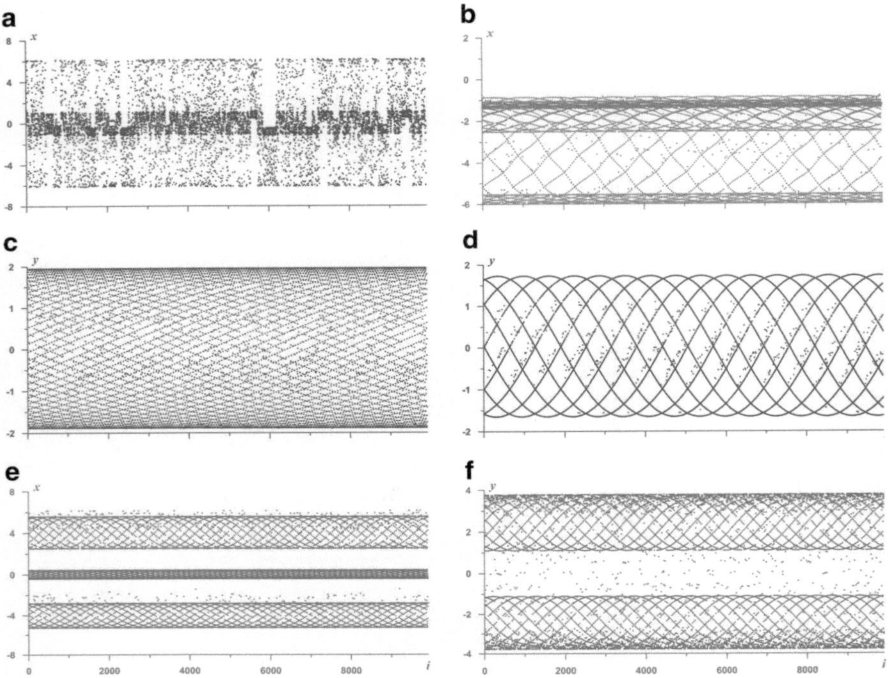

Fig. 2.50 Dependences of the coordinates $x(i)$ or $y(i)$ of the bottom layer for the case (1.1) in Fig. 2.49 and different initial positions. The variants correspond to the following characteristic motion regimes of bottom-layer vortices: (**a**) vortices in a chaotic layer near the separatrix; (**b**) vortices in the domain of nonlinear 1:2 resonance in lateral vortex domains; (**c**, **d**) quasi-regular behavior in the lateral vortex domains; (**e**) 1:3 resonance at the boundary of stochastic layer; (**f**) 1:2 resonance in the external domain

regimes in the closed domains of the inner separtrix and in the outer domain of the external separtrix. However, the regular barrier in the domain between separatrices hampers flip-overs from the neighborhood of the internal separatrix into the external domain.

Flip-over processes are vividly illustrated by Fig. 2.50, showing the dependences $x(i)$ or $y(i)$, where i is the number of the perturbation period for appropriate set of parameters. Characteristic types of vortex motions can be identified here. In the stochastic layer, with vortex coordinates varying irregularly, regimes can be clearly seen in which the vortex (a) lies in the right (left) closed domain—according to the position in the positive or negative domain of $x(i)$ values, (b) is situated in the upper (lower) closed domain—according to the small amplitude of variations of the y-coordinate, (c) moves within the inner or outer domain—according to the coordinate values with varying sign and typical minimal (in the former case) or maximal (in the latter case) amplitudes.

When trajectories are trapped by nonlinear resonance domains, two or three typical zones of variations in coordinates can be clearly seen and, overall, the

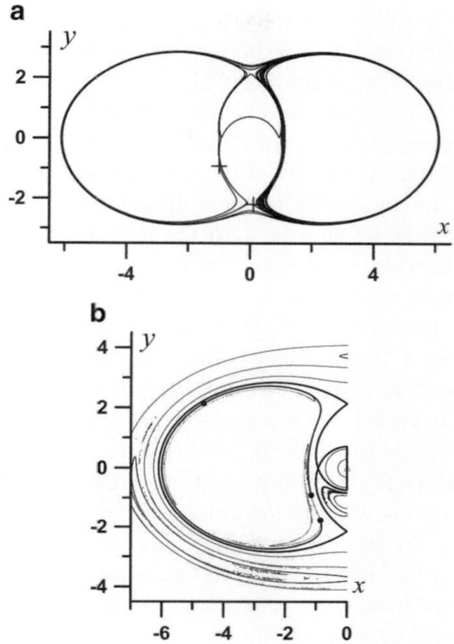

Fig. 2.51 Segments of trajectories from the (**a**) stochastic layer and (**b**) from the 1:2 resonance domain—variant (1.1) for Fig. 2.49. Symbols (+) mark the beginning and end of a part of stochastic trajectory and the position of the resonance trajectory (*black circles*) after each perturbation period $T = 2\pi/|\nu^*|$. Since the resonance trajectory never leaves the (**b**) vortex domain, only half of the figure is shown

behavior is of practically regular character. In regular domains, the vortex passes through the entire interval of coordinate variation and the number of regular segments of the curves corresponds to the ratio of rotation frequency of the trajectory to the perturbation frequency.

An interesting effect that can be identified by analyzing only one coordinate is that even practically regular trajectories are partially destroyed by high multiplicity resonances, which are too weak to be identified by such calculations. Obviously, the intervals of irregular behavior are very rare in such cases.

Quasi-regular segments of trajectories in the domains between flip-overs are shown in Fig. 2.51. Here, the vortex carries out several quasiregular revolutions along its trajectory in the right closed domain; next, it changes the type of its motion and carries out two revolutions in the external domain, followed by a revolution in the upper closed domain, after which it passes into the left closed domain. The same figure gives the segment of the trajectory belonging to the zone of 1:2 resonance in the lateral closed domain. The second trajectory shows that the vortex belonging to the 1:2 resonance domain, after a perturbation period, reaches a Poincaré section trajectory, belonging to a resonance stability island and, after an odd number of periods, the trajectory reaches the second stability island. In other words, during the motion in a moving coordinate system along a nonperturbed trajectory with frequency $\nu^*/2$, the marker moves along one of the two closed (maybe, partially destroyed) Poincaré section trajectories, belonging to this resonance domain. For the resonances of another order, the situation is similar, but each closed trajectory is visited by the vortex once in n periods, where n is the order of the resonance. In the case of fractional resonances, i.e., when the ratio of the rotation frequency of resonance trajectory to the perturbation frequency has the form $\nu(x_0)/\nu^* = m/n$, where m and n are integers, the situation is more complicated (see, for example, [443]).

Thus, the stochastic trajectory begins near the bottom hyperbolic point (its beginning is marked by a cross), next it is captured by the right vortex domain, where it carries out several revolutions, followed by a flip-over into the external domain. After two revolutions, the trajectory penetrates, along the internal part of the separatrix, into the top vortex domain, and, after one revolution in it, the trajectory changes over to the left vortex domain (the end of the trajectory is marked by a cross). The regular trajectory always remains in the left vortex domain, visiting an island of 1 : 2 resonance once per period. The beginning of this trajectory, marked by a black circle, belongs to the Poincaré section domain in the bottom stability island. Its position after one period is also marked by a black circle in the top island, and after one more period, the third black circle again lies in the bottom island. Only half of the phase portrait is shown for this trajectory, because the vortex never leaves the left vortex domain. Recall that the motions of the second bottom-layer vortex are symmetrical to the latter about the origin of coordinates.

In the case of reconnection of separatrices (variant (1.1) in Figs. 2.47 and 2.49), a vortex that has entered the near-separatrix stochastic layer carries out motions of all possible types, including possible passages from the internal into external domain and vice versa. Figure 2.51 gives two examples: a chaotic and a regular trajectory. The relationship $x(i)$ for this case given in Fig. 2.50 shows that flip-overs between motion regimes are frequent; therefore vortices do not penetrate deep into closed domains, though the real thickness of the stochastic layer is somewhat greater than that obtained for the given initial position of the vortex. The latter effect leads to the thicknesses of stochastic layers being somewhat less than those following from the analysis of undisturbed rotation frequencies. We can see that, in the case of reconnection of separatrices, the bottom-layer vortex can both move in the central domain (in a very thin near-separatrix layer) and penetrate into the external domain, thus making the behavior of the vortex system much more complicated.

The analysis given above shows that, as was the case with passive markers [380, 440, 443], the major effects of chaotization and the parameters at which they are realized can be derived from the undisturbed frequencies of bottom-layer vortices.

Let us consider one more effect, which has been analyzed in detail in [443], that is, a reconnection of separatrices of nonlinear resonances in a closed domain between separatrices. In the previous calculations, we deliberately chose the perturbation frequency such that the value of $\nu/3$ be slightly greater than the maximal rotation frequency in the domain under consideration (see Fig. 2.47). Let us change the perturbation frequency to a value slightly below this maximum. It can be seen that at $\nu = 0.0065$, in the closed domain between separatrices (see Fig. 2.48), 1:2 resonances are possible on different trajectories close to the trajectory with maximal frequency (one within the separatrix, and the other outside of it). Accordingly, when the resonance domains are wide enough, their separatrices can also be involved in a reconnection, resulting in chaotization in the central part of the domain. If overlapping with 1 : 3 resonance domains, lying near the separatrices, takes place, a chaotic flip-over can take place through this layer from the domain

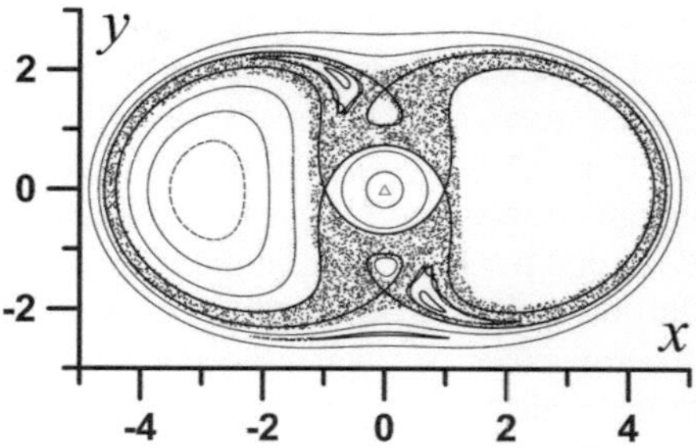

Fig. 2.52 Poincaré section for the left (in terms of the initial position) bottom-layer vortex, figure parameters 2.48 ($\nu^* = 0.0065$, $\varepsilon = 0.01$). Nonperturbed separatrices from Fig. 2.47 are given

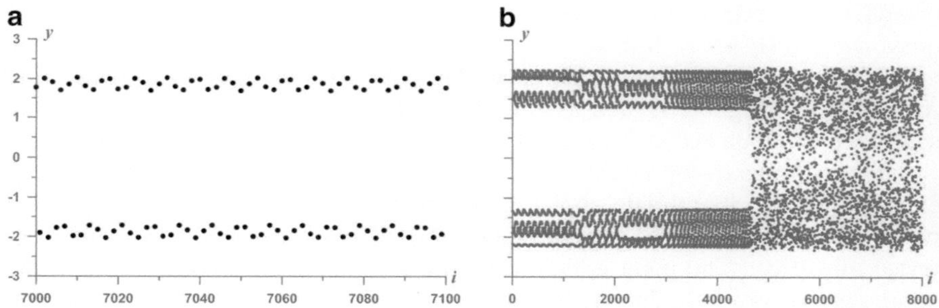

Fig. 2.53 Dependences $y(i)$ for a regular trajectory from the domain of regular behavior, corresponding to 1:2 resonance, in (**a**) the flow-through domain and (**b**) a vortex with the initial position from the stochastic layer in the flow-through domain. The parameters correspond to Fig. 2.52

of the internal separatrix into the external domain. It is important to note that the distance between 1:2 and 1:3 resonance domains is lesser in the internal part and greater in the external part of the closed flow-through domain, i.e., the extent of their overlapping is different.

The above effects are illustrated by Fig. 2.52. Here, in accordance with the analysis of dependence of nonperturbed rotation frequencies, we have chaotic behavior in the closed domain between the separatrices everywhere, except for two islands of regular behavior, located closer to the outer separatrix and corresponding to the 1:2 resonance. The regular barrier between the outer and inner domains can be destroyed by choosing the parameters that facilitate either reconnection of major separatrices or reconnection of nonlinear resonance separatrices in the barrier domain.

The same calculation makes it possible to attract attention to the "adhesion" phenomenon,[16] which is closely related with "Levy flights" [862]. Figure 2.53

[16] As far as we know, the effects of "adhesion" or "stickiness" of islands was first discussed by Zaslavsky [1007].

gives the dependences $y(i)$ for two markers: one from the domain of a stability island corresponding to $1:2$ resonance, not fully destroyed, and the other one from the chaotic domain. In the quasi-regular dependence (the left part of the figure), the marker visits one of the stability islands of this resonance once per period. Therefore, we see two trajectories: one for even periods, which corresponds to one island, and the second for odd periods, which corresponds to the second island.

The quasi-sinusoidal form of each dependence corresponds to the displacement of the marker along the trajectory of Poincaré section within the island. Here we see a small segment of this trajectory, but it does not change its form within 12,000 perturbation periods (i.e. up to infinitely).

The second dependence is more interesting. Here, the initial position of the marker has been chosen in the thin stochastic layer surrounding the domains of secondary resonances at the periphery of stability islands. At the initial, long enough (of the order of 5,000 perturbation periods) stage, this dependence looks like quasi-regular. However, it consists of several regular segments, and several times, at random moments, it makes flip-overs between domains with different motion regimes in the neighborhood of $1:2$ resonance. Next, the marker enters a stochastic sea, and the motion becomes strongly chaotic.

Let us consider one more example, where the stochastic layer can be appreciably extended by the appropriate choice of the perturbation frequency, its amplitude remaining unchanged. The variant (1.1) in Fig. 2.49 gives a good opportunity to reveal this effect. The point is that in that case, all resonance zones ($1:1$ in the bottom (top) domain, $1:2$ in the left (right) and in the external domains) are located not very far from the stochastic layer. Therefore, even an insignificant decrease in the perturbation frequency causes their approach to the stochastic layer (see Fig. 2.47).

In virtue of the estimate [1007]

$$\frac{\nu^*}{n} - \frac{\nu^*}{n+1} = \frac{\nu^*}{n(n+1)} = \nu(x_0|_{1:n}) - \nu(x_0|_{1:(n+1)})$$
$$\sim \frac{d\nu(x_0)}{dx_0}\left(x_0|_{1:n} - x_0|_{1:(n+1)}\right),$$

the distance along x coordinate between adjacent resonance domains

$$x_0|_{1:n} - x_0|_{1:(n+1)} \sim \frac{\nu^*}{n(n+1)} \Big/ \frac{d\nu(x_0)}{dx_0}$$

Will increase significantly, since with approaching the separatrix, the derivative of the rotation frequency with respect to the coordinate will increase, and the frequency itself will change insignificantly. As the result, the resonance domains and the stochastic layer will overlap, resulting in an increase in the width of the latter by the width of those domains.

With the assumption for the perturbation frequency $\nu^* = 0.01$, from the results of calculations given in Fig. 2.54, we see that the effect of extension of the stochastic layer indeed takes place.

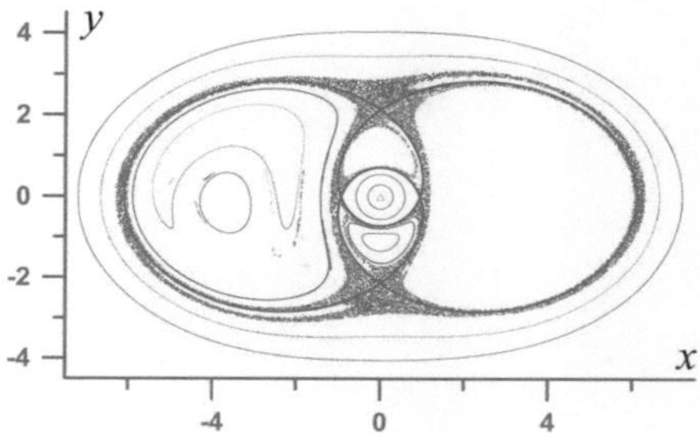

Fig. 2.54 Poincaré section for the left (in terms of the initial position) of the bottom-layer vortex, parameters of variant (1.1) of Fig. 2.47 ($\nu^* = 0.01$, $\varepsilon = 0.01$). Nondisturbed separatrices from Fig. 2.47 are given

The calculations show that the dynamic system under consideration allows the vortices to show practically all chaotization mechanisms of trajectories of bottom vortices known from the analysis of passive markers.

One more feature can be seen in the cases given in Figs. 2.43d and 2.44f. Those variants are of interest because they have no hyperbolic points but have a curve with a maximum in rotation frequency (this mechanism was considered above for another example) and, which is of particular importance, they have a curve (or two curves) with zero rotation frequency. The appropriate frequency dependence is given in Fig. 2.55a, c.

In this case, $s_0 = 0$; therefore, we consider the perturbation of the form

$$s(t) = \varepsilon \sin \nu t, \quad \omega(t) = \omega_0 (1 + \varepsilon \sin \nu t).$$

Analysis of Fig. 2.55a shows that, first, the trajectories with zero rotation frequency will clearly play the role of quasi-separatrices, since their neighborhoods, as well as those of ordinary separatrices, contain many domains of nonlinear resonances with rotation numbers $1 : n$, whose overlapping should lead to the appearance of a stochastic layer. Second, with the appropriate choice of the frequency (this is given in Fig. 2.55a), domains of nonlinear resonances (in this case, 1:2) will form within and beyond the trajectory with zero rotation frequency, and, if they overlap with the stochastic layer, the thickness of the latter will increase significantly. The Poincaré section given in Fig. 2.55b fully confirms our supposition. This figure shows a narrow stochastic layer in the neighborhood of an internal quasi-separatrix and a stochastic sea, associated with the overlapping of domains of internal and external 1:2 resonances in the neighborhood of the outer separatrix. Note two important facts:

- The vortices between and outside of quasi-separatrices move in different directions, i.e., in chaotic domains, vortices can change the direction of their motion and remain still for some time.
- As known from the theory of dynamic chaos, the thickness of the stochastic layer in the neighborhood of a hyperbolic point is larger than in other parts of separatrix. In our case, it is the same all over the length of the inner separatrix.

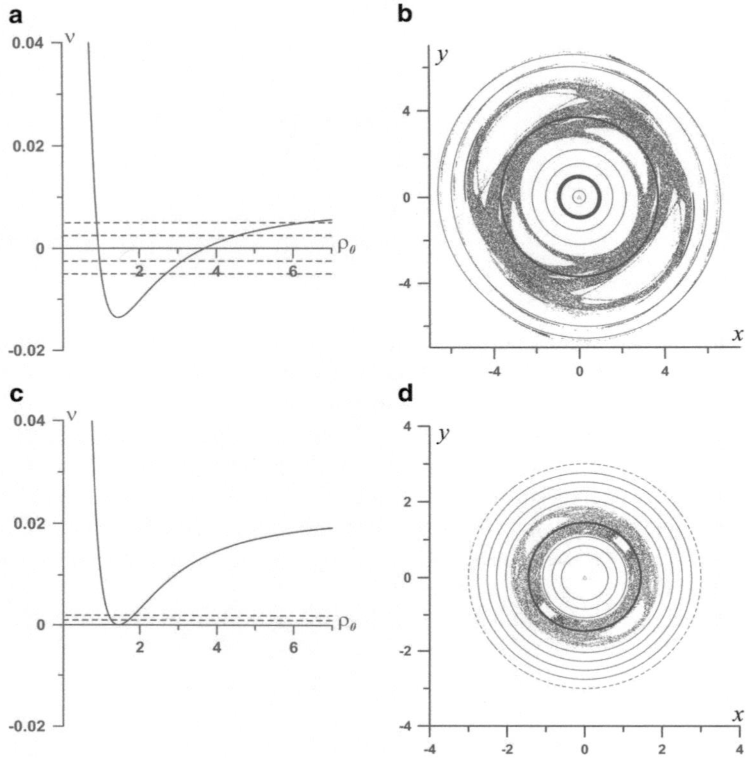

Fig. 2.55 Dependence of the rotation frequency (*left*) and Poincaré section (*right*) for the parameters of Fig. 2.44f ($\omega_0 = 0.10045$, $s_0 = 0$)—(**a**, **b**) and Fig. 2.43d ($\omega_0 = 0.270940$, $s_0 = 0$)—(**c**, **d**). The frequency dependences show the following levels: (**a**) $\nu^* = \pm 0.005$ and $\nu^*/2$, (**c**) $\nu^* = 0.002$ and $\nu^*/2$. The perturbation amplitudes: (**b**) $\varepsilon = 0.005$, (**d**) $\varepsilon = 0.003$

In particular, if a hyperbolic point exists at zero perturbation, then the thickness of the stochastic layer is of the order of $\sqrt{\varepsilon}$ near this point and ε in the rest part of the separatrix [522].

Note that the cases when the hyperbolic point exists in a stationary system and when it appears only at zero perturbation are essentially different. In the latter case, the stochastic layer is commonly exponentially thin. It is still unclear which case we have in this example. The natural question that requires special consideration is *what is the order of the thickness of the stochastic layer in the neighborhood of a separatrix that has no hyperbolic point?*

Calculations for the case of a single curve with zero frequency are given in Fig. 2.55c, d. In this case, the vortices move in the same direction both within and beyond this trajectory: we have only a minimum (zero) on the frequency

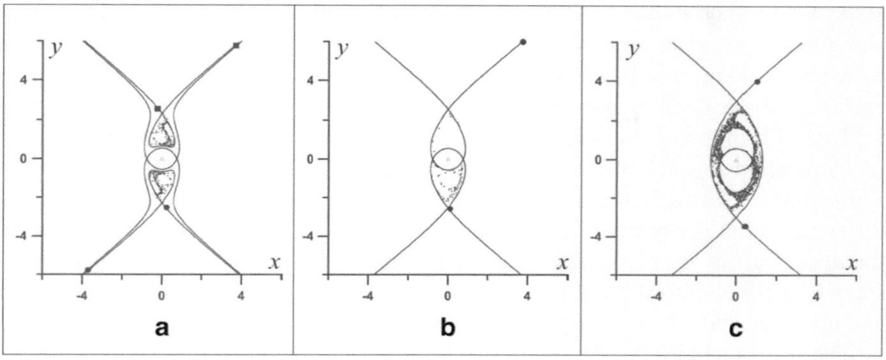

Fig. 2.56 Poincaré sections for $\omega_0 = 0$ and the following values of parameters: (**a**) $s_0 = 0.2$, $\nu^* = 0.03$, $\varepsilon = 0.05$ (the positions of both bottom-layer vortices are given, the *black circle* shows the initial and final positions of the left vortex, while the *square* shows the same for the right vortex); (**b**) $s_0 = 0.177590$, $\nu^* = 0.027560$, $\varepsilon = 0.05$ (a perturbed case of Fig. 2.45c); (**c**) $s_0 = 0.140$, $\nu^* = 0.030$, $\varepsilon = 0.033$

dependence. Resonances of the same order can exist on both sides of this curve, resulting in the overlapping of domains of large resonances (in our case, this is 1 : 2) and in the formation of a large chaotization domain. However, a question still to be answered is can a stochastic layer form because a reconnection of separatrices of resonances of the same order *in the neighborhood of a curve with minimal frequency at arbitrary frequency and amplitude*.

Finally, let us consider the case of motion close to variants in Fig. 2.45, where there is no rotational component in the external field. An important feature here is the presence of infinite vortex trajectories. In the steady case, for vortices that start at the initial moment from the x or y axis, we have several types of motion of bottom-layer vortices: they can rotate about the top vortex in the central domain; can stay in the capture domains, where they will rotate about elliptic points of the top and bottom closed domains or in the flow-through domain between separatrices; or they can drift away to infinity along the separatrices given in Fig. 2.45. However, if the vortices are originally far enough from the center in the first or third quadrant of (x, y) plane, then, after approaching the neighborhood of the domain bounded by separatrix whiskers, they will drift to infinity.

If in a nonstationary case, we take initial trajectories near separatrices (but far enough from the central domain) and choose a perturbation leading to effective chaotization, the incoming vortices will be captured by the vortex domains for some time, after which they will drift to infinity.

Figure 2.56 illustrates such situation. In Fig. 2.56a, the two bottom-layer vortices with the initial positions near the top right (black square) and bottom left (black circle) whiskers of the separatrix are captured by the top/bottom homoclinic loop, respectively, stay within it for about 180 perturbation periods, and next drift away along the appropriate branches of the separatrix. The latest calculated positions of the vortices are marked by the same symbols. Intermediate positions of vortices after each period are given by point markers.

To avoid confusion, in panels (b) and (c), we give the positions of only one vortex of the bottom layer, implying that the position of the second vortex is centrally symmetrical. Figure 2.56b gives a variant with a reconnection of separatrices. The difference from the previous case is that the vortex that was originally situated in the upper part of the plane is commonly captured by the bottom hole-like domain, stays in it for about 80 perturbation periods, passes into the top hole-like domain, and after one and a half revolutions in it, the vortex drifts to infinity.

The situation in Fig. 2.56c differs in the presence (in the stationary case) of a zone of flow-through motions of vortices between separatrices, this zone commonly serving as a barrier for chaotic trajectories. Nevertheless, this barrier can be destroyed with parameters chosen to correspond to a reconnection of separatrices of nonlinear resonances in the neighborhood of the trajectory with maximal rotation frequency $\nu^* = 0.016$. It would appear reasonable to use a perturbation with a frequency of, say, $\nu^* = 0.015$; however, calculations show the choice of $\nu^* = 0.03$ to be optimal for the following conditions to be satisfied simultaneously: (a) reconnection of the internal and external 1 : 2 resonances in the flow-through domain, (b) insignificant overlapping of external-resonance domains with the stochastic layer of the external separatrix, and (c) the formation of a dynamic trap. As can be seen from the figure, the vortex is captured by the separatrix layer of the external separatrix, after which it penetrates through a chaotic corridor into the stochastic layer of the inner separatrix and wanders within the chaotic domain during 1,665 periods with regular flip-overs between the central zone and both lunar domains, entering from time to time the neighborhoods of resonances in the flow-through domain and, finally, it leaves the trap and drifts to infinity.

2.3.3 The Case of Nonzero Total Momentum at $A = 2$

At $A = 2$, we have a three-vortex problem, whose full classification in the classical formulation (for an arbitrary distribution of intensities) is given in [31, 32, 103, 333,595,892]. In the two-layer formulation, comprehensive analysis of three-vortex motion is still far from complete. A particular case of two different intensities will be considered here, and the major portion of this section deals with the case where the intensities of bottom-layer vortices are equal ($\kappa_2^1 = \kappa_2^2$). The variant $\kappa_1^1 = \kappa_2^2$ (the intensity of the top-layer vortex is equal to that of one of bottom-layer vortices) will be analyzed briefly (as well as in the end of Sect. 2.3.1); detailed calculations will be given only for the particular case of zero total intensity ($Q = 0$).

The equations of motion (1.77)–(1.78) at $A=2$ become

$$u_1^1 = -\frac{1}{4\pi}\sum_{\alpha=1}^{2}\kappa_2^\alpha \frac{y_1^1-y_2^\alpha}{\left(r_{12}^{1\alpha}\right)^2}\left[1-\gamma r_{12}^{1\alpha}\mathrm{K}_1\left(\gamma r_{12}^{1\alpha}\right)\right],$$

$$u_2^\alpha = -\frac{1}{4\pi}\left\{\kappa_2^{3-\alpha}\frac{y_2^\alpha-y_2^{3-\alpha}}{\left(r_{22}^{\alpha(3-\alpha)}\right)^2}\left[1+\gamma r_{22}^{\alpha(3-\alpha)}\mathrm{K}_1\left(\gamma r_{22}^{\alpha(3-\alpha)}\right)\right]\right.$$
$$\left.+\kappa_1^1\frac{y_2^\alpha-y_1^1}{\left(r_{21}^{\alpha 1}\right)^2}\left[1-\gamma r_{21}^{\alpha 1}\mathrm{K}_1\left(\gamma r_{21}^{\alpha 1}\right)\right]\right\}, \tag{2.57}$$

$$v_1^1 = \frac{1}{4\pi}\sum_{\alpha=1}^{2}\kappa_2^\alpha \frac{x_1^1-x_2^\alpha}{\left(r_{12}^{1\alpha}\right)^2}\left[1-\gamma r_{12}^{1\alpha}\mathrm{K}_1\left(\gamma r_{12}^{1\alpha}\right)\right],$$

$$v_2^\alpha = \frac{1}{4\pi}\left\{\kappa_2^{3-\alpha}\frac{x_2^\alpha-x_2^{3-\alpha}}{\left(r_{22}^{\alpha(3-\alpha)}\right)^2}\left[1+\gamma r_{22}^{\alpha(3-\alpha)}\mathrm{K}_1\left(\gamma r_{22}^{\alpha(3-\alpha)}\right)\right]\right.$$
$$\left.+\kappa_1^1\frac{x_2^\alpha-x_1^1}{\left(r_{21}^{\alpha 1}\right)^2}\left[1-\gamma r_{21}^{\alpha 1}\mathrm{K}_1\left(\gamma r_{21}^{\alpha 1}\right)\right]\right\}. \tag{2.58}$$

In (2.57) and (2.58), $\alpha = 1, 2$.

2.3.3.1 Phase Portraits in Trilinear Coordinates

The relative motions satisfying (2.57)–(2.58) can be analyzed in trilinear coordinates (1.85), (1.87).

From (1.83) in the case (II), where $\kappa_2^1 = \kappa_2^2 = \kappa$; $\kappa_1^1 = \mu\kappa$, we have

$$\frac{L}{\kappa^2} = \mu\left[\left(r_{12}^{11}\right)^2 + \left(r_{12}^{12}\right)^2\right] + \left(r_{22}^{12}\right)^2, \tag{2.59}$$

whence it follows that L can change its sign only when

$$0 > \mu = -\frac{\left(r_{22}^{12}\right)^2}{\left(r_{12}^{11}\right)^2 + \left(r_{12}^{12}\right)^2} > -2, \tag{2.60}$$

while at $\mu \le -2$ and $\mu \ge 0$, it is sign-definite (negative in the former case and positive in the latter).

The condition (1.89) for the trilinear coordinates to belong to PD becomes

$$\left[(\mu+1)(t_2+t_3)-3\mu\right]^2 \le 4t_2t_3, \tag{2.61}$$

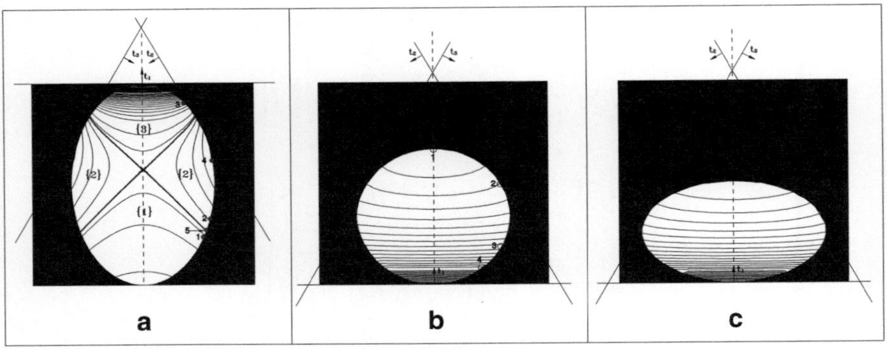

Fig. 2.57 Phase portraits (compact case) for three-vortex problem in trilinear coordinates in PD (2.62): (**a**) $\gamma^2 L = -5$, $\mu = -2.5$, (**b**) $\gamma^2 L = 2$, $\mu = 1$; (**c**) $\gamma^2 L = 2$, $\mu = 2$. As before, the *thick lines* in the fragment (**a**) are separatrices, separating the domains of existence of different types of solutions $\{1\}$, $\{2\}$, and $\{3\}$; their properties will be discussed in Sect. 2.3.2.3. The numbers mark the boundary points of PD corresponding to the initial conditions for numerical experiments, whose results are given in Figs. 2.69, 2.72 and 2.75

whence it readily follows that at $\mu \geq 1$ and at $\mu < -2$, all PD are compact, hence the relative motions are finite on finite supports. PD are infinite in interval $\mu \in [-2; 1)$.

In the case $\kappa_1^1 = \kappa_2^2 \equiv \kappa$, $\kappa_2^1 = \nu\kappa$, it is more convenient to eliminate the variable t_2 (rather than t_1) in formulas (1.88), in which case the expressions (2.59)–(2.61) will have the same form with μ replaced by ν, t_2 by t_1, and r_{22}^{12} by r_{12}^{12}. Clearly, the character of physical domains will remain qualitatively the same, though their axis of symmetry now will be the straight line $t_1 = t_3$, rather than $t_2 = t_3$. Moreover, in virtue of (**!**), the phase portraits (Hamiltonian contour lines) lose their symmetry at $\mu \neq 1$ (see, for example, Fig. 2.60).

This can be clearly seen in Figs. 2.57, 2.58, 2.59, 2.60 and 2.61, constructed for different fixed values μ and L. It is important to note that, unlike Sect. 2.2.2, where each phase portrait corresponded to a constant value of total momentum M (at $P_x = P_y = 0$), in this section, the total momentum being nonzero, the ratio between the angular and linear momentums in expressions (1.83) and (2.59) for integral invariant L will change at the passage from one trajectory to another.

Three examples of phase portraits of compact type for those values of μ are given in Fig. 2.57. PD in Figs 2.57a–c are specified by the conditions

$$\begin{aligned}
&\left(t_2+t_3-5\right)^2 \leq \frac{16}{9}t_2t_3, \quad t_1 \leq 0, t_2, t_3 \geq 0 \quad \text{at} \quad \gamma^2 L = -5, \mu = -2.5,\\
&\left(t_2+t_3-\frac{3}{2}\right)^2 \leq t_2t_3, \quad t_1 \geq 0, t_2, t_3 \leq 0 \quad \text{at} \quad \gamma^2 L = 2, \mu = 1,\\
&\left(t_2+t_3-2\right)^2 \leq \frac{4}{9}t_2t_3, \quad t_1 \geq 0, t_2, t_3 \leq 0 \quad \text{at} \quad \gamma^2 L = 2, \mu = 2, \qquad (2.62)
\end{aligned}$$

respectively.

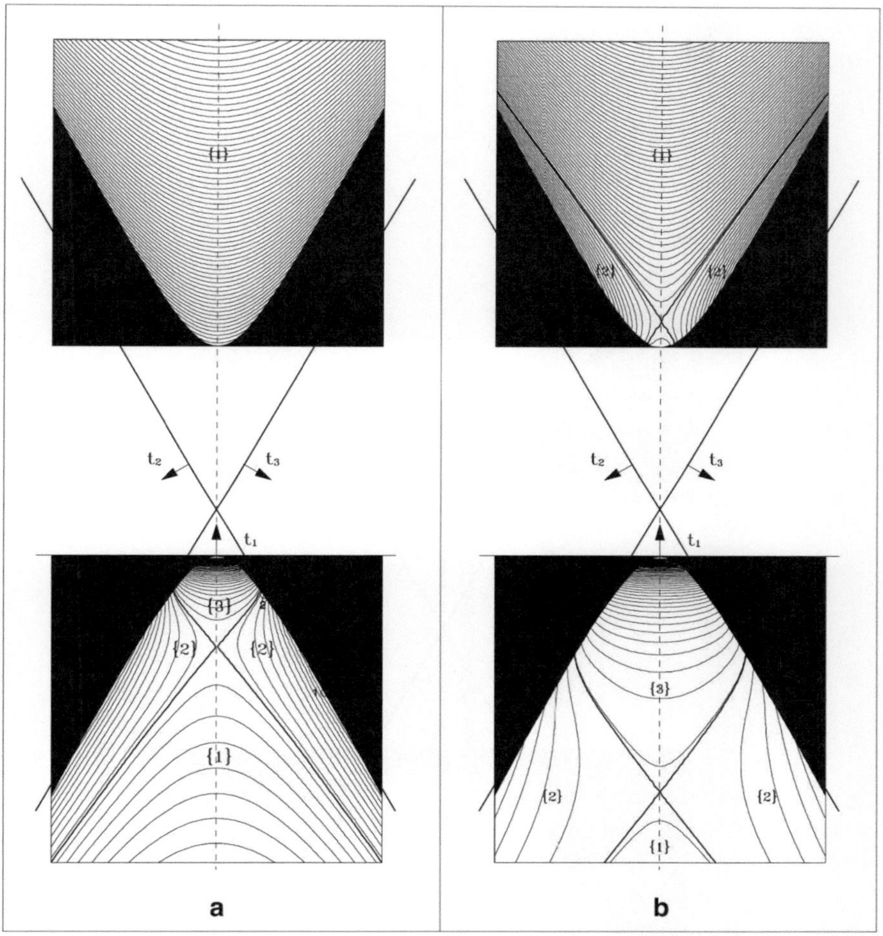

Fig. 2.58 Phase portraits of three-vortex problems in PD (2.63) at $\mu = -1.5$ and $L < 0$ (*bottom*), $L > 0$ (*top*): (**a**) $\gamma^2|L| = 2$, (**b**) $\gamma^2|L| = 0.65$. The numbers mark the PD points corresponding to the initial conditions for numerical experiments, whose results are given in Figs. 2.63, 2.67, 2.73 and 2.76

PD of noncompact type in Fig. 2.58 are described by the relationships

$$\left(t_2+t_3-9\right)^2 \leq 16t_2t_3, \quad \begin{cases} t_1 \leq 0, \ t_2, t_3 \geq 0, L < 0, \\ t_1 \geq 12; t_2, t_2 < 0, L > 0 \end{cases} \text{at } \mu = -1.5, \tag{2.63}$$

those in Fig. 2.59, by relationships

$$t_2t_3 \geq \frac{9}{4}, \quad \begin{cases} t_1 < 0, \ t_2, t_3 \geq 0; \gamma^2 L = -2, \\ t_1 \geq 6, \ t_2, t_2 < 0, \gamma^2 L = 2 \end{cases} \quad \text{at} \quad \mu = -1, \tag{2.64}$$

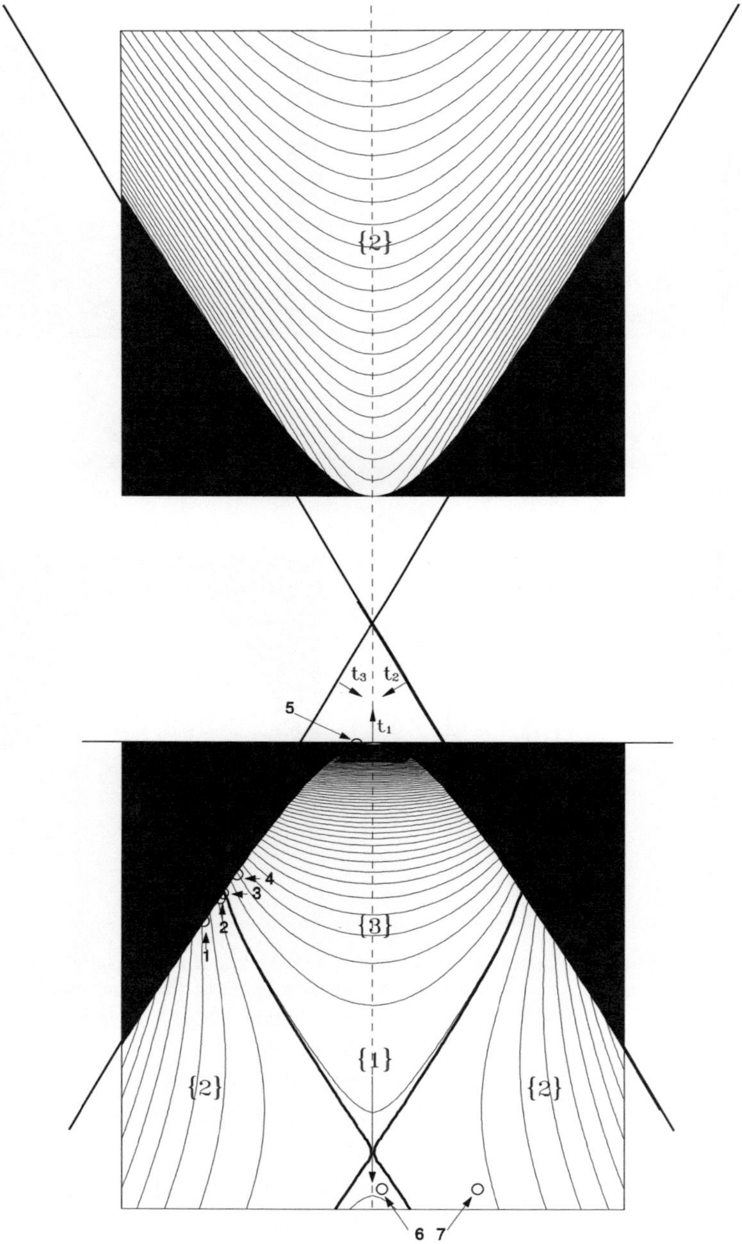

Fig. 2.59 Phase portraits in three-vortex problem in PD (2.64) at $\mu = -1$ and $\gamma^2 L = -2$ (*bottom*), $\gamma^2 L = 2$ (*top*). The numbers mark the PD points corresponding to the initial conditions for numerical experiments, whose results are given in Fig. 2.76

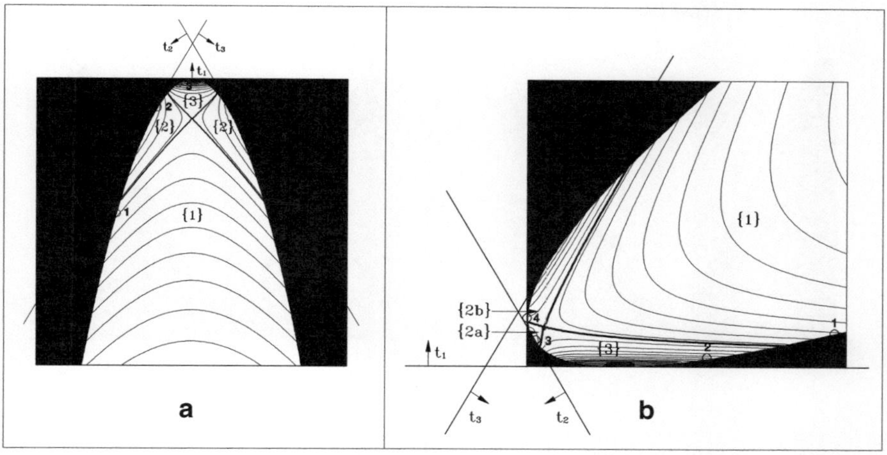

Fig. 2.60 Phase portraits of three-vortex problem in PD (2.65) at $Q=0$: (**a**) $\kappa_1^1=-2$, $\kappa_2^1=\kappa_2^2=1$, ($\mu=-2$) and $\gamma^2L=-2.89$; (**b**) $\kappa_1^1=\kappa_2^2=-1$, $\kappa_1^2=2$ $\nu=-2$ and $\gamma^2L=-25$. The numbers mark the PD points corresponding to the initial conditions for numerical experiments, whose results are given in Figs. 2.70 and 2.82

those in Fig. 2.60a, b (where $Q=0$, $L\le 0$):

$$\begin{gathered}\left(t_2+t_3-6\right)^2\le 4t_2t_3,\quad t_1\le 0, t_2, t_3\ge 0\\ \text{at}\quad \gamma^2L=-2.89;\, \kappa_1^1=-2,\, \kappa_2^1=\kappa_2^2=1\,(\mu=-2),\\ \left(t_2-2t_3+3\right)^2\le -8t_2t_3,\quad t_1\ge 0, t_2\le 0, t_3\ge 0\\ \text{at}\quad \gamma^2L=-25,\, \kappa_1^1=\kappa_2^2=-1,\, \kappa_2^1=2\,(\nu=-2)\end{gathered}\tag{2.65}$$

and those in Fig. 2.61a, b (where $L=0$):

$$\begin{gathered}\left(t_2+t_3-9\right)^2\le 16t_2t_3,\quad t_1\ge 0, t_2, t_3<0\quad \text{at}\quad \mu=-1.5,\\ \left(t_2+t_3+3\right)^2\le 16t_2t_3,\quad t_1\ge 0, t_2, t_3<0\quad \text{at}\quad \mu=-0.5.\end{gathered}\tag{2.66}$$

The phase portraits show a wide diversity of topological properties. Some such properties are discussed below along with a demonstration of the absolute trajectories of vortices and relative choreographies in the consideration of individual problems. However, first, we will find the existence conditions of singular points in the phase plane, which correspond to steady-state solution of the equations of motion (2.57)–(2.58).

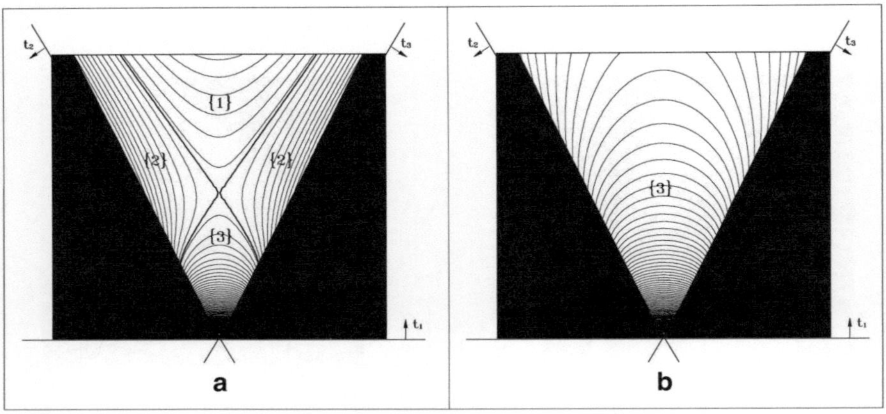

Fig. 2.61 Phase portraits in three-vortex problems in PD (2.66) at $L = 0$: (**a**) $\mu = -1.5$ and (**b**) $\mu = -0.5$

2.3.3.2 Analysis of Steady States

From Figs. 2.57, 2.58, 2.59, 2.60 and 2.61, we can conclude that the problems of the class considered here can have two types of singular points—stable elliptic and unstable hyperbolic points. The former always belong to PD boundary, while the latter coincide with separatrix self-crossing points. Obviously, elliptic and hyperbolic singularities correspond to rotational motions of solid-body configurations of three vortices—a collinear one in the former case and one in the form of an isosceles triangle (except for a nonsymmetrical variant in Fig. 2.60b, which will be considered separately below)—in the latter case. Those rotations take place about the vorticity center of the vortex structure with coordinates

$$(X_c, Y_c) = \left(\frac{\gamma P_x}{Q}, \frac{\gamma P_y}{Q} \right). \tag{2.67}$$

In their turn, the boundary elliptic singular points can also be divided into

- *polar*, where
  - two such points always exist for finite PD: one in the upper and one in the lower parts of the phase portrait (when the trilinear coordinate t_1 attains its maximal or minimal value, respectively),
  - for finite PD there is only one point, either in the upper part of the phase portrait (when $t_1 \leq 0$) or in its lower part (when $t_1 \geq 0$)
- and *lateral*.

If a *polar* point exists at $t_1 = 0$, it corresponds to a construction consisting of a top-layer vortex and theoretically merged bottom-layer vortices, while if $t_1 \neq 0$ for this point, it corresponds to *ordinary* or *inverse roundabout*.

Of particular interest is the existence of *lateral* elliptic points, which are phase images of *eccentric roundabouts*. It is worth mentioning that two important distinctions exist between such *eccentric roundabout* and the one considered in the end of Sect. 2.3.1, where the conditions $Q = P_x = P_y = 0$ are satisfied:

- the *rotation* center of the vortex structure in this case coincides with the center of *vorticity* (X_c, Y_c), while at $Q = 0$, the latter was situated at infinity;
- the rotation center now cannot belong to the segment in which vortices are concentrated (at $Q = 0$, it could always be situated only in the segment between the two bottom-layer vortices).

The expressions (2.67) become

$$(X_c, Y_c) = \begin{cases} \left(\dfrac{2(R-B)}{\mu+2}, 0\right) & \text{for lateral elliptic singular points,} \\ \left(0, \dfrac{2B\sin\varphi}{\mu+2}\right) & \text{for hyperbolic singular points.} \end{cases} \tag{2.68}$$

The meaning of variables B, R, and φ, where $\varphi \in [0; \pi/2)$, is elucidated by Figs. 2.62a, 2.63a, 2.66a and 2.67a, where, without loss of generality, we assume that, at the initial moment, the top-layer vortex $\binom{1}{1}$ is located in the origin of coordinates $(X, Y) = (0, 0)$, and that

(i) the collinear configuration belongs to the X-axis,
(ii) the triangular configuration is symmetrical about Y-axis.

It is obvious that for collinear configurations, we have

$$\begin{aligned} X_c \geq 0 \quad &\text{at} \quad \mu < -2, B \geq R \quad \text{or} \quad \mu > -2, B \leq R, \\ X_c \leq 0 \quad &\text{at} \quad \mu > -2, B \geq R \quad \text{or} \quad \mu < -2, B \leq R, \end{aligned} \tag{2.69}$$

and for triangular configurations,

$$\begin{aligned} Y_c \geq 0 \quad &\text{at} \quad \mu > -2, \\ Y_c \leq 0 \quad &\text{at} \quad \mu < -2. \end{aligned} \tag{2.70}$$

At $\mu \to -2$ (when the total intensity is zero), the position of the vorticity center tends to infinity.

We will mention some interesting cases of arrangement of coordinates (2.68).

- For a collinear construction, the vorticity center
  - belongs to the midpoint of the segment between vortices $\binom{1}{2}$ and $\binom{2}{2}$ at $\mu = 0$;
  - coincides with vortex $\binom{2}{2}$ at $\mu \neq 0$ and

$$B = \frac{2R(1+\mu)}{\mu}; \tag{2.71}$$

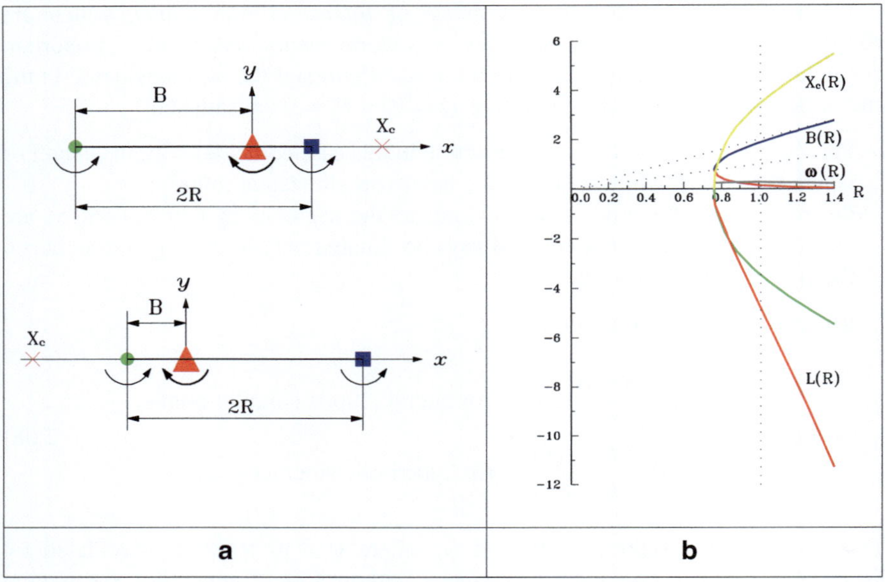

Fig. 2.62 (**a**) Schemes of the original arrangement of vortices for an *eccentric roundabout* at $B > R$ (*top*) and at $B < R$ (*bottom*). (**b**) $\mu = -2.5$: dispersion curve $B(R)$, satisfying (2.73), – *blue branch* at $B > R$ and *red branch* at $B < R$; the position of vorticity center $X_c(R)$ –*yellow branch* when $X_c > 0$ and *green branch* when $X_c < 0$ (see formula (2.69)); the value of the angular rotation velocity (2.77) of the vortex construction $\omega(R)$ – *light blue line*; and the value of the integral invariant $L(R)$ (2.59) – *brown line*. The *tilted dashed lines* are described by the equations $B = R$ and $B = 2R$. The intersection points of curves with *vertical dashed line* correspond to the parameters of numerical experiments at $\gamma^2 L = -5$, given in Fig. 2.72a

  - coincides with the midpoint of the segment between vortices $\binom{1}{1}$ and $\binom{2}{2}$ at $\mu \neq 2$ and

$$B = \frac{2R\mu}{\mu - 2}. \tag{2.72}$$

- For triangular construction, the vorticity center
  - lies within the triangle formed by vortices at $\mu > 0$;
  - lies beyond this triangle at $\mu < 0$, and, as mentioned above, at $\mu = -2$, it moves to infinity;
  - coincides with the center of the circumscribed circle of this triangle at $\mu = -2\cos 2\varphi$; the center of the circumscribed circle, in its turn, lies within the triangle at $\varphi > \pi/4$, when $0 < \mu < 2$, beyond the triangle at $\varphi < \pi/4$, when $-2 < \mu < 0$.

Let us find the conditions of existence of steady-state solutions of either collinear type in the form of an eccentric roundabout or in the form of an isosceles triangle.

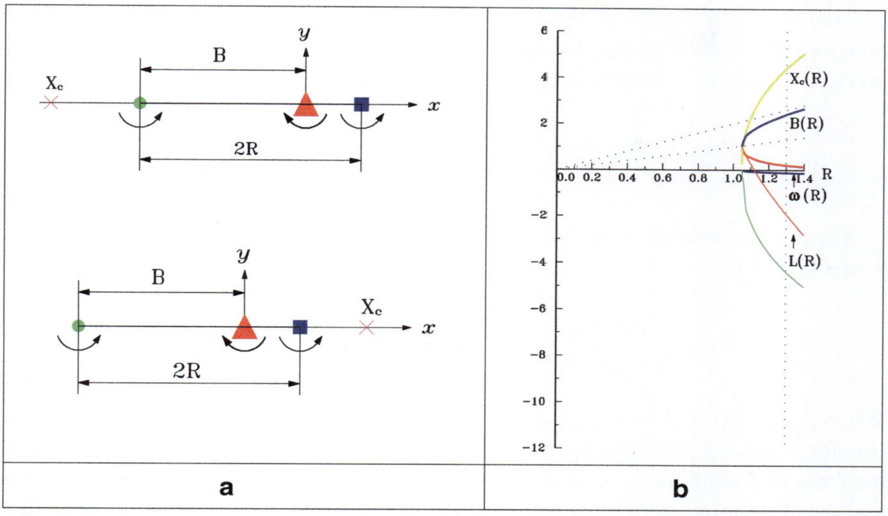

Fig. 2.63 The same as in Fig. 2.62, but with $\mu = -1.5$. The intersection points of curves with the *vertical dashed line* correspond to the parameters of numerical experiments at $\gamma^2 L = -2$, given in Fig. 2.73a

Eccentric roundabout. Using Eqs. (2.57) and (2.58), we can readily obtain a dispersion equation for the steady-state configuration in the form of a uniformly rotating collinear construction:

$$F(B,R;\mu) = \frac{1}{2R} + \frac{2R(1+\mu)}{B(2R-B)} + \mathrm{K}_1(2R) + \frac{(2R+B\mu)\mathrm{K}_1(2R-B) - [2R(1+\mu) - B\mu]\mathrm{K}_1(B)}{2(R-B)} = 0. \quad (2.73)$$

It can be seen that F is a monotonically decreasing function of both B and R, and

$$F \sim \begin{cases} 1/\varepsilon_1, \ \varepsilon_1 = \min(B,R) \text{ at } B,R \ll 1, \\ (1+\varepsilon_2)\left[1/\varepsilon_2 - \mathrm{K}_1(\varepsilon_2)\right], \ \varepsilon_2 = B \ll 1 \text{ at } R \gg 1, \\ (1+\varepsilon_3)\left[1/\varepsilon_3 - \mathrm{K}_1(\varepsilon_3)\right], \ \varepsilon_3 = 2R - B \ll 1 \text{ at } R \gg 1. \end{cases} \quad (2.74)$$

The expressions in square brackets (2.74) are always strictly positive; therefore, the necessary condition for the existence of solution to Eq. (2.73) can be written as the inequality

$$\mu < -1. \quad (2.75)$$

Thus, the coordinates of the rotation center for *stationary states in the form of an eccentric roundabout* have a narrower interval in terms of μ parameter than that given in (2.69) for *arbitrary collinear configurations*, namely:

$$\begin{aligned} X_c \geq 0 \quad &\text{at} \quad \mu < -2, B \geq R \quad \text{or} \quad -1 > \mu > -2, B \leq R, \\ X_c \leq 0 \quad &\text{at} \quad -1 > \mu > -2, B \geq R \quad \text{or} \quad \mu < -2, B \leq R. \end{aligned} \tag{2.76}$$

The angular rotation velocity of the linear vortex construction about the vorticity center (2.76) is

$$\omega = \frac{\gamma\kappa(\mu+2)}{4\pi(2R+B\mu)}\left[\frac{B+2R\mu}{2BR} - \mu \mathrm{K}_1(B) + \mathrm{K}_1(2R)\right]. \tag{2.77}$$

When $\mu = -2$ (i.e. $Q = 0$), the angular velocity becomes zero and the position of the rotation center (2.76) shifts into infinity. In this case, we have a nontrivial solution in the form of a configuration of three vortices lying along a straight line and moving with a constant translational velocity in the direction perpendicular to that line, like an ordinary heton (or a pair of vortices). Such structure (a discrete analogue of a modon [276, 418, 516, 518]) will be referred to as *triton*. In this case, the dispersion Eq. (2.73) becomes

$$F(B,R;-2) = \frac{B^2 - 2BR + 4R^2}{2BR(2R-B)} - \mathrm{K}_1(B) - \mathrm{K}_1(2R-B) - \mathrm{K}_1(2R) = 0. \tag{2.78}$$

The velocity of the translational motion of *triton* is

$$V = \frac{\kappa\gamma}{4\pi}\left[\frac{2(R-B)}{B(2R-B)} - \mathrm{K}_1(B) + \mathrm{K}_1(2R-B)\right]. \tag{2.79}$$

A nontrivial feature of this solution is that we have a **solid body** moving vortex construction in the presence of **nonzero** total momentum. This paradox is due to the fact that, according to (1.83), the contribution of momentum M to the vortex integral L vanishes.

If $R \to B$, we have:

- at $\mu < -2$ it becomes a degenerate case, in which the lateral elliptic point in the phase plane merges with the bottom polar point and the *eccentric roundabout* becomes an *ordinary* or *inverse roundabout* (depending on the chosen value of the integral invariant L), and the expression (2.77) takes the form

$$\omega = \frac{\gamma\kappa}{4\pi B^2}\left\{\frac{1}{2} + \mu[1 - B\mathrm{K}_1(B)] + B\mathrm{K}_1(2B)\right\}, \tag{2.80}$$

  coinciding with (2.34) at $A = 2$;
- at $\mu = -2$ a steady symmetrical *triton* (Eq. (2.79) has a solution $V = 0$) or, which is the same,—a static state of the *roundabout*.

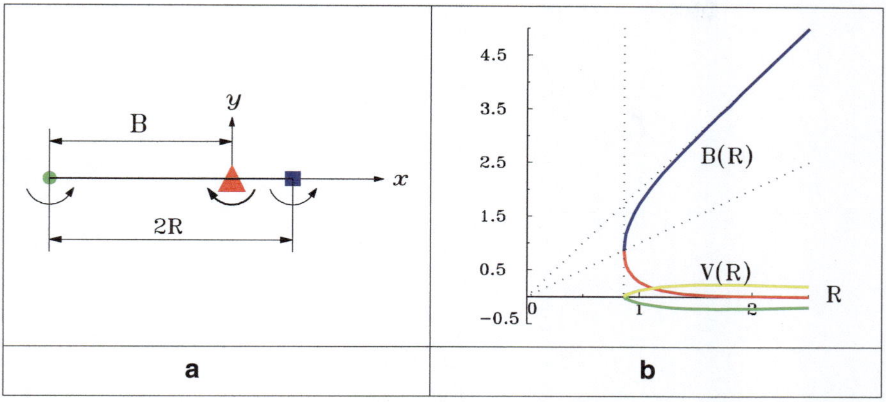

Fig. 2.64 The same as in Fig. 2.62, but for the collinear structure *triton* at $\mu = -2$. The curve $V(R)$ gives the distribution of the translational velocity of the vortex structure – *yellow branch* at $B > R$ and *green branch* at $B < R$. The position of the *vertical dashed line* is determined by the equality $B = R = R_0$, where R_0 is the solution of the equation $V = 0$ from formula (2.79)

Figures 2.62 and 2.63 demonstrate the character of solution of Eq. (2.73) at $\mu = -2.5$ and $\mu = -1.5$ within a wide range of values of integral L. In particular, they demonstrate the constant-sign properties of L at $\mu < -2$, mentioned in the previous section and the possible change of the sign of L at $\mu > -2$.

Note that the curve $X_c(R)$ is symmetrical about axis R, and $B(R)$ is symmetrical about the straight line $B = R$, and, with increasing R, the value of B asymptotically approaches 0 or $2R$ at $R > B$ and $R < B$, respectively, see (2.74). To illustrate those effects, the figures give auxiliary dashed half-lines $B = R$ and $B = 2R$. The angular velocity has a weak intermediate extremum; at $R = R_{min} = B$, it takes the value determined by the formula (2.80); and at $R \to \infty$, it tends to zero.

Figure 2.64 characterizes both the solution of Eq. (2.78) in R and the behavior of the translational velocity of the *triton* (2.79).

Figure 2.65 illustrates the motions of vortices in situations where both the dispersion Eq. (2.73) and one of relationships (2.71) or (2.72) are satisfied. In Fig. 2.65a the rotation is cyclonic and the vorticity center coincides with the vortex $\binom{2}{2}$, while in Fig. 2.65b the rotation is anticyclonic and the trajectories of vortices $\binom{1}{1}$ and $\binom{1}{2}$ coincide (the calculation was interrupted before they started overlapping). The last case is another example of generalized absolute choreography. Obviously, the collinearity condition for the configurations is identically satisfied, but the segments passing through three vortices are given in the figure only for two moments—for the initial (when the vortices are located on the X axis) and the final moment in each experiment.

Triangular stationary construction. In a similar way, we can find the relationships between geometric parameters of a stationary unstable vortex structure in the form of a uniformly rotating isosceles triangle, a case corresponding to a fixed hyperbolic singular point in the phase portrait.

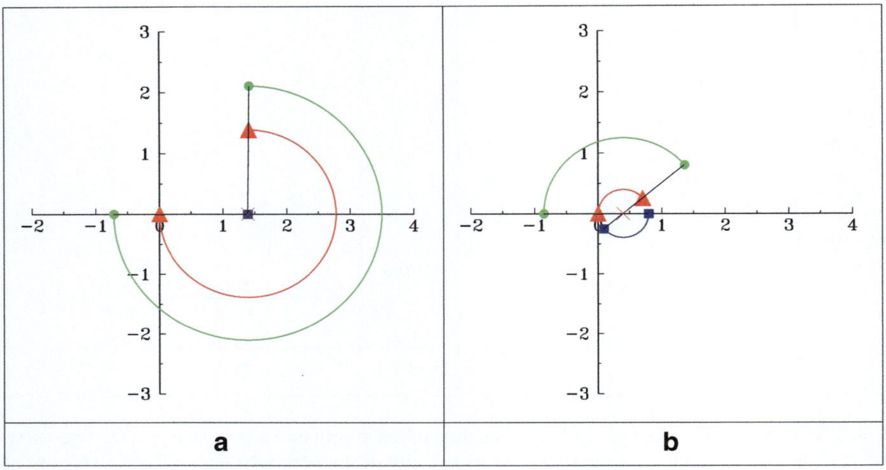

Fig. 2.65 Trajectories of absolute motion of vortex structures—*eccentric roundabouts*—when the conditions (2.71) (**a**) and (2.72) (**b**) are satisfied. The values of parameters: (**a**) $\mu = -1.5757$, $B = 0.7162$, $R = 1.0526$, $L = 0.5746$, (**b**) $\mu = -2.1598$, $B = 0.8551$, $R = 0.8235$, $L = -0.2211$

The dispersion equation, relating the length of the lateral side of the triangle B with the base angle of φ (see Figs. 2.66a and 2.67a), has the form

$$\frac{1+2B\cos\varphi \mathrm{K}_1(2B\cos\varphi)}{\cos^2\varphi} = -4(1+\mu)\left[1-B\mathrm{K}_1(B)\right]. \tag{2.81}$$

Since the expression in the left side (2.81) is always greater than unit, and $0 < 1-B\mathrm{K}_1(B) < 1$, we obtain that the necessary condition for the existence of solution of the dispersion equation becomes

$$\mu < -5/4. \tag{2.82}$$

Note that the comparison of conditions (2.75) and (2.82) provides an explanation for the fact that at $\mu = -1$ (Fig. 2.59), there are no lateral elliptic singular points, while there is a hyperbolic singular point.

The asymptotic properties of modified Bessel functions [725] determine the limitations for the base angle of the triangle at $B \gg 1$:

$$\arccos\varphi < \sqrt{-\frac{1}{4(1+\mu)}}. \tag{2.83}$$

The values of parameter μ must obviously belong to the half-line (2.82), on which the radical expression in (2.83) is always positive.

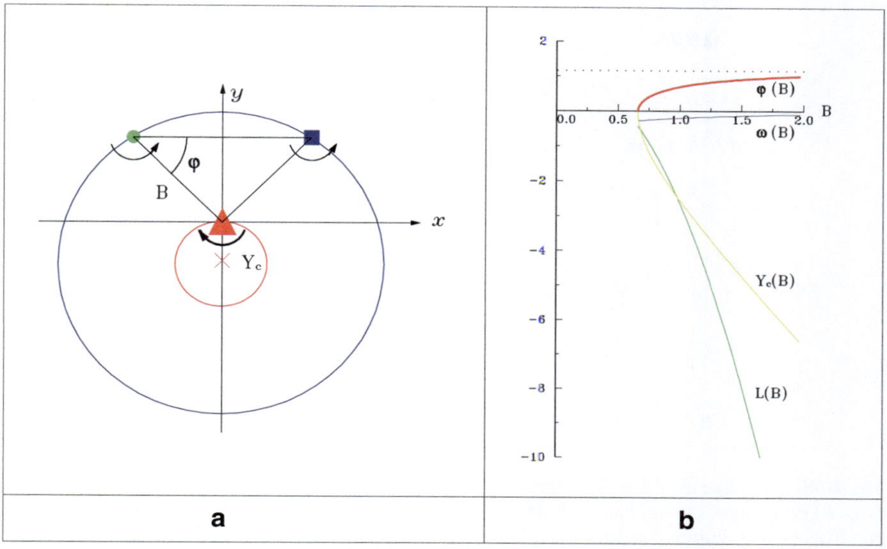

Fig. 2.66 (**a**) Scheme of the original arrangement of vortices for a solid-body rotating vortex structure in the form of an isosceles triangle at $\mu = -2.5$. The *solid circle lines* show the theoretical trajectories of vortices. (**b**) The appropriate dispersion curve $\varphi(R)$ (*red line*), satisfying (2.81); the value of the angular rotation velocity (2.84) of the vortex construction $\omega(R)$ (*light blue line*); the curve of the coordinates of the vorticity center $Y_c(R)$ (*yellow line*) and the value (2.59) of the integral invariant $L(B)$ (*green line*). The *horizontal dashed line* is the asymptote (2.83) for $\varphi(R)$

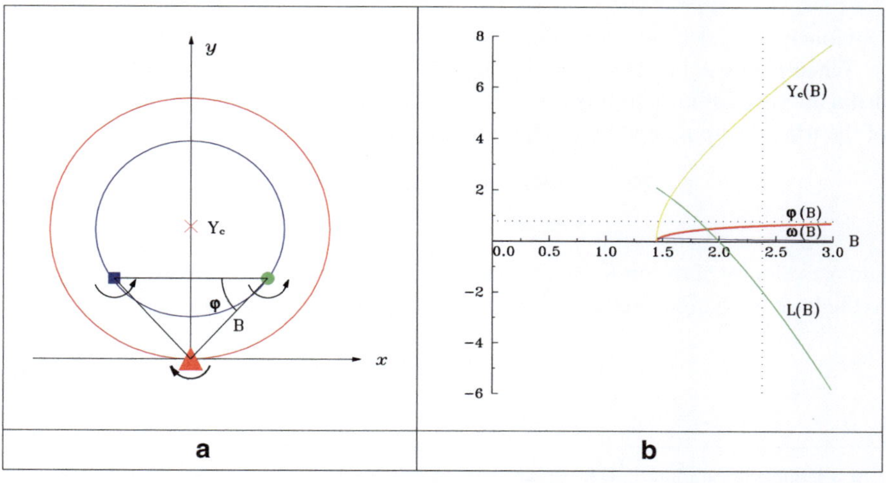

Fig. 2.67 The same as in Fig. 2.66, but with $\mu = -1.5$. The intersection points of the *vertical dashed line* with the curves determine the parameters of experiments in Fig. 2.73b

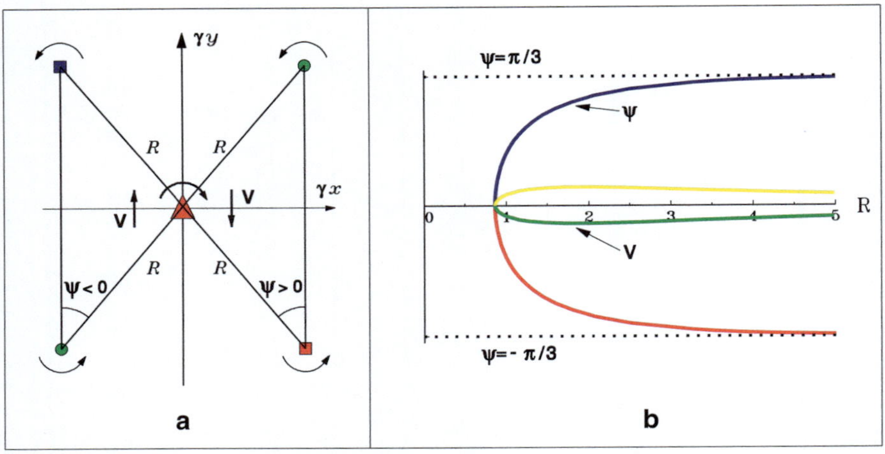

Fig. 2.68 The same as in Fig. 2.66, but with $\mu = -2$. In the dispersion curves (**b**), the *top branch* of $\psi(R)$ corresponds to the *bottom branch* of $V(R)$. The *dashed lines* $\psi = \pm\pi/3$ limit the domain of existence of solutions for Eq. (2.85)

The formula for the angular rotation velocity of the triangular structure about the vorticity center, whose coordinates $(0, Y_c)$ are determined by the bottom line of the formula (2.68), takes the form

$$\omega = \frac{\gamma\kappa(2+\mu)}{4\pi B^2}\Big[1 - B\mathrm{K}_1(B)\Big]. \tag{2.84}$$

Obviously, $\omega \le 0$ at $\mu \le -2$, and, in virtue of (2.82), $\omega \ge 0$ at $-5/4 > \mu \ge -2$.

When $\mu = -2$, i.e., at $Q = 0$, the angular velocity is zero and the vorticity center drifts away to infinity. In this case, the length of the lateral side and the base angle of the triangle are related by a dispersion equation

$$\frac{1 + 2R|\cos\psi|\mathrm{K}_1(2R|\cos\psi|)}{\cos^2\psi} = 4\big[1 - R\mathrm{K}_1(R)\big], \tag{2.85}$$

and the velocity of the uniform motion of the vortex structure in the direction parallel to the base of triangle (see Fig. 2.68a), can be described by the formula[17]

$$V = -\frac{\kappa\gamma\sin\psi}{4\pi R}\big[1 - R\mathrm{K}_1(R)\big]. \tag{2.86}$$

[17] In a homogeneous fluid [941], the steady-state solution of the type (2.85) corresponds to an equilateral triangle with an arbitrary side length. Expressions (2.79) and (2.86) are special cases of more general relationships obtained in [317] for the conditions of continuous stratification and with differential (nonuniform) rotation taken into account for the triangular configuration of vortices, of which two (with the same sign) lie in the same horizon, while the third vortex lies on the other horizon.

2.3.3.3 Classification of Motions of Triangular Vortex Structures: Trajectories of Absolute Motion, Choreographies

In this section, we consider the specific features of different types of vortex motion, on the one hand, by using the classification provided by phase portraits in Figs. 2.57, 2.58, 2.59, 2.60 and 2.61, and, on the other hand, by studying the behavior of absolute-motion trajectories of vortices and their relative choreographies with the use of numerical solution of the equations of motion (2.57)–(2.58).

When choosing the initial conditions in numerical experiments, coordinated with the analysis of phase portraits, we will rest upon the following considerations.

(i) If the phase curves of the class of motions under consideration reach PD boundary, and thus, the set of possible vortex configurations contains a collinear construction, we will assume that at $t=0$, all three vortices lie on the X-axis, the vortex $\binom{1}{1}$ is localized in the origin of coordinates, and the coordinates of vortices $\binom{1}{2}$ and $\binom{2}{2}$ satisfy the conditions of conservation of the integral invariant L, i.e.,

$$X_1^1 = Y_1^1 = Y_2^1 = Y_2^2 = 0,$$
$$X_2^1 = \begin{cases} \dfrac{X_2^2 - \sqrt{(X_2^2)^2\left[1-(1+\mu)^2\right] + \gamma^2 L(1+\mu)}}{1+\mu} \\ \text{at} \quad \mu \neq -1, \\ -\dfrac{\gamma^2 L}{X_2^2} \quad \text{at} \quad \mu = -1. \end{cases} \tag{2.87}$$

(ii) If phase trajectories do not reach PD boundary, we assume that at $t=0$ only the vortices $\binom{1}{1}$ and $\binom{1}{2}$ belong to the X-axis, and the vortex $\binom{2}{2}$ does not belong it, and the conditions

$$X_1^1 = Y_1^1 = Y_2^1 = 0,$$
$$X_2^1 = \begin{cases} \dfrac{X_2^2 - \sqrt{(X_2^2)^2\left[1-(1+\mu)^2\right] - (1+\mu)^2(Y_2^2)^2 + \gamma^2 L(1+\mu)}}{1+\mu} \\ \text{at} \quad \mu \neq -1, \\ -\dfrac{\gamma^2 L}{X_2^2} \quad \text{at} \quad \mu = -1. \end{cases} \tag{2.88}$$

are satisfied.

Considering (2.87)–(2.88), in the explanations to figures showing vortex trajectories, we will specify the parameters of the initial state by giving only the values of X_2^2 coordinate in the former case and the pair of coordinates (X_2^2, Y_2^2) in the latter case.

Analysis of phase portraits in Figs. 2.57, 2.58, 2.59, 2.60 and 2.61 shows that, at least in that relatively wide range of external parameters, not more than three qualitatively different motion types can exist:

- Type $\{1\}$—double capture, at which all three vortices are involved in joint rotation in the direction determined by the sign of their total intensity, or—at zero total circulation—by the intensity sign of the "strongest" vortex. Such motion can be conventionally determined by the scheme

$$\binom{1}{1}\binom{1}{2}\binom{2}{2}. \tag{2.89}$$

- Type $\{2\}$—simple capture, when a bottom-layer vortex merges with a top-layer vortex and one of the formulas holds

$$\binom{1}{1}\binom{1}{2}+\binom{2}{2} \quad \text{or} \quad \binom{1}{1}\binom{2}{2}+\binom{1}{2}. \tag{2.90}$$

- Type $\{3\}$—the dominating factor is the within-layer interaction between two bottom-layer vortices, and the motion of the vortex system is determined by the scheme

$$\binom{1}{2}\binom{2}{2}+\binom{1}{1}. \tag{2.91}$$

Characteristic examples of trajectories of absolute motion of all three types and the corresponding relative choreographies are given in Figs. 2.69 and 2.70, representing the cases of $Q \neq 0$ and $Q = 0$, respectively. The former figure shows choreographies constructed in a coordinate system uniformly rotating about the vorticity center in the anticyclonic direction, while the latter figure shows the motion in a coordinate system moving translationally with constant velocity in the negative direction of the y-axis.

Note that in the relative choreographies of motions of types $\{1\}$ and $\{3\}$, the bottom-layer vortices $\binom{1}{2}$ and $\binom{2}{2}$ move along the same trajectory, while staying at any moment on the opposite ends of its "diameter". However, in the former case (in the moving coordinate system) they move along closed trajectories in the clockwise direction induced by the top-layer vortex, while in the latter case, because of the predominant within-layer interaction, they move in the opposite direction (formulas (2.89), (2.91)). The vortex $\binom{1}{1}$ stays within the closed figure they describe for the motions of type $\{1\}$ and outside this figure for those of type $\{3\}$.

Figure 2.71 shows how the corresponding changes take place in the topological properties of those closed curves during the gradual movement of the initial coordinate of the image point in the phase plane along PD boundary from domain $\{1\}$ into domain $\{3\}$ in the case $\mu = -2$. The trajectories of absolute motion, which correspond to the choreographies 2.71a, 2.71g, 2.71j, are given in Fig. 2.70a–c, respectively. In Fig. 2.71e, the fixed points correspond to the stationary state of a

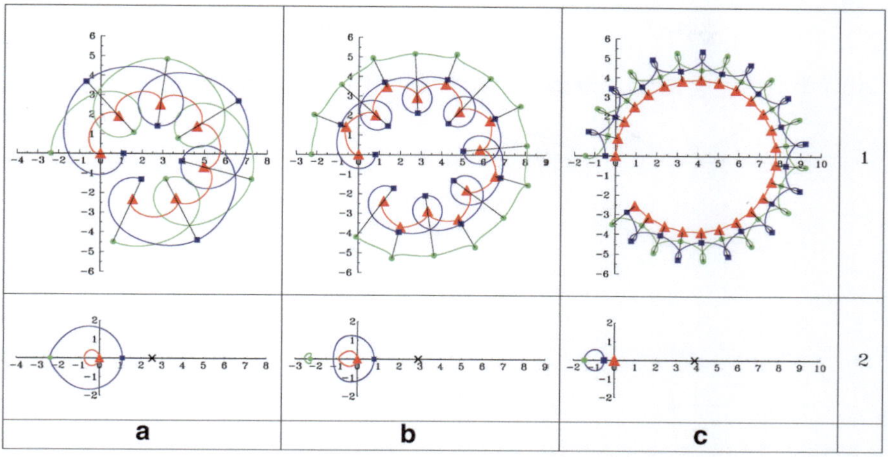

Fig. 2.69 Trajectories of absolute motion (**1**) and the corresponding choreographies (**2**) for motions of the types (**a**) – $\{1\}$, (**b**) – $\{2\}$, (**c**) – $\{3\}$ at $\gamma^2 L = -5$, $\mu = -2.5$ and the initial conditions determined by the formulas (2.87): (**a**) $X_2^2 = 1.1$, $t = (-7.2022;\ 1.8150;\ 8.3872)$, (**b**) $X_2^2 = 0.8$, $t = (-5.6143;\ 0.9600;\ 7.6543)$, (**c**) $X_2^2 = -0.5$, $t = (-0.5460;\ 0.3750;\ 3.1710)$. The *red lines* in the top of the figure show the trajectories of the top-layer vortex $\binom{1}{1}$, and the *blue* and *green lines* show the trajectories of bottom-layer vortices $\binom{1}{2}$ and $\binom{2}{2}$, respectively. The markers spaced by half-periods and the segments connecting them reflect synchronous (collinear) positions of vortices. The initial trilinear coordinates for the cases (a), (b), and (c) correspond to points with numbers 1, 2, and 3 in Fig. 2.57a. The crosses in the X-axis in the *bottom part* of the figure show the coordinates of vorticity centers $(X_c, 0)$. Adapted from [850] (Fig. 8) by permission of Cambridge University Press

triton (Fig. 2.74a). A specific feature of the figure is that at the initial moment, one of bottom-layer vortices lies strictly under the top-layer vortex. Figure 2.71b, c, h, i show the behavior of choreographies in transient regimes for the cases where the appropriate phase curves lie in the neighborhood of separatrices. In Figs. 2.71c, h, the choreographies are topologically similar. This is because the initial conditions for appropriate problems correspond to points on PD boundary near points where separatrices reach this boundary. Lying on different sides of the elliptic singularity, those are terminal points for phase curves very close to one another.

The series of Figs. 2.72a, 2.73 and 2.74a illustrates in greater detail the interactions of type $\{2\}$ for situations where the image point in the phase plane coincides with the lateral elliptic point: we see *steady* rotation of a solid-body collinear structure—an eccentric roundabout at $Q \neq 0$ (Fig. 2.72a and 2.73a) or a *steady* translational motion of triangular vortex configuration—*triton* $Q = 0$ (Fig. 2.74a). Clearly, the motions of this type correspond to relative choreographies, which degenerate into three fixed points.

Figures 2.72b, 2.73 and 2.74b illustrate the *unsteady* character of the triangular construction (the corresponding phase trajectory lies in the neighborhood of the separatrix): the vortices periodically lie in the vertices of an isosceles triangle, in the neighborhood of separatrix self-intersection, or on PD boundary. Note that in

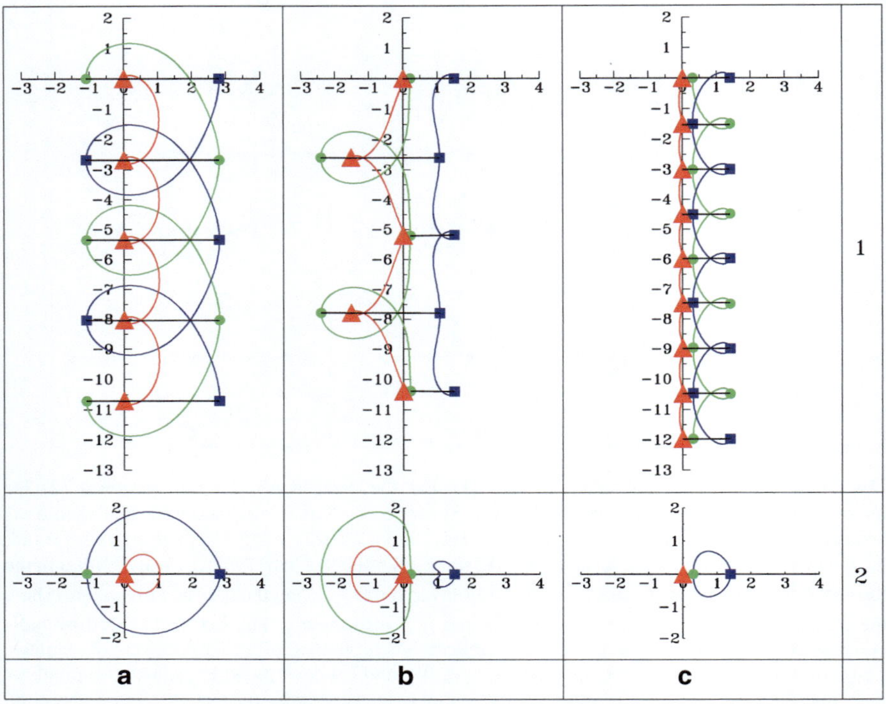

Fig. 2.70 The same as in Fig. 2.69, but with $\gamma^2 L = -2.89$, $\mu = -2$ and (**a**) $X_2^2 = 2.8$, $t = (-15.7889; 16.2768; 2.5121)$, (**b**) $X_2^2 = 1.9$, $t = (-4.5779; 7.4948; 0.0830)$, (**c**) $X_2^2 = 1.4$, $t = (-1.2561; 4.0692; 0.1869)$. The initial trilinear coordinates for the cases (a), (b), and (c) correspond to points with numbers 1, 2, and 3 in Fig. 2.60a. The vorticity center in this case lies at infinity

Fig. 2.72b, the relative choreographies were obtained in a coordinate system rotating about the vorticity center with a negative angular velocity, because $\mu < -2$ (see the formula for angular velocity (2.84)), and in Fig. 2.73b, where $\mu > -2$, in the system rotating with a positive velocity.[18] The choreographies in Fig. 2.74b were obtained in a coordinate system moving translationally with a constant velocity. Note that Figs. 2.72.1b, 2.73 and 2.74.1b show a part of trajectories of absolute motion of a vector tripod within one period (the configuration three times acquires collinear position and two times forms an isosceles triangle) for the case when the appropriate phase trajectory, while lying in domain $\{2\}$, belongs to the neighborhood of the separatix. The latter figures demonstrate an obvious property of instability of this state.

Let us consider some important special cases of motions of a three-vortex system.

[18]See also the behavior of curves $\omega(B)$ in Figs. 2.66 and 2.67.

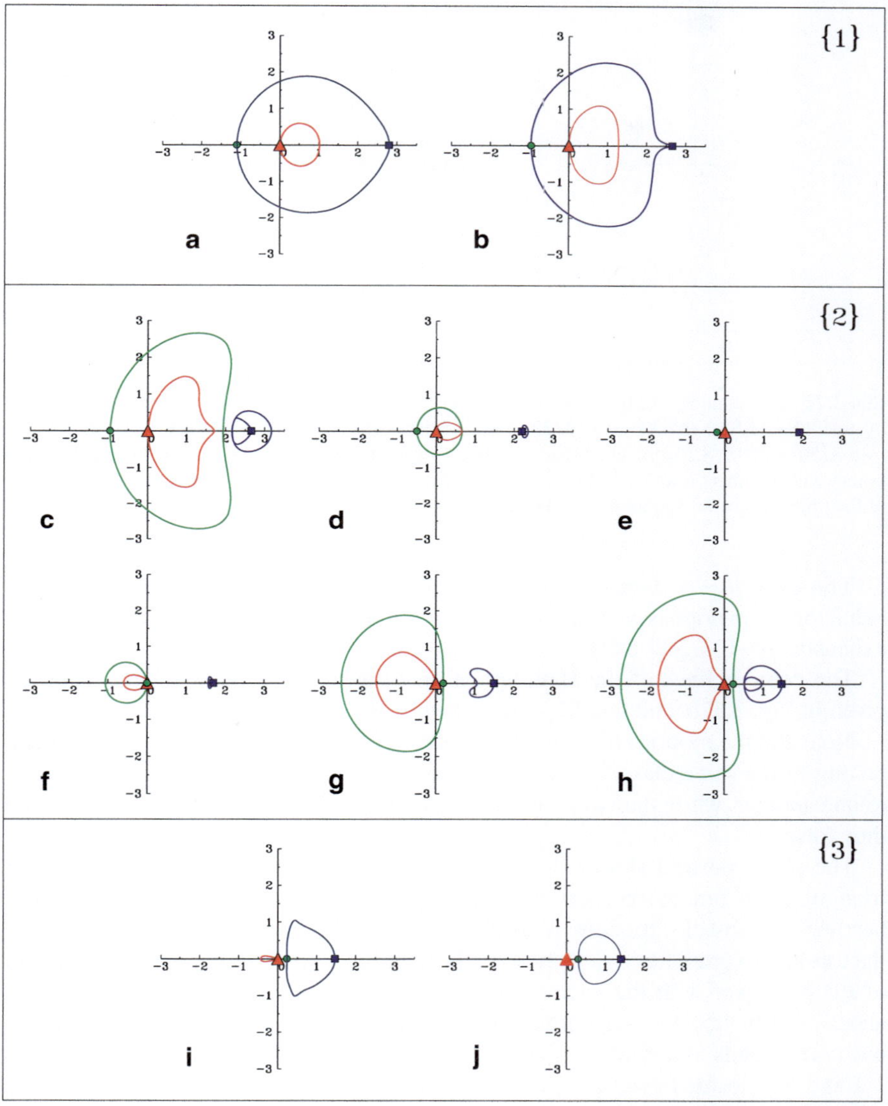

Fig. 2.71 Choreographies corresponding to motions of the types {1}, {2}, and {3} at $\mu = -2$: (**a**) $X_0 = 1.1$ (corresponds to Fig. 2.70a), (**b**) $X_0 = 0.9615$, $t = (-13.6257; 14.7054; 1.9193)$, (**c**) $X_0 = 0.9610$, $t = (-13.6182; 14.7009; 1.9173)$, (**d**) $X_0 = 0.5$, $t = (-7.5675; 10.0484; 0.5191)$, (**e**) $X_0 = 0.206$ (corresponds to Fig. 2.74a), (**f**) $X_0 = 0$, $t = (-3; 6; 0)$; (**g**) $X_0 = -0.2$ (corresponds to Fig. 2.70b), (**h**) $X_0 = -0.229$, $t = (-1.6013; 4.4924; 0.1089)$, (**i**) $X_0 = -0.230$, $t = (-1.5961; 4.4863; 0.1098)$, (**j**) $X_0 = -0.3$ (corresponds to Fig. 2.70c). The markers show the initial positions of the vortices

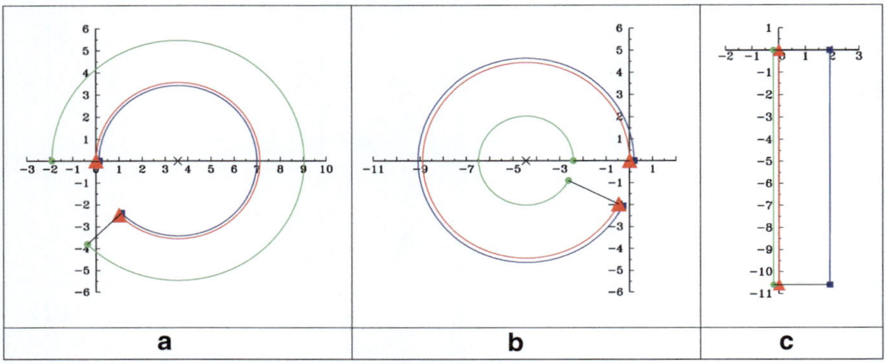

Fig. 2.72 The same as in Fig. 2.69 for motions of the type $\{2\}$ at the initial conditions specified by formulas (2.87): (**a**) $X_2^2 = 0.13233$, $t = (-2.5058; 0.0263; 5.4796)$, (**b**) $X_2^2 = 1.02777$, $t = (-6.8150; 1.5845; 8.2305)$. The trilinear initial coordinates for the cases (a) and (b) correspond to points with numbers 4 and 5 in Fig. 2.57a—in an elliptic singular point and in the neighborhood of the point where the separatrix reaches PD boundary

The case $\mu = 1$, when all three vortices have equal intensities. Note that the behavior of analogous vortex structures in a homogeneous fluid has been studied exhaustively [103, 595, 892].

The contour lines of the Hamiltonian in trilinear coordinates for this case are given in Fig. 2.57b, and the PD is described by the mean formula (2.62).

Note that the motions of vortices in this case are relatively simple and all of them belong to the same class $\{3\}$, when the bottom-layer vortices are rotating relative to one another, while the vortex $\binom{1}{1}$ can be either within or outside the construction they form.

The phase portrait shows all motions to be periodic. Moreover, each evolving triangular structure twice a period degenerates into collinear configurations and once becomes an isosceles triangle. A special case of the collinear state can be such were the top-layer vortex lies strictly above a bottom-layer vortex—in the tangency point of PD boundary with the lateral sides of the coordinate triangle: $t = (1.5, 0, 1.5)$ and $t = (1.5, 1.5, 0)$. The portrait contains two singular elliptic points. The top one corresponds to a stable steady state—a roundabout, when vortices with indices $\binom{1}{2}$ and $\binom{2}{2}$, located symmetrically about the fixed central top-layer vortex, rotate uniformly along a circle. The bottom point corresponds to a structure, in which the merged bottom-layer vortices rotate about a common center with theoretically infinite angular velocity.

Figure 2.75.1a–d illustrate the set of possible motions of such system in a series when at the passage from one experiment to another, the initial position of the image point of the phase space shifts along PD boundary from the top pole toward the bottom one.

Figure 2.75.1a gives a realization of a stationary roundabout. This is an example of absolute choreography. Clearly, the relative choreography (2.75.2a) in a coordinate

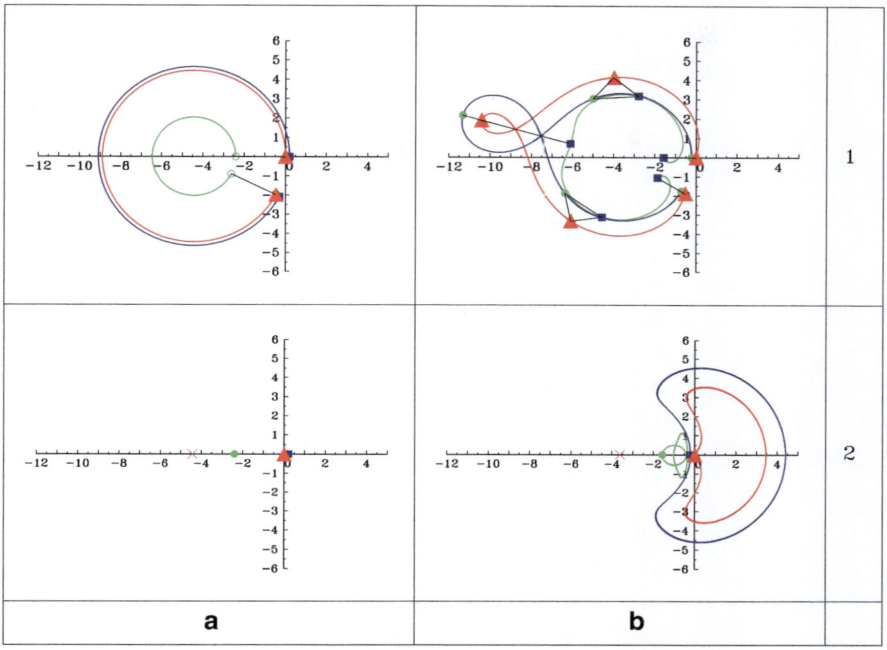

Fig. 2.73 Trajectories of absolute motion (**1**) and the corresponding relative choreographies(**2**) for motions of type $\{2\}$ at $\gamma^2 L = -2$, $\mu = -1.5$ and the initial conditions determined by formulas (2.87): (**a**) $X_2^2 = 0.1949$, $t = (-10.2416; 0.0855; 13.1561)$; (**b**) $X_2^2 = -0.2338$, $t = (-2.6901; 0.1230; 5.5671)$. In figure (**a**), linear segments show the collinear positions of vortices in the initial and final moments, and figure (**b**) shows their positions every quarter of the period (the configurations are collinear and isosceles-triangular). The trilinear initial coordinates for the cases (a) and (b) are shown by points with numbers 1, 2 in Fig. 2.58a. The crosses on the X-axis in the *bottom part* of the figure show the coordinates of vorticity centers $(X_c, 0)$

system rotating with the angular velocity of peripheral vortices can be determined by three fixed points. Figure 2.75.1b shows the initial stage of the evolution of the vortex structure in the case where at $t = 0$, the vortices $\binom{1}{1}$ and $\binom{2}{2}$ coincide. The system behaves in such a manner that the top-layer vortex aligns with bottom-layer vortices in turn and periodically. This can be vividly seen in the figures of absolute trajectories (Fig. 2.75.1b), and relative choreographies (Fig. 2.75.2b). Figure 2.75.1c, 1d give trajectories of absolute motion for situations where the vortex $\binom{1}{1}$ lies all the time outside the structure $\binom{1}{2}$–$\binom{2}{2}$. In the former case, bottom-layer vortices periodically stay in system vorticity center, while in the latter case, they always lie on the same side of it.

Clearly, as the initial point of phase trajectories is shifting toward the bottom pole, the bottom-layer vortices will be describing a figure approaching a circumference with decreasing (tending to zero) radius. A confirmation to this fact is also the phase trajectories in the bottom part of the Fig. 2.57b being practically horizontal lines.

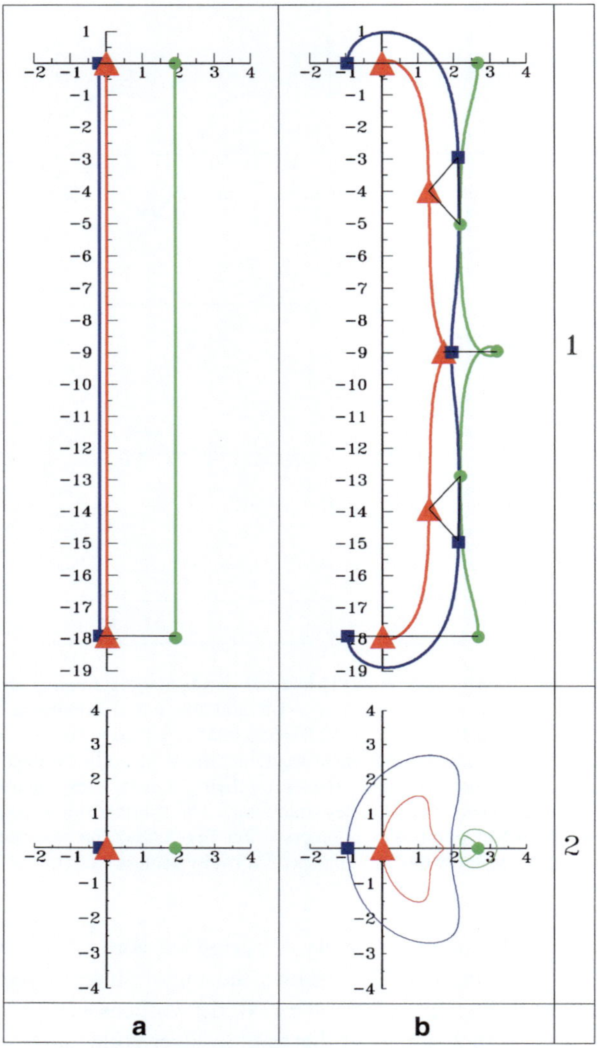

Fig. 2.74 (**1**) Stationary trajectories of three vortices: (**a**) a stable case (*triton*—elliptic singular point) and (**b**) an unstable case (in the form of an isosceles triangle—hyperbolic singular point) at $\gamma^2 L = -2.89$. Here (**a**) $X_2^2 = -0.2062$, $t = (-4.6321; 0.0883; 7.5438)$, (**b**) $X_2^2 = -0.9610$, $t = (-13.6182; 1.9173; 14.7009)$. (**2**) The corresponding relative choreographies in a coordinate system, which moves translationally and uniformly in the negative direction of the y-axis

Note that most cases considered above (Figs. 2.65, 2.69, 2.72 and 2.75) refer to the case of compact PD. Now let us consider the characteristic features of absolute motions that admit the existence of infinite trajectories.

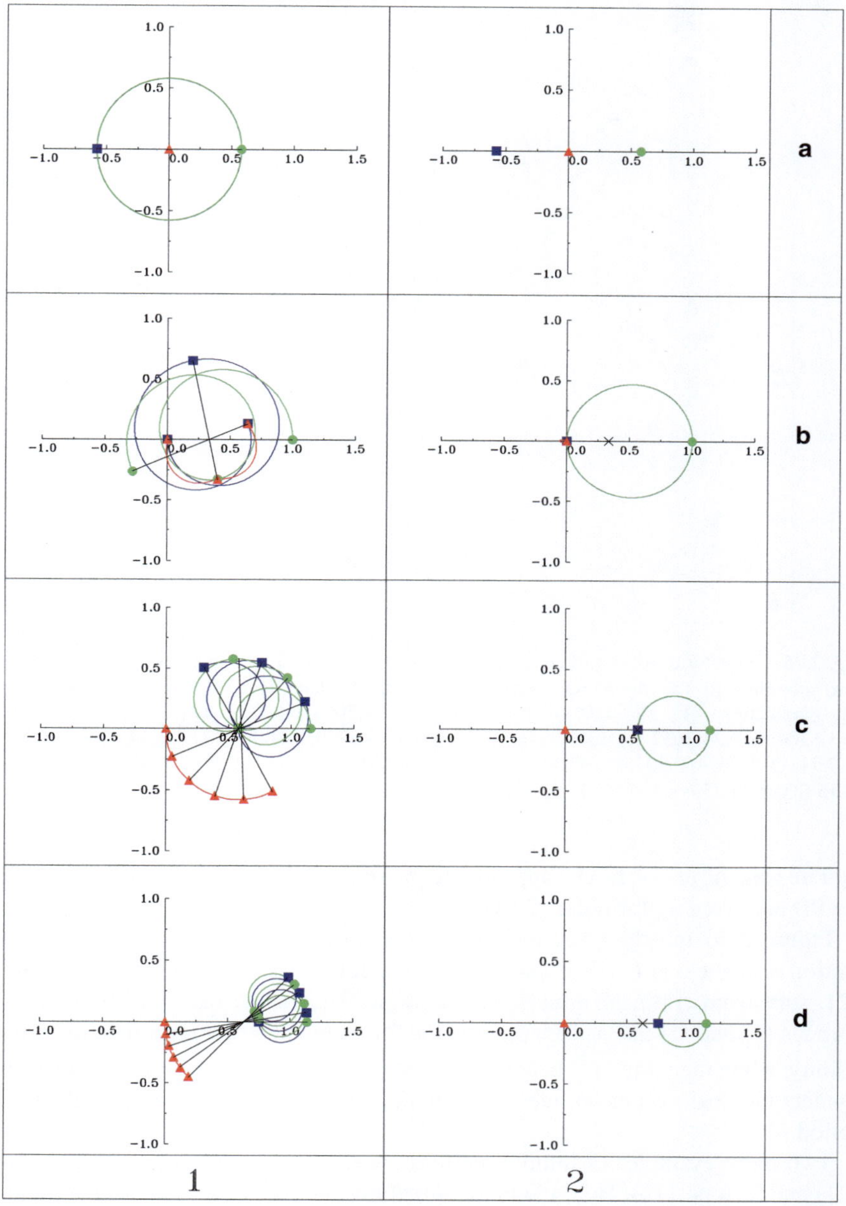

Fig. 2.75 (**1**) Trajectories of the absolute motion of three vortices in the case $\mu = 1$ and $L = 2$. (**2**) The corresponding relative choreographies in a coordinate system rotating in the cyclonic direction about the system vorticity center (its coordinates are marked by crosses). The initial conditions correspond to the following parameter values: (**a**) $X_2^2 = -1/\sqrt{3}$, $t = (2.0, 0.5, 0.5)$, point 1 in Fig. 2.57b; (**b**) $X_2^2 = 0$, $t = (1.5, 0.0, 1.5)$, point 2 in Fig. 2.57b; (**c**) $X_2^2 = 1/\sqrt{3}$, $t = (0.5, 0.5, 2.0)$, point 3 in Fig. 2.57b; (**d**) $X_2^2 = 0.75$, $t = (0.2227, 0.8438, 1.9335)$, point 4 in Fig. 2.57b

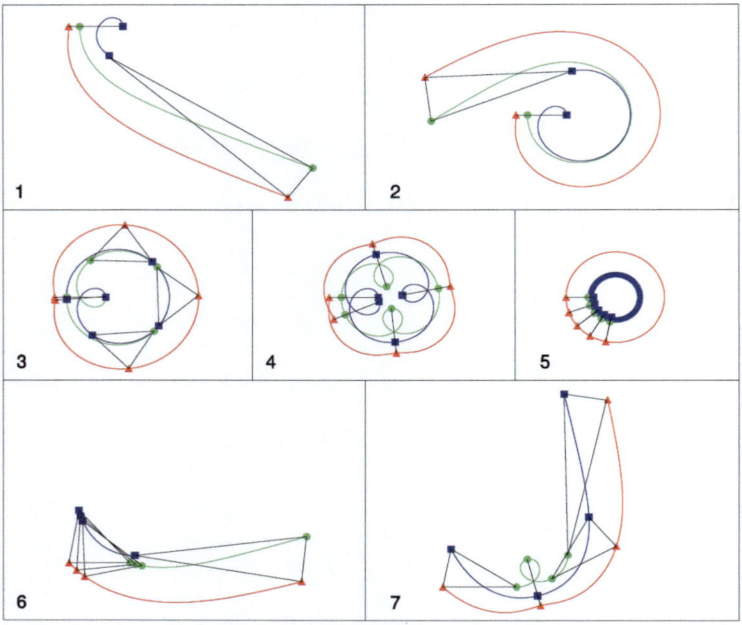

Fig. 2.76 Trajectories of absolute motion at $\mu = -1$, $\gamma^2 L = -2$ and the initial conditions corresponding to the numbered points in the phase plane (Fig. 2.58b) with the following parameters: (**1**) $X_2^2 = 0.45$, $t = (-4.7112; 7.4074; 0.3038)$, (**2**) $X_2^2 = 0.4810$, $t = (-3.8304; 6.4824; 0.3770)$, (**3**) $X_2^2 = 0.4811$, $t = (-3.3750; 6.0000; 0.3750)$, (**4**) $X_2^2 = 0.5$, $t = (-4.7112; 7.4074; 0.3038)$, (**5**) $X_2^2 = 0.9$, $t = (-0.0669; 1.8519; 1.2150)$, (**6**) $X_2^2 = 2.5$, $t = (-15.0000; 8.6250; 9.3950)$, (**7**) $X_2^2 = 3$, $t = (-15.0000; 4.5000; 13.5000)$

The case of $\mu = -1$. The appropriate phase portraits are given in Fig. 2.59, and the PD are given by formulas (2.64).

Figure 2.76 mostly gives the initial segments of trajectories of the absolute motion of vortices at $L < 0$. Experiments 1 and 2, which show a realization of regime $\{2\}$, start from collinear initial configurations. They show that mutual capture of vortices $\binom{1}{1}$ and $\binom{1}{2}$ takes place through the formation of a two-layer pair, drifting to infinity, while the vortex $\binom{2}{2}$ becomes stationary at $t \to \infty$. In those figures, segments connect the positions of vortices in the initial and final moments of the calculation period.

Two more examples of infinite motions are represented by experiments 6 (type $\{2\}$) and 7 (type $\{1\}$). Here the initial structures are triangles (not isosceles), such that the distances r_{22}^{21} (or trilinear coordinates t_1) for those two cases are equal. In this case, the coordinates Y_2^2 are derived from the condition of preservation of integral invariant L. The configurations accentuated in the first panel are as follows: (a) the initial, (b) in the form of an isosceles triangle (this corresponds to the position of the image point on the straight line $t_2 = t_3$), (c) antisymmetric with respect to the initial position (in which the sides of the triangle with lengths r_{12}^{12} and r_{12}^{11} change places) and (d) the final. The second fragment gives the same, except that in item (b), we have a collinear configuration (corresponding to the arrangement of the image point on PD boundary), and in item (c), we have a symmetric configuration. Thus, the possible class of vortex structures for motions of type $\{2\}$ includes the collinear state and does not include the configuration in the form of an isosceles triangle, while the motions of type $\{1\}$ show the opposite property.

Experiments 3, 4, and 5 demonstrate trajectories of finite type $\{3\}$[19] with dominating character of within-layer interaction between bottom-layer vortices $\binom{1}{2}$ and $\binom{2}{2}$.

Fragments 3–5 show trajectories in time segments $T/2$, $2T$, and $188T$ with instantaneous configurations spaced by intervals of $T/8$, $T/2$, and $12T$, respectively (T is the period of relative motion). Thus, the motions of type $\{3\}$, as well as before, involve the revolution of cyclonic bottom-layer vortices along loop-like trajectories about some variable center. Every quarter of the period, the vortex structure forms either a collinear or a isosceles-triangle configuration with alternative alternation of vortices $\binom{1}{2}$ and $\binom{2}{2}$.

Figure 2.77 gives both trajectories of finite-type absolute motion (left) and the corresponding choreographies (right). The lines (a) and (c) correspond to Figs. 2.76.1, 2, respectively. In both cases, the calculation time is extended (relative to the previous figure) to about $6T$. The origin of coordinates coincides with the vorticity center of the vortex structure. When choreographies were obtained, the coordinate system is subject to counterclockwise (Fig. 2.77a) and clockwise (Fig. 2.77c) rotational motion. It is obvious that in such case, a system must have some intermediate state in which no rotation of the coordinate system is required and thus the *relative* choreography becomes *absolute*. This is the situation given in Fig. 2.77b (the corresponding parameters are given in comments). In choreographies, the markers and the segments connecting them mark the collinear and triangular positions of vortices spaced by a quarter of the period.

Thus, in this case, as well as in the four-vortex problem with $P=0$ and $M\neq 0$ (Sect. 2.2.2.2), an absolute choreography corresponds to a phase trajectory in the class of finite motions $\{3\}$. It is important to note that this periodic solution shows a stable behavior: the vortex system in Fig. 2.77b has passed through more than 20 cycles.

More about choreographies. As mentioned above, the absolute motion of a finite-type vortex system can correspond to a discrete enumerable set of relative choreographies rather than a single choreography.

Indeed, varying the angular rotation velocity of the coordinate system, we can obtain an infinite number of closed vortex trajectories. Here we will consider this effect in the case of a three-body problem for motion of type $\{1\}$, when the relative motion involves the rotation of peripheral vortices of the bottom layer about the central top-layer vortex.

[19] Notwithstanding the initial conditions in problems 2 and 3 being very close (the markers with numbers 2 and 3 in Fig. 2.59 are practically indiscernible), the corresponding motions belong to different types, since the image points in the phase plane belong to trajectories lying in the neighborhood of the separatrix, but on different sides of it.

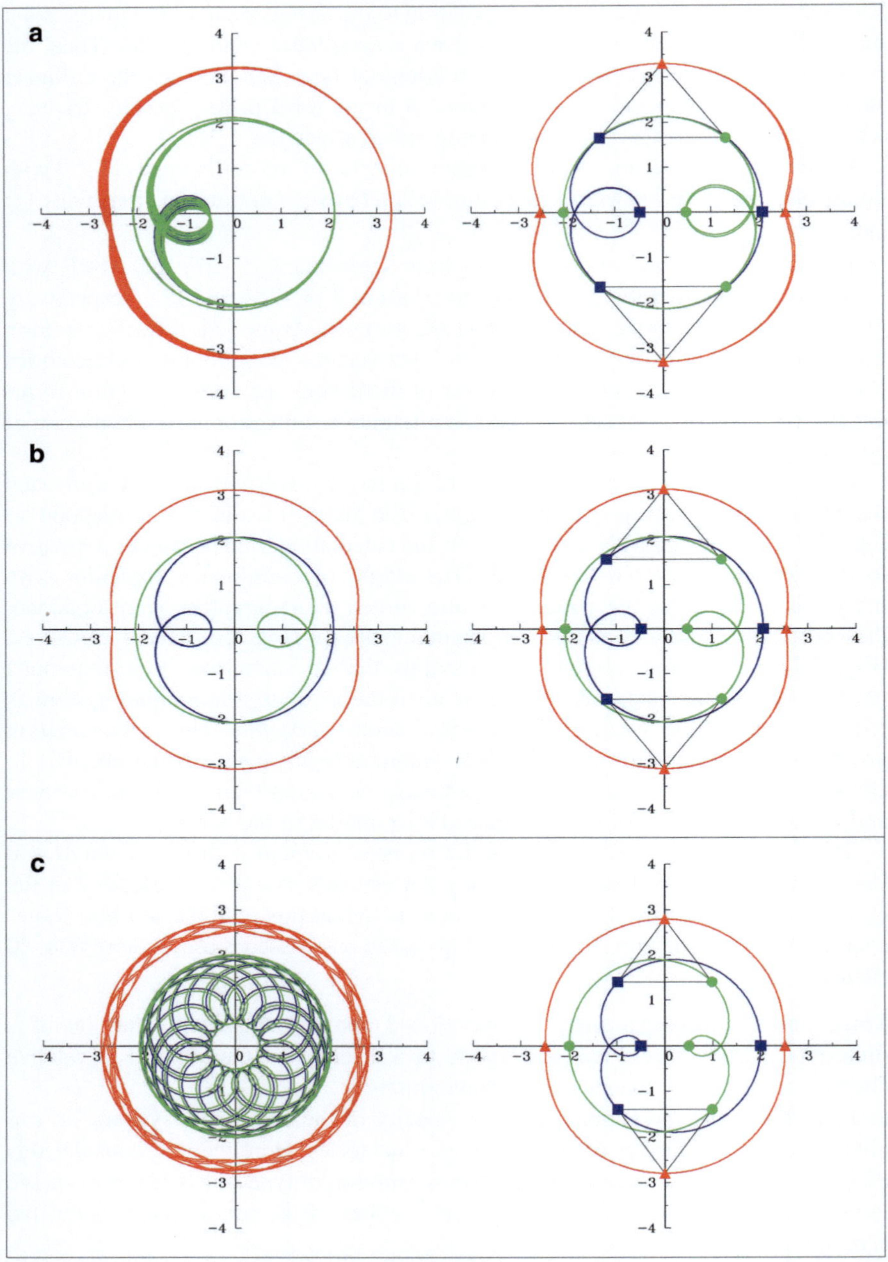

Fig. 2.77 The absolute trajectories of three vortices (*left*) and the corresponding choreographies (*right*) for motions of type $\{3\}$ at a collinear initial state and the following parameter values: **(a)** $X_2^2 = 0.4811, t = (-3.3750; 6.0000; 0.3750)$; **(b)** $X_2^2 = 0.4833, t = (-3.7722; 6.4218; 0.3504)$, **(c)** $X_2^2 = 0.5, t = (-4.7112; 7.4074; 0.3038)$

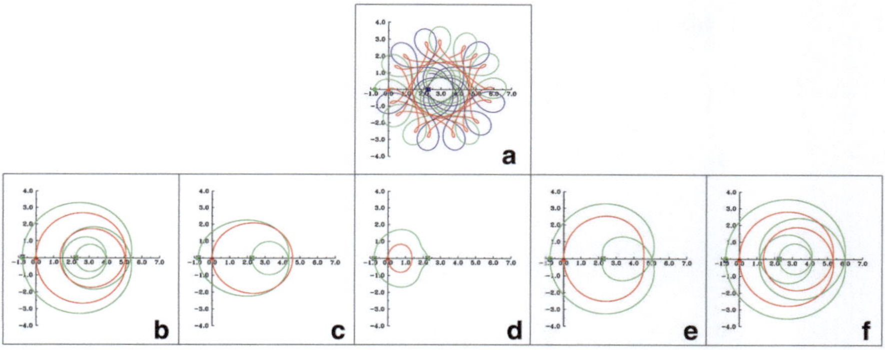

Fig. 2.78 Trajectories of absolute motion (**a**) and relative choreographies (**b–f**) for three-vortex problem with a collinear initial state and at $\mu = -1.5$, $\gamma^2 L = 0.65$, $X_2^2 = 2.8\sqrt{0.65}$, $t = (42.4361; -35.2800; -4.1561)$ and different values of angular rotation velocity of coordinate system: (**a**) $\omega = 0$ and $\omega = \omega_1^n = \omega_1^0 + n\Delta\omega$ at (**b**) $n=-2$, (**c**) $n=-1$, (**d**) $n=0$, (**e**) $n=1$, (**f**) $n=2$; $\omega_1^0 = 0.00002994$ and $\Delta\omega = 0.00010252$. The marked vortices and the segments connecting them in insets (**b–f**) track collinear configurations at $t=0$, $t=T_1/2$, $t=T_1$. Everywhere here, the vortices $\binom{1}{2}$ avd $\binom{1}{2}$ have the same relative trajectories (complex relative choreographies)

In this case, the absolute motion has two fixed internal scales $\Delta\omega_1$ and $\Delta\omega_2$ of angular velocities associated with periods T_1 and T_2 of satellite revolution of about the central vortex and the revolution of the entire system about the vorticity center, respectively. The choice of the scale of the angular velocity of additional rotation $\Delta\omega$ consists in the requirement that relationships of the form $\Delta\omega = \Delta\omega_1 N_1 = \Delta\omega_2 N_2$ must hold, where N_1 and N_2 are integers; this is always possible, because the relative motions are periodic.

With specified external parameters (the value of μ and the distances B and R between the vortices in the initial collinear configuration), this scale can be determined numerically. For the example given in Fig. 2.78, $\Delta\omega = 0.00010252$. This scale allows us to find on the ω-axis a discrete set of values ω_1^n $(n = 1, 2, \ldots)$ of the angular velocity of rotation of the coordinate system, which corresponds to closed trajectories. We well refer them to as first-order trajectories, as is reflected by their subscript. In the set of other choreographies, they can be distinguished by the fact that, in a rotating coordinate system, all three vortices come to the same position with a period of T_1, while every half-period, we have a collinear configuration, in which peripheral bottom-layer vortices change their places.

Again, the first-order choreographies contain the simplest one (ω_1^0, Fig. 2.78d), such that the top-layer vortex moves along an internal oval trajectory with period $T_1/2$, while bottom-layer vortices move along a common simple closed curve.[20] At $\omega = \omega_1^{-1}$ and $\omega = \omega_1^1$ (i.e., left and right of point $\omega = \omega_1^0$ on the ω-axis,

[20] An example of the simplest choreography is also the trajectory in Fig. 2.33b.

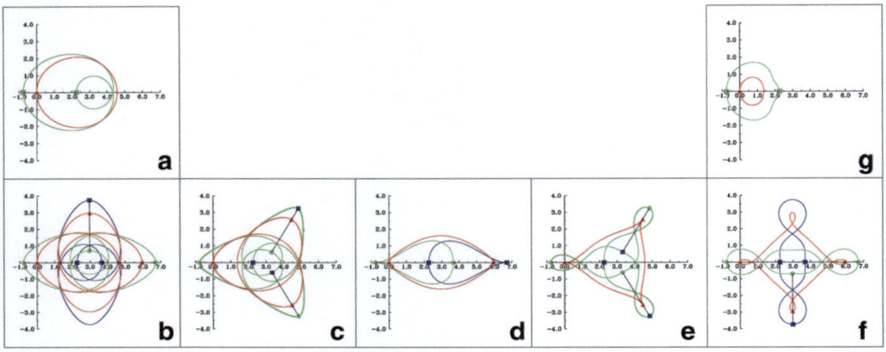

Fig. 2.79 A set of choreographies in the three-vortex problem at collinear initial state with the same parameters as in Fig. 2.78 for different values of the angular velocity of coordinate system rotation in the interval of angular velocities $[\omega_1^{-1}, \omega_1^0]$: **(a)** $\omega = \omega_1^{-1}$, **(b)** $\omega = \omega_4^1 = \omega_1^{-1} + \Delta\omega/4$, **(c)** $\omega = \omega_3^1 = \omega_1^{-1} + \Delta\omega/3$, **(d)** $\omega = \omega_4^2 = \omega_1^{-1} + \Delta\omega/2$, **(e)** $\omega = \omega_3^2 = \omega_1^{-1} + 2\Delta\omega/3$, **(f)** $\omega = \omega_4^3 = \omega_1^{-1} + 3\Delta\omega/4$, **(g)** $\omega = \omega_1^0$. The labeled vortices and the segments connecting them track the collinear configurations at the first local period of motion (at $t = 0$, $t = T_1/2$, $t = T_1$)

respectively), we have topologically similar trajectories (Fig. 2.78c, e), such that the top-layer vortex still describes a simple oval figure, while the bottom-layer vortices describe a self-intersecting loop. At the same time, vortices $\binom{1}{2}$ and $\binom{2}{2}$ during the first half-period cover half of both external and internal part of the loop and they complete their trajectories by the end of period. Note also that due to a change in the direction of rotation of the coordinate system in Fig. 2.78c all vortices move along their trajectories in the cyclonic direction, while those in Fig. 2.78e move in the anticyclonic direction (that is why we use superscripts with different signs). The subsequent drift in opposite directions from point ω_1^0 into the interval $\Delta\omega$ leads to the generation of a simple loop on the top-layer vortex trajectory and double loops on the trajectories of bottom-layer vortices (Fig. 2.78b, f).

In addition to first-order choreographies, there exists a continuum of more complex closed trajectories. Thus, each interval between adjacent points ω_1^n and ω_1^{n+1} contains a set of points with coordinates $\omega_k^m = \omega_1^n + m\Delta\omega/k$ (k and m are integers and $1 \le m \le k$), which correspond to appropriate choreographies. Fig. 2.79b–f give a series of higher order choreographies in the interval between ω_1^{-1} and ω_1^0 (insets 2.79a, g, respectively) at $k = 3$ and $k = 4$. In this series of experiments all relative trajectories are closed, i.e. they are choreographies. For $k = 3$ and all possible m (i.e. at $m = 0, 1, 2, 3$) we have complex choreographies, as the trajectories of vortices $\binom{1}{2}$ and $\binom{2}{2}$ coincide. In choreographies of the order k, each vortex returns to its original position only with a period of $T_k = kT_1$. Note that in the lower layer we observe simple choreographies at odd k and complex ones at even k, i.e., in the former case the vortices $\binom{1}{2}$ and $\binom{2}{2}$ move along the same trajectory, while in the latter case they are shifted by the angle π/k. For the identification of trajectories in the figures, the former is shown by the thicker line. Note that at larger k, the form of the choreography can be very complicated.

Now we return to the case of zero total intensity $\kappa_1^1 = \kappa_2^2 = -\kappa$, $\kappa_2^1 = 2\kappa > 0$, discussed in the end of Sect. 2.3.2.1, but in the case of nonzero total momentum.

A characteristic phase portrait of relative motion of three vortices for this case is shown in Fig. 2.60b.

Despite the equality $Q = 0$, in virtue of (**!**), the laws of interaction of the stronger cyclonic vortex $\binom{1}{2}$ with mutually equal anticyclones of the top and bottom layers are different in this case. It is also clear that the top-layer vortex $\binom{1}{1}$ must interact differently with the bottom-layer vortices $\binom{1}{2}$ and $\binom{2}{2}$, which now have different intensities (in terms of both their sign and magnitude). The difference between the interactions is reflected in the asymmetry of phase curves about the straight line $t_1 = t_3$, which, in this case, is the symmetry axis for only PD boundary. This fact plays its role in each identified type of motion, resulting, in particular in that domain $\{2\}$ divides into two subdomains $\{2a\}$ and $\{2b\}$.

The motions of types $\{1\}$ and $\{3\}$ are still determined by interactions according to formulas (2.89) and (2.91), respectively. Unlike the symmetrical case, the type $\{1\}$ now shows the counterclockwise rotation of all three vortices, caused by the intensity of the bottom-layer cyclone $\binom{1}{2}$, and type $\{3\}$, for the same reason, exemplifies a nonsymmetric within-layer interaction. The motions of type $\{2a\}$ correspond to the formula

$$\begin{pmatrix}1\\1\end{pmatrix}\begin{pmatrix}2\\2\end{pmatrix}+\begin{pmatrix}1\\2\end{pmatrix},$$

while for type $\{2b\}$, we have the law

$$\begin{pmatrix}1\\1\end{pmatrix}\begin{pmatrix}1\\2\end{pmatrix}+\begin{pmatrix}2\\2\end{pmatrix}.$$

Figure 2.82 exemplifies trajectories of absolute motion of vortices in this case, with the initial conditions, corresponding to the conditions of constant total momentum at any X_0, specified as

$$X_1^1 = 0, \quad X_2^1 = X_0/2, \quad X_2^2 = -\gamma P + X_0.$$

The time intervals in the figure are always equal to two periods of the relative motion of vortices. The markers, along with the horizontal segments drawn through them, show the positions of vortices every half period. The set of relative choreographies for this case is given in Figs. 2.80 and 2.81.

The elliptic singular point on the phase plane, which belongs to domain $\{3\}$, as well as in the previous case, corresponds to the two-layer pair from the top-layer vortex $\binom{1}{1}$ with the intensity $-\kappa$ and "merged" bottom-layer vortices $\binom{1}{2}$

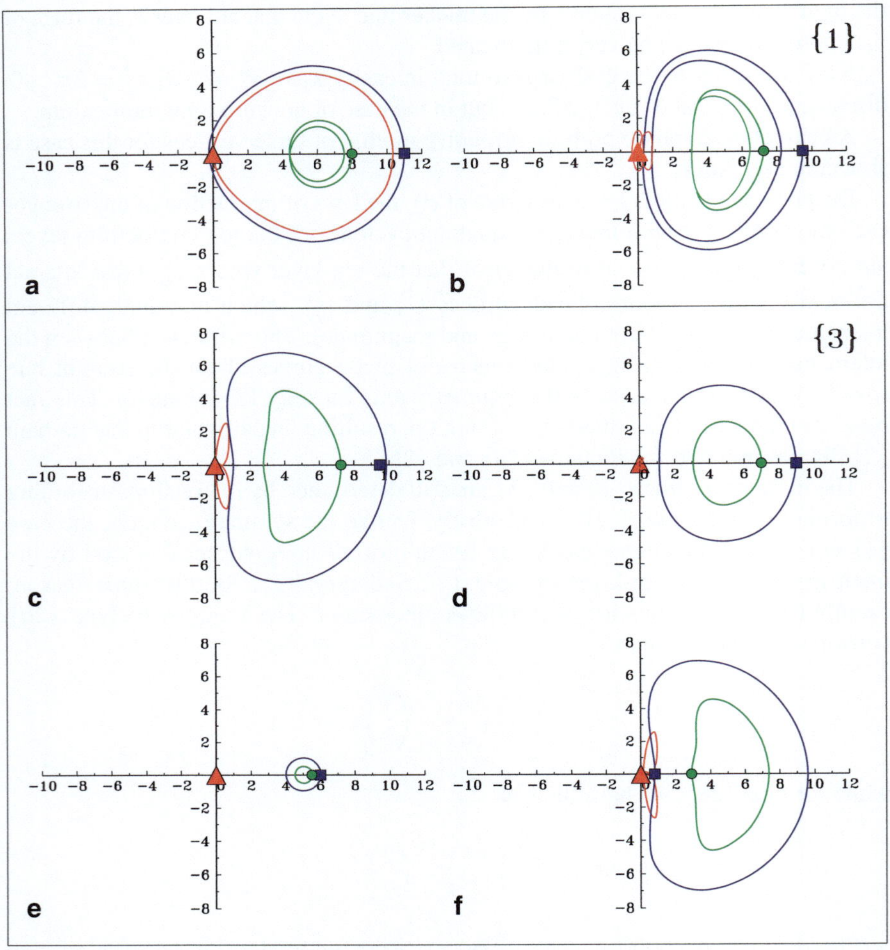

Fig. 2.80 Relative trajectories corresponding to motions of types{1} and {3} (shown in the figure): (**a**) $X_0 = 16$ (corresponds to Fig. 2.82a); (**b**) $X_0 = 14.481$, $t = (1.2047; -10.7867; 12.5820)$, (**c**) $X_0 = 14.479$, $t = (1.2037; -10.7822; 12.5785)$, (**d**) $X_0 = 14$ (corresponds to Fig. 2.82b), (**e**) $X_0 = 11$, $t = (0.06; -4.32; 7.26)$, (**f**) $X_0 = 5.792$, $t = (1.0624; -0.0753; 2.0129)$. Markers give the initial positions of vortices

and $\binom{2}{2}$ with intensities 2κ and $-\kappa$, respectively. Elliptic points from domains $\{2a\}$ and $\{2b\}$ correspond to structures *triton*, the dispersion equations for which coincide with (2.78), but the motion velocities differ from (2.79), becoming

$$V = -\frac{\kappa\gamma}{4\pi}\left\{\left[\frac{1}{B} - \mathrm{K}_1(B)\right] + 2\left[\frac{1}{2R-B} - \mathrm{K}_1(2R-B)\right]\right\}$$
$$\equiv \frac{\kappa\gamma}{4\pi} F[R, B(R)]. \tag{2.92}$$

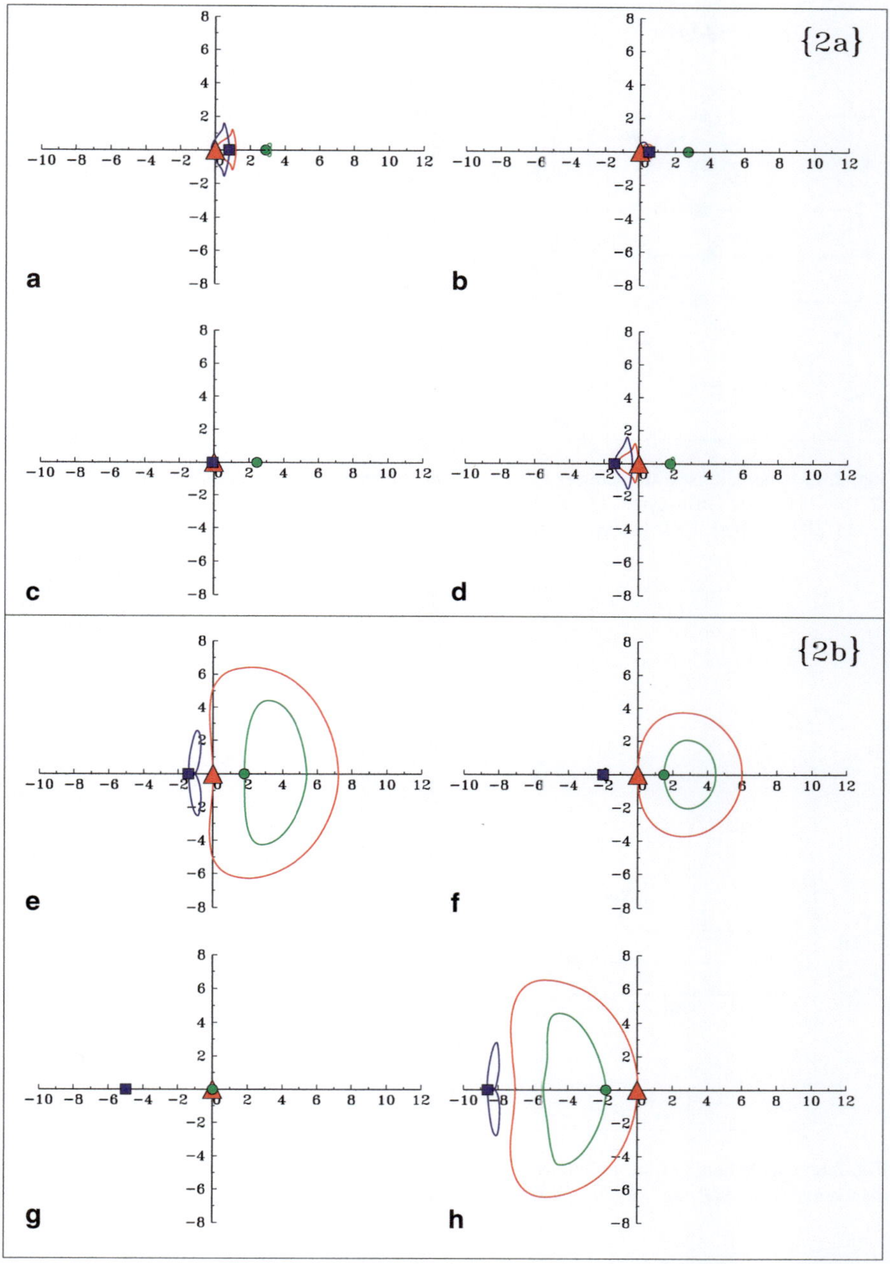

Fig. 2.81 The same as in Fig. 2.80, but for types $\{2a\}$ and $\{2b\}$: (**a**) $X_0 = 5.791$, $t = (1.0629; -0.0751; 2.0122)$, (**b**) $X_0 = 5.5$, (corresponds to Fig. 2.82c), (**c**) $X_0 = 4.912$, $t = (1.5533; -0.0009; 1.4476)$—triton, (**d**) $X_0 = 3.589$, $t = (2.4661; -0.2389; 0.7728)$, (**e**) $X_0 = 3.587$, $t = (2.4676; -0.2396; 0.7720)$, (**f**) $X_0 = 3$, (corresponds to Fig. 2.82d), (**g**) $X_0 = 0.0051$, $t = (5.9939; -2.9939; 0.0000)$—triton, (**h**) $X_0 = -3.613$, $t = (11.1188; -8.9021; 0.7833)$

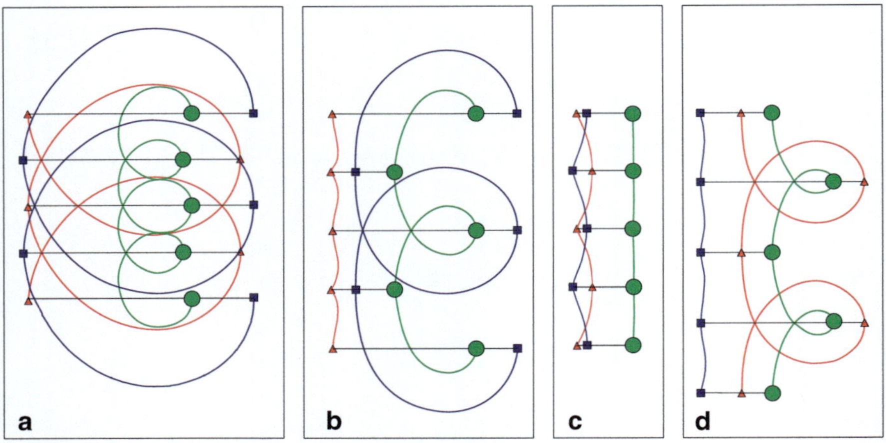

Fig. 2.82 Characteristic trajectories at $\gamma P = 5$ for different types of motions: (**a**) $X_0 = 16$, $t = (2.16; -14.52; 15.36)$—type $\{1\}$, (**b**) $X_0 = 14$, $t = (0.96; -9.72; 11.76)$—type $\{3\}$, (**c**) $X_0 = 5.5$, $t = (1.215; -0.030; 1.815)$—type $\{2a\}$, (**d**) $X_0 = 3$, $t = (2.94; -0.48; 0.54)$—type $\{2b\}$

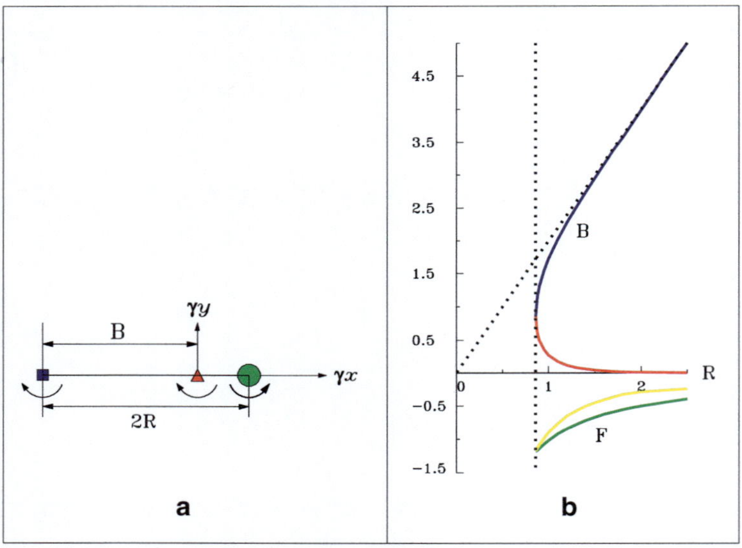

Fig. 2.83 (**a**) Scheme of the initial arrangement of vortices for collinear structure triton; (**b**) The dispersion curve $B(R)$ and function F (2.92), characterizing the translational velocity of triton

Figure 2.83, which is an analogue to Fig. 2.64, shows that the velocity in this case is a constant-sign function, attaining its maximum in terms of modulus at $B = R = R_{min}$. Here, as well as in Fig. 2.64, the full (dashed) lines correspond to the branches on which $B < R$ ($B > R$).

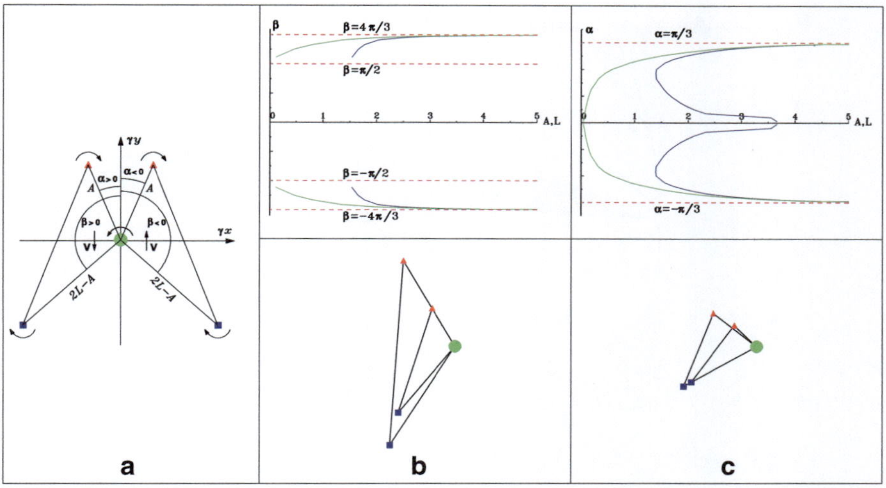

Fig. 2.84 The characteristics of an unstable triangular construction of vortices: (**a**) layout of vortices; (**b**) $\psi_1 = \pi/6$: dispersion curves ($R(\psi_2)$—the *full line* and $B(\psi_2)$—the *dashed line*) as solutions of Eqs. (2.93) and (2.94)—*top*; examples of stationary vortex configurations at $\psi_2 = 141°$ and $\psi_2 = 148°$—*bottom*; (**c**) $\psi_2 = 2\pi/3$: dispersion curves $R(\psi_1)$ (*full line*) and $B(\psi_1)$ (*dashed line*)—*top*; examples of stationary vortex configurations at $\psi_1 = 45°$ and $\psi_1 = 50°$—*bottom*

The hyperbolic singular point on the phase plane corresponds to a triangular (unstable) vortex structure, whose parameters are to be determined. The conditions for its uniform, translational, solid-body motion can be reduced to a system of two dispersion relationships:

$$\frac{\sin\psi_1}{B}\Big[1 - B\mathrm{K}_1(B)\Big] - \frac{\sin\psi_2}{C}\Big[1 + C\mathrm{K}_1(C)\Big] \\ - \frac{B\sin\psi_1 - C\sin\psi_2}{D^2}\Big[1 - D\mathrm{K}_1(D)\Big] = 0, \tag{2.93}$$

$$\frac{\cos\psi_1}{B}\Big[1 - B\mathrm{K}_1(B)\Big] + \frac{\cos\psi_2}{C}\Big[1 + C\mathrm{K}_1(C)\Big] = 0, \tag{2.94}$$

where $C = 2R - B$, $D = \sqrt{B^2 + (2R-B)^2 - 2B(2R-B)\cos(\psi_2 - \psi_1)}$.

Thus, (2.93)–(2.94) establish relationships between parameters B, R, ψ_1 and ψ_2 (see Fig. 2.84). Clearly, we always have $B \le R$. Asymptotically $B \sim R \to \infty$ at $\psi_2 = 2\psi_1 = 2\pi/3$. Again, from (2.94) it follows that the conditions $0 < \psi_1 < \pi/2 < \psi_2 < \pi$ are necessary for the existence of triangular stationary states.

Figure 2.84 illustrates the general form of a vortex structure, the dispersion curves, and gives examples of configurations obtained by solving the system of Eqs. (2.93) and (2.94).

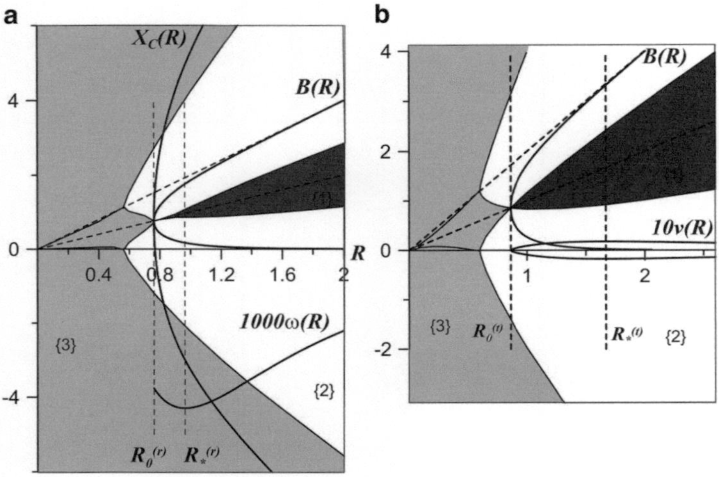

Fig. 2.85 Plots of functions: (**a**) $B(R)$, $\omega(R) \cdot 1000$ and $X_c(R)$ for an eccentric roundabout with $\mu = -2.5$, (**b**) $B(R)$ and $v(R) \cdot 10$ for a triton with $\mu = -2$. The *tilted dashed lines* $B = R$ and $B = 2R$ give the axis of symmetry of dispersion curves (2.73) and (2.78) and their asymptotics at $R \gg 1$. When $\mu = -2.5$, *vertical dashed lines* $R_0^{(r)}$ and $R_*^{(r)}$ correspond to the R value, characteristic to the static state and to the case of maximum angular velocity of the roundabout, correspondingly. When $\mu = -2$, the line $R_0^{(t)}$ means the same for the static state, and the line $R_*^{(t)}$ - for the case of maximum translation velocity of the triton motion. The notations {1}, {2}, and {3} for the domains marked by colors (*dark gray*, *white*, and *light gray*, respectively) have the same meaning as in Figs. 2.57a and 2.60a. Panel (**a**) adapted from [850] (Fig. 5) by permission of Cambridge University Press

In the analysis of the possible types of motion of triangular vortex structures, it can be of use to construct diagrams of possible types of their interaction in terms of distances between vortices at collinear arrangement in the initial moment. Examples of such diagrams are given in Fig. 2.85 for the cases of $\mu = -2.5$ and $\mu = -2$. Their analysis provides exhausting knowledge about the character of motion of vortex structures. In the following section, we will repeatedly address this figure.

2.3.3.4 Analysis of Weakly Perturbed Tripolar Collinear States

Let us return to the discussion of stable stationary states of the type of *eccentric roundabout* and *triton*. Both those states are partial solutions for type {2} motions, and they take place when additional conditions (2.73) and (2.78) for distances between vortices are satisfied. The appropriate curves on the plane of variables (R, B) are given, in particular, in Fig. 2.85.

At final deviations from stationary states, we have the trajectories given in Figs. 2.69b, 2.72b for the case $\mu = -2.5$ and 2.70 b, 2.71c–h for the case $\mu = -2$. Here we consider the case of small initial deviations from those steady states.

The elliptic type of the fixed points corresponding to the solutions on phase portraits considered here 2.57a and 2.60a suggests their stability with respect to small perturbations.

First, let us consider the case of eccentric roundabout. In the coordinate system rotating with the angular velocity (2.81), all three vortices will be fixed. Now let us determine the character of *perturbed* motions of vortices in the neighborhood of this stationary position.

Let us write an expression for the Hamiltonian (1.81) (where we now assume $A_1 = 1$, $A_2 = 2$, $h_1 = h_2 = 1/2$) in a coordinate system rotating with the angular velocity (2.81)

$$\begin{aligned}\mathscr{H} = -\frac{1}{8\pi}\Big\{\mu\Big[F_1\left(\sqrt{(x_2^1-x_1^1)^2+(y_2^1-y_1^1)^2}\right)\\ +F_1\left(\sqrt{(x_2^2-x_1^1)^2+(y_2^2-y_1^1)^2}\right)\Big]+F_2\left(\sqrt{(x_2^2-x_2^1)^2+(y_2^2-y_2^1)^2}\right)\\ -8\pi\frac{\omega}{4}\Big[(x_2^1)^2+(y_2^1)^2+\mu\left((x_1^1)^2+(y_1^1)^2\right)+(x_2^2)^2+(y_2^2)^2\Big]\Big\}, \quad (2.95)\end{aligned}$$

with notations

$$F_1(r) = \ln(r) + \mathrm{K}_0(r), \qquad F_2(r) = \ln(r) - \mathrm{K}_0(r). \qquad (2.96)$$

Note that the square bracket at ω in (2.95) is the total momentum, i.e., it is constant. Let us consider small perturbations relative to stationary configurations satisfying the dispersion Eq. (2.73). We represent the coordinates of vortices as

$$\begin{aligned}&x_2^1 = -B+\varepsilon_2^1, \quad y_2^1 = \delta_2^1,\\ &x_1^1 = \varepsilon_1^1, \quad y_1^1 = \delta_1^1, \qquad (2.97)\\ &x_2^2 = 2R-B+\varepsilon_2^2, \quad y_2^2 = \delta_2^2,\end{aligned}$$

where perturbations $\varepsilon_j^\alpha, \delta_j^\alpha \ll 1$ are chosen such that the total momentum (which is equal to zero in the coordinate system with the origin in the vorticity center) remained unchanged, i.e., the perturbed parts of the components of momentum satisfy the conditions

$$\Delta P_x = \frac{1}{2}\left[\varepsilon_2^1+\mu\varepsilon_1^1+\varepsilon_2^2\right] = 0, \quad \Delta P_y = \frac{1}{2}\left[\delta_2^1+\mu\delta_1^1+\delta_2^2\right] = 0. \qquad (2.98)$$

In this case

$$\Delta M = \frac{1}{2}\left[(\varepsilon_2^1)^2+(\delta_2^1)^2+\mu\left((\varepsilon_1^1)^2+(\delta_1^1)^2\right)+(\varepsilon_2^2)^2+(\delta_2^2)^2\right] = C_1 \qquad (2.99)$$

and

$$\begin{aligned}\Delta L &= 4Q\Delta M \\ &= \left\{\mu\left[(X_1)^2+(Y_1)^2+(X_3)^2+(Y_3)^2\right]+(X_2)^2+(Y_2)^2\right\} = C_2, \end{aligned} \tag{2.100}$$

where C_1, C_2 are constants, and the following notations are introduced for the projections of relative displacements of vortices

$$\begin{aligned} X_1 &= \varepsilon_1^1 - \varepsilon_2^1, \quad Y_1 = \delta_1^1 - \delta_2^1, \\ X_2 &= \varepsilon_2^2 - \varepsilon_2^1, \quad Y_2 = \delta_2^2 - \delta_2^1, \\ X_3 &= \varepsilon_2^2 - \varepsilon_1^1, \quad Y_3 = \delta_2^2 - \delta_1^1. \end{aligned} \tag{2.101}$$

Expanding the expression for Hamiltonian (2.95) to the second order in the coordinates of relative perturbed motion, we obtain

$$\begin{aligned}\mathscr{H} = -\frac{1}{8\pi}\cdots-\frac{1}{16\pi}\Big\{&\left[\mu F_{1x_2^1x_2^1}(B)+F_{2x_2^1x_2^1}(2R)\right](\varepsilon_2^1)^2 \\ &+\left[\mu F_{1x_2^2x_2^2}(2R-B)+F_{2x_2^2x_2^2}(2R)\right](\varepsilon_2^2)^2+\mu\left[F_{1x_1^1x_1^1}(B)+F_{1x_1^1x_1^1}(2R-B)\right](\varepsilon_1^1)^2 \\ &+2F_{2x_2^1x_2^2}(2R)\varepsilon_2^1\varepsilon_2^2+2\mu F_{1x_2^1x_1^1}(B)\varepsilon_2^1\varepsilon_1^1+2\mu F_{1x_2^2x_1^1}(2R-B)\varepsilon_2^2\varepsilon_1^1 \\ &+\left[\mu F_{1y_2^1y_2^1}(B)+F_{2y_2^1y_2^1}(2R)\right](\delta_2^1)^2+\left[\mu F_{1y_2^2y_2^2}(2R-B)+F_{2y_2^2y_2^2}(2R)\right](\delta_2^2)^2 \\ &+\mu\left[F_{1y_1^1y_1^1}(B)+F_{1y_1^1y_1^1}(2R-B)\right](\delta_1^1)^2+2F_{2y_2^1y_2^2}(2R)\delta_2^1\delta_2^2 \\ &+2\mu F_{1y_2^1y_1^1}(B)\delta_2^1\delta_1^1+2\mu F_{1y_2^2y_1^1}(2R-B)\delta_2^2\delta_1^1\Big\}. \end{aligned} \tag{2.102}$$

The subscripts x_j^α and y_j^α denote derivatives with respect to this variables. We neglect both constant zero-order terms (including the term at ω), and all identically zero terms, in particular, first-order terms (because of the stationary character of unperturbed configuration), and terms with mixed derivatives with respect to x_j^α and y_j^α that vanish at $y_j^\alpha = 0$.

It can be readily shown that the derivatives satisfy the relationships:

$$\begin{aligned} F_{nz_1^1z_1^1}(Z) &= F_{nz_2^2z_2^2}(Z) = F_{nz_2^1z_2^1}(Z) \\ &= -F_{nz_1^1z_2^1}(Z) = -F_{nz_1^1z_2^2}(Z) = -F_{nz_2^1z_2^2}(Z) \\ &= -\frac{1}{Z^2} + \mathrm{K}_0(Z) + \frac{\mathrm{K}_1(Z)}{Z} \equiv \Phi(Z), \end{aligned} \tag{2.103}$$

where $n=(1,2)$, z_j^i denote either x_j^i or y_j^i, and Z denote B, $2R$, or $2R-B$.

Now, considering (2.101) and (2.103), the quadratic part in the expansion of the Hamiltonian can be written as a function of the distance between perturbed positions of vortices

$$\begin{aligned} \mathscr{H}_2(X_1,Y_1;X_2,Y_2;X_3,Y_3) = \frac{1}{2}\Big\{ & a\Big[(X_1)^2+(Y_1)^2\Big] \\ & +b\Big[(X_2)^2+(Y_2)^2\Big] + c\Big[(X_3)^2+(Y_3)^2\Big]\Big\}, \end{aligned} \tag{2.104}$$

where

$$[a,b,c] = -\frac{1}{8\pi}\Big[\mu\Phi(B), \Phi(2R), \mu\Phi(2R-B)\Big]. \tag{2.105}$$

To reduce this to a dynamic system with one degree of freedom, we apply Poisson brackets in the form

$$\begin{aligned} \{f,g\} = 2\Bigg\{ & \frac{\partial f}{\partial x_2^1}\frac{\partial g}{\partial y_2^1} - \frac{\partial f}{\partial y_2^1}\frac{\partial g}{\partial x_2^1} + \frac{1}{\mu}\left[\frac{\partial f}{\partial x_1^1}\frac{\partial g}{\partial y_1^1} - \frac{\partial f}{\partial y_1^1}\frac{\partial g}{\partial x_1^1}\right] \\ & + \frac{\partial f}{\partial x_2^2}\frac{\partial g}{\partial y_2^2} - \frac{\partial f}{\partial y_2^2}\frac{\partial g}{\partial x_2^2}\Bigg\} \end{aligned} \tag{2.106}$$

to first integrals.

We will take into account the obvious consideration that a configuration of three vortices is always either a triangle or a segment, i.e., the projections of the distances between vortices are related by the expressions

$$X_1 = \xi X_2, \quad X_3 = (1-\xi)X_2, \qquad Y_1 = \zeta Y_2, \quad Y_3 = (1-\zeta)Y_2. \tag{2.107}$$

The integrals (1.79) yield the equality

$$\mu B X_1 + 2RX_2 + \mu(2R-B)X_3 = 0 \tag{2.108}$$

for all $\varepsilon_j^\alpha = 0$. However, since, with the condition (2.73) taken into account, we have

$$\{\mathscr{H}_2, \mu B X_1 + 2RX_2 + \mu(2R-B)X_3\} = 0, \tag{2.109}$$

then the equality (2.108) also holds at $\varepsilon_j^{\alpha} \neq 0$, implying that

$$\xi = \frac{2R(1+\mu) - \mu B}{2\mu(R-B)}. \tag{2.110}$$

A similar expression for ζ can be derived from the trivial expression for the Poisson bracket for the quadratic part in the development of the Hamiltonian with an invariant $Y_1 - \zeta Y_2$:

$$\{\mathscr{H}_2, Y_1 - \zeta Y_2\} = \Big[-\zeta\Big(a\xi + 2b + c(1-\xi)\Big) + \Big(a\xi + \mu b - c(1-\xi)\Big)\Big] X_2 = 0, \tag{2.111}$$

hence we have

$$\zeta = \frac{(1+\mu)a\xi + \mu b - c(1-\xi)}{\mu\Big(a\xi + 2b + c(1-\xi)\Big)}. \tag{2.112}$$

From expressions (2.110) and (2.112), it follows that ξ and ζ can be considered constant in the context of development of the Hamiltonian to the second order. Now, using (2.100), we introduce two variables e_1 and e_2, which constitute a canonical pair and allow the action–angle variables to be determined in a natural way:

$$\frac{\Delta L}{\mu+2} = (e_1)^2 + (e_2)^2, \tag{2.113}$$

where

$$(e_1)^2 = \frac{\mu(X_1)^2 + (X_2)^2 + \mu(X_3)^2}{\mu+2},$$
$$(e_2)^2 = \frac{\mu(Y_1)^2 + (Y_2)^2 + \mu(Y_3)^2}{\mu+2}, \tag{2.114}$$

and the factor $1/(\mu+2)$ is introduced to make the Eq. (2.113) an equality of positive values, since $L \leq 0$ at $\mu \leq -2$. Now, using (2.107), we express the original variables and the quadratic part of the Hamiltonian in terms of variables (2.114):

$$\left(\frac{X_1}{\xi}\right)^2 = \left(\frac{X_3}{1-\xi}\right)^2 = \left(X_2\right)^2 = \frac{\mu+2}{\mu\xi^2 + 1 + \mu(1-\xi)^2}(e_1)^2,$$
$$\left(\frac{Y_1}{\zeta}\right)^2 = \left(\frac{Y_3}{1-\zeta}\right)^2 = \left(Y_2\right)^2 = \frac{\mu+2}{\mu\zeta^2 + 1 + \mu(1-\zeta)^2}(e_2)^2, \tag{2.115}$$

and

$$\mathscr{H}_2(e_1,e_2) = \frac{\mu+2}{2}\left[\frac{a\xi^2+b+c(1-\xi)^2}{\mu\xi^2+1+\mu(1-\xi)^2}(e_1)^2 + \frac{a\zeta^2+b+c(1-\zeta)^2}{\mu\zeta^2+1+\mu(1-\zeta)^2}(e_2)^2\right]. \tag{2.116}$$

The dynamic equations in the new variables become

$$\begin{aligned} \dot{e}_1 &= \{\mathscr{H}_2, e_2\} = \tilde{\omega} e_2, \\ \dot{e}_2 &= \{\mathscr{H}_2, e_1\} = -\tilde{\omega} e_1, \end{aligned} \tag{2.117}$$

where

$$\tilde{\omega} = \frac{a\xi\zeta+b+c(1-\xi)(1-\zeta)}{\sqrt{\left(\mu\zeta^2+1+\mu(1-\zeta)^2\right)\left(\mu\xi^2+1+\mu(1-\xi)^2\right)}}. \tag{2.118}$$

Now, reducing the system of Eq. (2.117) to a second-order equation

$$\ddot{e}_1 = -\tilde{\omega}^2 e_1,$$

we see that $\tilde{\omega}$ is the frequency of the perturbed motion.

Thus, the solution of the system of Eq. (2.117) describes periodic oscillations with a frequency of (2.118) along elliptic orbits with the principal semiaxes ratio of

$$\lambda = \frac{(e_1)_{max}}{(e_2)_{max}} = \sqrt{\frac{\mu\xi^2+1+\mu(1-\xi)^2}{\mu\zeta^2+1+\mu(1-\zeta)^2}}. \tag{2.119}$$

From (2.115) it follows that in the differential relative coordinates, we have three different ellipses with the appropriate ratios:

$$\frac{(X_1)_{max}}{(Y_1)_{max}} = \frac{\xi}{\zeta}, \quad \frac{(X_2)_{max}}{(Y_2)_{max}} = 1, \quad \frac{(X_3)_{max}}{(Y_3)_{max}} = \frac{1-\xi}{1-\zeta}. \tag{2.120}$$

The shape of curves in the coordinates of relative motion x_j^α and y_j^α can be obtained by using (2.101) and (2.115):

$$\begin{aligned} \varepsilon_2^1 &= -\frac{\mu X_1+X_2}{\mu+2}, \quad \varepsilon_2^2 = \frac{-\mu X_1+(1+\mu)X_2}{\mu+2}, \quad \varepsilon_1^1 = \frac{2X_1-X_2}{\mu+2}, \\ \delta_2^1 &= -\frac{\mu Y_1+Y_2}{\mu+2}, \quad \delta_2^2 = \frac{-\mu Y_1+(1+\mu)Y_2}{\mu+2}, \quad \delta_1^1 = \frac{2Y_1-Y_2}{\mu+2}. \end{aligned} \tag{2.121}$$

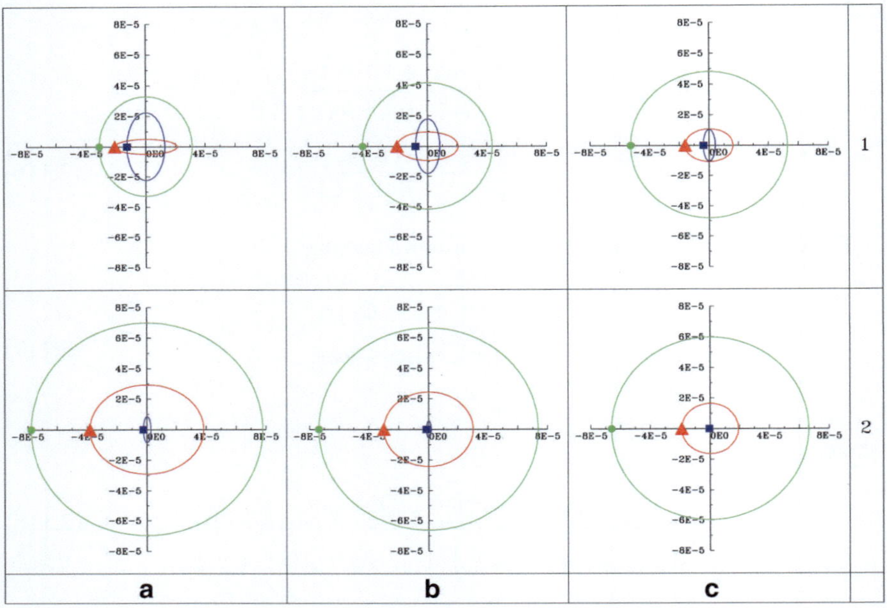

Fig. 2.86 Trajectories of perturbed motions of point vortices in variables $\varepsilon_j^\alpha(t)$ and $\delta_j^\alpha(t)$ (the centers of ellipses are shifted to the origin of coordinates) with the initial conditions (2.123) at $\varepsilon_2^1 = -\varepsilon_2^2 = 2.5\cdot 10^{-5}$, $\varepsilon_1^1 = \delta_1^1 = \delta_2^1 = \delta_2^2 = 0$: (**1**) $R_0 = 0.87$, (**a**) $\mu = -2.1$, $B_0 = 0.2724$, (**b**) $\mu = -2.5$, $B_0 = 0.1221$, (**c**) $\mu = -3.5$, $B_0 = 0.0780$, (2) $R_0 = 1.27$, (**a**) $\mu = -2.1$, $B_0 = 0.5021$; (**b**) $\mu = -2.5$, $B_0 = 0.0521$, (**c**) $\mu = -3.5$, $B_0 = 0.0271$. The markers on the initial collinear configuration show the correspondence between trajectories and vortices in accordance with Fig. 2.62. Adapted from [850] (Fig. 6) by permission of Cambridge University Press

So, we obtain that, in the assumed approximation, the perturbed relative motions of vortices of the eccentric roundabout take place along elliptic orbits with semiaxes ratios

$$\frac{\left(\varepsilon_2^1\right)_{max}}{\left(\delta_2^1\right)_{max}} = \frac{\mu\xi+1}{\mu\zeta+1}, \quad \frac{\left(\varepsilon_2^2\right)_{max}}{\left(\delta_2^2\right)_{max}} = \frac{\mu(1-\xi)+1}{\mu(1-\zeta)+1}, \quad \frac{\left(\varepsilon_1^1\right)_{max}}{\left(\delta_1^1\right)_{max}} = \frac{2\xi-1}{2\zeta-1} \tag{2.122}$$

and a frequency of (2.118).

Figure 2.86 gives trajectories of relative motion in coordinates $\varepsilon_j^\alpha(t)$ and $\delta_j^\alpha(t)$ under the initial conditions

$$\begin{aligned}
(x_2^1, y_2^1) &= (-B_0 + \varepsilon_2^1, \delta_2^1),\\
(x_1^1, y_1^1) &= (\varepsilon_1^1, \delta_1^1),\\
(x_2^2, y_2^2) &= (2R_0 - B_0 + \varepsilon_2^2, \delta_2^2),
\end{aligned} \tag{2.123}$$

where B_0 and R_0 satisfy the Eq. (2.73), and the lower branch of curve in Fig. 2.85a is taken for all six cases. The values of ε_j^α, δ_j^α are given in underline to Fig. 2.86.

It is obvious that in virtue of (2.121), the perturbation of the stationary position of at least one vortex induces appropriate deviations of the other two vortices, which is the idea of this figure. To facilitate the comparison of corresponding ellipses, their centers in the figure are placed in the same point $(0,0)$. As shown in the figure, with the given external parameters, the trajectory of vortex $\binom{1}{2}$ is nearly circular and has the maximal amplitude, while other vortices have the following properties: $(\varepsilon_1^1)_{max} > (\delta_1^1)_{max}$ and $(\varepsilon_2^2)_{max} < (\delta_2^2)_{max}$. With an increase in the absolute value of top-layer vortex intensity, the oscillation amplitudes of vortices $\binom{1}{1}$ and $\binom{2}{2}$ decrease. Note the interesting situation shown in the panel 2.86.1c, where $(\delta_1^1)_{max} = (\delta_2^2)_{max}$.

In the case of triton from (2.121) and (2.122), we obtain

$$\xi = \zeta = \frac{1}{2}, \tag{2.124}$$

i.e., all three ellipses are similar; therefore, can take X_2, Y_2 as canonica variables and write the Hamiltonian as

$$\mathscr{H}_2(X_2,Y_2) = \frac{1}{2}\left(\frac{a}{4} + b + \frac{c}{4}\right)\left(X_2{}^2 + Y_2{}^2\right), \tag{2.125}$$

where, in accordance with (2.105),

$$[a,b,c] = \frac{1}{4\pi}\left[\Phi(R), \frac{1}{2}\Phi(2R), \Phi(2R-B)\right]. \tag{2.126}$$

Now, we readily obtain the frequency of the quasi-elliptic motions of the vortices constituting the perturbed triton:

$$\tilde{\omega}(R,B,-2) = \frac{a}{4} + b + \frac{c}{4}. \tag{2.127}$$

The same expression can be derived taking the limit of Eq. (2.118) for $\mu \to -2$, if we take into account that at the same time, $\xi, \zeta \to 1/2$.

It is important to note that the frequencies (2.118) and (2.127), generally speaking, have nonzero limits $\tilde{\omega}_e$ at zero amplitudes ($\varepsilon_j^i = \delta_j^i = 0$). The dependencies of the limiting frequencies (eigenfrequencies) of the perturbed motion on the size of configuration R are illustrated by Fig. 2.87. The absolute values of frequencies increase with increasing R and $|\mu|$ at all $\mu < -1$; the direction of rotation changes its sign at $\mu = -2$. Clearly, the conditions (2.73) or (2.78) hold on all curves. The equality $\tilde{\omega}_e(R,B,\mu) = 0$ will be attained only in degenerate *static* states at $B = R_0^{(t)}$ or $B = R_0^{(r)}$. In this case, for a roundabout, according to (2.68), we have $X_c = Y_c = 0$.

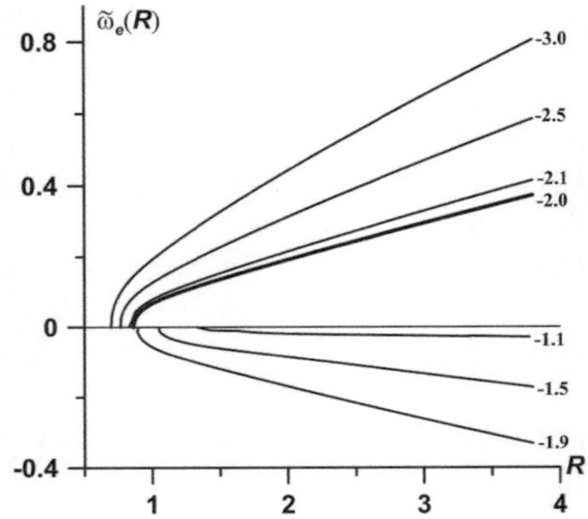

Fig. 2.87 Eigenfrequencies of unperturbed motions $\tilde{\omega}_e$ at different values of parameter μ as functions of the initial size of configuration R. The *thick line* shows the case of $\mu = -2$ (triton). Adapted from [850] (Fig. 7) by permission of Cambridge University Press

2.3.3.5 Regular Advection Near Stationary Configurations

We will focus now on the advection of fluid particles [27, 28] in a velocity field induced by the stationary vortex structures described above. In the stationary state, the trajectory of motion of material particles coincide with the streamlines of horizontal motion, and the simplest way to study the motion of particles is to add zero-intensity point vortices to the system. By analogy to the case considered above, we will pay special attention to studying the behavior of rotation frequency Ω of liquid particles in a vortical flow, an important characteristic of the chaotic properties of advective processes [380, 440, 443].

We will start our analysis from the case $B = R = R_0^{(r)}$, i.e., a symmetrical roundabout.

Figure 2.88 shows the phase portraits for the top and bottom layers at $\mu = -2.5$. The lateral cyclonic formations in the top layer are obviously induced by peripheral bottom-layer vortices. The separatrix in the top layer has a heteroclinic structure. A heteroclinic separatrix, embracing the point vortices, forms in the bottom layer. The dashed streamline in the bottom layer corresponds to a minimum of the turnover frequency. Dependences on the initial position taken on the axis $y = 0$ for turnover frequency Ω and azimuthal velocity v of fluid particles are also given here. Because of symmetry, only the right part of the figure is given (for positive abscissa). Within the separatrix, in the vicinity of point vortices, we have indefinitely growing turnover frequency and azimuthal velocity, which corresponds to the singularity in point vortices. The situation beyond the separatrix is also of great interest. The turnover frequency has a maximum near the separatrix and a minimum at a considerable distance from it (the inset in the bottom figure), while the velocity of particles has only a minimum, which is somewhat closer to the center than the frequency minimum. The behavior of the velocity and turnover frequency with distance from the center corresponds to a solid-body rotation of the roundabout (the fluid is immovable at infinity in the absolute coordinate system).

Figure 2.89 gives both layers' phase portraits for the ($R = 0.8200$, $B = 0.3712$) situation, which appreciably differs from symmetry. The left loop of the outer

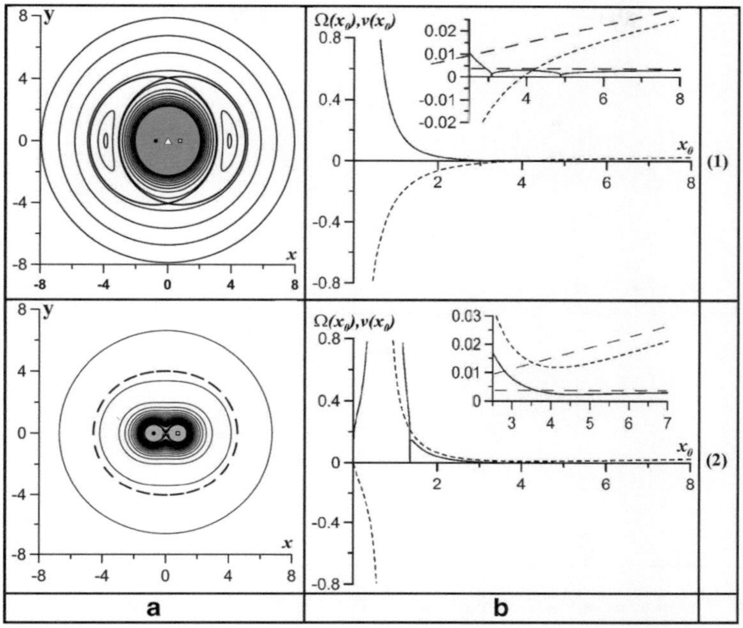

Fig. 2.88 (**a**) Isolines of flow function in a coordinate system rotating with an angular velocity of (2.77) at $\mu = -2.5$ and $B = R = R_0^{(r)} = 0.7609$ in the top (**1**) and bottom (**2**) layers. The separatrices are shown by thick lines. The isolines are drawn with a velocity step of 0.005 for the top layer and 0.01 for the bottom layer. The cross marks the rotation center, coinciding with the central vortex of the top layer. The closed *dashed line* is the trajectory of the bottom-layer particle with minimal turnover frequency. (**b**) Dependencies of turnover frequency $\Omega(x_0)$ (*solid lines*) and the azimuthal velocity $v(x_0)$ (*short-dash lines*) of fluid particles originating from the x-axis, as functions of the distance x_0 from the origin of coordinates. Insets with extended vertical coordinates demonstrate the behavior of curves in critical domains. The *long-dash straight lines* show asymptotes corresponding to solid-body rotation with the angular velocity of (2.77). Adapted from [444] (Fig. 2) by permission of Cambridge University Press

separatrix in the top layer has increased due to the effect of rotation of the vortex system as a whole, and the turnover frequency of particles along the trajectory in this loop has dramatically decreased. In the bottom layer, the separatrix embracing the point vortices is now asymmetric (its left loop has decreased), and an additional external asymmetric heteroclinic separatrix, containing the internal separatrix in its large right loop, has formed. In the external domain, the fluid is still asymptotically immovable.

The patterns of streamlines for the case $R = 1.5$ at small distance between vortices $\binom{1}{1}$ and $\binom{1}{2}$ are given in Fig. 2.90 where $B = 0.0246$. In this case, domains of closed motion of fluid particles (induced vortices), separated from the separatrix by a domain of another closed current, form in the bottom layer above and below the main separatrix. The vorticity center now lies at a greater distance from the vortex structure and outside the internal separatrix. Now the external separatrix

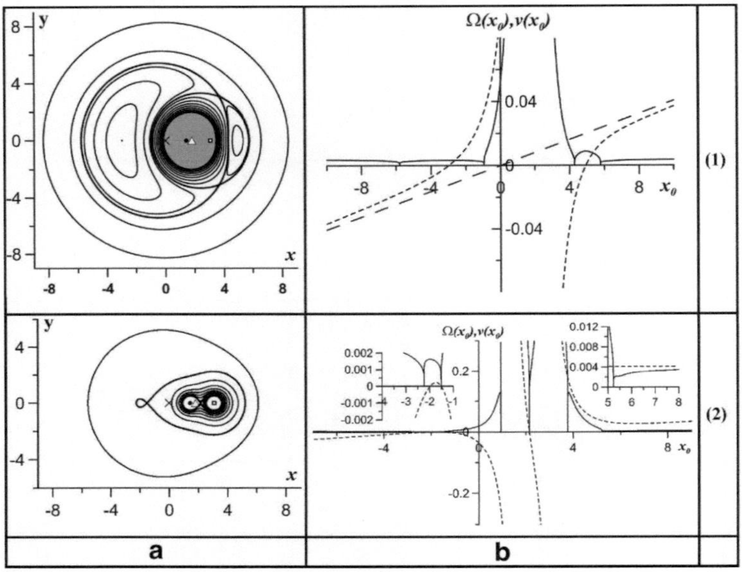

Fig. 2.89 The same as in Fig. 2.88, but with $B = 0.3712$, $R = 0.8200$. The isolines are drawn with a step of 0.010 for the top layer and 0.025 for the bottom layer. Adapted from [444] (Fig. 3) by permission of Cambridge University Press

can be interpreted as a homoclinic structure with connected whiskers, and in phase portraits, the vicinity of the roundabout can be isolated in both layers, moving along separatrix whiskers which encompass the rotation center. The situation in the top layer shows almost no qualitative changes.

In frequency relationships, the boundary of the closed flow can be clearly seen here in insets near the values $x_0 = -9$ and $x_0 = 11$. Note that a closed flow, separating the internal and external separatrices still persists in this case. This flow is also identified based on the frequency relationship, and, as can be clearly seen in the inset, these two flows are separated by the external separatrix.

Consider an important particular case corresponding to *reconnection* of a separatrix of induced vortices with the external separatrix of roundabout. This situation is shown in Fig. 2.91. The two separatrices have merged heres and the domain of the external closed flow has disappeared. In such situations, under a perturbation of a stationary configuration, it is very likely that a wide domain of chaotization will appear in the vicinity of such a merged separatrix [443, 849]. Unlike the situation in Fig. 2.90b, there is no domain of external closed flow for frequency relationships, which we do not give here. No significant changes are observed in the top layer either; therefore, we simply show the separatrices of the top layer, superimposed on the phase portrait of the bottom layer. The comparison of separatrix configurations gives an idea about the shape of the fluid column captured by the roundabout, and allows us to conclude that when R is large enough, this column is localized in the rectangular box drawn in Fig. 2.91.

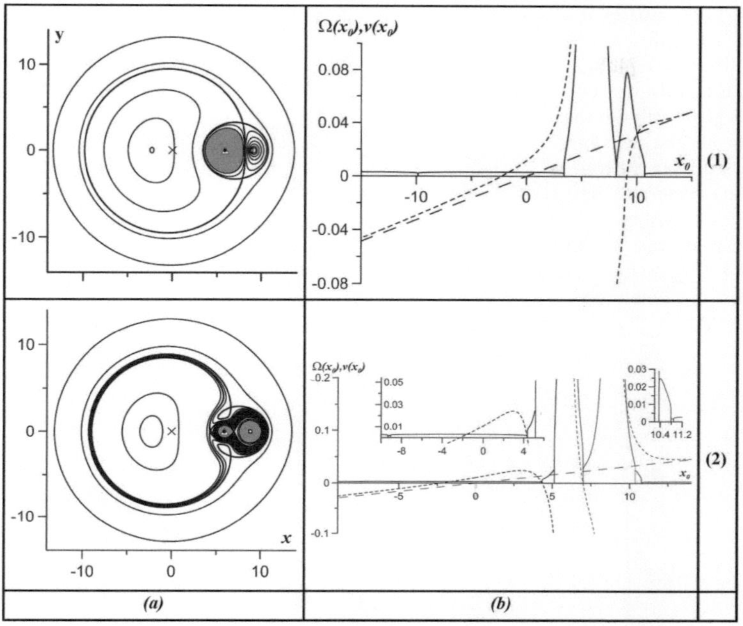

Fig. 2.90 The same as in Fig. 2.88, but with $B = 0.0246$, $R = 1.5000$. Isolines are drawn with a step of 0.001 for both the bottom and top layers. Panel (2) adapted from [444] (Fig. 4) by permission of Cambridge University Press

To analyze the effect of parameter μ on the character of phase portraits of an eccentric roundabout, we consider a series of dispersion curves at different μ (Fig. 2.92). They are qualitatively similar to one another with the limiting value $R_0^{(r)}$ increasing from zero at $\mu \to -\infty$ to infinity at $\mu = -1$, when $B = R$. Curves X_c are also plotted here, displaced by the amount of B for all named values μ except $\mu = -2$ (triton) when $|X_c| = \infty$.

Calculations show that the topological properties of phase portraits, shown in Figs. 2.88, 2.89, 2.90 and 2.91 for the case of $\mu = -2.5$, may also form at other values of parameter μ. However, because of the special role of the value $\mu = -2$, the crossing of this level is accompanied by changes in the phase picture which are illustrated in Fig. 2.93, by two examples with $\mu < -2$ and $\mu > -2$. Thus, when $\mu = -4$ (Fig. 2.93a), the structure of bottom-layer separatrices is the same as with $\mu = -2.5$ (Fig. 2.91), and the only difference is that now the rotation center is lacalized outside the left loop of the inner separatrix. However, when $\mu = -1.5$ (Fig. 2.93b), the rotation centerchanges its place to the right half plane, the motion of the vortex structure changes its direction, and the vortex atmosphere, surrounding the point vortices, appears within the loop composed of separatrix whiskers, embracing the rotation center.

The further analysis of bifurcations of phase portraits will be carried out for the case of triton ($\mu = -2$).

Fig. 2.91 The same as in Fig. 2.88, but with $B = 0.0119$, $R = 1.7418$. Streamlines with a step of 0.005 are given only for the bottom layer. The *gray lines* are separatrices of the top layer. The structure of the boxed domain will be described in the text. Adapted from [444] (Fig. 5) by permission of Cambridge University Press

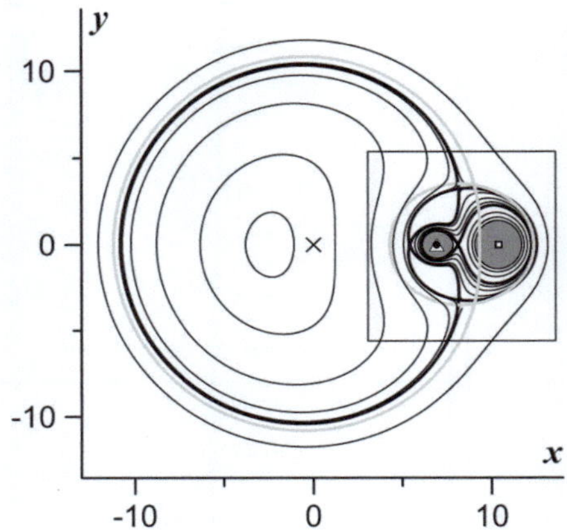

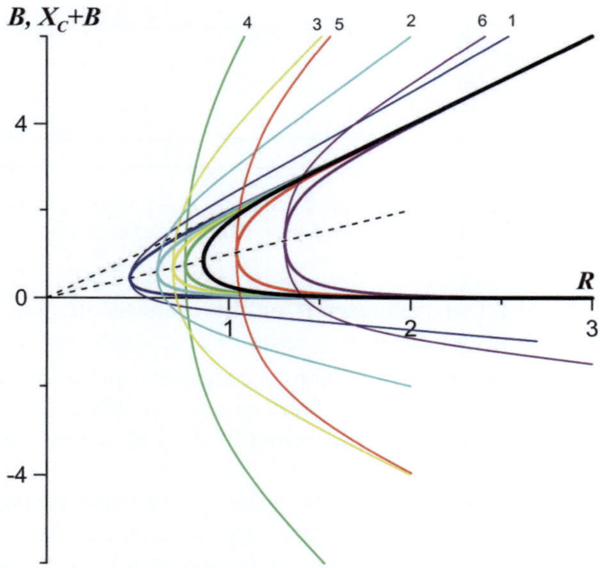

Fig. 2.92 Dispersion curves $B(R)$ (within the angle between the straight lines $B = 0$ and $B = 2R$) and dependences $[X_c+B](R)$ for an eccentric roundabout with: (**1**) $\mu = -7.5$ (*dark blue*); (**2**) $\mu = -4.0$ (*light blue*); (**3**) $\mu = -3.0$ (*yellow*); (**4**) $\mu = -2.5$ (*green*); (**5**) $\mu = -1.5$ (*red*); (**6**) $\mu = -1.2$ (*magenta*). The curve corresponding to the triton at $\mu = -2.0$ is marked by a *black line*. The inclined *dashed line* $B = R$ is the axis of symmetry of curves (2.73) and (2.78), while the line $B = 2R$ and the axis R are their asymptotics at $R \gg 1$. Adapted from [444] (Fig. 6) by permission of Cambridge University Press

As before, we begin with the limiting symmetric state $B = R = R_0^{(t)} = 0.8602$, which in this case, is static. Figure 2.94a gives phase portraits and 2.94a gives frequency characteristics of fluid particle motion. The latter, by virtue of symmetry, are given only for positive values of the initial positions along the axis x or y.

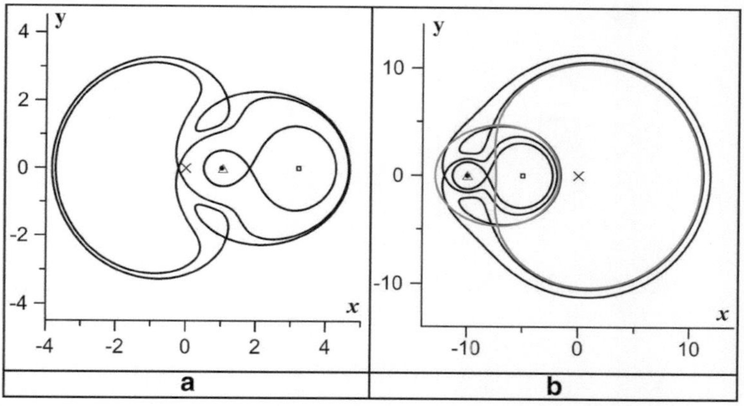

Fig. 2.93 Separatrices of stream-function field in the bottom layer in a coordinate system rotating with the angular velocity (2.77): (**a**) $R = 1.1$, $\mu = -4.0$ and (**b**) $R = 2.5$, $\mu = -1.5$. In (b), for comparison, the *grey line* is the top-layer separatrix. Adapted from [444] (Fig. 7) by permission of Cambridge University Press

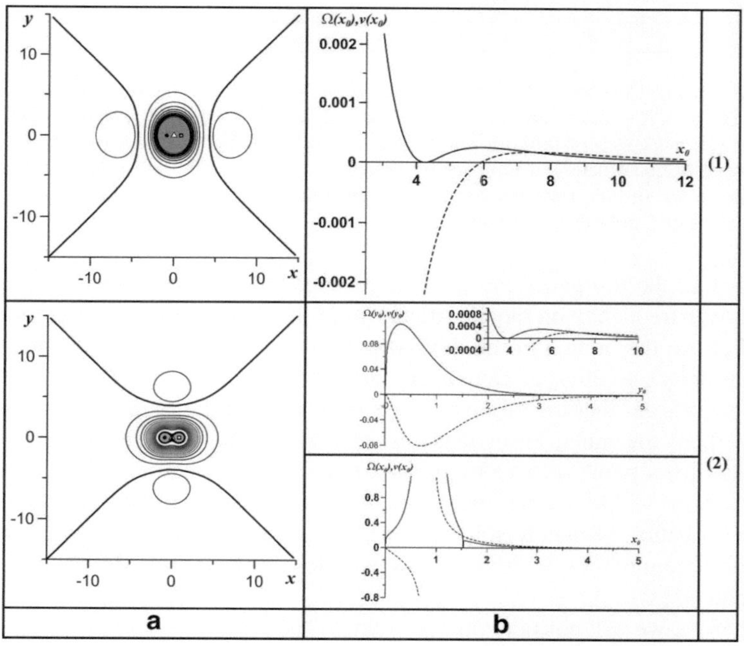

Fig. 2.94 (**a**) Isolines of stream-function in a fixed coordinate system with $B = R = R_0^{(r)} = 0.8602$ (static position) in the top **(1)** and bottom **(2)** layers. Separatrices are shown by *thick lines*. Isolines are drawn with a step of 0.005 for the top layer and 0.01 for the bottom. (**b**) Turnover frequencies Ω (*solid lines*) and asimuthal velocities v (*short-dash lines*) of fluid particles that have started from the x or y axes as functions of distance x_0 or y_0 to the origin of coordinates. Insets with extended vertical coordinates elucidate the behavior of curves in critical domains. Adapted from [444] (Fig. 8) by permission of Cambridge University Press

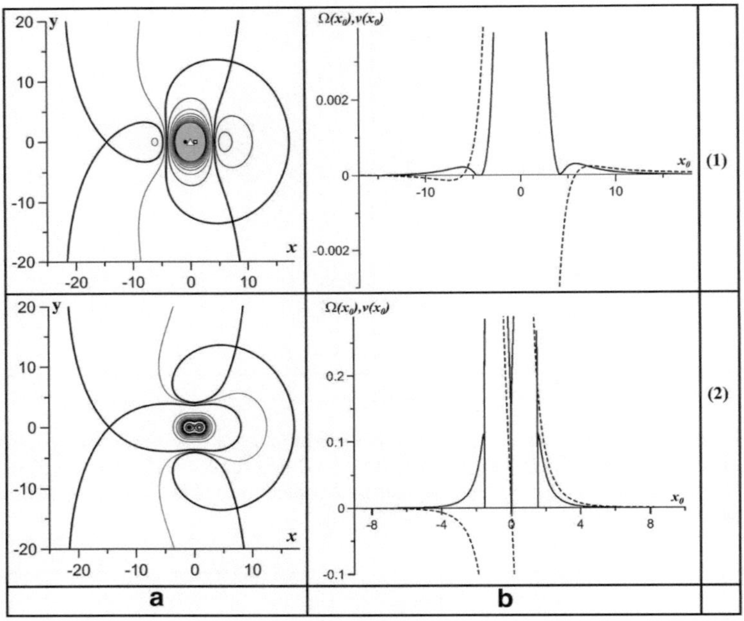

Fig. 2.95 (**a**) The same as in Fig. 2.94, but in the coordinate system moving translationally with velocity (2.78) at $R = 0.860188$, $B = 0.8565$. The isolines are drawn with a step of 0.002 for the top layer (**1**) and a step of 0.01 for the bottom layer (**2**). (**b**) Dependences (on the distance x_0 to the origin of coordinates) of turnover frequencies Ω (*solid lines*) and azimuthal velocities V (*short-dash lines*) of fluid particles that have started from the x-axis. Adapted from [444] (Fig. 9) by permission of Cambridge University Press

In particular, the top panel and the inset in Fig. 2.94.2b shows the dependence of the turnover frequency on the coordinate y_0 in the vicinity of the top elliptic point located above the branch of the quasi-separatrix for the bottom layer (this branch runs to infinity), where such frequencies are much lower than in the top layer.

It is significant that the turnover frequencies outside the separatrices embracing point vortices are much lower than inside these separatrices. In induced vortices (separated from point vortices by two separatrices), the frequencies are also much less, a fact that will play a significant role in the analysis of possible chaotic regimes after perturbation of steady solutions. In the same figure, dashed lines show the behavior of azimuthal velocities of fluid particles that have started from appropriate coordinate axes.

As before, we will consider pairs of values B and R referring to the lower branch of the dispersion curve. Here, $B \leq R$, the triton's velocity is negative, and the vortex structure in the figure moves downward. In the coordinate system moving with the triton, the slip flow is directed upward.

Figure 2.95 gives the case of $R = 0.860188$, i.e., nearly limiting one. Even with the deviation from the symmetrical position of vortices as small as that, considerable deformation takes place in the phase portrait. The separatrices of induced vortices

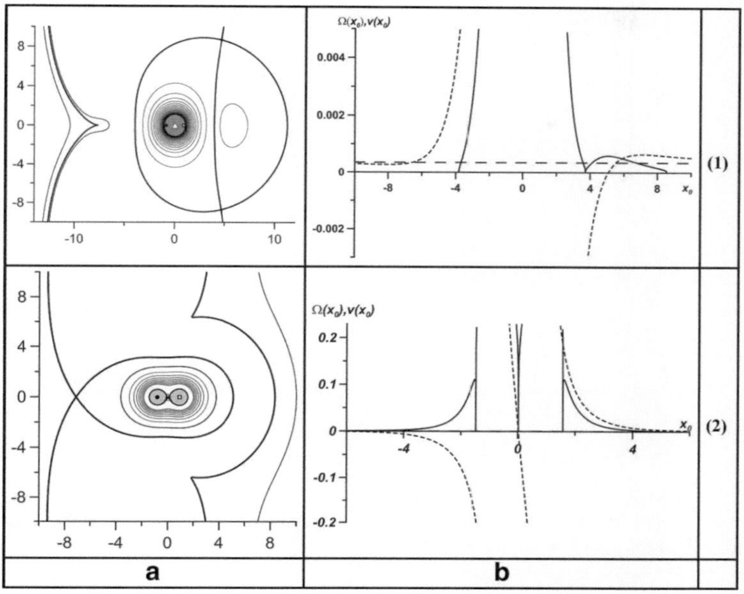

Fig. 2.96 (**a**) The same as in Fig. 2.94 at $\mu=-2$ and (**1**) $R=0.86023$, $B=0.8458$ (*top layer*) and (**2**) $R=0.86050$, $B=0.8234$ (*bottom layer*). The isolines are drawn with a step of 0.01 for both layers. (**b**) Turnover frequencies (*full lines*) and azimuthal velocities (*fine-dash lines*) of fluid particles, which were originally located on the x-axis as functions of the distance x_0 from the origin of coordinates. Panel (**a**) adapted from [444] (Fig. 10) by permission of Cambridge University Press

take on classical homoclinic form. In the top layer, the left induced vortex detaches from the main separatrix, while the two vortex domains on the right are joined by a common separatrix. Three isolated vortex domains form in the lower layer.

An interesting feature is the appearance of two narrow zones of flow-through currents in the bottom layer and one such zone in the top layer, where the separatrices of induced vortices closely approach the main separatrices. Though the velocities of fluid particles are not large here, they are still far in excess of the velocity of the triton itself. These zones can be associated with the jet flows generated by the triton.

The dependences of turnover frequencies for initial positions on the x-axis in the bottom layer are practically the same in this figure and in Fig. 2.94. The only difference is that the frequency vanishes on the external separatrix rather than in infinity, the left induced vortex has separated from the main separatrix, and the rotation frequency in it has decreased. The differences for the fluid particles originating from the y axis are also insignificant.

As the triton asymmetry increases further, the vortex domains not associated with point vortices degrade with increasing R and next disappear completely, first in the top and next in the bottom layer.

Figure 2.96a gives examples of phase portraits for the top and bottom layers for the values of R and B at which the appropriate vortex domains in the bottom layer

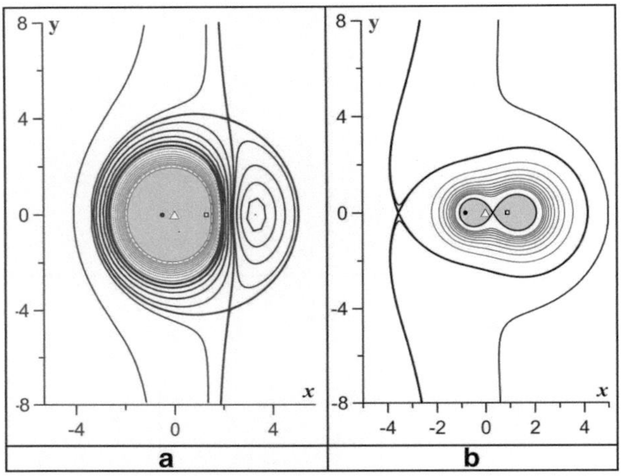

Fig. 2.97 Isolines of stream-function at $\mu = -2$ and $R = 0.90$, $B = 0.4923$: (**a**) in the top layer, velocity step 0.005, (**b**) in the bottom layer, velocity step 0.01. Adapted from [444] (Fig. 11) by permission of Cambridge University Press

and the left vortex in the top layer disappear. Naturally, the domains of jet flows are also not pronounced. The circulation ring in the top layer associated with the right point vortex in the bottom layer is considerably shifted right from the domain occupied by the triton.

As can be seen from Fig. 2.96b, the law of variations in the turnover frequency of bottom-layer particles has changed only slightly relative to the previous case, reflecting only the fact of disappearance of induced vortices. In the top layer, the right induced vortex has disappeared, while the left one has not changed significantly (Fig. 2.96b).

The subsequent increase in the asymmetry of triton because of the increase in parameter R in a wide interval does not lead to qualitative changes in phase portraits; however, as can be seen from Fig. 2.97, we have evolutionary changes in the picture. With this set of external parameters, the effect of vortex structure on the background flow is more significant in the left part in the top layer and in the right part in the bottom layer. The frequency relationships for this configuration will be analyzed in the following section.

As R grows further, the structure of the stream function field in the top layer shows practically no change, so we will consider a number of phase portraits only for the bottom layer (Fig. 2.98), demonstrating a series of new bifurcations.

In panels 2.98a and 2.98b, we essentially observe the process of revival of the separatrix that has disappeared before and the formation of new finite domains of motion of fluid particles, tending to a two-layer, practically vertical, vortex structure, consisting of an anticyclone in the top layer and the left-hand cyclone in the bottom layer.

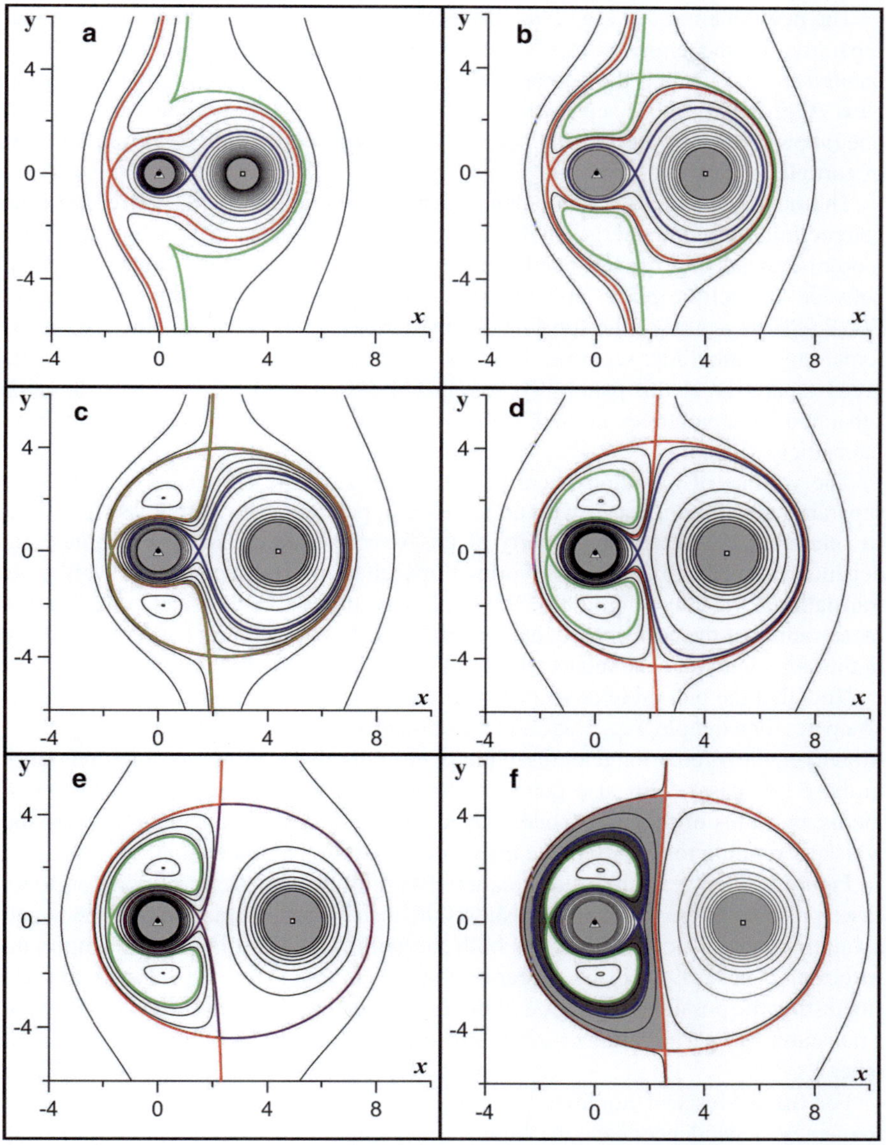

Fig. 2.98 Isolines of stream-function in the bottom layer at parameter values as follows: (**a**) $R = 1.546$, $B = 0.03523$, $V = -0.01750$, (**b**) $R = 2.0$, $B = 0.009255$, $V = -0.01699$, (**c**) $R = 2.1881$, $B = 0.005516$, $V = -0.01628$, (**d**) $R = 2.4$, $B = 0.003131$, $V = -0.01539$, (**e**) $R = 2.470$, $B = 0.002608$, $V = -0.01412$, (**f**) $R = 2.70$, $B = 0.001438$, $V = -0.01200$. Here, V is the triton velocity (2.79). Adapted from [444] (Fig. 12) by permission of Cambridge University Press

The new separatrix in Fig. 2.98b, to a considerable extent embraces the external separatrix of the entire vortex structure, while in Fig. 2.98c we observe their *reconnection*, at which they completely merge and acquire heteroclinic character. Next (Fig. 2.98d), those separatrices split in the left hyperbolic point and the first one (shown in green) becomes a boundary of a two-cell domain of closed motions of particles.

The next *reconnection* phenomenon can be seen in Fig. 2.98e. Here, the blue heteroclinic structure merges with the external (red) separatrix, while the right loop disappears and the left one becomes a homoclinic structure. Water exchange between the left, tripolar, and the right, monopolar, domains ceases. Next, in Fig. 2.98f, we again see the splitting of separatrices, after which the blue separatrix, remaining homoclinic, separates from the red one, while its whiskers close on the green separatrix. In this panel, two domains of closed regular flows in the domains separated by separatrices in the left loop are shown by grey shading with different intensities.

The volume of the atmosphere of the vortex structure associated with the red separatrix steadily increases with an increase in parameter R (from 2.98a to 2.98f). In general, a characteristic property of the baroclinic model is the nonmonotonic dependence of the size of triton vortex atmosphere on parameter R, as well as its translational velocity (Fig. 2.85). We think it is impossible to exactly evaluate the mean radius of the atmosphere, but calculations show its minimal size to lie nearly in the same interval of R values where the triton's velocity is maximal.

Note that the picture of advection near the triton and the roundabout are similar: compare, for example, Fig. 2.98c and the domain enclosed by the square in Fig. 2.91. Moreover, the bifurcation features of phase portraits also have appropriate analogues in those two cases. This also points to the nonmonotonic character of behavior of the mean radius of the atmosphere surrounding it. The main distinction is that the whiskers running to infinity in the triton are closed in the roundabout.

Figure 2.99 gives kinematic characteristics of fluid particle motion for some cases presented here. Figure 2.99a shows the distribution of rotation frequency and azimuthal velocity of particles for both the bottom and top layers. An important feature in panel 2.99.1a is the appearance of dips in frequency and velocity profiles, suggesting the possible realization of interesting features of chaotization [443] when a nonstationary perturbation of triton is introduced (analogous effects can be seen at other R).

For the analysis of nonlinear resonances, it is necessary to know the type of frequency dependence, i.e., the sign of the appropriate curvature [443]. In the case given in Fig. 2.99a, we have asymmetry for the initial values taken right and left from the inner separatrix. In this situation, analysis of reconnections should be carried out with the phase of nonstationary perturbation taken into account. We will also mention important facts of existence of frequency maximums in the domains between the inner and outer separatrices in the presented dependences, the facts that suggest that reconnections are possible in the presence of nonstationary perturbations [443].

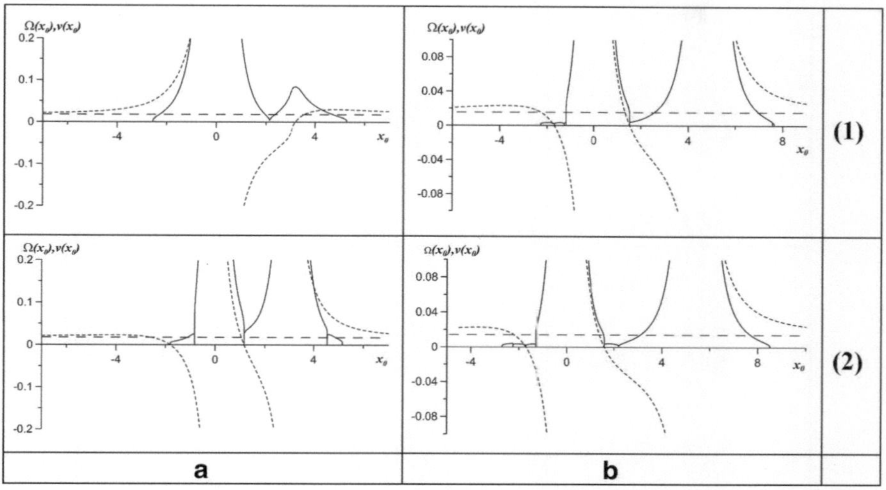

Fig. 2.99 Turnover frequencies and azimuthal velocities of fluid particles as functions of their initial position on the x-axis: (**a**) for the top (**1**) and bottom (**2**) layers for the phase portrait in Fig. 2.98a, b for the bottom layer and phase portrait in Fig. 2.98d—(**1**) and for phase portrait in Fig. 2.98f—(**2**). The *horizontal dashed line* corresponds to triton velocity (2.79)

To identify transport corridors in this situation, one should search for the possible appearance of simultaneous reconnections of separatrices of nonlinear resonances in adjacent domains of different flows. In particular, Fig. 2.99.1b, 2b show that such situation is very likely, since the maximal frequencies in adjacent through-flows have similar values: in [443], it is shown that reconnection of nonlinear resonances takes place exactly when the corresponding frequency of nonstationary perturbation is nearly maximal.

At further increase in R, no new bifurcations are found to appear; it is only in the limit at $R \to \infty$ that reconnection of the green and blue separatrices takes place, at which they degenerate and tend to acquire a common circular shape. Those effects are demonstrated in Fig. 2.100, where $R = 10$. Clearly, the size of the vortex atmosphere has increased significantly.

It is important to mention the localized character of singularities. In the majority of vortex domains adjacent to the outer separatrix, the azimuthal velocity and frequency are small (see Fig. 2.100). Because of this, we can expect that sufficiently pronounced effects of chaotization will realize at small perturbations of stationary construction.

On the other hand, the domains where the derivative of frequency with respect to the initial position of the trajectory (small nonlinearity parameter) can serve as barriers for chaotic advection, i.e., separate the well mixed domain around the centers of singular vortices from the zone of chaotic transport near the separatrix [443].

In the following section, we will consider some mechanisms of chaos initiation in the problem under consideration.

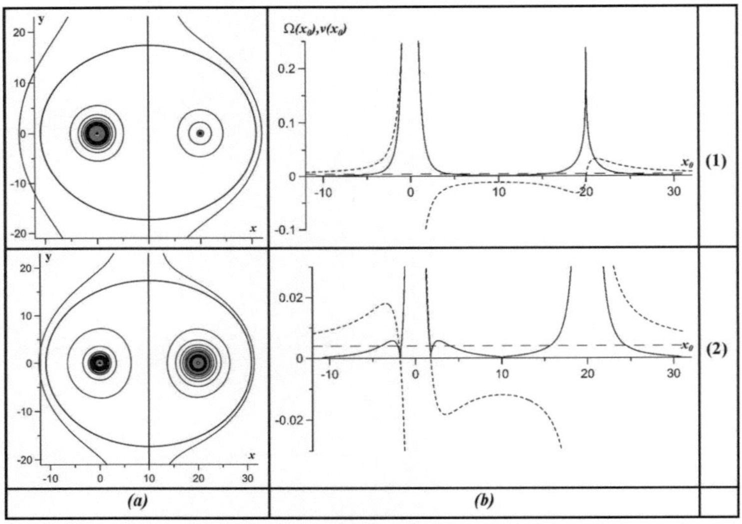

Fig. 2.100 (**a**) Phase portraits, (**b**) Turnover frequencies Ω (*solid lines*) and azimuthal velocities v (*short-dash lines*) of fluid particles at $R = 10.00$ and $B = 0.0000000046$. The isolines are drawn with a step of 0.01 for both layers: (**1**) – the top layer, (**2**) – the bottom layer. The isolines are drawn with a step of 0.01 for both layers. The notations are the same as in Fig. 2.91. Adapted from [444] (Fig. 13) by permission of Cambridge University Press

2.3.3.6 Chaotic Advection Near Stationary Configurations

Above we considered the motion of a vortex system in both stationary and nonstationary cases and evaluated the characteristic frequencies of the perturbed motion. It has been shown that perturbed stationary configurations either move translationally with constant mean velocity or rotate about a rotation center with constant mean angular velocity. Thus, passing to a coordinate system moving with appropriate translational or angular velocity, we can study nonstationary motion of the vortex system near some mean stationary state.

Under small perturbations, the motion of fluid particles will be similar to the motion in the velocity field induced by a stationary mean state of vortex system, as considered in the previous section. In a perturbed case, the velocity field will be nonstationary but periodic with a period of relative motion. In such case, the so-called *separatrix chaos* can form in *stochastic layer* (a narrow domain in the neighborhood of the separatrix of induced velocity field). At greater perturbations with appropriately chosen perturbation frequencies, the *separatrix chaos* can pass into *global chaos* or, in other words, into a *stochastic sea* [1007].

As shown above, in the unperturbed case, a considerable amount of fluid particles have characteristic rotation frequencies, comparable with the frequency of relative motions (the frequency of nonstationary perturbation). In such case, chaotization is very likely to take place in domains remote from separatrices because of overlapping of nonlinear resonance domains [443, 1007]. Barriers consisting of

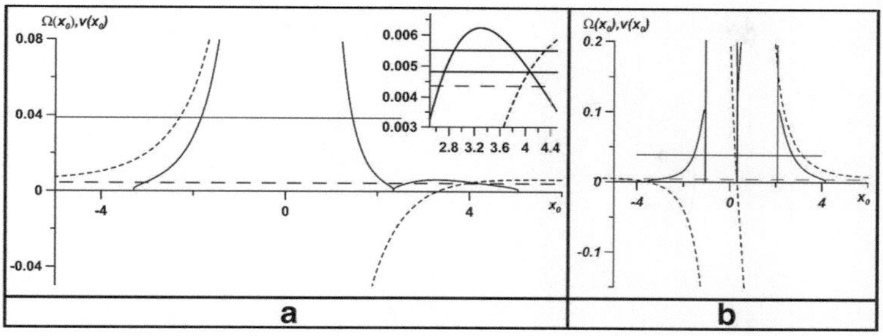

Fig. 2.101 Dependencies of turnover frequencies Ω (*solid lines*) and azimuthal velocities v (*short-dash lines*) of fluid particles on their initial position on the x axis for the top (**a**) and bottom (**b**) layers. The *long-dash horizontal line* is triton velocity. The *solid horizontal lines* in panels (**a**) and (**b**) show the frequency $\tilde{\omega}$ of perturbed motion. The *solid horizontal lines* in the inset show $\tilde{\omega}/8$ and $\tilde{\omega}/7$ of the perturbed motion frequency. The curves correspond to the phase portrait in Fig. 2.97 where $\mu = -2$ and $R = 0.90$, $B = 0.4923$. Adapted from [444] (Fig. 14) by permission of Cambridge University Press

regular trajectories (KAM-tori) can exist between the domains of separatrix chaos and the domains of overlapping of nonlinear resonances of main series [330, 331]. Such barriers can be destroyed without an increase in the perturbation amplitude due only to the appropriate choice of the frequency at which reconnection of separatrices takes place [443]. Below we consider two examples of perturbed configurations, illustrating the global separatrix chaos and the destruction of a regular barrier for chaotic transport due to reconnection of separatrices.

We limit our consideration to the situation of a perturbed triton; in the case of an eccentric roundabout, the scenarios of chaotization are similar.

For the sake of simplicity, we will consider the perturbations that preserve the zero component of momentum P_y. In this case, we can consider the class of initial configurations with symmetrical separation of the peripheral bottom-layer vortices, i.e., replace R by $R+\Delta R$ and B by $B+\Delta R$.

Now we return to Fig. 2.97, which gives an example of streamlines for motions of fluid particles in a velocity field induced by a triton at $R = 0.90$, $B = 0.4923$. As shown in [380, 440, 443], for the chaotization of trajectories in the vortex domain to be maximal, the rotation frequencies of undisturbed trajectories and the frequency of nonstationary perturbation must be commensurable.

In the case under consideration, the perturbation frequency can be evaluated by the formula (2.127), yielding $\omega \approx 0.0385$. To construct the possible scenario of chaotization of trajectories, we need to know the turnover frequencies of fluid particles in the vortex domain in the absence of perturbation. The appropriate dependences of the initial position x_0 of the particle on the x-axis are given in Fig. 2.101. The solid horizontal line shows the frequency of perturbed motion. We can that the perturbation frequency and the turnover frequencies in the top and bottom layers are of the same order of magnitude in a considerable portion of the vortex domain.

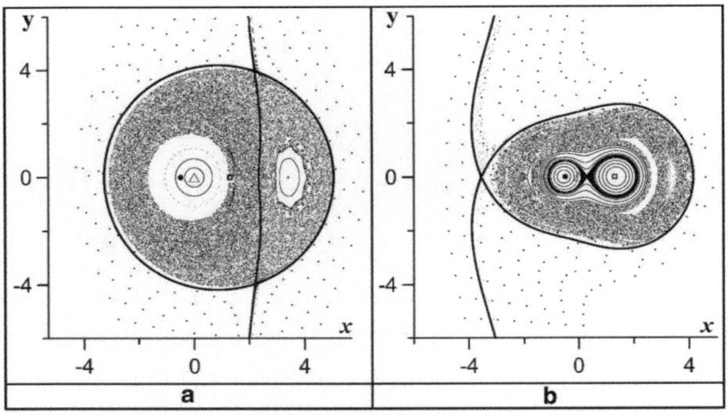

Fig. 2.102 Poincaré sections for unperturbed triton at $\mu = -2$, $B = 0.4923$, $R = 0.9$ and $\Delta R = 0.05$ for the top (**a**) and bottom (**b**) layers. The phase portrait for the unperturbed case is given in Fig. 2.97. Adapted from [444] (Fig. 15) by permission of Cambridge University Press

The intersection of the horizontal line, indicating to the perturbed motion frequency, with the plot of rotation frequency of perturbed motion of fluid particles indicates the trajectory on which nonlinear resonance 1:1 takes place. As a rule, this is the largest resonance in the system [380, 440, 443]. Maximal chaotization can be expected to occur in the vicinity of such a trajectory.

Figure 2.102 gives appropriate Poincaré sections. To begin, we consider the top layer (Fig. 2.102a). In the vortex domain containing the central vortex, the intersection takes place at the initial coordinates of $x_0 \approx -2.0$ and $x_0 \approx 1.5$. The Poincaré sections in this example correspond to the external problem in the terminology of [469] and have been calculated in the following manner: 11 markers were placed within the segment $[-2.0 < x_0 < 4.0, y_0 = -6]$ and 11 more markers were placed in a small segment in the vicinity of the separatrix at $y_0 = -6$; the positions of these markers were output once during each period of perturbation.

The behavior of markers, initially remote from the separatrix, is regular; they move around the vortex domain almost in the same manner as unperturbed trajectories do. The markers placed near the separatrix can penetrate into the vortex domain, where they demonstrate some features of fractal behavior: some markers penetrate into the vortex domain, while others do not.

The markers that have penetrated into the vortex domain demonstrate chaotic behavior within the separatrix, and a boundary of the chaotization domain can be clearly seen in the left part of the vortex domain. This boundary corresponds to the position of the 1:1 resonance, i.e., $x_0 \approx -2.0$, which corresponds to a hyperbolic point of nonlinear resonance, and $x_0 \approx 1.5$, which is smaller than the position of the nonperturbed resonance trajectory by the width of the resonance domain. The situation is somewhat more complicated in the right part of the vortex domain, which is subject to the effect of the right vortex of the bottom layer. In the frequency dependence for this domain, we see that 1:1 resonance cannot take place.

The inset in Fig. 2.101 shows the levels corresponding to frequencies $\tilde{\omega}/7$ and $\tilde{\omega}/8$. The largest among the realizable resonances has the ratio of 1:7 and consists of seven islands, i.e., in a coordinate system rotating with a frequency of $\tilde{\omega}$, there seven elliptic and seven hyperbolic points [1007]. The domains of resonances 1:7 and 1:8 have a lesser width than the 1:1, therefore, the boundary of chaotization zone lies further from the elliptic point of the right vortex domain. The analysis of the shape of this boundary shows it to lie between the domains of resonances 1:7 and 1:8; i.e., we see partial overlapping of these resonances.

The situation in the bottom layer is radically different (Fig. 2.102b). First, because of the singularity and the point vortices belonging to this layer, the turnover frequency of fluid particles very rapidly decreases in the vicinity of the internal separatrix, embracing the point vortices. The second significant distinction is that the external vortex domain has no elliptic point and is located between two separatrices. Since the turnover frequency vanishes on separatrices, there exists a trajectory between them, corresponding to a maximum of turnover frequency.

The two intersections of the curve of turnover frequency and the horizontal line corresponding to the perturbation frequency lie on two different trajectories. Note that the two symmetric domains with turnover frequency maxima with negative and positive x_0 correspond to the same trajectories; therefore, it is sufficient to consider only one of these domains. The problem in which closed streamlines lie between separatrices and the frequency dependence has a single maximum has been considered in detail in [443], where one of the separatrices was at infinity. Such a situation admits the existence of two sets of nonlinear resonances with the same multiplicity and, accordingly, reconnections of separatrices of these resonances are possible. Without going into detail, we note that the model demonstrates all effects revealed in cited paper.

Thus, in the lower layer, the curve of turnover frequency intersects with the level of perturbation frequency very close to the separatrix, in both the internal and external vortex domains; hence, we have a very narrow separatrix stochastic layer. One more nontrivial intersection takes place in the external vortex domain on the trajectory passing through points $(x_0 \approx -1.55, y_0 = 0)$ and $(x_0 \approx 2.75, y_0 = 0)$.

As a result, we have an island of regular behavior, corresponding to resonance $1:1$ at $x_0 \approx 2.75$, and the boundary of stochastic sea passes via points $(x_0 \approx -1.52, y_0 = 0)$ and $(x_0 \approx 2.55, y_0 = 0)$, i.e., it is spaced apart from the position of resonance trajectory by the width of the resonance domain. Moreover, we see a partially destroyed 1:2 resonance of two islands of regular behavior, i.e., these largest nonlinear resonances partially overlap in this case.

The degree of chaotization in the external domain will be appreciable, the perturbation of the triton being relatively small; in other words, the perturbation is not large for the triton, but sufficiently large from the viewpoint of the approaching nonlinear resonance. The stochastic layer in the vicinity of the internal separatrix is very narrow, since, because of singularity, the derivative of frequency with respect to the initial position of the trajectory is very large in this domain and hence, the thickness of the stochastic layer is very small here. Because of the singularity,

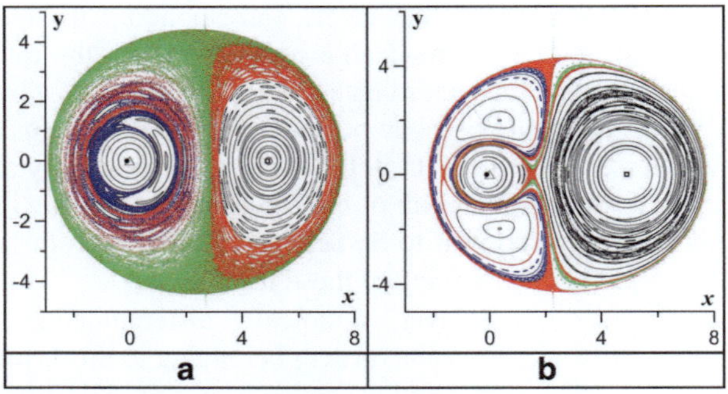

Fig. 2.103 Poincaré sections at $\mu = -2$, $B = 0.003131$, $R = 2.4$ and $\Delta R = 0.1$ for the top (**a**) and bottom (**b**) layers. The figures correspond to perturbed behavior of triton, whose unperturbed phase portrait is given in Fig. 2.98d, the separatrices are shown separately in Fig. 2.106a. Adapted from [444] (Fig. 16) by permission of Cambridge University Press

vicinities of point vortices should be regular [779]. Thus, the internal separatrix and the external chaotic domain are separated by a barrier hampering the transport of fluid particles from the central part of the vortex atmosphere to its periphery.

An important point is the capture and release of part of its mass by the vortex atmosphere. The capture of fluid from the incoming flow is possible only within a very narrow domain near the separatrix. The release of fluid particles also takes place in a narrow domain near the separatrix on the opposite side of the vortex atmosphere. Tentative estimates show that, on the one hand, this is a very slow process, but, on the other hand, all the fluid in the external domain of chaotization will be, sooner or later, carried out from the vortex atmosphere.

In the Poincaré section, it can be seen that the capture of fluid from the oncoming flow is possible only in a very narrow domain near the separatrix. The release of fluid particles is also possible only in a narrow domain near separatrix on the opposite side of the vortex atmosphere. Tentative estimates show that, on the one hand, this process is very slow but, on the other hand, the entire fluid in the outer domain of chaotization will be washed out from the vortex atmosphere after some time.

Not meaning to fully analyze the scenarios of chaotization in vortex atmospheres of perturbed structures, we will illustrate two other important effects. First, we consider the effect of stratification.

As it was shown above, the chaotic domain can occupy a considerable portion of the vortex atmosphere in both layers. Figure 2.103 gives an example where the degree of chaotization in the bottom layer is small, while that in the top layer is considerable. Such difference was obtained through the choice of sufficiently small perturbation amplitude, corresponding to a large frequency of perturbed oscillations. However, the effect is primarily determined by the form of dependence of rotation frequencies of fluid particles in the velocity field induced by stationary configuration.

In studies [380, 440, 443, 779], it was shown that when perturbations are not too small, the chaotization of phase portrait domains not adjacent to separatrices, is determined by the extent of overlapping of nonlinear resonances, which, in

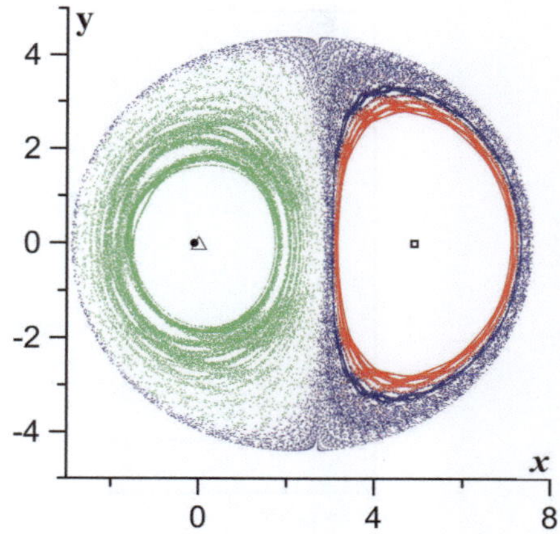

Fig. 2.104 Three-color illustration of the trap effect of the *red marker* in the top layer (Fig. 2.103a). Here, *red markers* corresponds to 40,000 first perturbation periods, *blue markers* – next 20,000 periods, *green markers* – next 20,000 periods. See details in the text. Adapted from [444] (Fig. 17) by permission of Cambridge University Press

its turn, is governed by the width of the domains of those resonances and the distance between neighboring resonance domains. Those parameters are determined by the derivative of turnover frequency with respect to the action; this derivative is proportional to the derivative of frequency with respect to coordinate. The derivative of frequency is determined, on the average, by the value of frequency in the elliptic point (the frequency is zero on the separatrix) and the size of the corresponding vortex domain. Thus, in domains with singularities, the maximal frequency tends to infinity, hence the derivative is large enough.

This means that a regular domain always exists in the neighborhood of singularity, even at considerable perturbations [779]. If the size of such vortex domain is large enough, the derivative of the frequency becomes smaller at some distance from the singularity, and the overlapping of domains of nonlinear resonances is possible here. The boundary starting from which they overlap is determined by their width and hence, by the perturbation amplitude. An important point is also that the derivative of frequency can be too small in some domains (the nonlinearity parameter is small [1007]), and such domains, as shown in [443], serve as barriers for chaotic advection.

Consider Figs. 2.103 and 2.104 from this viewpoint. We start from the bottom layer. Nonlinear resonances realize in the neighborhood of trajectories for which the condition $m\tilde{\omega}+n\omega \sim 0$ is satisfied and the width of resonance domains abruptly drops with increasing n and m, of which the system is more sensitive to m [380, 1007].

Figure 2.105 shows the levels of frequencies $\tilde{\omega}/n$, corresponding to a series of largest resonances. With increasing n, the resonance trajectory approaches the separatrix and the distance between the neighboring resonance domains decreases. In this case, if the derivative of frequency with respect to coordinate is not too small, the widths of resonance domains decrease with increasing n slower than the distances between them. Accordingly, starting from some n, overlapping begins and the neighborhood of overlapping resonances becomes chaotic [179, 1007]. Because

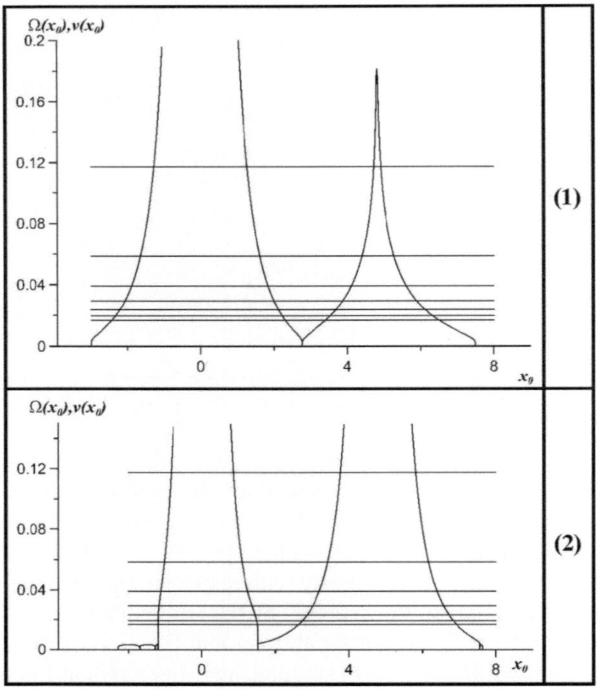

Fig. 2.105 Rotation frequency of fluid particles versus the initial position on the x-axis in the stationary case (triton) at $R = 2.4$, for the top (**1**) and bottom (**2**) layers. The *horizontal lines* show the values of frequencies $\tilde{\omega}/n$ (n varies from 1 to 7) for the case of Fig. 2.103. Adapted from [444] (Fig. 18) by permission of Cambridge University Press

of the appearance of additional separatrices, the singular vortex domain in the left part of the vortex atmosphere of the bottom layer is small and the overlapping forms in a narrow domain near the separatrix.

The right singular vortex domain is much greater, and the derivative of frequency with respect to coordinate can reach values of the order of 1. In this domain, nonlinear resonances overlap. We can see such domain in the neighborhood of trajectories with initial coordinate close to $(x_0 = 2.35, y_0 = 0)$. However, this vortex domain is large enough for the derivative to become small, i.e., a regular barrier forms between this, partially chaotized, domain and the separatrix. Such barriers were studied in detail in [443].

In the induced regular domains of the left part of vortex atmosphere and in the flow-through domains, the turnover frequencies are much less, so only resonances of very high multiplicity can realize here, because such resonances are not wide enough to overlap. Thus, we see narrow separatrix stochastic layers labeled by red markers. Additionally, we see regular behavior with many nonlinear resonances in the right vortex domain (black markers—one for each trajectory). In accordance with Fig. 2.105.2, resonances 1:1, 1:2, 1:3, etc. can be easily identified. Clearly, the figure does not show all resonances. Domains of regular resonances 2:1 (at double perturbation frequency, i.e., $m = 2, n = 1$), 1:1, 1:2, 1:3 can be seen in the left loop, while the resonance domain 1:4 is already assimilated by the stochastic layer. For the induced vortices, the domains of resonances with high multiplicity are too narrow. Resonances of very high multiplicity (blue and green markers), as well as regular trajectories and KAM-tori, exist in flow-through zones.

Analysis of the behavior of turnover frequency allowed us to elucidate the weak chaotization in the bottom layer. Note that the marker that sweeps the stochastic layer near the outer separatrix corresponds to the external problem, i.e., it has been trapped from the external flow (several its positions can be seen in the bottom part of the figure) and spent a long time in the stochastic separatrix layer, after which it was carried out into the external flow. This example shows that vortex atmosphere can not only acquire, but also loose mass. It will be shown below that an increase in perturbation amplitude and a decrease in frequency are accompanied by a considerable increase in the extent of chaotization, in particular, because of the widening of the weakly chaotized domain in the right loop.

In the top layer, we have two loops—the left, singular one and the right, regular one. In the right loop, because of the finite magnitude of maximal rotation frequency $\Omega_{max} = 0.1813$, the derivative of frequency with respect to coordinate is small enough even in the neighborhood of the elliptic point, and the resonance 1:1 is situated near this point. Further, the picture is largely similar to that we see in the right loop of the bottom layer. However, because of the smaller size of the vortex domain, the regular barrier in the top layer is very thin and permeable. Because of this, the stochastic separatrix layer merges with the overlap domains of nonlinear resonances to become a stochastic sea [1007]. In the left part, the singular vortex domain is greater than the analogous domain in the bottom layer, though it is somewhat less than the right singular domain.

These circumstances cause two important distinctions. On the one hand, because of the lesser derivative of the frequency with respect to coordinate, the widths of all resonance domains are much greater than those in the bottom layer, but, on the other hand, the frequency function here contains no domain with small values of the derivative, hence there are no conditions for the formation of a regular barrier. So the overlap domain of nonlinear resonances also merges with separatrix stochastic layer. It is important that the resonance domains with low orders still partly overlap, i.e., there exist barriers between them whose permeability steadily decreases with the distance from the separatrix.

Thus, regular trajectories (labeled by black markers) can also be seen in Fig. 2.103. An interesting feature is that secondary resonances can be seen to surround the weakly destroyed 1:1 resonance in its neighborhood. Here we could find resonances of the orders of 1:3 and 1:7 (the point here is the rotation frequencies around the resonance domain 1:1 and the perturbation frequency associated with the effect of other primary resonances on this domain). Additionally, the red marker shows a vast domain of a stochastic sea. Obviously, a fluid particle can penetrate from the right boundary into the domain of the left boundary sweeping a very large part of the vortex atmosphere.

In such cases, the so-called dynamic traps form [442], associated with the low–permeability layers mentioned above. The domain swept by the blue marker corresponds to such trap: this marker stays fairly long near the separatrix of 1 : 1 resonance, after which it detaches and penetrates through several barriers into the domain external for this trap. The trap can be assumed the blue domain itself. A second trap, surrounding it, is separated from it only by a low-permeability barrier and it consists of a series of higher permeability barriers in which the markers entangle, carrying out a series of Levy flights [442].

The effect of traps can be illustrated more vividly by the auxiliary Fig. 2.104, showing the behavior of a particle marked in red in the main figure, though three colors are used here, alternating depending on the particle's residence time in the corresponding domains. The red color corresponds to the most closed trap, from where the particle started and where it stayed for almost 40,000 perturbation periods. The positions of the particle after its leaving this domain are given in blue; the particle stays here for about 20,000 perturbation periods, though it spends most of the time in the neighborhood of the red trap. So, the domain of maximal density of blue markers can be referred to as a semi-open trap, from which the particle escapes periodically, visiting the left loop in the neighborhood of the outer separatrix. However, in some moment, the particle will be captured by the left outer trap, where it will stay for 20,000 perturbation periods. At the last stage, the positions of this marker in Fig. 2.104 are shown in green.

Returning to Fig. 2.103a, we add that the green color in it refers to a marker that originally lied beyond the atmosphere (external problem). It is captured by the vortex domain, sweeps a considerable portion of the stochastic sea, which is much greater than that in the bottom layer, and next it is again carried out into the external flow.

Mechanisms that have their effect on chaotizaton of vortex atmosphere were discussed above. Now, we will consider one more interesting mechanism of chaotization.

The variant of reconnection near rotation frequency peak between two separatrices, which was considered above, has been studied well, e. g., in [443]; however, the system under study may show not only reconnections of separatrices of nonlinear resonances induced by perturbation, but also reconnections of nonperturbed separatrices accompanying changes in the size of configurations [849]. Examples of such reconnection for the bottom layer are given in Figs. 2.98e, and 2.106b (only separatrices are given in the latter case). The comparison of all three panels in Fig. 2.106 vividly demonstrates the process of disappearance of the domain separating the inner and outer separatrices during reconnection.

The dependences of turnover frequency near the reconnection are given in Fig. 2.107, where full horizontal lines are drawn at levels $\omega/2$, $\omega/4$, and $\omega/5$. Importantly, the perturbation frequency in this case is rather large, and the domains of major nonlinear resonances 1:1, 1:2, 1:3 form too close to the singularity, the fact that in accordance with the results in [779], facilitates their abrupt narrowing.

So, the boundary of chaotization domain is determined by the position of 1:4 resonance. Another feature of the frequency dependences is the existence of a gentle segment of separatrix in the domain of the right bottom vortex, in which case,

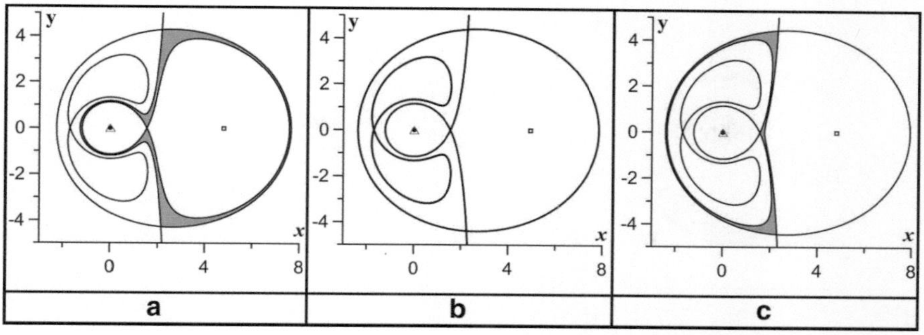

Fig. 2.106 Separatrices at (**a**) $R = 2.4$, (**b**) $R = 2.469630$ (reconnection) and (**c**) $R = 2.5$. The domain separating the inner and outer separatrices is shown in *gray*. Adapted from [444] (Fig. 19) by permission of Cambridge University Press

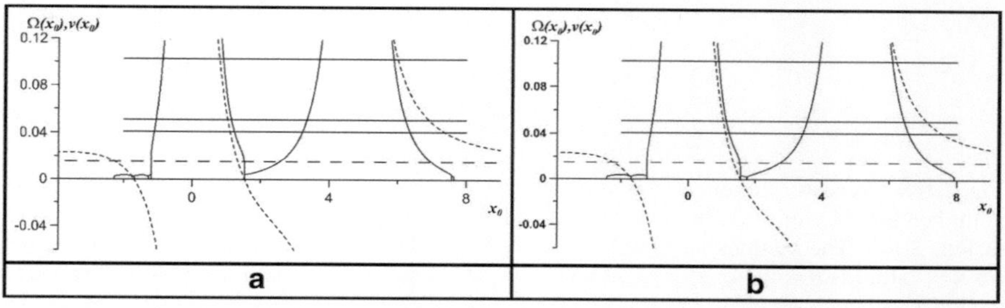

Fig. 2.107 Rotation frequencies and azimuthal velocities versus the initial position of a bottom-layer particle on the x-axis: (**a**) $R = 2.4$, (**b**) $R = 2.5$. They correspond to separatrices in Fig. 2.106a, c, respectively. Adapted from [444] (Fig. 20) by permission of Cambridge University Press

as shown in [443], we can expect the formation of a regular barrier preventing fluid particles from penetrating into the inner chaotization domain from the near-separatrix stochastic layer. With appropriately chosen perturbation amplitude, we indeed can see such barrier, which is later destroyed because of reconnection of separatrices. Such scenario of chaotic mixing is given in Fig. 2.108.

The results of calculations in the time intervals before and in the moment of reconnection are given for the internal (the markers are initially located in the vortex domain of the inner separatrix, surrounding singular domain) and external (the markers are initially located outside the vortex domain in the neighborhood of separatrix) problems.

In these figures, we can clearly see a barrier of regular behavior, which significantly hampers the penetration of external markers into the right vortex domain, where internal markers mix due to overlapping of nonlinear resonances of lower orders. The existence of such barrier is due to the small size of the domain of quasilinear behavior of the dependence of rotation frequency. Similar situation was studied in detail in [443]. We will mostly focus of the external problem. Recall [298, 1007] that the width of the separatrix stochastic layer is maximal near the hyperbolic point.

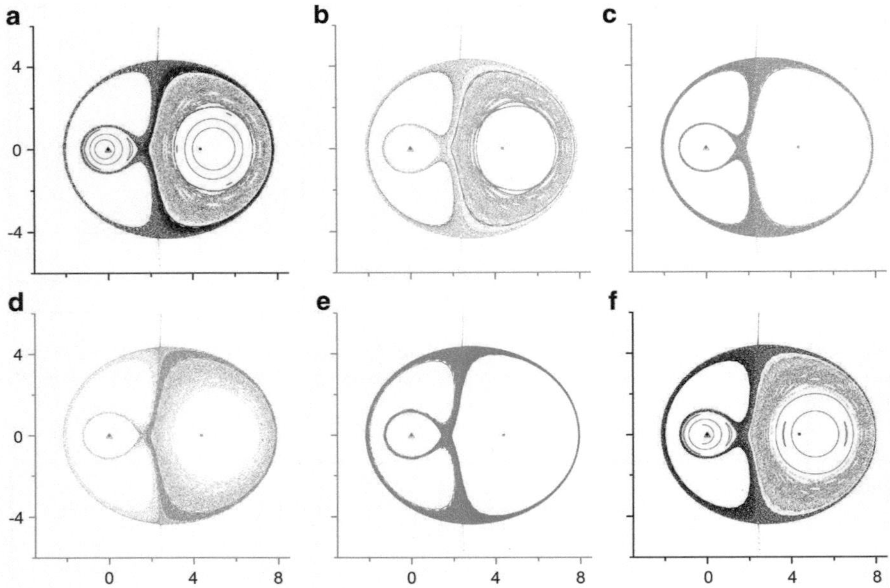

Fig. 2.108 Poincaré section at $\Delta R = 0.185$ for the bottom layer. The figures correspond to perturbed behavior of triton in the neighborhood of reconnection of the unperturbed triton shown in Fig. 2.106. The sections are given successively for the values: (**a**) $R = 2.41$, (**b**) $R = 2.42$, (**c**) $R = 2.43$, (**d**) $R = 2.44$, (**e**) $R = 2.45$, (**f**) $R = 2.46$. The *blue* and *black markers* correspond to the inner and outer problem, respectively. Panels (**a**), (**d**) and (**f**) adapted from [444] (Fig. 21) by permission of Cambridge University Press

So, at $R = 2.41$ (2.108a), the internal sepatrix and the inner part of the outer separatrix (along which the external markers penetrate into the vortex domain) are separated by a thin layer with closed trajectories. Exchange of markers takes place within this layer, i.e., the stochastic layers of those separatrices merge. However, hyperbolic points are spaced apart and the width of the stochastic layer on the side of the right vortex domain of the internal separatrix is the same as in the absence of the external separatrix. The barrier has low permeability: for example, in Fig. 2.108b, at $R = 2.42$, only one marker out of 10 penetrates into the mixed domain in the right loop of the internal separatrix and stays there for a long time, thus suggesting a low permeability from inside. In the moment of reconnection, which is somewhat shifted because of perturbation and takes place at $R \approx 2.44$, the separatrices merge and hyperbolic points now lie on the same trajectory, resulting in an increase in the thickness of the stochastic layer.

In the Poincaré section (Fig. 2.108), corresponding to the value $R = 2.44$, we see that outer markers penetrate into the right vortex domain, but they also leave it easily and stay for a long time near the barrier. At further increase in R, we again see a domain separating separatrices, each stochastic layer becomes thinner (they merge within the separating domain), and barrier permeability becomes low, as can be seen in the sections in Fig. 2.108e, f at $R = 2.45$ and $R = 2.46$, respectively (the latter figure shows inner markers as well).

Note that the examples of chaotization of vortex atmosphere for a perturbed triton are quite typical, for example, at the further increase of triton size. The barrier considered above must increase because of the stretching of the dependence of rotation frequency and extension of the plateau on this dependence (see details in [443]), and reconnections of separatrices are observed at several values of parameters. Thus, at $R \approx 2.188$, the separatrix of induced vortices located above and below the x-axis will penetrate into the domain lying within the outer separatrix.

Note that the numerous problems relating to chaotic transport as applied to the ocean and atmosphere are widely represented in the reviews [223, 394, 441, 442, 470, 566, 800, 895, 971, 981, 1018].

2.4 Heton Structures in a Three-Layer Fluid

Formulas (1.108) readily yield expressions for the velocities of stationary vortex structures in a three-layer fluid. Assuming for simplicity $h_1 = h_2 = h_3 = h$, $\Delta\rho_1 = \Delta\rho_2 = \Delta\rho$ and $\gamma_1 \equiv \gamma$ (in this case, we have $\gamma_2 = \sqrt{3}\gamma$) and introducing dimensionless distance $L = l\gamma$, we obtain that at $\kappa_1^1 = -\kappa_3^1 \equiv \kappa$, $\kappa_2^1 = 0$, the translational velocity of 1-modal heton (Fig. 1.5.1b) with nonzero kinetic and angular momentums is

$$v = \frac{\gamma\kappa}{3\pi L}\left[1 - \sqrt{3}L\mathrm{K}_1(\sqrt{3}L)\right], \tag{2.128}$$

(cf with (2.8)).

A 2-modal heton with an *arc-shaped* axis, consisting of vortices with intensities $\kappa_1^1 = \kappa_3^1 = \kappa$, $\kappa_2^1 = -2\kappa$ (see Fig. 1.5.2c), moves with the velocity

$$v = \frac{\gamma\kappa}{6\pi L}\left[1 - \frac{3L}{2}\mathrm{K}_1(L) + \frac{\sqrt{3}L}{2}\mathrm{K}_1(\sqrt{3}L)\right]. \tag{2.129}$$

The comparison of (2.128) and (2.129) shows that with increasing m, the velocity of the m-modal heton 175 decreases because of the increasing degree of horizontal localization.

However, if the 2-modal heton has zero momentum (Fig. 1.5.2b), it has no translational motion and the angular velocity of its rotation is

$$\omega = \frac{\gamma^2\kappa}{6\pi L^2}\left[\frac{3}{2} - 2\sqrt{3}L\mathrm{K}_1(\sqrt{3}L) + \frac{3}{2}L\mathrm{K}_1(2L) - \frac{\sqrt{3}}{2}L\mathrm{K}_1(2\sqrt{3}L)\right]. \tag{2.130}$$

Note that the vortex structure (2.130) combines the properties of *heton* (zero total intensity) and an analogue of *ordinary (two-layer) roundabout*. Therefore, this structure can be naturally referred to as *three-layer roundabout*.

Let us consider some examples of solutions describing uniformly rotating vortex structures in a three-layer fluid:

- *Tripole*, localized in the top ($\kappa_1^1 = -2\kappa$, $\kappa_1^2 = \kappa_1^3 = \kappa$) or bottom ($\kappa_3^1 = -2\kappa$, $\kappa_3^2 = \kappa_3^3 = \kappa$) layer, whose peripheral vortices rotate with an angular velocity

$$\omega = \frac{\gamma^2\kappa}{6\pi L^2}\left[\frac{3}{2} + 3L\mathrm{K}_1(L) + \sqrt{3}L\mathrm{K}_1(\sqrt{3}L) - \frac{3}{2}L\mathrm{K}_1(2L) - \frac{\sqrt{3}}{2}L\mathrm{K}_1(2\sqrt{3}L)\right]. \tag{2.131}$$

- *Tripole*, localized in the middle layer ($\kappa_2^1 = -2\kappa$, $\kappa_2^2 = \kappa_2^3 = \kappa$), with angular rotation velocity

$$\omega = \frac{\gamma^2\kappa}{6\pi L^2}\left[\frac{3}{2} + 4\sqrt{3}L\mathrm{K}_1(\sqrt{3}L) - 2\sqrt{3}L\mathrm{K}_1(2\sqrt{3}L)\right]. \tag{2.132}$$

- Symmetric *roundabout*, consisting of vortices with intensities $\kappa_1^1 = \kappa_1^2 = \kappa$ in the top layer (or $\kappa_3^1 = \kappa_3^2 = \kappa$ in the bottom layer) and a central layer $\kappa_2^1 = -2\kappa$ in the middle layer. The angular rotation velocity is

$$\omega = \frac{\gamma^2\kappa}{6\pi L^2}\left[\frac{3}{2} - 2\sqrt{3}L\left(\mathrm{K}_1(\sqrt{3}L) + \mathrm{K}_1(2\sqrt{3}L)\right)\right]. \tag{2.133}$$

- Symmetric *reverse roundabout*, consisting of two vortices of the middle layer $\kappa_2^1 = \kappa_2^2 = \kappa$ and one vortex of the top layer $\kappa_1^1 = -2\kappa$ (or bottom layer $\kappa_3^1 = -2\kappa$). In this case, the angular rotation velocity is

$$\omega = \frac{\gamma^2\kappa}{6\pi L^2}\left[\frac{3}{2} - 2\sqrt{3}L\mathrm{K}_1(\sqrt{3}L) - \frac{3}{2}L\mathrm{K}_1(2L) - \frac{\sqrt{3}}{2}L\mathrm{K}_1(2\sqrt{3}L)\right]. \tag{2.134}$$

- Symmetric *roundabout*, consisting of a vortex $\kappa_1^1 = -2\kappa$ in the top layer (or $\kappa_3^1 = -2\kappa$ in the bottom layer) and two vortices with intensities $\kappa_3^1 = \kappa_3^2 = \kappa$ in the bottom layer (or, in the second case, $\kappa_1^1 = \kappa_1^2 = \kappa$—two vortices in the top layer). The angular velocity is

$$\omega = \frac{\gamma^2 \kappa}{6\pi L^2}\left[\frac{3}{2} - 3L\mathrm{K}_1(L) + \sqrt{3}L\mathrm{K}_1(\sqrt{3}L) + \frac{3}{2}L\mathrm{K}_1(2L) - \frac{\sqrt{3}}{2}L\mathrm{K}_1(2\sqrt{3}L)\right]. \tag{2.135}$$

- Vortex structure with a tilted axis, for which $\kappa_1^1 = \kappa_3^1 = \kappa$, $\kappa_2^1 = 0$, supplementing (2.128) and determining a *top* with angular velocity

$$\omega = \frac{\gamma^2 \kappa}{3\pi L^2}\left[1 - \frac{3L}{2}\mathrm{K}_1(L) + \frac{\sqrt{3}L}{2}\mathrm{K}_1(\sqrt{3}L)\right]. \tag{2.136}$$

We will also mention the existence of *heton* nonmodal structures with a tilted axis, which are localized either in the top two layers ($\kappa_1^1 = -\kappa_2^1 = \kappa$, $\kappa_3^1 = 0$), or in the bottom two layers ($\kappa_3^1 = -\kappa_2^1 = \kappa$, $\kappa_1^1 = 0$), Moving with the velocity

$$v = \frac{\gamma \kappa}{6\pi L}\left[1 - \sqrt{3}LK_1(\sqrt{3}L)\right] \tag{2.137}$$

and the corresponding nonmodal structures of the type of *top* ($\kappa_1^1 = \kappa_2^1 = \kappa$, $\kappa_3^1 = 0$ and $\kappa_3^1 = \kappa_2^1 = \kappa$, $\kappa_3^1 = 0$), rotating with the angular velocity

$$\omega = \frac{\gamma^2 \kappa}{3\pi L^2}\left[1 - \sqrt{3}L\mathrm{K}_1(\sqrt{3}L)\right]. \tag{2.138}$$

The properties of hetons (2.137) were studied in [218, 329].

The studies in [327] have shown that the interaction between 1-modal hetons (Fig. 1.5.1a) is analogous to the interaction between hetons in a two-layer model, see Figs. 2.25 and 2.26. In the case of interaction between 2-modal hetons (Fig. 1.5.2a), trajectories of type $\{2\}$ become nonsymmetric about the y-axis because of the appearing asymmetry in the interaction between layers. However, asymptotically (at $t \to \infty$), a 2-modal heton with an arc-shaped axis will form (see Fig. 1.5.2c and formula (2.129)). Such heton is the simplest example of a quasi-stationary vortex cluster, if the centers of the top and bottom layers are spaced small distance a apart. When $a \ll l$, the interaction between top and bottom layer vortices leads to their mutual rotation, i.e., to the formation of a vortex top, whose interaction with a vortex with opposite sign in the middle layer leads to a displacement of the system as a whole—a heton cluster. Heton clusters are typical examples of a vortex structure in multilayer models [327, 329].

The dynamics of *distributed* vortices under the three-layer MKD-model was studied in [844–846], and some its aspects will be touched upon in Sect. 3.4 of this book.

Chapter 3
Dynamics of Finite-Core Vortices

Abstract In this chapter, we consider the stability of an isolated finite-core (or distributed) two-layer vortex with respect to relatively small and finite perturbations. An analogy between a distributed heton and A-symmetrical structure of discrete hetons is demonstrated. The specific features of the nonlinear stage of evolution of unstable vortices, and the interaction between two distributed hetons or antihetons are considered. The model is shown to be promising for the description of deep-convection processes,water mass mixing in the ocean, and the formation of new quasistationary vortex structures. We study the effect of external flow and of an isolated hill on heton motion. The results obtained for a three-layer, quasigeostrophic model are given; in particular, specific features of the dynamics of meddies are studied. The role of baroclinicity in the formation of the kinematic and thermohaline structure of the ocean is analyzed.

The key features of finite-core vortices, as opposed to discrete ones, are the possible instability of a solitary vortex under small perturbations of its shape and a tendency toward merging of vortex patches of the same sign once they approach each other closely enough. Different approaches to the stability problem of barotropic and baroclinic (including two- and three-layer) vortices are discussed in papers [8, 47, 48, 51, 56, 61, 70, 77–81, 88, 121, 141, 144, 148, 149, 151, 152, 175, 180, 184, 202, 225, 231, 233–236, 240, 248, 269, 274, 280, 289, 290, 293, 311, 313, 339, 340, 343, 356, 358, 363, 371–374, 387, 412, 427, 436, 477, 531, 544–547, 567, 570, 604, 608, 613, 614, 617–619, 639, 657, 660, 664, 666, 684, 687, 688, 730, 755, 756, 787, 802, 806, 820, 832, 833, 839, 840, 842, 844–846, 851, 859, 866, 867, 899, 900, 904, 905, 954, 966, 980, 1009], the effect of vortex merging is also discussed in many publications, such as [21, 63, 117, 118, 122, 134, 142, 143, 158, 182, 193, 197, 240, 267, 286–288, 309, 310, 353, 354, 357, 396, 447, 512–514, 519, 523, 529, 531, 549, 564, 564, 568, 572, 573, 575, 581, 586, 589–591, 602, 603, 652, 680, 685, 686, 704, 711, 723, 728, 731, 732, 747, 757, 760, 829, 830, 834, 910, 919–921, 938, 944, 945, 948, 955, 963, 967, 972, 983, 997–999, 1024].

M.A. Sokolovskiy and J. Verron, *Dynamics of Vortex Structures in a Stratified Rotating Fluid*, Atmospheric and Oceanographic Sciences Library 47,
DOI 10.1007/978-3-319-00789-2_3,

This chapter is based on materials of the authors' papers [263–266, 325, 443, 444, 477, 562, 839–847, 849–853, 859, 860, 919–921, 944, 945, 948].

3.1 Studying the Linear Stability of a Two-Layer Vortex

As shown in Sect. 1.2.5, the stability analysis of an axisymmetric two-layer vortex of the type (1.47)–(1.49), subject to small perturbations (1.52), (1.68) of vortex patch boundaries, can be reduced in a linear approximation to solving the spectral problem (1.69) and (1.70). The approach to stability analysis proposed in Sect. 1.2.5 is of particular convenience for piecewise constant distributions of potential vorticity typical of CDM.

3.1.1 *A Vortex with a Vertical Axis: Two Circular Vortex Patches*

Let

$$\Pi_1(\rho) = \bar{\Pi}_1 H(a_1 - \rho), \quad \Pi_2(\rho) = \bar{\Pi}_2 H(a_2 - \rho), \tag{3.1}$$

where $\bar{\Pi}_1$ and $\bar{\Pi}_2$ are constants, and $H(\rho)$ is a unit Heaviside function. Now, writing the system (1.69)–(1.70) at $\rho = a_1$ and $\rho = a_2$, we obtain a system of four linear algebraic equations in $\varepsilon_j(a_k)$, $j, k = 1, 2$. Equating its determinant to zero, one finds the dispersion equation

$$\Big[(\delta - 2X)(\delta - 2Y) - Z\Big]\Big[\delta - S(a_2)\Big]\Big[\delta - T(a_1)\Big] = 0, \tag{3.2}$$

where

$$\begin{aligned}
X &= \frac{1}{2}\Big\{S(a_1) + \bar{\Pi}_1\Big[h_2 J_0^{(m)}(a_1, a_1) + h_1 J^{(m)}(a_1, a_1)\Big]\Big\},\\
Y &= \frac{1}{2}\Big\{T(a_2) + \bar{\Pi}_2\Big[h_2 J^{(m)}(a_2, a_2) + h_1 J_0^{(m)}(a_2, a_2)\Big]\Big\},\\
Z &= h_1 h_2 \bar{\Pi}_1 \bar{\Pi}_2 \Big[J^{(m)}(a_1, a_2) - J_0^{(m)}(a_1, a_2)\Big]^2,\\
S(b) &= P(b) - h_2 Q(b), \quad T(b) = P(b) + h_1 Q(b),
\end{aligned}$$

while $J^{(m)}$ and $J_0^{(m)}$ are determined by expressions (1.62)–(1.63), where $m \geq 1$ is the number of the mode.

The roots of the quadratic polynomial in (3.2) have the form:

$$\delta = X + Y \pm \sqrt{D}, \quad D = (X - Y)^2 + Z, \tag{3.3}$$

therefore, the instability condition is the inequality $D < 0$, whence, in particular, it follows that this requires that $\bar{\Pi}_1 \bar{\Pi}_2 < 0$, implying the opposite rotation in the layers.

Since, in dimensionless variables, $h_1 + h_2 = 1$, we assume $h_1 = h$, then $h_2 = 1 - h$. Introducing the notations

$$\Delta = \frac{(1-h)\bar{\Pi}_2}{h\bar{\Pi}_1}, \quad \delta_m = \delta(m),$$

we consider the case of a cylingrical two-layer vortex $a_1 = a_2 = 1$. Now (3.3) takes the form of a dispersion relationship

$$\delta_m = \frac{\bar{\Pi}_1}{2(1-h)} \left\{ (1-h)\left(1 - L_1(\gamma) - \frac{1}{2m}\right)(1+\Delta) + h\left[\left(\frac{1-h}{h}\right)^2 + \Delta\right] \right.$$
$$\left. \times \left(L_1(\gamma) - L_m(\gamma)\right) \pm \sqrt{\Phi_m(\gamma, h)} \right\}, \tag{3.4}$$

where

$$\Phi_m(\gamma, h) = \left\{ (1-h)\left(L_1(\gamma) - \frac{1}{2m}\right)(1-\Delta) + h\left[\left(\frac{1-h}{h}\right)^2 - \Delta\right] \right.$$
$$\left. \times \left(L_1(\gamma) - L_m(\gamma)\right) \right\}^2 + 4(1-h)^2 \Delta \left(\frac{1}{2m} - L_m(\gamma)\right)^2$$

and

$$L_m(\gamma) = \mathrm{I}_m(\gamma)\mathrm{K}_m(\gamma), \ m \geq 1.$$

Heton with Vertical Axis

In a particular case of heton, i.e., at $\Delta = -1$, the expresson (3.4) becomes simpler

$$\delta_m = \frac{\bar{\Pi}_1}{2(1-h)} \left[(1-2h)\left(L_1(\gamma) - L_m(\gamma)\right) \pm \sqrt{\Phi_m(\gamma, h)} \right], \tag{3.5}$$

and now

$$\Phi_m(\gamma, h) = \left(L_1(\gamma) - L_m(\gamma)\right)\left[L_1(\gamma) - (1-2h)^2 L_m(\gamma) - \frac{2h(1-h)}{m}\right]. \tag{3.6}$$

$L_m(\gamma)$ functions monotonically decrease with increasing γ and m, so

$$L_m(\gamma) \to \frac{1}{2m} \quad \text{at} \quad \gamma \to 0 \quad \text{and} \quad L_m(\gamma) \to \frac{1}{2\gamma} \quad \text{at} \quad \gamma \to \infty.$$

Then, $\delta_1 = 0$, and we obtain neutral-stability curves for $m \geq 2$ by equating the second multiplier in (3.6) to zero, whence we have

$$h = \frac{1}{2}\left[1 \pm \sqrt{\left(\frac{1}{2m} - L_1(\gamma)\right)\left(\frac{1}{2m} - L_m(\gamma)\right)}\right]. \tag{3.7}$$

The Eq. (3.7) was used to construct neutral-stability curves shown in the top part of Fig. 3.1, while its bottom part gives the dependence of $|\Im\delta_m|$ on parameter γ from (3.5) for different values of the top layer depth h. In the top Fig. 3.1, the mth mode becomes unstable above each mth curve, and thus, if $n > 2$, then at $\gamma_n(h) < \gamma(h) < \gamma_{n+1}(h)$, unstable modes will have numbers $m = 2, 3, \ldots, n$. From (3.7) and Fig. 3.1, it can be seen that heton has the minimal stability at $h = 0.5$. In this case $\gamma_2\Big|_{h=0.5} = \gamma^* = 1.705$. For high modes, we have

$$\gamma_m \sim m, \tag{3.8}$$

where γ_m is the root of the equation

$$L_1(\gamma) = \frac{1}{2m} \tag{3.9}$$

(see Eq. (3.7)). The validity of the asymptotic estimate (3.8) can be seen from the following set of solutions (3.9), obtained at $h = 0.5$:

$$\gamma_3 = 2.840, \quad \gamma_4 = 3.893, \quad \gamma_5 = 4.919, \quad \gamma_6 = 5.934. \tag{3.10}$$

The analysis of the top part of Fig. 3.1 shows that the most unstable are hetons with equal layer thicknesses. At $\gamma < \gamma^*$, all perturbation modes are neutrally stable. At $\gamma > \gamma^*$, instability conditions become valid for the second mode, and higher modes become unstable as γ increases further. Note that the explicit dependence of the stability properties of oceanic rings on parameter γ is discussed, in particular, in [666, 671].

At a fixed value of stratification parameter γ and a simultaneous increase in the relative thickness h_j of the layer with number j ($j = 1, 2$), the stability factor of each mode will also increase (symmetrically with respect to j). As can be seen from the bottom part of Fig. 3.1 and Table 3.1, a decrease in the top-layer thickness h is accompanied by an increase in heton stability, while the entire spectrum of increments of unstable modes shifts into the domain of greater γ.

We will model the nonlinear stage of the process of collapse of unstable vortices with the help of CDM, whose algorithm is briefly described in Sect. 1.2.3. With this in view, we will take 120 markers (fluid particles) on each, originally circular, contour with unit radius. The values of potential vorticity of vortex patches in the layers will be defined such that the dimensionless time unit will correspond to a half of the rotation period of the top-layer vortex. From (1.47), we have

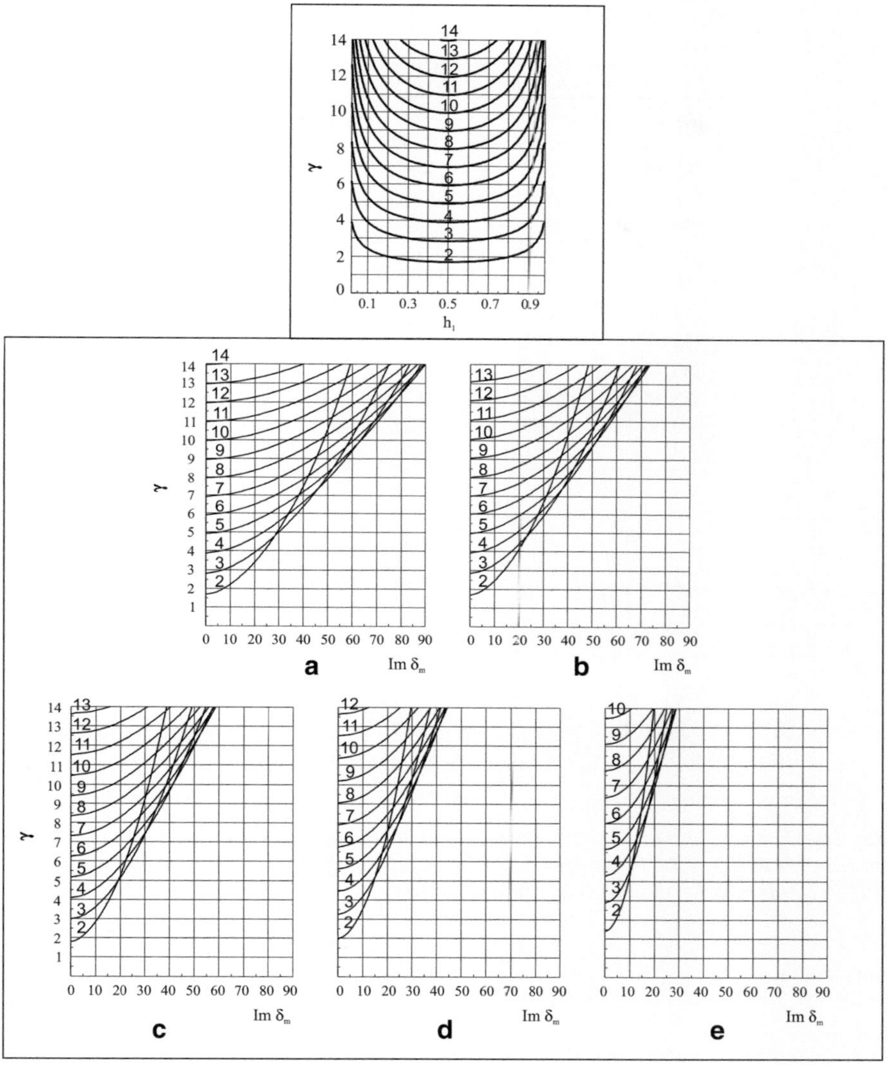

Fig. 3.1 Heton neutral-stability curve on the plane of parameters (h, γ) (*top*) and the appropriate relationships between the absolute values of growth increments of unstable modes $|\mathrm{Im}\delta_m|$ and parameter γ at $h = 0.5$ (**a**), $h = 0.4$ (**b**), $h = 0.3$ (**c**), $h = 0.2$ (**d**), $h = 0.1$ (**e**) (*bottom*). Figures at curves are the values of m for those curves

$$V_1\Big|_{r=1} = \bar{\Pi}_1 W(h, \Delta, \gamma),$$

where

$$W(h, \Delta, \gamma) = \frac{1}{2}h(1+\Delta) + \Big[h(1+\Delta) - 1\Big] L_1(\gamma),$$

Table 3.1 The absolute values of growth increments in unstable modes of heton at $\gamma = 14$

m	$h = 0.5$	$h = 0.4$	$h = 0.3$	$h = 0.2$	$h = 0.1$
2	59.17	48.31	38.73	29.57	19.70
3	74.82	61.09	48.96	37.35	24.82
4	83.52	66.17	54.60	41.59	27.48
5	88.23	71.99	57.59	43.73	28.58
6	90.01	73.70	58.59	44.25	28.38
7	89.40	72.84	57.99	43.38	26.92
8	86.70	70.54	55.82	41.19	24.06
9	82.08	66.62	52.27	37.65	19.33
10	75.55	61.09	47.25	32.53	10.80
11	66.92	53.75	40.46	25.17	
12	55.73	44.14	31.16	12.50	
13	40.00	30.30	15.78		
14	6.67				

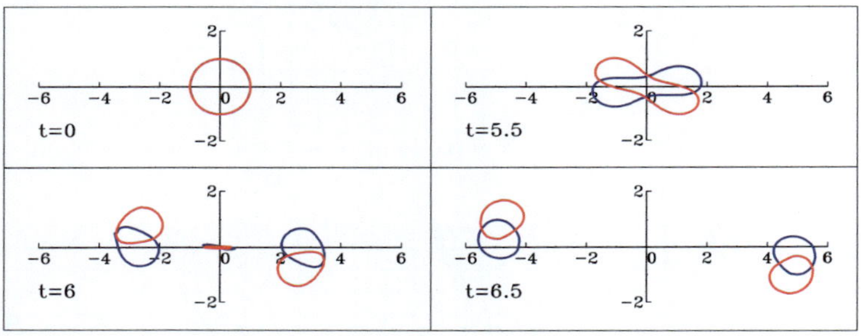

Fig. 3.2 Configurations of contours of the *top* (*full lines*) and *bottom* (*dashed lines*) layers for an unstable heton at $\gamma = 2.4$ in the specified moments of dimensionless time. Adapted from [851] (Fig. 3a) by permission of Cambridge University Press

and hence,

$$\bar{\Pi}_1 = 2\pi/W.$$

The value of $\bar{\Pi}_2$ is determined from the condition $\Delta = -1$, i.e., $h\bar{\Pi}_1 + (1-h)\bar{\Pi}_2 = 0$.

In all calculations that follow (except for Sect. 3.1.2), we will consider the case of equal thicknesses ($h = 0.5$).

Figures 3.2 and 3.3 give the results of calculations where the most unstable modes are those with numbers 2 and 3, respectively (see the bottom part of Fig. 3.1). The initial configurations of the vortex patches are assumed circular; the sources of their perturbations are errors in the approximation of the scheme and rounding. Calculations show that, under the effect of this "noise," the original vortex remains axisymmetric for some time, next deforms, and, finally, falls into two-layer pairs with tilted axes, running away in radial directions. It is worth mentioning that at

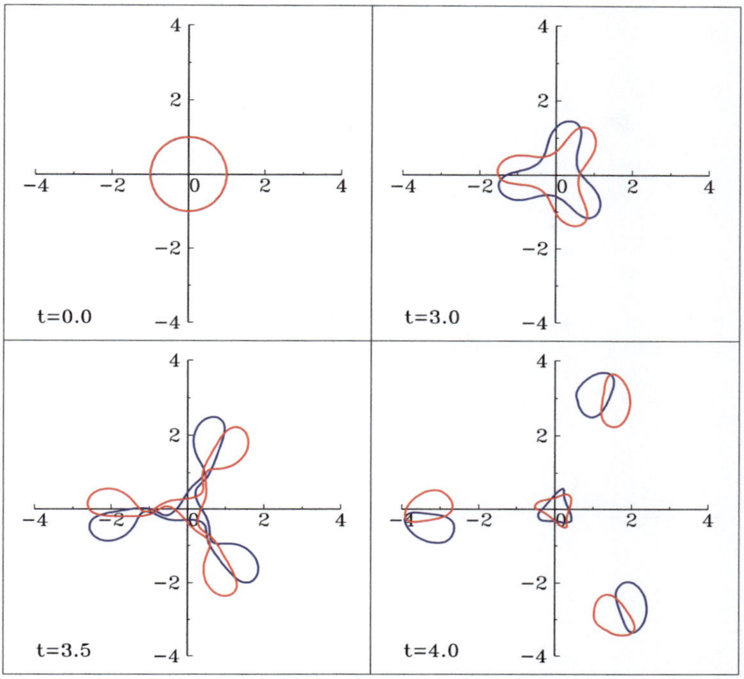

Fig. 3.3 The same as in Fig. 3.2, but at $\gamma = 4$. Adapted from [851] (Fig. 3b) by permission of Cambridge University Press

$m = 3$, the collapse will be incomplete—the central part of the vortex will remain in the same place (as shown below, this property will appear always when $m > 2$).

With an increase in parameter γ (a weaker stratification), the process of an unstable heton decomposition becomes more complex. An interesting example of the so-called "cascade instability" is given in Fig. 3.4, where $\gamma = 14$ and, at the initial moment, a weak harmonic perturbation of the form (1.68) with $\varepsilon_1 = \varepsilon_2 \equiv \varepsilon = 0.02$ and $m = 9$ is superimposed onto the circular contours of vortex patches. In this numerical experiment, the decomposition of the vortex begins earlier and consists of three stages:

- first, a set of nine pairs of hetons moving apart will form,
- next, one more set of nine smaller hetons will start,
- next, a tendency toward the generation of one more vortex front can be seen.

In this case, as mentioned above, a small, practically immobile core persists in the central part.

As can be seen from table 3.3, at $\gamma = 14$ and $h = 0.5$, the unstable modes correspond to $m = 2, 3, \ldots, 14$, and the most unstable one is the mode with $m = 6$. A decrease in h is accompanied by a decrease in both the increments of all modes and the number of unstable modes of higher order. The actual realization of a mode

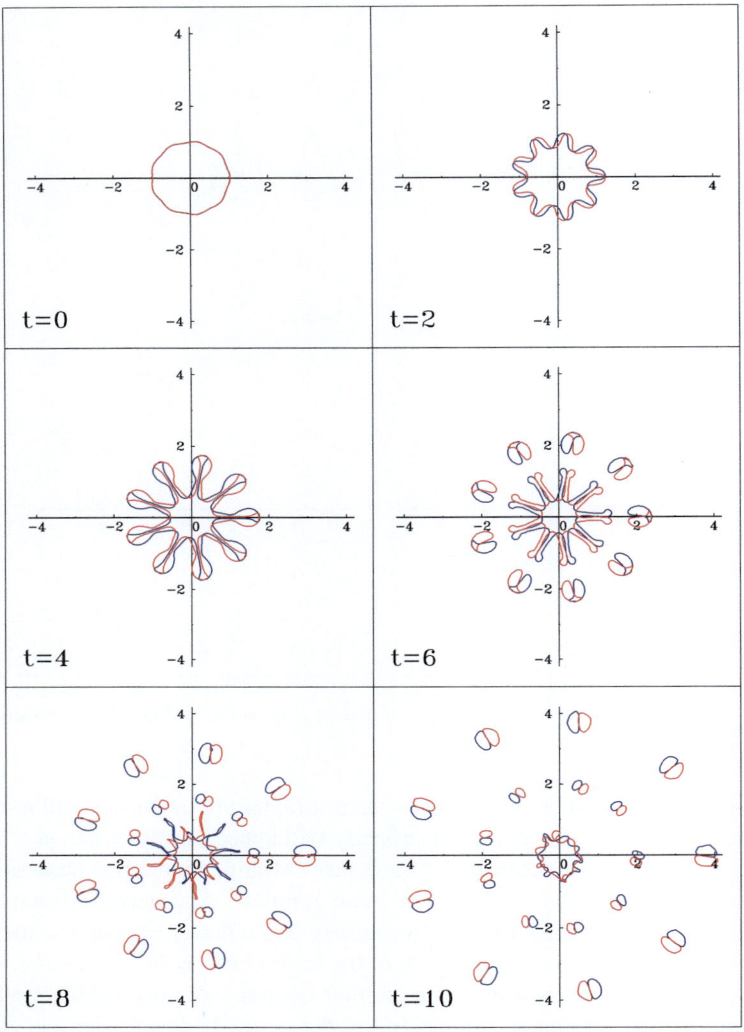

Fig. 3.4 An illustration of the process of cascade instability of a weakly disturbed heton at $\gamma = 14$, $\varepsilon = 0.02$ for $m = 9$. The contours of the *top* (*bottom*) layers are shown in *red* (*blue*)

is determined not only by its growth increment, but also by the specified initial perturbation. However, numerical calculations show that at large enough γ, the initial perturbations in the form of lower harmonics generally lead to the formation of unstable modes of higher order.

For example, in Fig. 3.5, where $m = 4$ for the initial perturbation, complex multicascade vortex structures form at the nonlinear stage: in the final calculation

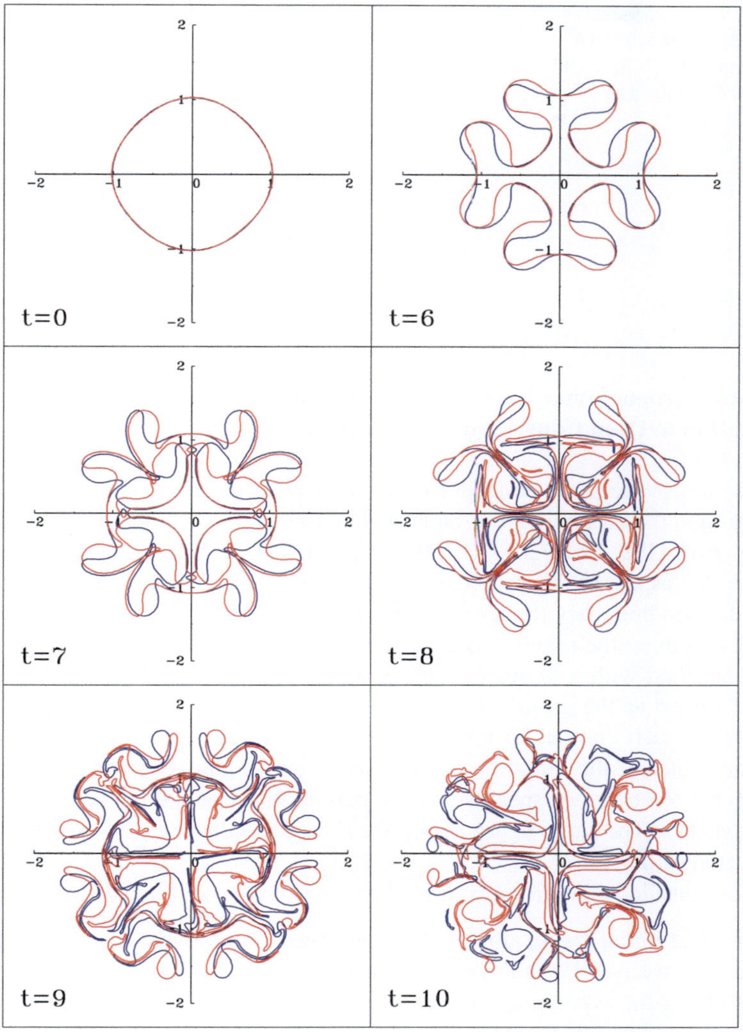

Fig. 3.5 Evolution of an unstable heton at $\gamma = 14$ and at the initially specified perturbation with $\varepsilon = 0.02$ for $m = 4$. The contours of the *top* (*bottom*) layer are given in *red* (*blue*)

moment ($t = 10$), at least three vortex pairs octets with different characteristic scales reached the frontal area. It is important that they took part in the "leapfrog" type of interaction.

The illustrations of the destruction of unstable hetons vividly show their important feature, i.e., the extension of the outer boundary of the domain filled by whirling fluid. Since any heton is associated, one way or another, with a thermal (or density) anomaly, we can speak of the motion of a warm or cold patch front.

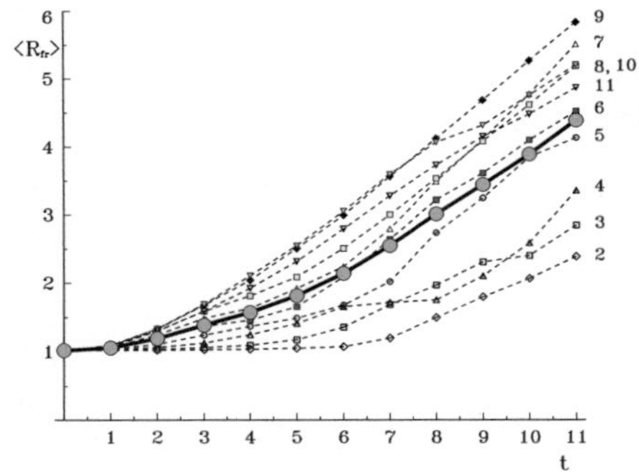

Fig. 3.6 Time dependences of $\langle R_{fr} \rangle$ at $h = 0.5$, $\gamma = 14$, $\varepsilon = 0.02$ and $m = 2, 3, \ldots, 11$. The *thick line* shows an "averaged" law

"Ballistic" Propagation; Law of Vortex Domain Boundary: Application to Deep Convection in the Ocean

The idea to use the heton concept for the description of the dynamic stage of deep convection in the ocean was first suggested by Legg and Marshall in [526], and next developed in [199, 200, 220–222, 229, 325, 326, 395, 525, 527, 540]. Here we will consider one aspect of this problem, associated with the time law of propagation of vortex domain boundary (temperature patch).

In determining the required dependence, we will track the motion of the vortex front, identified with a dummy particles with radius-vector $\langle R_{fr} \rangle$. Its coordinates are determined as the arithmetic mean of the coordinates of 40 fluid particles (20 particles on each contour in the top and bottom layer patches) the most distant from the center of the initial heton. The set of curves thus constructed is given in Fig. 3.6 for ten cases of specified initial perturbations of the boundaries of heton vortex patches. The same figure shows the time dependence of the weighted mean distribution of $\langle R_{fr} \rangle$.

Let us consider some features of the behavior of the constructed curves:

- after some initial stage, almost all dependencies (and, what is of prime importance, the average dependence) obeys a ballistic (linear) law;
- the slope of the asymptotic line correlates with the number of perturbation mode (contrary to some deviations from a linear dependence).

Such deviations are due to the specific features of the nonlinear stage of development of each mode. In particular, the almost horizontal intermediate plateau on the curve with $m = 4$ is due to the alteration of the modal structure of the vortex (see Fig. 3.5); however, with time, the curve acquires a nonzero slope. The case of $m = 9$, for which the front motion velocity becomes very nearly constant, will be considered below as a reference case.

It is interesting, that the quasi-linear behavior was observed for all three leading fronts during the heton cascade collapse (Fig. 3.7). It should be noted that we used the shorter time steps when calculating the evolution of vortex patches than it is given in Fig. 3.4. The position of the trailing edge was determined using the same

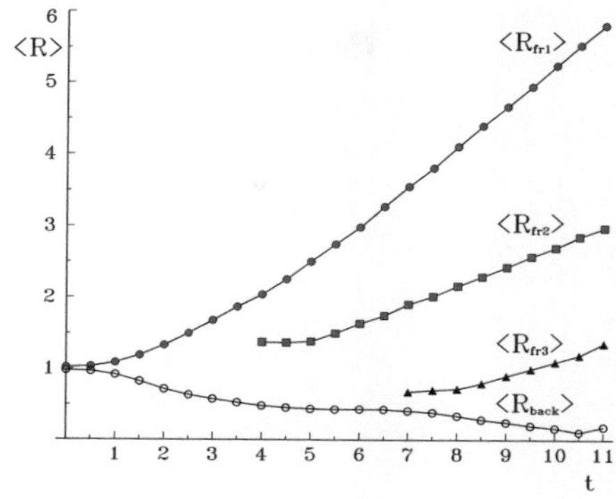

Fig. 3.7 The same as in Fig. 3.6 for the first $\langle R_{fr1} \rangle$, second $\langle R_{fr2} \rangle$, and third $\langle R_{fr3} \rangle$ front edges and the back $\langle R_{back} \rangle$ edge for the reference experiment at $h = 0.5$, $\gamma = 14$, $\varepsilon = 0.02$ and $m = 9$ (Fig. 3.4)

algorithm as that used for the leading edges, but with the help of markers at minimal distance from the center. The results of this experiment (as well a many others) suggest the stabilization of $\langle R_{back} \rangle$.

In the numerical experiments considered above, the initial perturbations of vortex patch contours are regular. Now suppose that the small perturbation magnitudes A_{jk} (where still $j = 1, 2$, and the integral parameter $k = 1, 2, \ldots, 120$ is the number of the marker in the contour) are random (they are produced by a random number generator). An example of heton evolution for this variant is given in Fig. 3.8, where, to make it more vivid, the contours of vortex patches of the top and bottom layers are shifted apart. Clearly, the fluid particles scatter along random directions.[1] Figure 3.9, where curve "1" characterizes the function of front velocity, averaged over 10 realizations under random perturbations of contours as compared with the velocity of the reference front (curve "2"), suggests that the ballistic; scatter law is valid in the general case as well. Wirth [983], who analyzed the results of numerical experiments studying the process of baroclinic instability of zonal flows in the ocean, and Dewar [217, 218], who studied the regularities in the propagation of a heat patch in both the open ocean and a coastal zone, came to the same conclusions.

Note that convective temperature anomalies play a significant role in the heat balance of the ocean [571]. In ocean general circulation models, the mesoscale heat transfer is generally assumed diffusive (proportional to the local temperature gradient) [294, 570, 668, 965]. In the models using constant diffusion coefficient, the mean radius of temperature anomaly increases as $\langle R \rangle \sim t^{1/2}$. The growth of the mean radius in the models based on a nonlinear diffusion coefficient proportional to temperature gradient is even weaker: $\langle R \rangle \sim t^{1/3}$. As shown above, the heton theory and numerical experiments convincingly suggest the linear law $\langle R \rangle \sim t$, leading to

[1] The values of stratification parameters in most calculations considered here was chosen for qualitative comparison of this figure (and those similar to it, where perturbations of contours were determined by a set of random amplitudes) with the results of laboratory experiments in the work [395], assuming $\gamma = 14$.

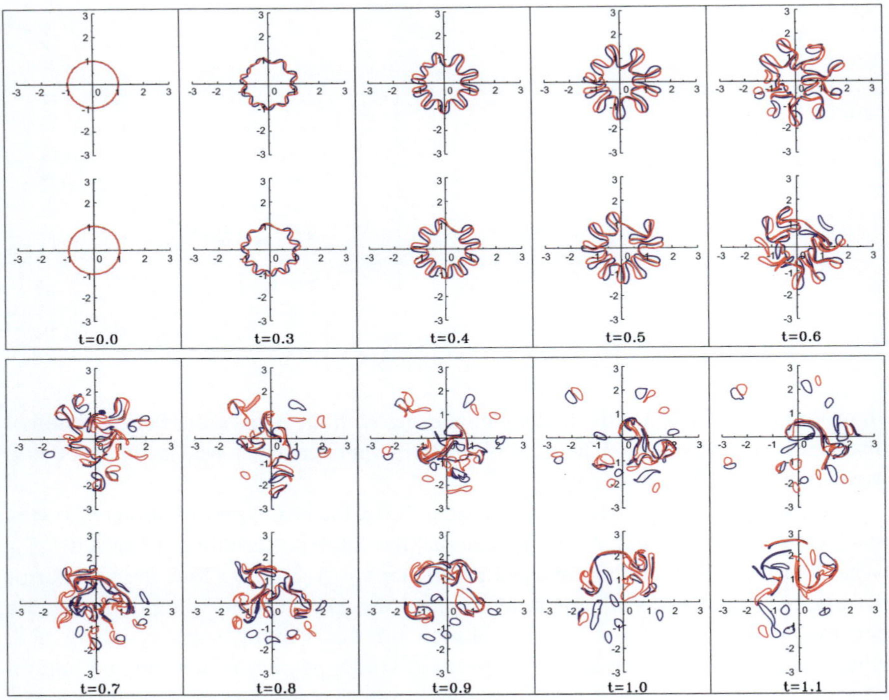

Fig. 3.8 Evolution of heton with random initial perturbations of contours at $h = 0.5$, $\gamma = 14$ and $|\varepsilon_{jk}| \leq 0.02$. The *top* and *bottom lines* of each frames show synchronous contours of vortices for two different realizations

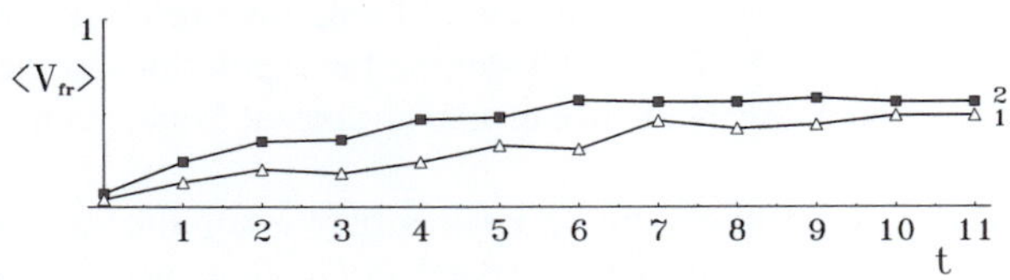

Fig. 3.9 The mean propagation velocities of the fronts at $h = 0.5$ and $\gamma = 14$: *1* a heton averaged over ten realizations with random perturbation of contours with $|\varepsilon_{jk}| \leq 0.02$; *2* the reference heton at $m = 9$, $\varepsilon = 0.02$

more efficient heat transfer as compared with the diffusion one. The problem of construction of physically sound parameterizations, taking into account the non-diffusion character of heat transfer by hetons, has not been solved completely [324, 525, 864]. The present results give grounds for optimism.

Analogy with *A*-Symmetric Structure of Discrete Hetons

An evident analogy exists between the behavior of a distributed heton and a system of discrete hetons with vertical axes, uniformly distributed over a circumference (Sect. 2.2.1). Indeed, returning to Figs. 2.2–2.4, we notice the following circumstances:

- If *the effective radius*[2] of the circumference R is greater than R^S_{min}, then the originally vertical heton axes tilt, and the newly formed two-layer pairs will move away along radial directions. The distributed dipole structures that form as the result of collapse of an unstable finite-size heton behave in a similar manner.
- The asymptotics (3.8), which implies a linear dependence between the critical values of stratification parameter γ_m and the numbers of high unstable modes m, is still valid in discrete case up to proportionality factor

$$R^S_{min}(m) \sim \alpha m, \quad \alpha \approx 2.7.$$

However, one should take into account the fundamental difference between the two models:

  – for discrete vortices at $R > R^S_{min}(m)$, a system of m radially scattering pairs always form,

  while

  – for a distributed heton, the value of γ_m determines only the lower boundary of the domain in which the mode with number m becomes unstable, and the conditions of realization of this mode are not necessarily preferable.

- Nevertheless, this analogy enables a mathematical explanation to be given for the possible separation mechanism of distributed pairs: **for a newly formed vortex pair to start moving away from the center of the original vortex, it is necessary that its local vorticity center fall beyond the separatrix bounding the domain of finite motions of the appropriate system of discrete vortexes**.
- The zone of "irregular scatter" (Fig. 2.4) in the case of a distributed heton corresponds to a domain of intense thread formation and "dissipation" of small-size vortex structures.
- An important feature of the analogy is also a correlation between the existence of a domain of capture of discrete hetons in the Fig. 2.4, on the one hand, and the corresponding establishment of $\langle R_{back}\rangle$ (Fig. 3.9) after the collapse of an unstable distributed heton at small R, on the other hand.

---

[2]Note that we used variables of the type $R = \gamma r$ as space variables in the problems of the theory of discrete vortices. Thus, an increase or a decrease in the distance can be interpreted as an appropriate change in the stratification parameter γ, which, in its turn, is responsible for the state of stability of a distributed two-layer vortex.

Noncompensated Two-Layer Vortex

All previous results referred to hetons, purely baroclinic two-layer structures, satisfying the compensation condition $\Delta = -1$. The presence of a nonzero barotropic component of motion, which manifests itself in a weaker potential vorticity of the bottom-layer patch, leads to the phenomenon of *localization of instability domains* for each mode (Fig. 3.10). In this figure, boundaries of infinite instability areas are black colored, and those ones of finite instability areas (two boundaries for each m) have their special colors.

At the same time, we can see an increase in the degree of instability of vortices with a thicker top layer. The arrows in stability diagrams point to the top and bottom boundaries of the instability domains of the appropriate modes, where it is necessary for better understanding of the picture. As can be seen from Fig. 3.10, when the contribution of the barotropic component is significant, the instability domains of different modes (at least, lower ones) may not intersect, resulting in the formation of intermediate domains with neutral vortex stability. Because of this, the instability zone becomes discontinuous with respect to parameter γ. This effect is shown in Fig. 3.11, where the successive increase in the stratification parameter causes changes in regimes: (a) instability of the second mode, (b) neutral stability, (c) instability of the third mode. In this case, the contribution of the strong barotropic component manifests itself in that, at the realization of the mth unstable mode, the behavior of the top-layer vortex is stable, and m satellites separate from the periphery of the core of the bottom-layer vortex patch (after a short period of active filamentation). The entire two-layer structure rotates in the direction determined by the top-layer anticyclone. It is worth noting that the originally round vortex patch of the top layer, as well as the already formed core of the bottom layer, in the process of their evolution, take on a quasi-elliptic shape at $m = 2$ and a quasi-triangular shape at $m = 3$.

3.1.2 Annular Two-Layer Vortex: Four Circular Vortex Patches

Let us consider models that more adequately reflect the kinematics of oceanic rings. The main mechanism of the formation of rings in the currents of the type of Gulf Stream and Kuroshio is known to be the meandering of current with subsequent looping of the meander. In this case, the vortex that has formed still has a jet structure, which is typical of the current, with a maximum of azimuthal velocity shifted from the external boundary toward the center. Staying in the class of piecewise-constant distributions of potential vorticity, we can simulate the ring current in each layer by a set of two nested vortex patches (an analogue of Carton–McWilliams vortex [152]). Such study within the framework of barotrophic model was carried out by Kozlov and Makarov in [472, 473]; here, their results are generalized to the case of two-layer fluid.

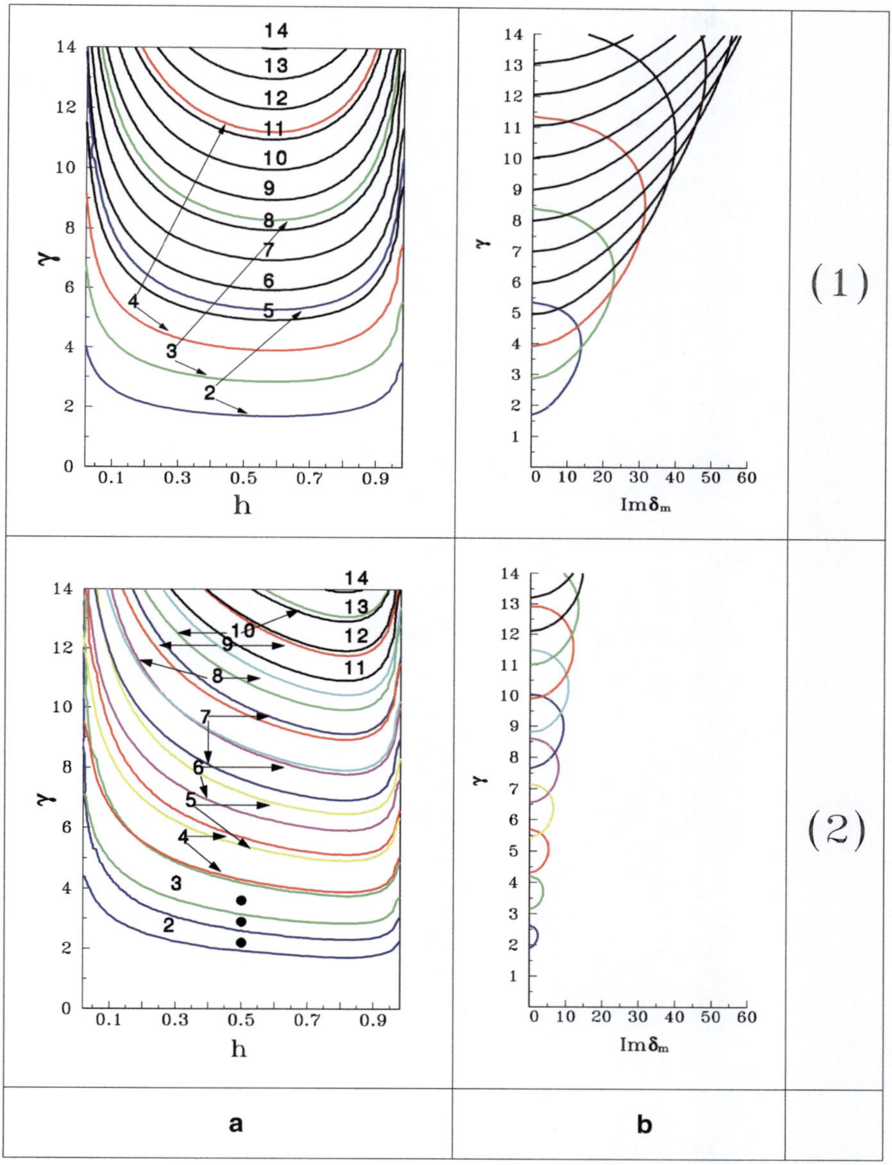

Fig. 3.10 (**a**) Neutral-stability curves for a noncompensated two-layer vortex on the plane of parameters (h, γ) and (**b**) the appropriate relationships between the absolute values of increments of unstable modes $|\mathrm{Im}\delta_m|$ and the parameter γ at $h = 0.5$, as well as at (*1*): $\Delta = -0.5$ ($m = 2$ *blue*, $m = 3$ *green*, $m = 4$ *red*, $m \geq 5$ *black*) and (*2*): $\Delta = -0.05$ ($m = 2$ *blue*, $m = 3$ *green*, $m = 4$ *red*, $m = 5$ *yellow*, $m = 6$ *light magenta*, $m = 7$ *light blue*, $m = 8$ *cyan*, $m = 9$ *brown*, $m = 10$ *light green*, $m \geq 11$ *black*). Figures near the curves correspond to the numbers m of unstable modes

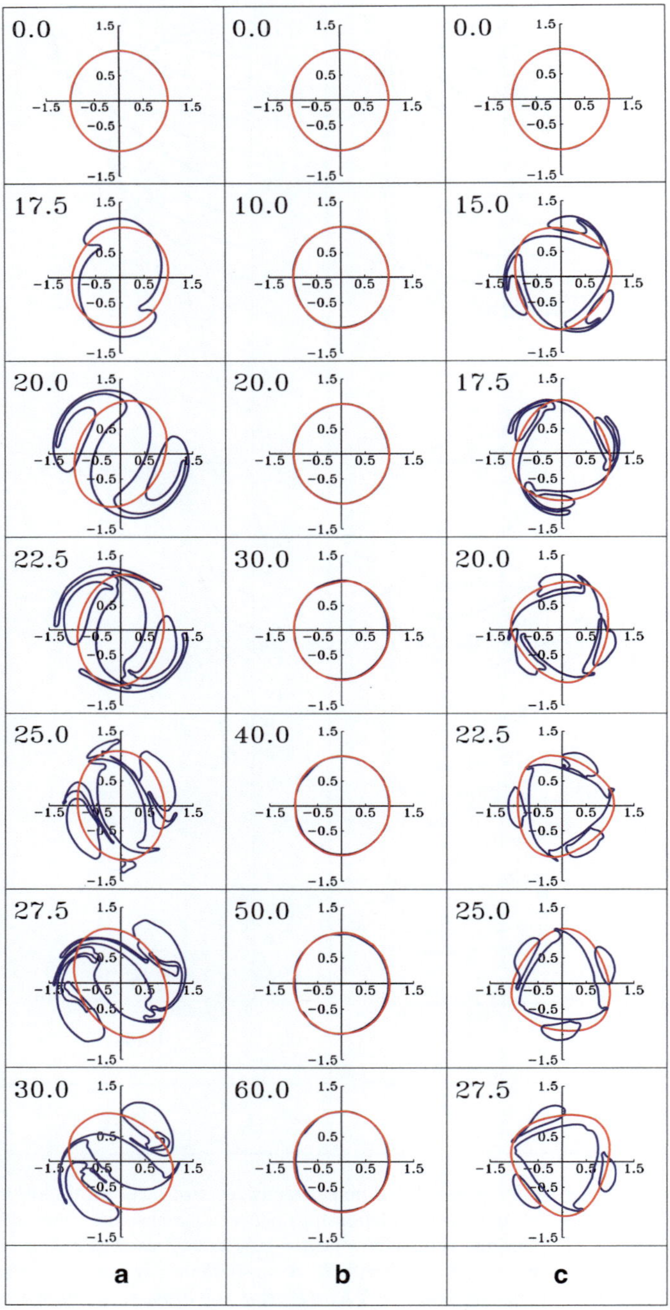

Fig. 3.11 The evolution of noncompensated two-layer vortex at $\Delta = -0.05$: (**a**) $\gamma = 2.2$, (**b**) $\gamma = 2.9$, (**c**) $\gamma = 3.6$

Studying the Stability of Rings

Instead of (3.1) we assume

$$\Pi_1(\rho) = \bar{\Pi}_{11} H(a_1 - \rho) + \bar{\Pi}_{21} H(a_2 - \rho),$$
$$\Pi_2(\rho) = \bar{\Pi}_{21} H(a_1 - \rho) + \bar{\Pi}_{22} H(a_2 - \rho). \tag{3.11}$$

Without loss of generality, we assume $a_1 = \alpha$, $a_2 = 1$ and consider two particular cases:

$$a)\, \bar{\Pi}_{21}/\bar{\Pi}_{11} = \bar{\Pi}_{22}/\bar{\Pi}_{12} = -\alpha^2, \quad b)\, \bar{\Pi}_{11} + \bar{\Pi}_{21} + \bar{\Pi}_{12} + \bar{\Pi}_{22} = 0. \tag{3.12}$$

Acting by analogy with (1.46)–(1.49), we obtain that the velocity field generated by such two-layer vortex has an azimuthal component of the type

$$V_1(\rho) = PV_{bt} - (1-h)QV_{bc}, \quad V_2(\rho) = PV_{bt} + hQV_{bc}, \tag{3.13}$$

where $P = h\bar{\Pi}_1 + (1-h)\bar{\Pi}_2$, $Q = \bar{\Pi}_2 - \bar{\Pi}_1$. The variables V_{bt} and V_{bc}, which characterize the barotropic and baroclinic velocity components, have the form

$$V = \begin{cases} V^{\alpha} - \alpha^2 V^1, & \text{in the case } \; a), \\ V^1 - V^{\alpha}, & \text{in the case } \; b), \end{cases}$$

and

$$V_{bt}^{\alpha} = \frac{\alpha}{2} \begin{cases} \rho/\alpha, & \rho \le \alpha, \\ \alpha/\rho, & \rho \ge \alpha; \end{cases} \qquad V_{bc}^{\alpha} = \alpha \begin{cases} \mathrm{K}_1(\gamma\alpha)\mathrm{I}_1(\gamma\rho), & \rho \le \alpha, \\ \mathrm{I}_1(\gamma\alpha)\mathrm{K}_1(\gamma\rho), & \rho \ge \alpha. \end{cases}$$

According to formulas (3.12)–(3.13), in the barotropic case, the condition *a*) from (3.12) implies complete isolation, within a unit circle, of the vortex that has maximum azimuthal velocity at $\rho = \alpha$, and the condition *b*) implies zero velocity at $\rho \le \alpha$, its practically linear growth at $\alpha \ge \rho \ge 1$, and further decrease of $V_j \sim \rho^{-1}$, $j = 1, 2$. The extent of the effect of baroclinicity on the distributions of (3.12)–(3.13) is shown in Fig. 3.12, where typical profiles of azimuthal velocities, standardized by $\bar{\Pi}_{11}$ are given and denotation $\kappa = \bar{\Pi}_{21}/\bar{\Pi}_{11}$ is introduced.

Since (3.11), as well as any axially symmetric distribution represents a stationary solution, we can examine its stability, assuming instead of (1.68),

$$\varepsilon_{jk}(\theta, t; \rho) = \varepsilon_{jk}(\rho_j) e^{im(\theta - \delta t)}, \quad m \ge 1, \tag{3.14}$$

where now $j, k = 1, 2$.

The corresponding spectral problem in this case reduces to solving a system of eight algebraic equations. Their structure is such that, equating the system determinant to zero, we can always restrict ourselves to the analysis of a fourth-order

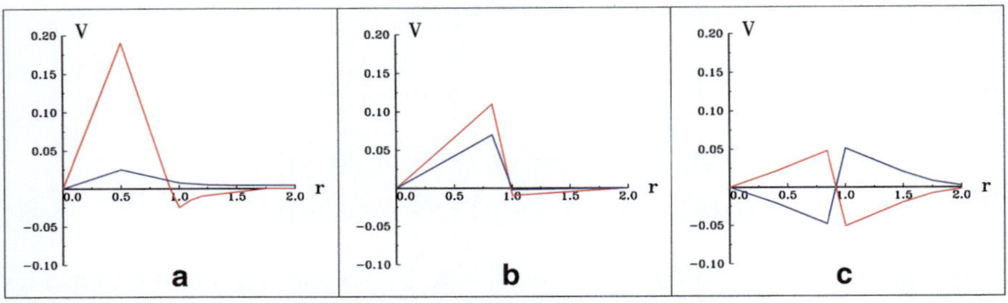

Fig. 3.12 Profiles of azimuthal velocities V_1 (the *top layer*, the *full line*) and V_2 (the *bottom layer*, the *dashed line*) in a ring two-layer vortex at (**a**) $\kappa=0$, $\gamma=2$, $h=0.5$, $\alpha=0.5$, (**b**) $\kappa=0.5$, $\gamma=2$, $h=0.25$, $\alpha=0.825$, (**c**) $\kappa=-1$, $\gamma=3$, $h=0.5$, $\alpha=0.84$. They characterize the initial conditions for the numerical experiments, represented by Figs. 3.13–3.15

eigenvalue problem. As an example, let us consider the dispersion relationships that are obtained in the particular case of $\kappa=0$ (a passive bottom layer):

$$\delta_m = \frac{1}{2}\left[-(b_1+b_2)\pm\sqrt{(b_1-b_2)^2-4b_3^2}\right], \tag{3.15}$$

where

$$b_1 = \begin{cases} h(1-\alpha^2)/2+(1-h)\left[(1-2\alpha^2)L_{11}^{(1)}+\alpha L_{12}^{(1)}\right]-h/2-(1-h)L_{11}^{(m)}, \\ (1-h)\left(L_{11}^{(1)}-L_{12}^{(1)}/\alpha\right)+h/2+(1-h)L_{11}^{(m)}; \end{cases}$$

$$b_2 = \begin{cases} \alpha(1-h)\left(L_{12}^{(1)}-\alpha L_{22}^{(1)}\right)+\alpha^2\left[h/2+(1-h)L_{11}^{(m)}\right], \\ h(1-\alpha^2)/2-(1-h)\left(\alpha L_{12}^{(1)}-L_{22}^{(1)}\right)-h/2-(1-h)L_{22}^{(m)}; \end{cases}$$

$$b_3 = \left[h\alpha^m/2+(1-h)L_{12}^{(m)}\right]\begin{cases} \alpha^2, \\ 1. \end{cases}$$

Here, the top line everywhere corresponds to the case *a*), while the bottom line, to the case *b*); denotations $L_{11}^{(m)} = m\mathrm{I}_m(\gamma\alpha)\mathrm{K}_m(\gamma\alpha)$, $L_{12}^{(m)} = m\mathrm{I}_m(\gamma\alpha)\mathrm{K}_m(\gamma)$, $L_{22}^{(m)} = m\mathrm{I}_m(\gamma)\mathrm{K}_m(\gamma)$ are introduced here.

At $h=1$, when the top layer extends throughout the depth, the dispersion relationships (3.15) aquire their limiting barotropic form [472, 473].

At $\gamma \sim O(1)$ and any $\alpha \le 1$ in both cases there exists a finite (bounded by the value of M_a or M_b, respectively) number of exponentially growing modes with numbers m, such that $2 \le m \le M_a$ and $3 \le m \le M_b$.

For the general case of $\kappa \neq 0$, the dispersion relationships are more cumbersome and we do not give them here.

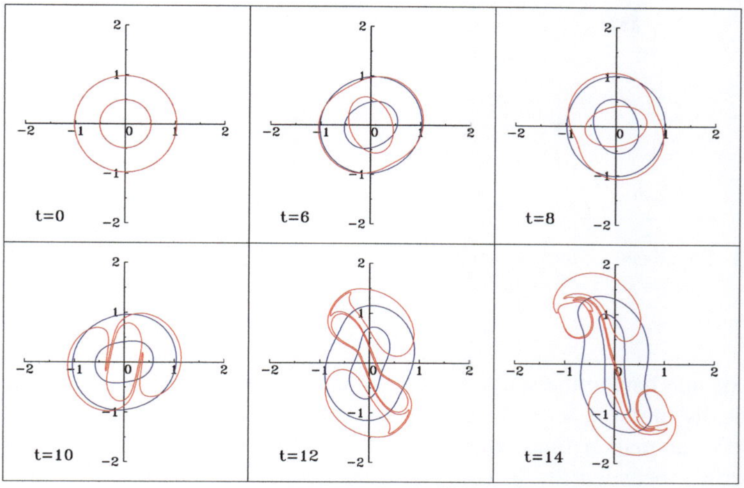

Fig. 3.13 The process of formation of a *mushroom-shaped* structure from a two-layer ring vortex at $\kappa = 0$, $\gamma = 2$, $h = 0.5$, $\alpha = 0.5$

According to stability analysis, the three velocity profiles given in Fig. 3.12a–c correspond to instability conditions of the second, fifth, and sixth modes, respectively.

Modeling the Transformations of an Oceanic Ring into Smaller Vortex Structures

The work [473] proposed a model of formation of a mushroom-shaped current [255, 256, 607] as the result of development of a shear instability of a vortex with a compensation envelope when instability conditions hold for the mode with $m = 2$. During its decomposition, the vortex produces two pairs moving apart in different directions; one of them obviously will be brought by the current that has generated the ring, while the other—along with a part of the neck connecting them—forms a mushroom-shaped configuration, which can be often seen in satellite photographs and apparently is located only in the surface oceanic layer.

In his paper [956], where an impulse two-layer model of a mushroom-shaped current is proposed, Voropayev writes: "No reliable data on mushroom-shaped currents in the water mass of stratified ocean are available yet; therefore, we will consider only currents in the top oceanic layer." Here, we will also extend the barotropic model [473] to a two-layer model where nonzero vorticity is concentrated only in the top layer ($\kappa = 0$).

Figure 3.13 gives the contours of an evolving ring domain at $\gamma = 2$, $h = 0.5$ and $\alpha = 0.5$ in the given time moments. To accelerate the process, we chose the contours at the initial moment to have the shape of ellipses slightly elongated along

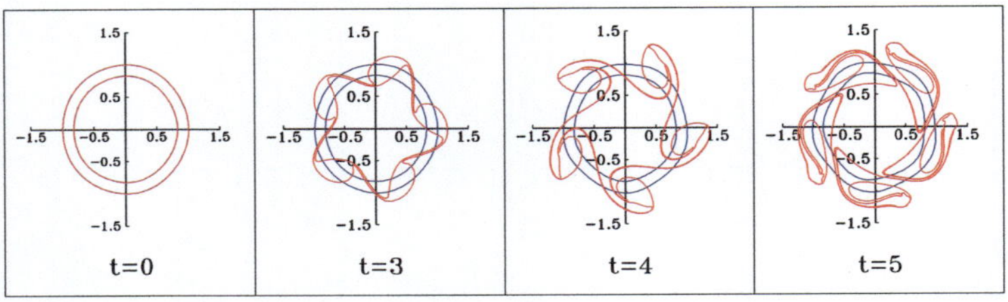

Fig. 3.14 The process of formation of cyclonic ringlets at the periphery of an anticyclonic ring at $\kappa = 0.5$, $\gamma = 2$, $h = 0.25$, $\alpha = 0.825$

the x-axis and having the semiaxes ratio of $a/b = 1.05$, which is associated with a theoretically unstable second eigen mode. In this case, the contours of the bottom layer serve as marker lines, and their deformations, as well as any motion below the interface between the layers, are induced by top-layer vorticity. The surface vortex structure at the first stage is involved in the rotation, which becomes slower as the ring decays. Next, two two-core mushroom-shaped vortex structures will form, moving away in opposite directions. Clearly, the cause of their formation is the shear instability of the loop-shaped jet flow in the top layer.

The following example can be regarded as the simplest model of generation of the so-called *ringlets* [411, 651]—small-size cyclonic satellite vortices, identified at the periphery of an anticyclonic vortex "82B". The two-layer model of vortex "82B" proposed in [672] and assuming the interface between layers to follow the 10° isotherm, allows us to take approximately $\gamma = 2$, $h = 0.25$. The character of velocity profiles given in [411] allows the barotropic component to be considered quite moderate, so we assume $\kappa = 0.5$. To obtain a configuration consisting of five ringlets and similar to that described in [411], we assign the value of 0.825 to the parameter α, predicted by the stability analysis and ensuring the predominance of the fifth mode. The profiles of azimuthal velocities in both layers at the initial moment are given in Fig. 3.12b. Moreover, here and in the next example, a small initial perturbation of contours with amplitude of 0.005 is specified, corresponding to the most unstable mode.

Figure 3.14, where the vortices of the top and bottom layers are artificially shifted relative to each other, shows how the external envelope of a ring in the surface layer transforms in the process of evolution into a construction of five smaller cyclones distributed almost uniformly. The entire vortex structure rotates in the anticyclonic direction under the effect of the central core. As can be clearly seen in Fig. 3.14, the formation of peripheral cyclonic vortices is accompanied by an intrusion of fluid from the compensating envelope into core domain. As noted by Nof [651], this is an indispensable condition of the appearance of ringlets.

The mechanism of barotropic-baroclinic instability used in the model can be considered as a possible cause of detachment of cyclonic vortices from a warm anticyclonic ring.

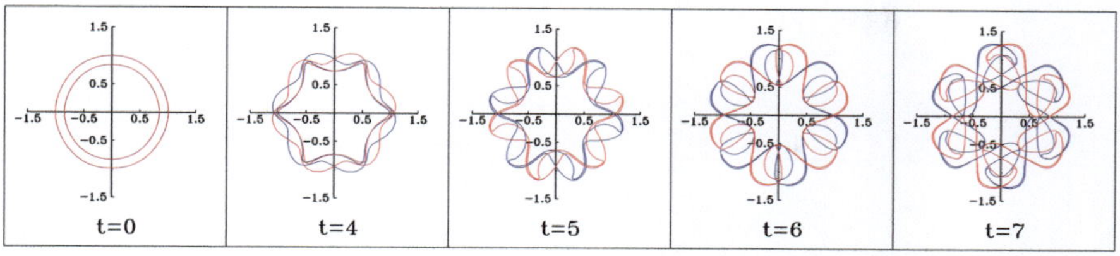

Fig. 3.15 Evolution of a heton-type ring at $\kappa = -1$, $\gamma = 3$, $h = 0.5$, $\alpha = 0.84$

Now let us consider an example (Fig. 3.15) of nonlinear evolution of an unstable ring of a heton structure; for this ring, we set $P = 0$ and specify the condition (3.12b); the profiles of its azimuthal velocities are given in Fig. 3.12c.

Calculations show that the outer contours in both layers at the first stage acquire the shape of hexagonal structures, shifted with respect to one another, and thus, six separate mini-hetons will form. The inner contours first form a vertically homogeneous configuration, whose shape is similar to that of the separatrix of the contour line field of the Hamiltonian for a hexagon consisting of discrete hetons (Fig. 2.2b); next, vortex threads actively form from the angular parts of this figure and wind onto the forming satellite vortices. The newly formed hetons with tilted axes tend to scatter, but next (in the last calculation fragment), they become practically aligned, and their vortex patches in each layer again tend to merge to form ring structures. Note that the stable vortex polar hexagon is observed in Saturn's atmosphere [9, 17].

Thus, this calculation demonstrates the possibility in principle of periodical transformation of an axisymmetric circular ring into a structure with six axes of symmetry, which next again transforms into a two-layer ring. Note that a similar situation was observed in numerical experiments [484] with a barotropic model, where, because of Helmholtz instability, the stream flow breaks to form a chain of vortices which later merge again, forming a continuous wave-like flow.

In particular, the latter two figures show that, during the nonlinear evolution of rings, small-scale polygonal vortex structures, either with or without a central vortex, can form. This fact characterizes an important applied aspect of the model of discrete vortices (Sects. 2.2 and 2.3).

3.2 The Impact of Finite Perturbations

A particular result of Sect. 3.1 is the establishment of conditions for the linear instability of an axisymmetric heton, comprising two vortex patches with constant vorticities $\bar{\Pi}_1$ and $\bar{\Pi}_2$ with opposite signs. It is shown that in the case of equal layer thicknesses, the necessary and sufficient condition of heton instability is $\gamma > 1.705$. At finite deviations from axial symmetry, theoretical (analytical) analysis of stability

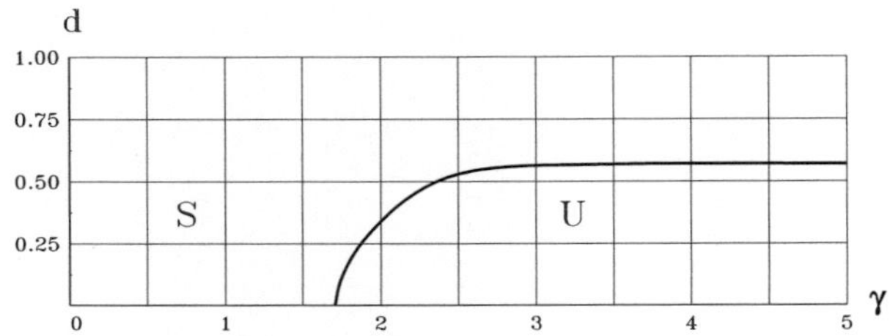

Fig. 3.16 Diagram of stable (*S*) and unstable (*U*) states of a heton with a tilted axis on the plane (γ, d). Adapted from [851] (Fig. 6) by permission of Cambridge University Press

becomes impossible. In that case, numerical experiments can be used to study the stability. Let us consider the simplest variants of asymmetry:

- two separate circular vortex patches—a heton with a tilted axis (an analogue of mode $m = 1$),
- two vortex patches, which originally had an elliptic shape—a heton with a vertical axis (an analogue of mode $m = 2$).

3.2.1 Heton with a Tilted Axis: Two Initially Circular Patches

A possible mechanism of formation of a heton with a tilted axis in the ocean is the baroclinic instability of the potential vorticity anomaly. The schematic representation of the development of such process, which was given in Fig. 1.3 in the preface, was taken from paper [633], studying the effects of isopycnic and diapycnic mixing of oceanic currents.

First, we elucidate the terminology. When, in the context of this problem, we speak about *circular* vortex patches, this means just that the patch contours are approximated by circumferences in the *initial* moment. While interacting with one another, the vortices are permanently changing their shapes, adapting to stationary one (see the Sect. 3.2.2). This note also refers to the term *elliptic* vortices, which will be discussed in the Sect. 3.2.3.

Let $2d$ be the distance between the centers of circular vortex patches comprising the heton and belonging to different layers. Intuition suggests (and calculations confirm) that, at large d, vortex patches behave practically as discrete vortices and the appropriate two-layer pair moves rectilinearly and uniformly in the direction normal to the straight line passing through their centers.

Numerical experiments show that such behavior of distributed hetons can also take place at moderate values of d, as well as at $d < 1$, i.e., at the initial overlapping of the top- and bottom-layer vortex patches. The boundary between stable and unstable regimes on the parameter plane (γ, d) is given in Fig. 3.16. The general result of many calculations can be formulated as follows: (1) at $\gamma < 1.705$, all motions of hetons with tilted axes are stable, (2) in the interval $1.705 \leq \gamma < 2.5$, the boundary line is a function of two variables: γ and d, (3) next, the instability domain

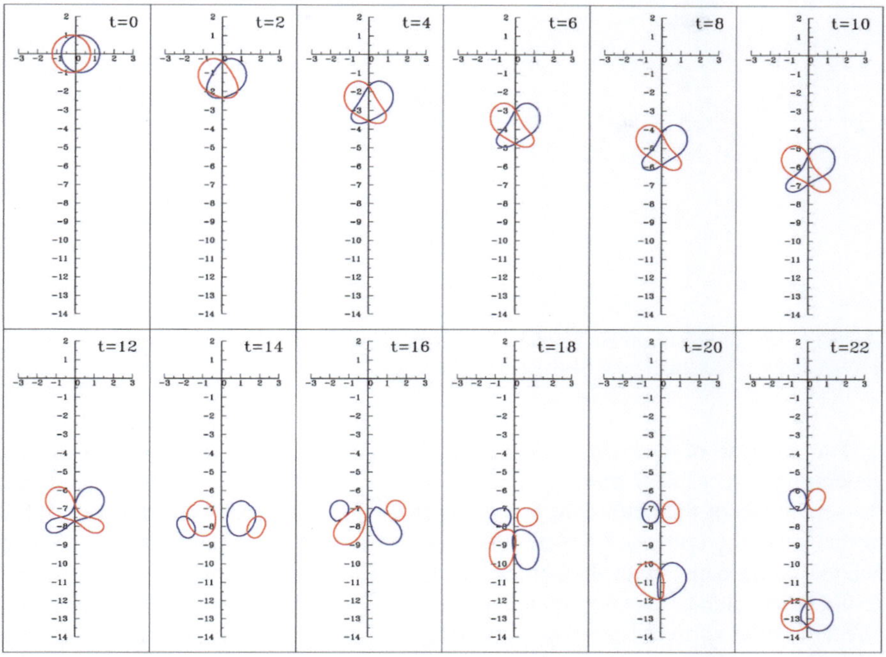

Fig. 3.17 Evolution of a heton with tilted axis at $\gamma = 1.9$, $h = 0.5$, $d = 0.24$

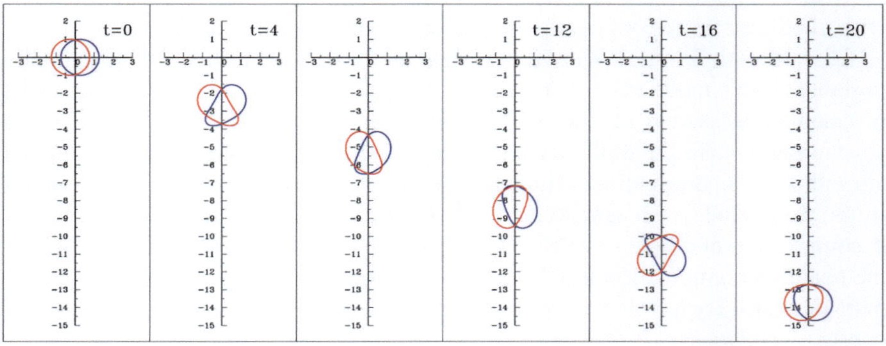

Fig. 3.18 The same as in Fig. 3.17 at $d = 0.25$

extends to the entire semi-infinite strip $[\gamma > 2.5; 0.0 < d < 0.6]$. This boundary, as shown in several examples below, is conventional (we will explain what we mean by "conventional"), but gives a physically sound characteristic of the stability state of hetons with tilted axes. Let us consider some examples.

Figures 3.17 and 3.18 show the behavior of hetons with tilted axes near the interface of regimes on different sizes of it at $\gamma = 1.9$, when the only unstable mode of an identical two-layer vortex with vertical axis is the second one (see Fig. 3.1).

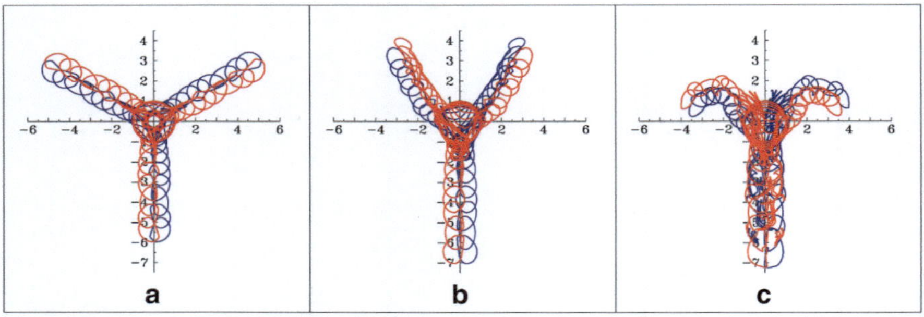

Fig. 3.19 Sets of combined configurations of contours for moments $t = 0, 1, \ldots, 11$ at $\gamma = 4$, $h = 0.5$ and $d = 0.00$ (**a**), $d = 0.01$ (**b**) and $d = 0.02$ (**c**)

The interior of domain U shows a "competition" between the first mode, generating a rectilinear motion of the pair, and the second mode, which induces the separation of the heton into two equal two-layer pairs. As seen from Fig. 3.17, in this case, the collapse of vortex patches consists of four major stages: (1) they become asymmetric with their thickness increasing near the rear front, (2) as the result, each patch breaks up into unequal parts, (3) smaller vortices rotate under the effect of larger vortices from the opposite layer, (4) two new two-layer pairs, moving apart, form in this case, the stronger one moving in the direction determined by the tilt of the axis of the original heton.

With a slight increase in d (Fig. 3.18), we pass into domain S, where the heton with tilted axis carries out purely translational motion without destruction.

Characteristic features of heton evolution under the condition that the maximally unstable is the mode $m = 3$ are given in Figs. 3.19 and 3.20. The first series of calculations shown in Fig. 3.19 is aimed to study small deviations from the axisymmetric state. In all three experiments, in the initial moment, heton contours are subject to a perturbation, symmetrical about the vertical axis and corresponding to the third mode with amplitude 10^{-3}. Figure 3.19a gives a general picture of decomposition of a heton, which had a vertical axis in the initial moment. Clearly, the resulting configuration has three axes of symmetry, along which the two-layer pairs that have formed due to the instability mechanism move away with equal and constant velocities.

In the next two fragments (b) and (c), the initial state corresponds to weak tilts of axes with $d = 10^{-2}$ and $d = 2 \cdot 10^{-2}$, respectively. Since the specified tilt initiates the motion of initial vortex patches in the negative direction of the y-axis, we see, at the initial stage of evolution, the redistribution of mass in favor of the pair moving in this direction. This, in its turn, leads to the capture of part of the swirling fluid from two-layer vortices moving away in lateral directions. They lose their compensation property and, hence, the property of linear motion. This tendency becomes stronger

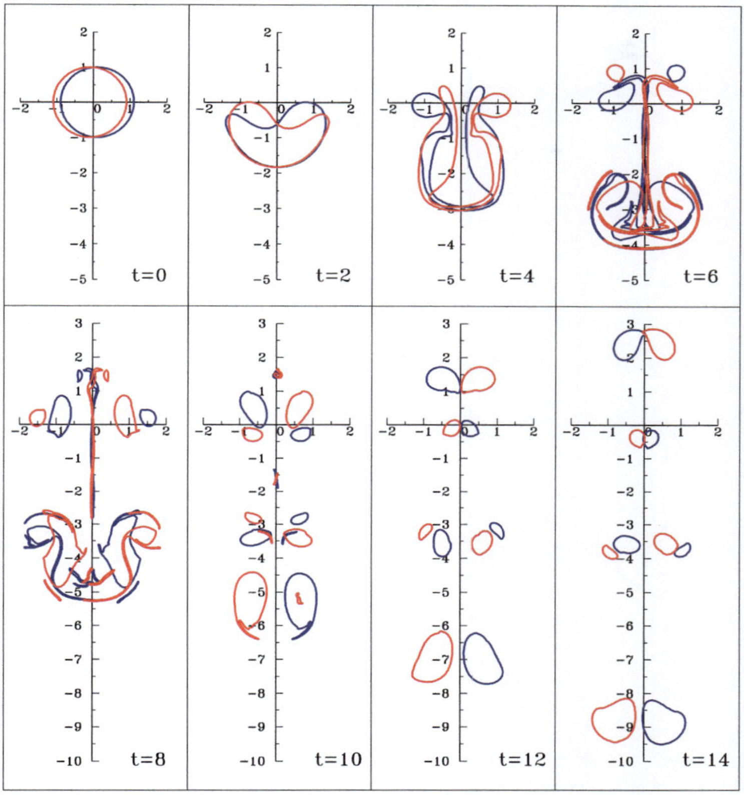

Fig. 3.20 Evolution of a heton with tilted axis at $\gamma = 4$, $h = 0.5$ and $d = 0.1$

as d increases. On the other hand, at the intermediate stage, the increasing vortex patches of the stronger pair actively form filaments, after which they aquire quite compact shape.

Figure 3.20 with $d = 10^{-1}$, where the emerging trends can be seen quite clearly, gives a more detailed picture of the process. The ancestors of the lateral pairs in Fig. 3.19a are now four small vortex patches, which, after some transformations, form pairs moving away in different directions along the y-axis (the upper part of the figure). The vortices of the stronger pair experience much more complex transformations. After the actual separation from the rest of the vortex structure in the time interval between $t = 4$ and $t = 6$, this heton, before it acquires its relatively stationary shape, frees itself from a considerable number of vortex threads and two two-layer vortices (the central part of the figure). The calculation was stopped at $t = 14$, when the symmetry of the figure showed distortion because of the accumulated computational errors. We can expect that the vortex patches in both layers in the system of those four vortices will merge later to form a single pair with a tilted axis, which will start moving in the positive direction of the y-axis.

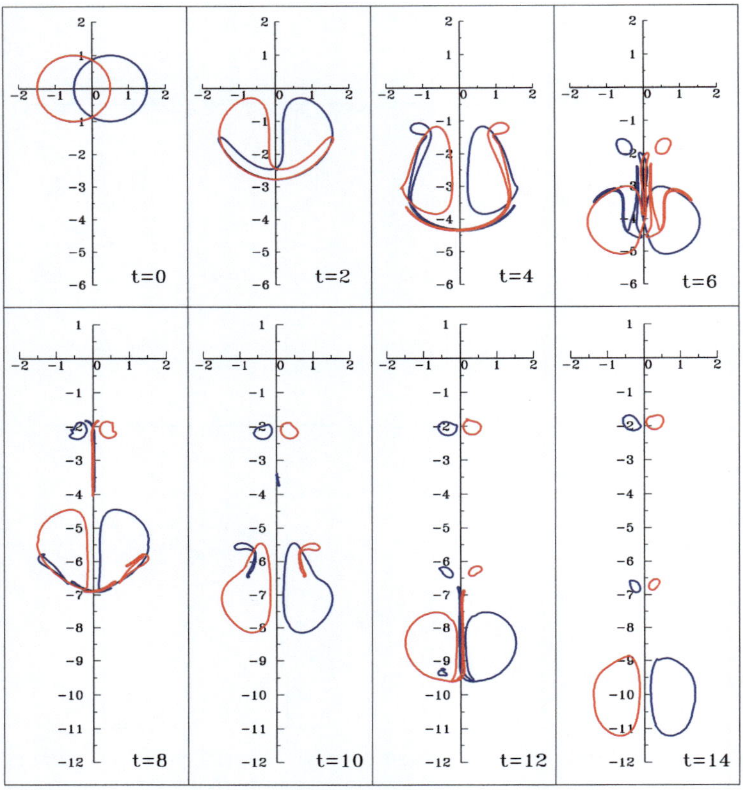

Fig. 3.21 The same as in Fig. 3.19, but at $d = 0.5$

A fivefold increase in d (Fig. 3.21) facilitates the separation process considerably, resulting in that the stronger pair now has a higher velocity. This experiment gives an example suggesting that if a heton has a maximally unstable mode $m > 2$, then the presence of a finite tilt of its axis at $d < d^*$ will lead to a change in the direction of scatter of the newly formed two-layer pairs along the straight line perpendicular to the plane of the tilted axis.

Finally, Fig. 3.22 demonstrates the fact that, when the tilt of the axis is considerable ($d = 0.6$), the heton, after setting aside some excessive mass at the initial stage of motion, will remain practically stable two-layer pair at any γ. The above mentioned "conventional character" of the boundary between the regimes U and S is due to the existence of an intermediate narrow interval in d, where a short-time filamentation process of vortex patches can be seen.

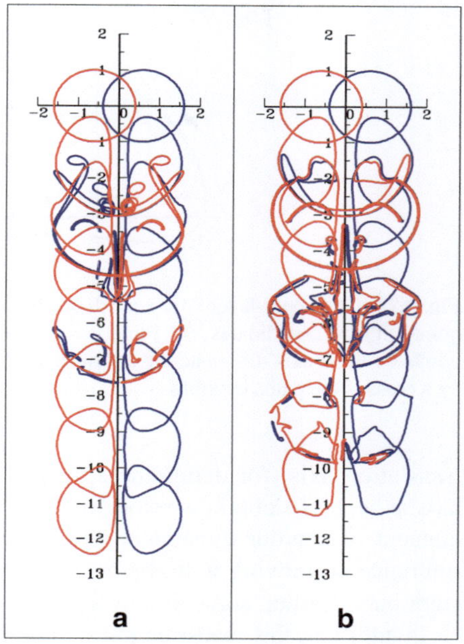

Fig. 3.22 Sets of combined configurations of contours for moments $t = 0, 2, \ldots, 14$ at $h = 0.5$, $d = 0.6$ and (**a**) $\gamma = 4$, (**b**) $\gamma = 14$

3.2.2 Stationary Translation Hetonic V-States[3]

Figures 3.18–3.22 show, that with time, vortex patches moving transitionally, try to take a stationary shape different from the circular one and having elements of a double symmetry, when two-layer vortex structures move with tilted axes and zero potential vorticity.

Theoretical study of different translating or rotating stationary states was the subject of many works [3, 4, 23, 27, 39, 129–131, 159, 166, 167, 173, 174, 180, 185–192, 210, 231, 232, 238, 247–249, 303, 334, 414, 414, 416, 416, 418, 420–423, 431–433, 458–462, 464–466, 473, 480, 505–509, 516–519, 532, 550–555, 560–562, 580, 588, 592, 598, 601, 611, 611, 615, 628, 629, 631, 653, 657, 660, 661, 664, 665, 679, 691, 696, 700, 701, 705, 706, 730, 733, 746, 767, 785, 790–794, 810, 840, 845, 846, 872, 877, 880, 909, 913, 986, 1025, 1027, 1028].

The subject of interest in this subsection is translating finite-core f-plane doubly symmetric hetons. This study is a logical continuation of the Makarov and Kizner' work [561] where a few families of stationary translating and orbital barotripic dipoles were constructed and their stability investigated.

The double symmetry of a heton means that the layer thicknesses are equal, and the shapes of the two vortex patches arranging a heton possess symmetry about the

[3]Most part of this Section and, in particular, Figs. 3.23–3.29, adapted from [562] by permission of Cambridge University Press

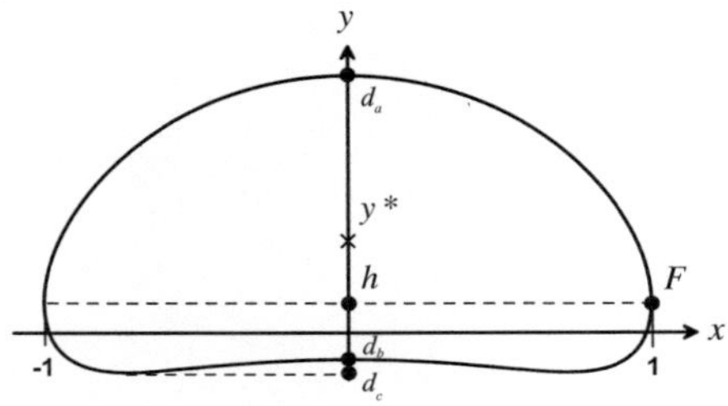

Fig. 3.23 The upper-layer vortex patch of a two-layer doubly symmetric heton. Here F is the front point with the coordinates $x = 1$ and $y = h$. The symbols d_a, d_b, d_c, h and y^* next to the marked points on the y-axis designate the ordinates of these points (see the text); the point $(0, y^*)$ marked by a cross is the patch centroid

translation axis (for definiteness, this will be the x-axis) and about the transverse axis as well. To obtain a heton solution, an iterative procedure (which fits into the concept of contour dynamics) will be used, based on the fact that, in a frame of reference co-moving with the heton, the contours bounding the vortex patches are stationary. Earlier, some doubly symmetric hetons were obtained by Polvani [704].

In addition, this investigation will define more exactly and generalize the results of our preliminary study of the steady motion of distributed hetons with tilted axes by example of initially circular vortex patches (Sect. 3.2.1, Fig. 3.16).

We shall construct a two-parameter family of doubly symmetric hetons, choosing for the identification of a heton its half-length $\gamma^* = D^*/\lambda$ (cf. with formula (1.21)), in the translation direction, relative to the Rossby radius λ and the non-dimensional distance h from the front point of its upper-layer vortex patch to the translation axis (Fig. 3.23). The properties of the hetons found will be discussed both using the parameters γ^* and h, and in terms of the mean radius of a vortex patch (again relative to the Rossby radius). and the non-dimensional horizontal separation $2d$ between the centroids of the patches. Next, classification of the doubly symmetric heton equilibria will be carried out on the (γ^*, h) and $(\gamma, 2d)$ planes and their bifurcations will be examined in detail. Practically, d is not always the best parameter to characterize the geometry of a dipole (e.g. Sects. 3.2.1 and [237,561]). Our computations suggest that a more convenient parameter for this purpose can be the ordinate h of the front point F normalized by a new length scale D^*, the patch half-length in the x-direction. Therefore, we compared this approach with an alternative algorithm, in which the input parameters for the construction of a heton solution were parameters γ and d. The comparison showed that the set of solutions obtained with the procedure based on h and D^* is wider than the set obtained with the alternative $(\gamma, 2d)$-based algorithm, although, when the alternative procedure converges, the solutions obtained with the two methods are identical.

A relative equilibrium is an ensemble of vortices that can move as a whole, translationally or rotationally, at either a constant linear or angular velocity, but keeps unchanged the shapes and mutual arrangement of the vortices. So, we will understand a relative equilibrium like a stationary state when considered in a co-moving frame of reference. Capturing of such a heton is based on the fact that, in a co-moving frame of reference, the total velocity in any point of a contour bounding a vorticity patch should be tangent to the contour. The relative velocity is a difference of the velocity in a fixed frame of reference and the translation speed of the heton. Putting it differently, the stationarity condition consists in vanishing of the scalar product of two vectors, the velocity of a particle located at a boundary contour of a vortex patch, and a normal (to the contour) vector at this point.

At this stage, since a stationary heton comprises only one uniform-vorticity patch in each layer, the superscript α in expressions for velocities (1.35) can be omitted. Then, for a heton travelling along the x-axis at a constant velocity U, the stationarity condition for the each boundary point with number $k = 1, 2, \ldots, K_1$ of a vortex patch in layer j is

$$v_j^k(x_j^k, y_j^k)\dot{x}_j^k - \left[u_j^k((x_j^k, y_j^k) - U\right]\dot{y}_j^k = 0, \quad j = 1, 2,$$

where the velocity components u_j^k and v_j^k are determined by (1.35). Here, for definiteness, we assume that the positive vortex patch resides in the upper layer ($\Pi_1 = 1$ and $\Pi_2 = -1$) and look for hetons with $U > 0$, i.e. those propagating rightward or standing in place (referred also to as static). Owing to the $y \to -y$ and $t \to -t$ symmetry of the f-plane model given by Eqs. (1.11) and (1.12), a symmetry transformation about the x-axis of the solutions found will yield hetons propagating leftward.

Along with the stability to (generally) asymmetric perturbations, a particular case is considered, where the perturbations do not violate the antisymmetry of the upper- and lower-layer flows about the x-axis, inherent in the unperturbed state. For this purpose we use a modified version of the CDM code, in which this symmetry is preserved (this approach was earlier used by Polvani [704]). The characteristic feature of this version is that, at any time step, only the coordinates of the boundary contours in the upper layer are computed via Eq. (1.35). The shapes of the boundary contours in the lower layer are obtained as mirror images (with respect to the x-axis) of those in the upper layer; the upper-layer velocity field, however, is calculated as induced by the vortex patches in both layers. This version of the algorithm is referred below to as mirror-symmetric.

In the CDM simulations, parameter γ is varied in a wide range, whereas the initial areas of the vorticity patches are taken equal to π. Such unification allows us to use fixed time and length (along a contour) steps Δt and ΔC, respectively, for all simulations. As soon as a heton solution at a given pair of parameters γ^* and h is found, one can determine its y^* and upper-layer patch area S, and then calculate parameters γ and d as

$$\gamma = \gamma^* \sqrt{S(\gamma^*, h)/\pi}, \quad d = y^* \sqrt{S(\gamma^*, h)/\pi}. \tag{3.16}$$

Note that the conditions $U \geq 0$ and $d \geq 0$ (or $y^* \geq 0$) are equivalent.

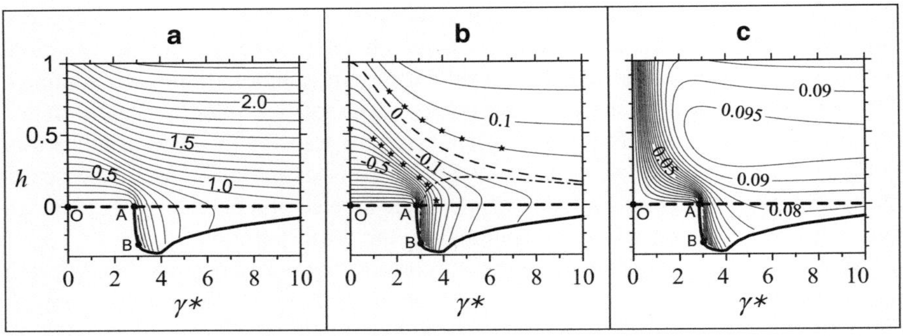

Fig. 3.24 Isolines of selected characteristics of translating doubly symmetric hetons on the (γ^*, h) plane. (**a**) Relative separation $2d$ between the vorticity centroids (see the second of relationships (2.14)). (**b**) The ratio d_b/d_a, which at $d_b/d_a \leq 0$ can serve as a measure of the overlap of the upper and lower vortex patches; the *thin dashed line* (at which $d_b/d_a = 0$) demarcates the overlapping and non-overlapping hetons; *dash-dotted line*, at which $d_c - d_b = 0$, indicates the upper bound of the region occupied with hetons composed of beanlike patches; *asterisks* represent Polvani's [704] solutions (6 of the 12 solutions corresponding to $d_b/d_a = 0.05$ do not fit into the plot frame). (**c**) The heton translation speed U. In all of the panels, the region filled with isolines is the region where the rightward translating heton solutions can be constructed. The rectilinear segment *OA* of the *bold dashed line* (where $h = 0$) corresponds to static, circular purely baroclinic (thus static) hetons that are limiting to the translating hetons as $h \to \infty$ at $\gamma < \gamma_A$; the remainder of the *bold dashed line* (at $\gamma < \gamma_A$) represents static circular fully baroclinic solutions that exist along with the rightward translating hetons whose front points lie on the x-axis. The *bold solid line* is the borderline of the region where hetons with $h < 0$ exist (the $h = \Phi(\gamma^*)$ line). The region between the *bold dashed line* and the *bold solid line* is occupied by the cross-over solutions, in which the centroid and the front point of a vortex patch are located on opposite sides of the x-axis; static cross-over hetons correspond to the *solid bold line* segment *AB*

The iterative procedure used here virtually repeats the one described in [561] devoted to barotropic dipole equilibria. As in the barotropic case, the procedure allows the construction of translating and rotating relative equilibria; in particular, asymmetric translational hetons can be constructed, composed of patches that differ in potential vorticity, area and shape. Also symmetric about the x-axis but asymmetric about the y-axis hetons can be obtained. A solution of this kind, which we call "dehiscent", appears as a couple of quasi-elliptic vortex patches (with generally non-zero separation) whose principal axes make an angle which does not exceed 90°; such a solution requires suitable initialization of the iterative procedure.

Doubly symmetric solutions were constructed with γ^* varying in the range $10^{-13} < \gamma^* \leq 10$ at a step 0.2, and h varying in the range $-0.34 \leq h \leq 1$ at a step 0.02; for computational reasons, the value $\gamma^* = 10^{-13}$ was used instead of $\gamma^* = 0$. Overall, over 2,800 stationary solutions were constructed. In Fig. 3.24, some characteristics of translating hetons found are shown as contour lines on the (γ^*, h) plane. The shapes of a number of hetons overlaid with the isolines of the upper-layer co-moving streamfunction, $\psi_1 + Uy$, are shown in Fig. 3.25.

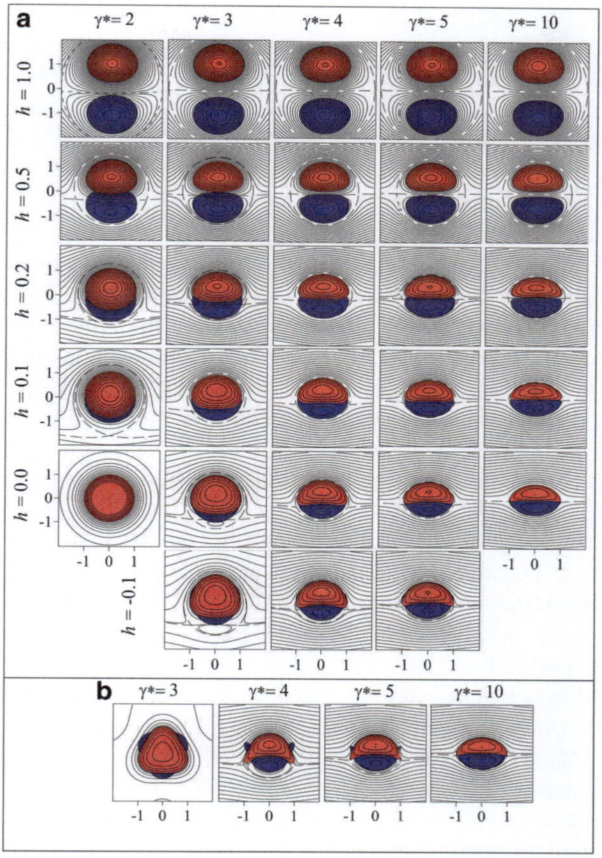

Fig. 3.25 Examples of doubly symmetric heton solutions computed at different combinations of parameters γ^* and h (indicated on the *top* and *left*). (**a**) Hetons at $h \geq -0.1$. (**b**) Heton solutions near the boundary $h = \Phi(\gamma^*)$ of the region were the solutions exist (Fig. 3.24). *Red* and *blue domains* are the vortex patches in the upper and *lower layers*, respectively; thin *solid lines* are the streamlines in the *upper layer* (given at a step 0.01); *thin dashed lines* are the separatrix streamlines in the *upper layer*

As expected, doubly symmetric hetons exist for, apparently, any positive values of parameters γ^* and h (Figs. 3.24 and 3.25a). A result that is a little more surprising is the existence of solutions with negative h on the right of point A whose coordinates are $\gamma_A^* \approx 2.840$ and $h < 0$h (Figs. 3.24 and 3.25b). To distinguish between the solutions with $h > 0$ and $h < 0$, we shall refer to them as to ordinary and cross-over hetons, respectively (because of the unusual shape of the latter).

For any γ^*, there is a minimal value of h allowing the existence of rightward or static doubly symmetric hetons. The borderline of the region, where such heton solutions exist, consists of two parts, a rectilinear piece OA, where $0 \leq \gamma^* \leq \gamma_A^* \approx 2.840$ and $h = 0$, and a curvilinear piece, where $\gamma^* > \gamma_A^*$ and $h < 0$. The shape of this curvilinear part (the bold solid line in Fig. 3.24) is estimated numerically as some nonlinear function $h = \Phi(\gamma^*) < 0$ of parameter γ^*. In a solution with a negative h, the centroid and the front point of a vortex patch are located on opposite sides of the x-axis, so the two patches overlap in a somewhat exotic, cross-over manner (see the lowest row in Fig. 3.25a at $3 \geq \gamma^* \geq 5$ and Fig. 3.25b). Note that cross-over solutions have never been discussed in the literature before. Below we shall see that the cross-over hetons are all unstable.

The behaviour of parameters d and d_b/d_a is of particular interest, because Makarov and Kizner [561] used the separation to identify barotropic dipoles, whereas Polvani [704] employed parameter d_b/d_a to compute finite-core hetons. With this approach, however, Polvani did not obtain the cross-over solutions. At $h > 0$ and $\gamma^* < \gamma_A^*$, the separation d and the ratio b/a monotonically depend on h at any given γ^* (panels 3.24a, 3.24b). At $\gamma^* > \gamma_A^*$ and $\Phi(\gamma^*) \leq h \leq \infty$, the dependences of d and d_b/d_a on h are not monotonic; for example, at a fixed $\gamma^* > \gamma_A^*$, parameter d_b/d_a reaches its minimum at a positive h (Fig. 3.24b). This explains why the use of parameter h in constructing the family of stationary solutions is preferred over that of d or d_b/d_a.

The thin dashed line in Fig. 3.24b indicates the dipoles in which $d_b = 0$; below this line, d_b/d_a is negative. In the hetons occupying the region beneath the dash-dotted line, the patches are beanlike, according to Polvani's [704] terminology (see e.g. Fig. 3.25, $h = -0.1; 0; 0.1$ and $\gamma^* = 4, 5$). Noticeably, the thin dashed line goes above the dash-dotted line. This means that we cannot expect that the overlap of the upper and lower patches can be present in the front and rear parts of a heton and, at the same time, be absent in its central part. Thus, in non-overlapping hetons, necessarily, $d_b/d_a \geq 0$. In overlapping hetons, i.e. where $d_b/d_a < 0$, this ratio can serve as a reasonable measure of the overlap. However, as Fig. 3.25 suggests, this measure is meaningful only in application to ordinary hetons (at $h \geq 0$). The overlap in ordinary hetons is most pronounced at $h = 0$ and $0 < \gamma^* \leq \gamma_A^*$, and it decreases as h increases and γ^* goes to the right of γ_A^* (panels 3.24b and 3.25a). When γ^* varies from 0 to approximately 1 and $0 < h \leq 10$, the translation speed depends monotonically on each of the arguments γ^* and h, but outside this region U becomes a two-valued function of h (at any given γ^*) reaching its maximum in the vicinity of the line $h = 0.5$ (Fig. 3.24c). This appears quite natural if we recall the dependence of the translation speed on the separation in point-vortex hetons [350].

A well-known heton solution is given by a pair of circular vortex patches of the same radius positioned one exactly above the other ([477] term such a heton axially symmetric). Such trivial solutions correspond to $h = 0$ and exist at any γ^* (Fig. 3.24, bold dashed line). In a solution of this kind, $\psi_{BT} = 0$ identically, therefore its translation speed is necessarily zero (see, e.g. [427–429]), i.e. the heton is static. Numerically, a circular fully baroclinic solution with $h = 0$ is easily obtained in the range $0 < \gamma^* \leq \gamma_A^*$ (segment OA in Fig. 3.24); it is limiting to the translating solutions with the same γ^*. In contrast, at $\gamma^* > \gamma_A^*$ and $h = 0$, we managed to obtain static, fully baroclinic solutions only when the initial guesses were already close to circular; otherwise, the procedure converged to translating hetons with incomplete overlap of the patches (Fig. 3.25a, the panels in the row $h = 0$).

It is instructive to compare the hetons obtained by Polvani [704] with ours. Polvani produces two tables with several parameters characterizing two series of solutions, namely 12 hetons at $d_b/d_a = 0.05$ and 8 hetons at $d_b/d_a = -0.3$, referred below to as series 1 and 2, respectively. Using some of these parameters, for each of the 20 Polvani's solutions we determine the relative separation $2d$ and, by setting $2d$ and γ (which differs from Polvani's γ by a factor $\sqrt{2}$), construct a heton solution via our alternative $(\gamma, 2d)$ procedure (see below).

Next, we calculate d_b/d_a and the translation speed U, as well as h and γ^* for our solutions, and compare our d_b/d_a and U with the d_b/d_a and the properly normalized translation speed of Polvani's hetons. For series 1, the discrepancy in d_b/d_a grows from 0.02 to 1 % with increasing γ, while the discrepancy in U remains approximately 0.05 % for the whole range of γ values. For series 2, with increasing γ, the discrepancy in d_b/d_a grows from 0.00 to 0.62 % as long as the vortex patches remain convex, and for the only beanlike heton in this series, the discrepancy in d_b/d_a equals to 2.44 %; the discrepancy in U increases from 0.08 to 0.16 %. Owing to the pretty good agreement, we conclude that Polvani's [704] solutions are members of the family of our doubly symmetric heton solutions (the 2 % discrepancy in d_b/d_a in the only beanlike heton is apparently due to the fact that this solution is close to the convergence bound of Polvani's algorithm, where the accuracy of his computations is the lowest). Because parameters h and g^* for our counterparts to Polvani's hetons are known, we can show these solutions in Fig. 3.24b (asterisks); in the figure caption, we loosely refer to them as to Polvani's hetons.

Classification of the doubly symmetric rightward translating and static hetons (i.e. with $d \geq 0$) on the (γ^*, h) and $(\gamma, 2d)$ planes is summarized in Fig. 3.26. Comparison of Fig. 3.26a, b convinces us that the (γ^*, h) plane is indeed better suited for the construction and classification of the heton equilibria. For instance, on the $(\gamma, 2d)$ plane (Fig. 3.26b), the boundaries of the region of existence of cross-over hetons intertwine due to the non-monotonic dependence of $2d$ upon h. On the other hand, the $(\gamma, 2d)$ plane appears to be a more natural environment for the discussion of the heton stability.

In the limit of $\gamma^* \to 0$ equation (1.11) at $j = 1$ and $j = 2$ describe two independent two-dimensional flows. Thus, the two vortex patches in a limiting heton do not interact and are circular, so the heton is static ($U = 0$) and the separation between the centres of the patches is $d = h$. On the other hand, for any fixed γ^*, when $h \to \infty$, the separation $2d$ goes to infinity and the barotropic interaction between the patches becomes the leading one. This is because at the periphery of a vortex patch, as the distance from a patch increases, the barotropic flow induced by this patch dies out more slowly than the baroclinic flow. Therefore, in this limit too, the PV patches tend to become circles and $U \to 0$.

Another class of effectively barotropic limiting solutions is achieved as $\gamma^* \to \infty$. As (1.18) suggests, $h \to 0$ in the limit of $\gamma \to \infty$, hence, as $\gamma^* \to \infty$. However, generally, p does not vanish in this limit; so one may say that the flow in the two layers undergoes barotropization as $\gamma^* \to \infty$. In other words, although the two vorticity patches reside in different layers, the flows in the upper and lower layers tend to become identical as $\gamma^* \to \infty$. Accordingly, the vortex patches constituting a

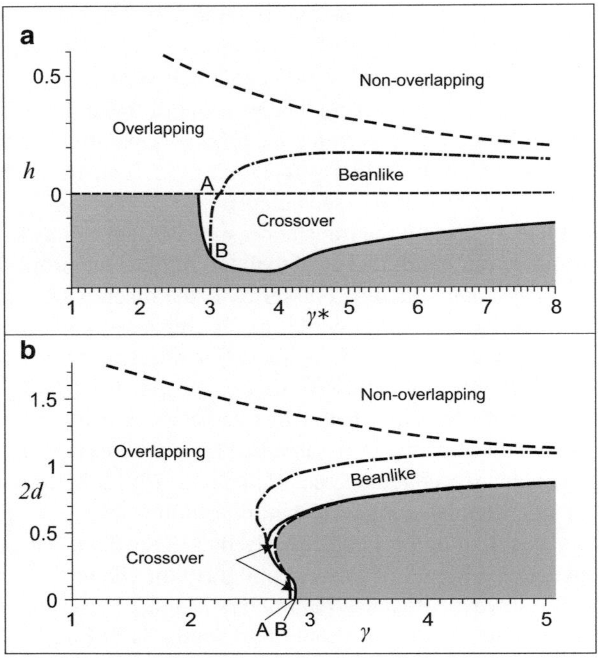

Fig. 3.26 Classification of the doubly symmetric hetons with $d \geq 0$. it (**a**) Regions in the (γ^*, h) plane occupied by solutions of different types. (**b**) Regions in the $(\gamma, 2d)$ plane occupied by solutions of different types. In it (**a**) and (**b**), *thin dashed lines* demarcate the regions of overlapping and non-overlapping hetons; beanlike hetons are found *beneath* or to the *right* of the *dash-dotted lines*; *straight bold dashed lines* represent static, purely baroclinic solutions; *curved bold solid lines* represent the solutions with $h = \Phi(\gamma^*)$; segment *AB* represents the static cross-over hetons. *Shaded* in (**a**) is the region on the (γ^*, h) plane, where no rightward translating hetons are found; its mapping on the $(\gamma, 2d)$ plane is *shaded* in (**b**); *curved bold dashed line* in (**b**) represents translating hetons with $h = 0$. Translating cross-over solutions fill the region between the *curved bold solid line* and the *straight dashed line* in (**a**) and the two regions between the *curved bold solid line* and the *curved bold dashed line* in (**b**)

heton tend to take the same shape as in the single-parameter family of barotropic uniform-vorticity dipoles first described by Pierrehumbert [696]; this fact was mentioned by Polvani [704]. In particular, at $h = 0$, the limiting shape to which the incompletely overlapped heton tends is the same as of the Sadovskii [785] barotropic dipole in which the boundaries of the patches have two vortices each and a common rectilinear boundary segment on the translation axis. (In contrast, the flow induced by a purely baroclinic circular heton solution mentioned above degenerates to zero with growing γ^*). Obviously, the translation speed of the limiting two-layer heton with a “barotropized” streamfunction is half that of a truly barotropic dipole of the same shape.

The line $h = \Phi(\gamma^*)$ bounding from below the region on the (γ^*, h) plane, in which cross-over heton equilibria exist (bold solid line in Figs. 3.24 and 3.26a), consists of two parts, the segment AB where the coordinates of point B are $\gamma_B^* \approx 3.022$ and $h_B \approx -0.269$, and a half-infinite segment where $\gamma^* > \gamma_B^*$. The hetons corresponding to the segment AB (as other cross-over hetons) are not fully overlapping. Thus, the barotropic flow associated with such a boundary solution is non-zero; yet these hetons do not move. This can happen only if $d = 0$, i.e. if the tendencies for self-propulsion generated by different parts of a cross-over heton extinguish each other. The shape of a heton of this kind resembles a star of David (Fig. 3.25b, $\gamma^* = 3$); it is composed of a pair of curvilinear convex Δ-like patches with a common centroid, around which the lower patch is turned by 180° relative to the upper one.

All of the boundary solutions with $\gamma^* > \gamma_B^*$ and $h = \Phi(\gamma^*)$ are translating since in these hetons $d \neq 0$. Here, in each layer, the edges of the vortex patches coincide with the separatrices of the co-moving streamfunction. The same feature is inherent in the two-layer counterpart of the Sadovskii dipole, to which our boundary solution apparently tends as $\gamma^* \to \infty$ (because, seemingly, $\Phi(\gamma^*) \to 0$ as $\gamma^* \to \infty$). The boundary solutions at $\gamma^* > \gamma_B^*$ cannot be achieved numerically, but we managed to approach them quite closely; in this way the shape of the line $h = \Phi(\gamma^*)$ was estimated. Examples of the boundary solutions are shown in Fig. 3.25b.

The common property of the solutions considered above is that $y^* \geq 0$, i.e. $d \geq 0$, so these hetons either travel rightward ($y^* > 0$) or are static, i.e. stand in place ($y^* = 0$). As mentioned, for each solution with $y^* > 0$, there exists a mirror-symmetric (about the x-axis) leftward-moving counterpart with $y^* < 0$. Thus, on the (γ^*, h) plane, the line given by the pair of equations $h = -\Phi(\gamma^*)$ and $h = \Phi(\gamma^*)$ (*bold line* in Fig. 3.27) bounds from above and below the region, in which any pair (γ^*, h) determines two solutions that differ not only in the direction of their translation, but also in the shape of the vortex patches.

Most sharp gradients of the characteristics of heton equilibria are observed in the vicinity of the point A in which, as we saw, a family of static circular purely baroclinic solutions with $h = 0$ (represented by a dashed line in Figs. 3.24 and 3.27a) bifurcates from the translating hetons (namely, from those whose front points are on the x-axis). Another set of hetons that bifurcates from the translating solutions in point A is the family of cross-over hetons. Static cross-over solutions, a subclass of the family of cross-over solutions with $d = 0$, are represented in Fig. 3.27a by segment BAB' of the solid line bounding the region where both rightward- and leftward-translating hetons exist; the point labelled B' is symmetric to B about the line $h = 0$.

Contour lines of γ^* on the (h, U) plane (panel 3.27b) and of h on the (γ^*, U) plane (panel 3.27c) give additional information on bifurcations. On the (h, U) plane, all of the isolines of γ^* with $\gamma^* < \gamma_A^* \approx 2.840$ smoothly pass through the point with the coordinates $h = 0$ and $U = 0$. In contrast, the isolines with $\gamma^* > \gamma_A^*$ break when they run into the straight-line segment labelled BB' in panel 3.27b. Recall that, on the (γ^*, h) plane, the static cross-over solutions with negative and positive h correspond to the segments AB and AB', respectively (see Figs. 3.24, 3.26a, and 3.27a). On the (γ^*, U) plane, however, these two kinds of solutions cannot be discriminated between because they fall into the same segment on the γ^*-axis (in Fig. 3.27c, the left end of this segment is labelled A, while the right end is deliberately labelled with two labels, B and B'). The isolines of h, which run into this segment, break off

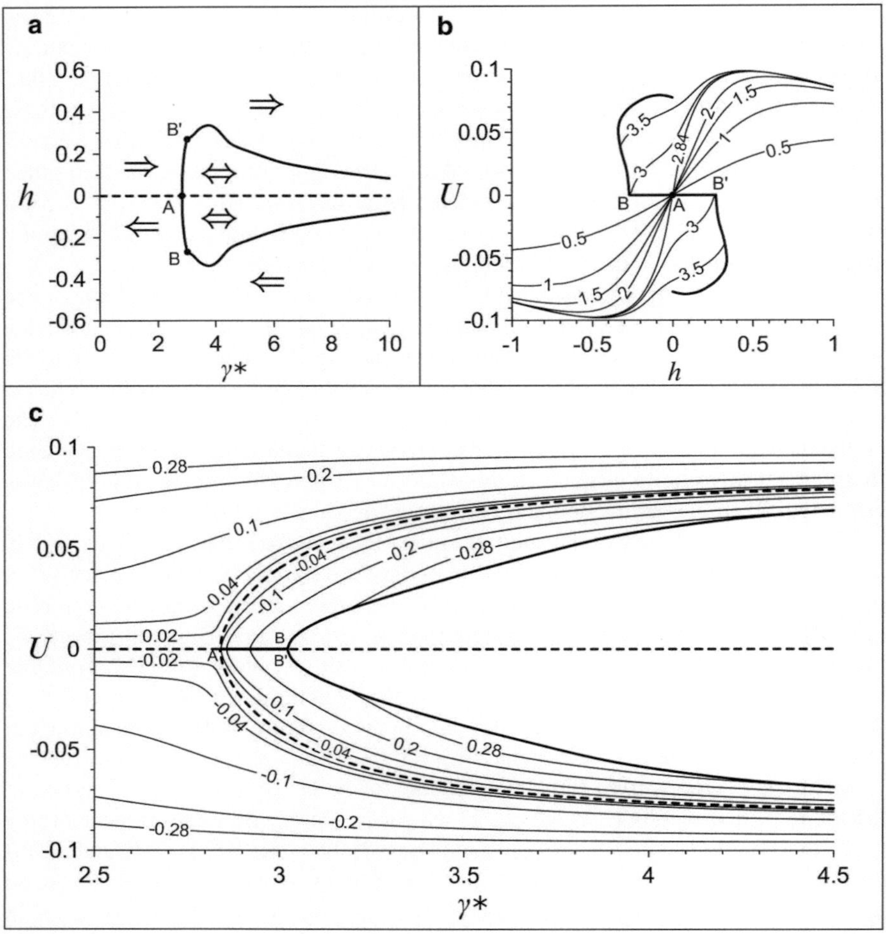

Fig. 3.27 Bifurcations of heton equilibria. (**a**) Regions on the (γ^*, h) plane, where the hetons travelling rightwards or leftwards exist; *arrows* label the directions of the heton translation. Shown in *bold* is the line $h = \pm\Phi(\gamma^*)$ bounding the region on the (γ^*, h) plane where both rightward and leftward hetons exist (labelled by *bidirectional arrows*); points A and B are the same as in Figs. 3.24 and 3.26, and B' is symmetric to B about the γ^*-axis. (**b**) Contour lines of γ^* as a function of h and U (*thin lines*). (**c**) Contour lines of h as a function of γ^* and U (*thin lines*). In panels (**a**) and (**c**), *dashed lines* represent the hetons with $h = 0$ (some of them are static and others are not); in panels (**b**) and (**c**), symbols A, B and B' label the same solutions as in (**a**). In all of the panels, the *bold lines AB* and AB' represent the static cross-over hetons, and the remaining *bold lines* represent translating boundary cross-over hetons (but ordinary solutions are also present on the *bold lines*)

here. The isoline $h = 0$, when looked from left to right, splits in point A into three branches, upper, mid and lower, at which the translation speed is positive, zero and negative, respectively. Note that, from point B (or B') rightward on the γ^*-axis of the (γ^*, U) plane, only axisymmetric fully baroclinic hetons exist (Fig. 3.27c).

We tested the stability of doubly symmetric hetons numerically based on CDM simulations with the above-described stationary solutions taken as initial states. In these tests, the numerical noise (coming from the discretization and round-off errors) served as permanently acting perturbations. Such perturbations are, generally, asymmetric. In the special case where the stability of doubly symmetric hetons to symmetric perturbations was studied, we employed the mirror-symmetric version of the algorithm. In principle, even small local perturbations of the shape of the vortex patches comprising a heton may affect the overall heton trajectory. However, as long as the variations in the heton shape remain small, such a response should not be interpreted as a manifestation of instability, since a relative equilibrium is considered.

Within the linear stability framework, instability of a vortical steady state to initial perturbations implies exponential growth with time of the perturbation magnitude. In contrast, in the nonlinear case, the growth of a perturbation can be limited even if the steady state is unstable. If, however, a small initial perturbation causes significant disruption of the shape of a vortical system and violates its structural integrity, one may speak of the structural instability (not to be confused with the structural instability with respect to changes in the parameters of a dynamical system). Although the numerical noise in a CDM simulation acts permanently, it affects the heton effectively in the same way as initial perturbations do, inducing a certain oscillation that in some circumstances may remain small (which is indicative of stability), but may grow in others (instability). In particular, the stability or instability of a heton can be judged by examining the variation in time of the perimeter of one of its vortex patches, say, $P_1(t)$, the perimeter of the upper-layer vortex. We express this variation as the percentage relative to the initial perimeter, $\delta P(t) = 100[P_1(t) - P_1(0)]/P_1(0)$. This approach works reliably not only with symmetric perturbations, but with asymmetric perturbations as well, because, in response to slight asymmetric perturbations (asymmetric numerical noise), a heton which is close to the stability bound develops almost antiphase oscillations of the upper and lower patches. The oscillations persist as long as the heton retains its structural identity. Thus, the central question is how strong the oscillations are.

It is known that static fully baroclinic circular hetons ($h = d = 0$) are linearly stable if $\gamma < \gamma_{cr} \approx 1.705$, while at $\gamma > \gamma_{cr}$ the mode-2 perturbation becomes unstable, and its growth rate increases with increasing γ. This fact was established analytically [274, 343, 477]. Therefore, in order to design a reasonably accurate procedure for the distinction between stable and unstable hetons, we start with the purely baroclinic circular hetons near the critical point $\gamma < \gamma_{cr} \approx 1.705$ and $d = 0$. As seen in Fig. 3.28a, b, at $\gamma = 1.705$, parameter δP barely oscillates. The slow lowering (within 0.025 % by $t = 4{,}000$) of the dashed curve corresponding to $\gamma = 1.705$ is due to a gradual shrinking of the patches typical of contour dynamics computations; the shrinking is so slight that the number of nodes at the contour bounding the vortex patch remains unchanged by $t = 4{,}000$. When we increase γ by 0.00025, thus entering into the zone of linear instability, δP performs a weak oscillation whose amplitude is of the order of 0.05 and 0.075 % in the asymmetric and symmetric case, respectively. Taking one more 0.00025 step rightward, we obtain a stronger oscillation

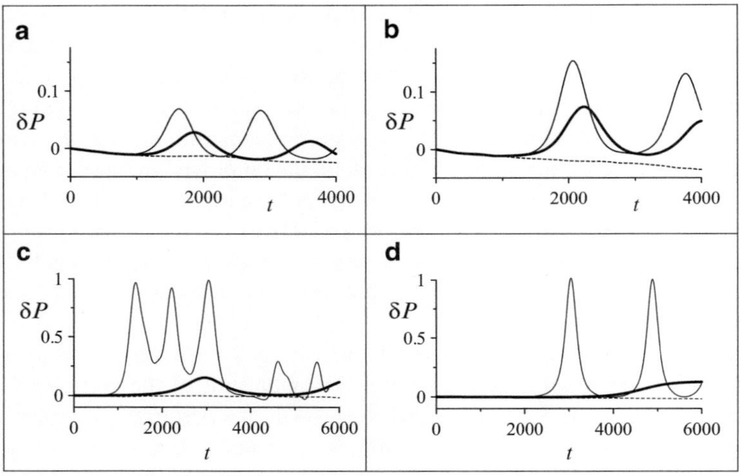

Fig. 3.28 Determination of the stability bounds. Noise-induced oscillations of the perimeter of the *upper-layer* vortex patch presented as the percentage relative variation $\delta P(t) = 100[P_1(t) - P_1(0)]/P_1(0)$. (**a**) Effect of asymmetric noise on the static circular hetons (i.e. with zero separation) at $\gamma = 1.705$ (*dashed line*), $\gamma = 1.70525$ (*bold line*) and $\gamma = 1.7055$ (*thin solid line*). (**b**) Same as in (**a**), but for symmetric noise. (**c**) Effect of asymmetric noise on the translating hetons with $2d = 0.3$ at $\gamma = 1794$ (*dashed line*), $\gamma = 1.79425$ (*bold line*) and $\gamma = 1.7945$ (*thin solid line*). (**d**) Effect of symmetric noise on the translating hetons with $2d = 0.3$ at $\gamma = 1.804$ (*dashed line*), $\gamma = 1.80425$ (*bold line*) and $\gamma = 1.8045$ (*thin solid line*)

(the oscillations induced by the one- and two-step increase of γ are shown in panels 3.28a, 3.28b by bold and thin solid lines, respectively). The presence of a quasi-periodic regime at supercritical (in the sense of the linear stability) values of γ should not be surprising, since, in the simulations, we are dealing with nonlinear processes.

These preliminary computations suggest the following procedure for the high-resolution determination of the stability bounds. First, we make a rough estimate of the position of the stability bound on the $(\gamma, 2d)$ plane and choose a reasonably thin domain enclosing the tentative bound. Then we fix d and, starting from a definitely stable solution within this domain, gradually increase γ with a step of 0.00025; alternatively, we can fix γ and decrease d with the same step (the choice of a proper version depends on the slope of the tentative bound). The process continues until the amplitude of the oscillations in δP reaches the level of 0.05 or 0.075 % in the asymmetric or symmetric case, respectively, or exceeds this level (bold solid lines in panels 3.28c, 3.28d). The subsequent increase of γ (decrease of d) leads to the growth in order of the magnitude of the amplitude (compare the positions of the bold and thin solid lines in Fig. 3.28). The penultimate value of γ (or d) in the above-described process is taken as the stability bound for the given d (or γ).

The shapes of the stability bounds on the $(\gamma, 2d)$ plane shown in Fig. 3.29 are based on 42 points estimated in our stability tests as explicated above. For reasons of symmetry, in these diagram, only the region $d \geq 0$ is shown. The bounds of

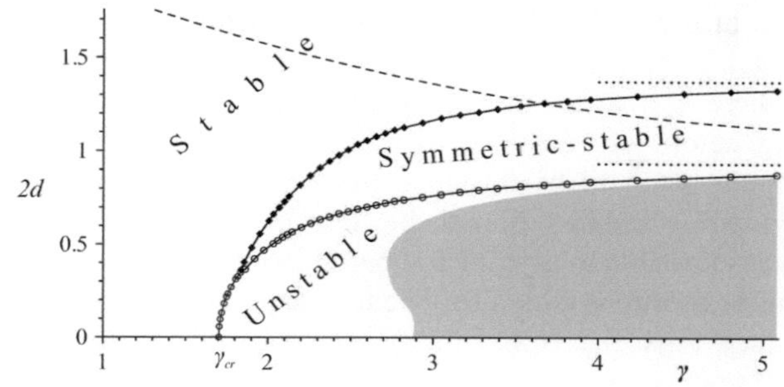

Fig. 3.29 Stability bounds on the $(\gamma, 2d)$ plane. *Diamonds* and *empty circles* mark the bounds of stability to asymmetric and symmetric perturbations, respectively. The region above the upper stability bound represents unconditionally stable hetons; the solutions belonging to the region between the two stability bounds are stable to symmetric perturbations and unstable to asymmetric perturbations; the region below the lower stability bound represents solutions unstable to the both types of perturbations. The asymptotic levels, to which the stability bounds approach as $\gamma \to \infty$, are given by *dotted straight-line* segments. *Thin dashed line* and *shaded region*, are the same as in Fig. 3.26b

stability to asymmetric and symmetric perturbations are marked with diamonds and empty circles, respectively. In the interval from $\gamma = 1.705$ to $\gamma = 2.5$, the stability bounds are determined through varying the parameter γ, while at $\gamma > 2.5$, parameter d is varied. It might seem that, within the interval from $\gamma = 1.705$ to approximately $\gamma = 1.8$, the two bounds have a common segment. This is, however, just a graphical effect due to the thickness of the lines in Fig. 3.29.

Thus, we can conclude that the hetons comparable in size to the Rossby radius and smaller, i.e. at $\gamma < \gamma_{cr} \approx 1.705$, are stable in both senses. When the critical size γ_{cr} is exceeded, the heton stability becomes monotonically dependent on the separation, i.e. the larger the heton, the higher the separation required for the stability; and the separation needed for the stability to symmetric perturbations is smaller than that required for the stability to arbitrary perturbations (Fig. 3.29). These results are in qualitative agreement with the observations made earlier with two- and three-layer β-plane modons bearing smooth baroclinic riders [427].

As $\gamma \to \infty$, i.e. when the flow in the two layers undergoes barotropization, the two stability bounds approach two different asymptotic levels. In this limit, the stability properties of hetons must be the same as in barotropic dipoles. Indeed, the estimated asymptotic threshold separations for the stability to asymmetric (arbitrary) and to symmetric perturbations are quite close to $2d \approx 1.374$ and $2d \approx 0.936$, respectively (see Fig. 3.29, where these asymptotes are shown by dotted lines). The first of these bounds, $2d \approx 1.374$, corresponds to Dritschel's [237] estimate of the linear stability bound for barotropic dipoles, and the second bound, $2d \approx 0.936$, is the separation between the centroids in the Sadovskii dipole, i.e. the minimum possible separation in barotropic dipoles which were all shown to be stable to symmetric perturbations [561]. Note that, due to the different choice of the length scale, the stability bounds provided here and in the work by Makarov and Kizner [561] differ by a factor $\sqrt{\pi}$.

As indicated in Fig. 3.29, all non-overlapping hetons are stable to symmetric perturbations and, at $\gamma < 3.66$, they are stable to arbitrary perturbations as well. At $\gamma > 3.66$, there is a sector on the $(\gamma, 2d)$ plane, where moderately separated nonoverlapping hetons are unstable to arbitrary, asymmetric perturbations. For a non-overlapping heton to be unstable, the separation $2d$ must necessarily be below the asymptotic stability bound, i.e. less than 1.374 (Fig. 3.29). Beanlike hetons are always unstable to asymmetric perturbations; however, at approximately $\gamma > 2.55$, beanlike solutions with a sufficient separation can be stable to symmetric perturbations (see Figs. 3.26b and 3.29). All cross-over hetons are unstable to perturbations of both types.

The issues of bifurcations and behaviour of perturbed circular hetons, in our opinion, are interrelated. Therefore, we shall speak very briefly upon the nonlinear evolution of perturbed purely baroclinic circular hetons in the supercritical region $\gamma > 1.705$.

Kinematically, the oscillation of an initially circular baroclinic heton consists of synchronous periodical elongation of its vortex patches (this behaviour was first observed by Helfrich and Send [343]). Slightly perturbed hetons behave in a similar way. They are represented by the points $(\gamma, 2d)$ adjacent to the "common" segment of the stability bounds (Fig. 3.29) to the right of it. Remarkably, on the $(\gamma, 2d)$ plane, along this part of the bound of stability to symmetric perturbations, the family of dehiscent solutions bifurcates from the doubly symmetric solutions. A dehiscent solution is obtained by initializing the iterative procedure by two separated ellipses whose principal axes make an angle of $\sim 20°$; such a solution possesses symmetry about the x-axis, but is asymmetric about the y-axis. Bifurcations in dynamical systems are usually related to the emergence of instabilities. Therefore, the bifurcation of the dehiscent hetons serves as one more piece of evidence in favour of the correctness of our determination of the stability bounds for the doubly symmetric solutions.

The limiting value of γ, at which the increase in the oscillation amplitude does not affect the integrity of an initially circular purely baroclinic heton (i.e. does not lead to structural instability), is $\gamma \approx 1.752$. When the value 1.752 is overstepped, the elongation of the vortex patches becomes accompanied by some cross-clamping, so the whole structure resembles in shape two crossed dumbbells. This process culminates in splitting the heton into two hetons that run in opposite directions. As our simulations suggest, the two emerged translating hetons are not steady, but oscillate around some stable steady states that can be depicted by one of our stationary solutions.

Continuing to increase γ, we reach the next critical point, $\gamma \approx 2.840$, starting from which, mode 3 becomes unstable [274, 343, 477]. In the second supercritical region, next to 2.840, the growth rate of mode 2 exceeds that of mode 3. Therefore, as before, a perturbed heton splits into two translating hetons. However, when γ becomes larger than approximately 4.75, mode 3 begins to prevail, and a slightly perturbed heton takes the form of a pair of curvilinear Δ-like patches turned by some angle relative to each other, that is, becomes similar in shape to a static cross-over solution shown in Fig. 3.26b (leftmost panel). In this case, with the passage of time, three translating hetons detach from the parent structure [343, 477]. As established above, the family of static cross-over solutions of the star-of-David kind bifurcates from the circular baroclinic hetons just at point A with the coordinates $\gamma_A^* = 2.840$ and $h = 0$ (Figs. 3.24 and 3.26a). On the $(\gamma, 2d)$ plane, the coordinates of point A calculated via (3.16) are equal to those on the (γ, h) plane, i.e. $\gamma^* = \gamma_A^* = 2.840$ and $d_A = h_A = 0$. This is because the solution corresponding to point A is a circular purely baroclinic heton with a unit radius. Thus, the bifurcation occurs exactly at the second critical point, where mode 3 becomes linearly unstable. In light of the theory of bifurcations in dynamical systems, such a coincidence is hardly accidental.

Aiming at the determination of the main parameters governing the stability of two-layer modons, we have considered a simplified model, namely, a two-parameter family of translating and static (i.e. motionless) doubly symmetric finite-core heton equilibria. In terms of potential vorticity, such a heton can be viewed as a pair of uniform vortex patches, of which one (say, the positive vortex patch) resides in the upper layer of a two-layer rotating fluid, and the other (the negative one) in the lower layer. The shapes of the patches are supposed to be symmetric about the translation axis and the transverse axis.

We obtained heton solutions numerically with an iterative procedure, which implements the condition that, in a frame of reference co-moving with the heton, the bounding contours of the vortex patches are stationary. The parameters identifying a doubly symmetric translating heton can be chosen in different ways. For instance, the mean radius γ of the vortex patches (relative to the Rossby radius) and the non-dimensional horizontal separation d between their centroids (scaled with the mean radius) have a clear geometrical and physical meaning and are suitable for the discussion of the heton stability. However, at the stage of searching for the heton equilibria, this way of parameterization has some technical limitations. More convenient in this context appear the heton's half-length γ^* in the translation direction (again, relative to the Rossby radius) and the non-dimensional distance h from the front point of the upper-layer vortex patch to the translation axis (scaled with the half-length). Following this path, we managed to obtain hetons in a wide range of shapes, including a novel class of solutions, the so-called cross-over hetons formed by overlapping beanlike vortex patches whose centroids and front points are (generally) located on opposite sides of the translation axis. The barotropic flow associated with a cross-over solution is non-zero, because hetons of this type are not fully overlapping; yet among the cross-over solutions there is a subclass of static hetons, i.e. with $d = 0$. We examined the bifurcations of the solutions found, as well as the properties of the hetons in the limits of $\gamma^* \to 0$, $\gamma^* \to \infty$ and $h \to \infty$.

Stability of the heton equilibria was tested with fine-resolution contour-dynamics simulations, in which the permanent numerical noise served as a perturbing factor. Two kinds of stability were examined separately, the stability to arbitrary perturbations that do not preserve the symmetry of the initial state, and the stability to symmetric perturbations, which do not violate the initial symmetry. In addition, we discussed the interdependence between instability and bifurcations.

We found that the hetons comparable in size with the Rossby radius, or smaller ($\gamma < \gamma_{cr} \approx 1.705$), are stable in both senses. When the critical size γ_{cr} is exceeded, the heton stability becomes dependent on the separation; and the larger the heton, the higher the separation required for stability. The latter agrees with the numerical results of Kizner et al. [427], who carried out just a few simulations with two- and three-layer β-plane modons bearing smooth baroclinic riders and considered the modon stability in terms of the degree of overlap. The β-effect forces a modon to travel eastward and symmetrize about the translation axis to within an additive constant in the background potential vorticity field [347, 413, 427, 428].

Thus, there must be some similarity between the behaviour of the β-plane modons and f-plane hetons subjected to symmetric perturbations [424, 425]. Indeed, we have found that non-overlapping and moderately overlapping hetons are stable to symmetric perturbations. Moreover, sufficiently separated non-overlapping hetons ($2d > 1.374$) of any size are stable to arbitrary perturbations. Hetons whose separation is smaller than 1.374 can be unstable if they are constituted by large vortex patches (with $\gamma > 3.66$). In the parameter plane $(\gamma, 2d)$, such unstable hetons fill a relatively narrow sector above the region occupied by beanlike solutions. The characteristic feature of the beanlike hetons is the concave shape of their vortex patches. Beanlike and cross-over hetons belong to the category of overlapping solutions. Beanlike hetons are unstable to asymmetric perturbations, but some of them can be stable to symmetric perturbations. The cross-over hetons, whose shapes are the most complex among all of the solutions found, are unstable in both senses.

The results presented above offer some new perspective in the investigation of the dynamics of two-layer modons. Now that the two-parameter family of doubly symmetric hetons is constructed and the stability bounds (with respect to symmetric and asymmetric disturbances) are known, a possibility arises to examine the presumably different oscillatory regimes and transitions typical of hetons moved off from equilibrium states determined by different pairs $(\gamma, 2d)$.

3.2.3 *Heton with a Vertical Axis: Two Initially Elliptic Vortex Patches*

As noted above, the introduction of *small* ellipticity to the initial configurations of vortex patches creates the conditions for the second mode to becomes unstable. Let us suppose now that the ellipses that compose a heton have arbitrary semiaxes a and b along the axes x and y, respectively. Now, for an "equivalent" ellipse (with an area equal to that of the unit circle, i.e., at $ab = 1$), the parametric specification of coordinates of contour points

$$x = a\cos\varphi = a\sqrt{\frac{1+\cos 2\varphi}{2}}, \quad y = b\cos\varphi = \frac{1}{a}\sqrt{\frac{1-\cos 2\varphi}{2}}$$

yields

$$\rho = \sqrt{x^2+y^2} = \sqrt{\frac{\chi^2+1}{2\chi}} \cdot \sqrt{1 + \frac{\chi^2-1}{\chi^2+1}\cos 2\varphi},$$

where $\chi = a/b$.

Before going to the analysis of numerical calculations, we list the available results of studying the behavior of elliptic vortex patches. The classic Kirchhoff solution [416] for an elliptic vortex patch with semiaxes a and b and with vorticity ω, which rotates as a solid body with constant angular velocity $\Omega = \omega ab/(a+b)^2$ has been known since 1876. Seventeen years later, Love [544] showed this solution to be unstable at $\chi = a/b > 3$. Mitchell and Rossi [615] gave a full quantification by parameter χ of the regimes of linear and nonlinear stability of the elliptic vortex.

Note, that Thomson (Lord Kelvin) in [904] proposed an energy-based argument for determining the stability of inviscid vortical flows. Recently, Luzzatto-Fegiz and Williamson [550, 553] proposed a new approach based on the construction of "Imperfect–Velocity–Impulse" (IVI) diagrams, which represents a rigorous and practical implementation of Kelvin's argument for determining the stability of vortical flows.

Kirchhoff's solution has many generalizations. In particular,

- Chaplygin [166,598], and later Kida [414], Dritschel [236], Dhanak and Marshall [219], Legras and Dritschel [529] and others showed that the introduction of external velocity field, linearly depending on coordinates (this is analogous to an affine transformation of coordinates that converts ellipses into ellipses) enables one to construct a solution specifying the behavior of a pulsing vortex rotating with varying velocity at the value of χ varying over time.
- Polvani and Flierl [706] introduced the notion of generalized Kirchhoff vortex for a system of N embedded elliptic patches. They studied the stability of such solution and constructed diagrams of stable and unstable states in the space of external geometric parameters.
- Kozlov [459–461] generalized the problem of elliptic vortex to a case with the effect of "involvement" taken into account through the introduction of "effective" bottom friction. This mechanism initiates the cyclone-anticyclone asymmetry that can be seen in the ocean or the atmosphere (at changes in the sign of ω). In [459], the conditions are determined under which the model yields the particular cases of Kirchhoff, Chaplygin, and Kida.

Here, we will use CDM to study the possible evolution scenarios of *two-layer vortices with elliptic shape*. It is assumed in all cases that, in the initial moment, the ellipses of the top and bottom layers are identical, are aligned strictly one above the other, and have opposite signs of potential vorticity (i.e., an elliptic heton with vertical axis).

Figure 3.30 gives a diagram of possible states of a two-layer elliptic heton in the plane of parameters (γ, χ), constructed based on the analysis of many numerical calculations of the evolution of vortex patches. Figure 3.30 shows that even small deviations of the initial configurations of vortex patches from circular ones ($\chi = 1.1$) give essentially new results, especially at $\gamma > 3$. The experiments given in the left and right parts of Fig. 3.31, have been carried out at the same values of γ as those given in Figs. 3.2 and 3.3, where, in the axisymmetric case, the second and third modes form, respectively, and the scatter directions of the newly formed two-layer pairs are determined in a random manner.

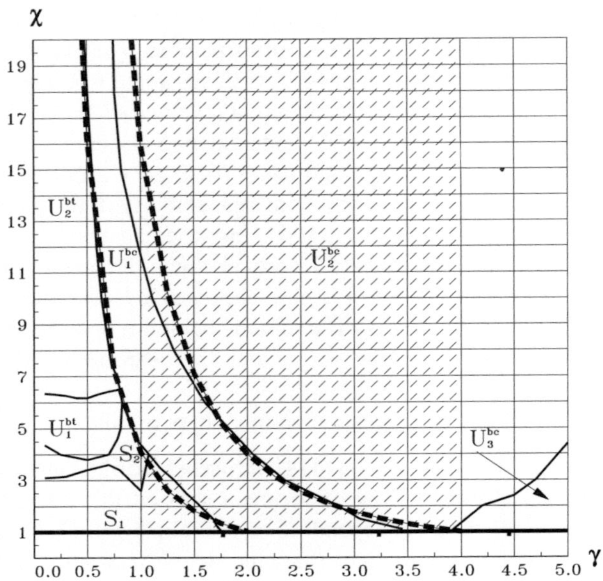

Fig. 3.30 Diagram of possible states of an elliptic heton with vertical axis in the rectangular domain ($\gamma \in (0, 5], \chi \in [1, 20]$). Denotations for subdomains: S_1 stable states, S_2 stable states with partial loss of mass due to the casting of vortex threads, U_1^{bt} decomposition of elliptic patches into asymmetric parts because of barotropic instability, U_2^{bt} decomposition of elliptic patches into symmetric parts because of barotropic instability, U_1^{bc} decomposition of elliptic vortex patches because of baroclinic instability into two hetons with tilted axes moving apart, U_2^{bc} cascade baroclinic instability with small-scale hetons moving away along both axis of the ellipse, U_3^{bc} cascade baroclinic instability with small-scale hetons moving away along both axes of the ellipse. The *thick dashed lines* can be described by equations $\gamma\sqrt{\chi} = 2$ and $\gamma\sqrt{\chi} = 16$ (see the text). The marked rectangular domain corresponds to mesoscale oceanic vortices. The *ticks* on the *thick straight line* $\chi = 1$ separate the domains (*left* to *right*): the stability of the axisymmetric heton, instability of the mode with $m = 2$, $m = 3$, and $m = 4$, respectively. Adapted from [859] (Fig. 1) by permission of Springer

The main distinctions of the realization of the second mode ($\gamma = 2.4$) in the considered case with Fig. 3.2 consists in: *(a)* the scatter direction is now determined by the orientation of the initial vortex ellipses, *(b)* the time interval of the passage to a nonlinear stage of evolution becomes shorter. At $\gamma = 4$, the differences from the scenario in Fig. 3.3 are of deeper, qualitative character. Indeed, here we see "competition" between mechanisms responsible for the formation of the second (due to the initial elliptic perturbation of the shape of vortex patches) and third (due to the larger value of γ) modes. The process of interaction between vortices can be conventionally divided into several stages. At the first stage ($t = 2, 3$), two pairs moving apart are formed. However, the newly formed vortex patches are still rather large (as compared with circular vortices with radius R_d), and, at the second stage, they break up into three unequal parts ($t = 4 \div 7$). At the third stage, we see six well-formed two-layer pairs: two leading and two tailing ones (moving apart in opposite directions—a cascade process) and two pairs moving toward one another

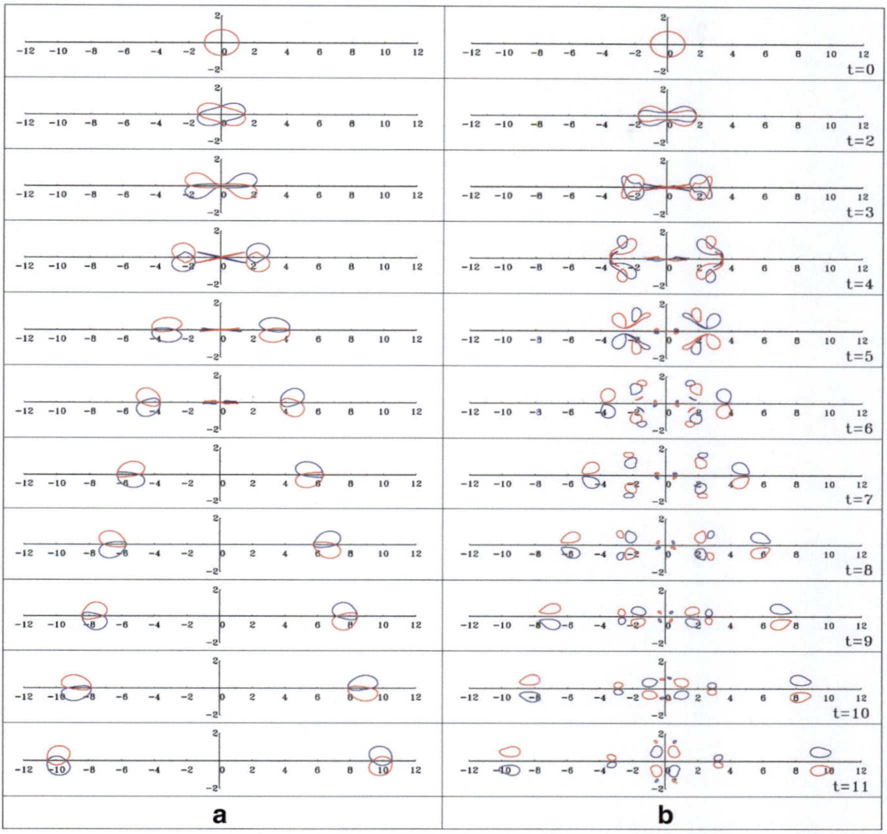

Fig. 3.31 Synchronous configurations of unstable elliptic hetons with vertical axes at $\chi = 1.1$ and $\gamma = 2.4$ (*left*, type U_1^{bc}) and $\gamma = 4$ (*right*, type U_3^{bc})

($t = 8, 9$). Finally, at the fourth stage ($t \geq 10$), the latter pairs change their partners and the new hetons with tilted axes thus formed move apart in opposite directions now along the y-axis.

The results of calculations at $\chi = 2$ are given in the series of Figs. 3.32–3.33. Figure 3.32 shows a relatively "stiff" passage from stable to unstable scenario (from S_1 to U_1^{bc}) with an increase in parameter γ by 0.01. The analogy mentioned in section "Analogy with A-Symmetric Structure of Discrete Hetons" can be seen very clearly here. Indeed, the dumbbell-shaped structure, which is most typical of the stable vortex patches in the left part of the figure, characterizes a trend toward the splitting of each patch into two parts. The entire configuration, while staying within the domain limited by the appropriate separatrix (for the problem of four point vortices—see Fig. 2.25a), does not decompose, but the rotation of vortex patches becomes much slower when their wider head parts are near separatrix

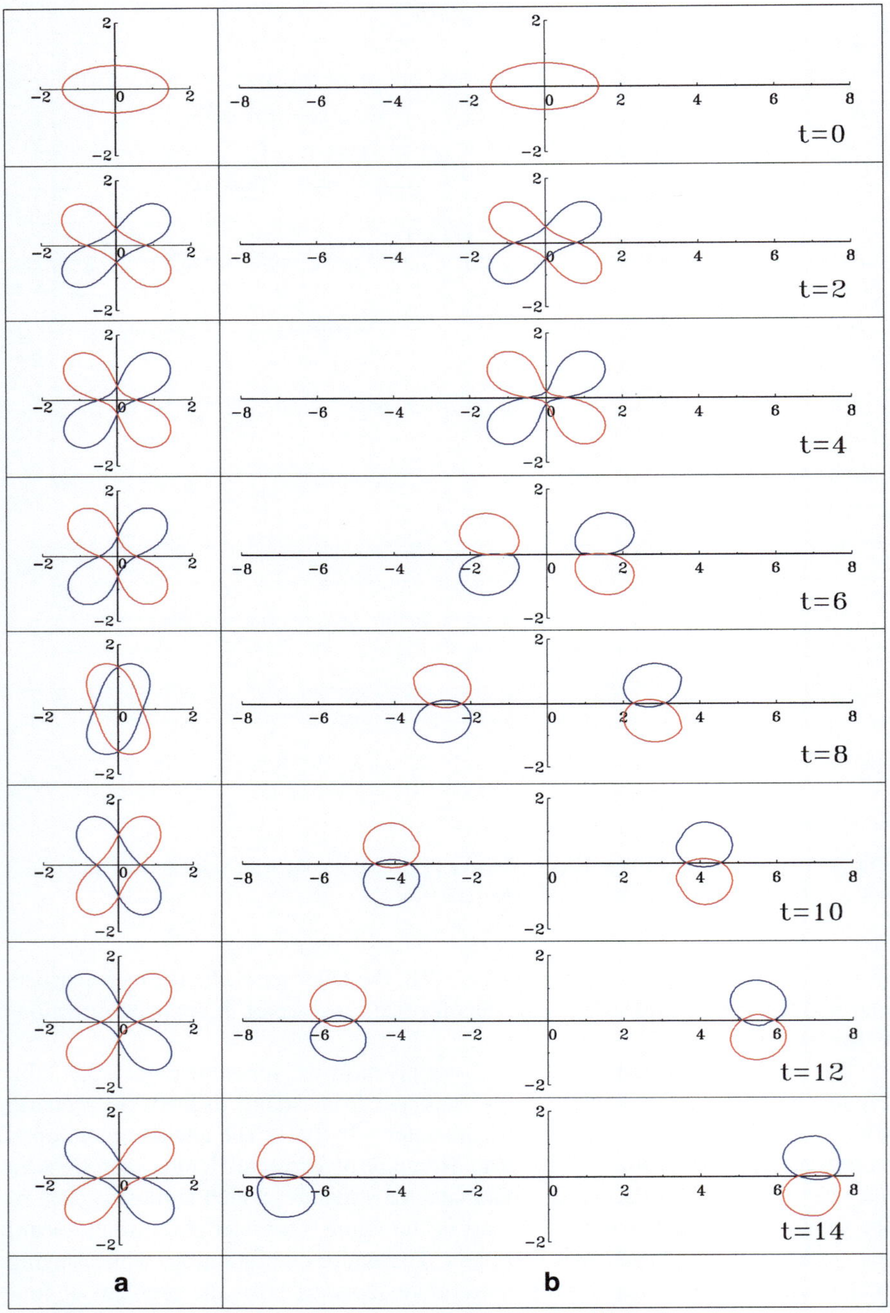

Fig. 3.32 Synchronous configurations of (**a**) a stable ($\gamma = 1.56$, type S_1) and (**b**) an unstable ($\gamma = 1.57$, type U_1^{bc}) elliptic heton with vertical axis at $\chi = 2$. Adapted from [859] (Fig. 2) by permission of Springer

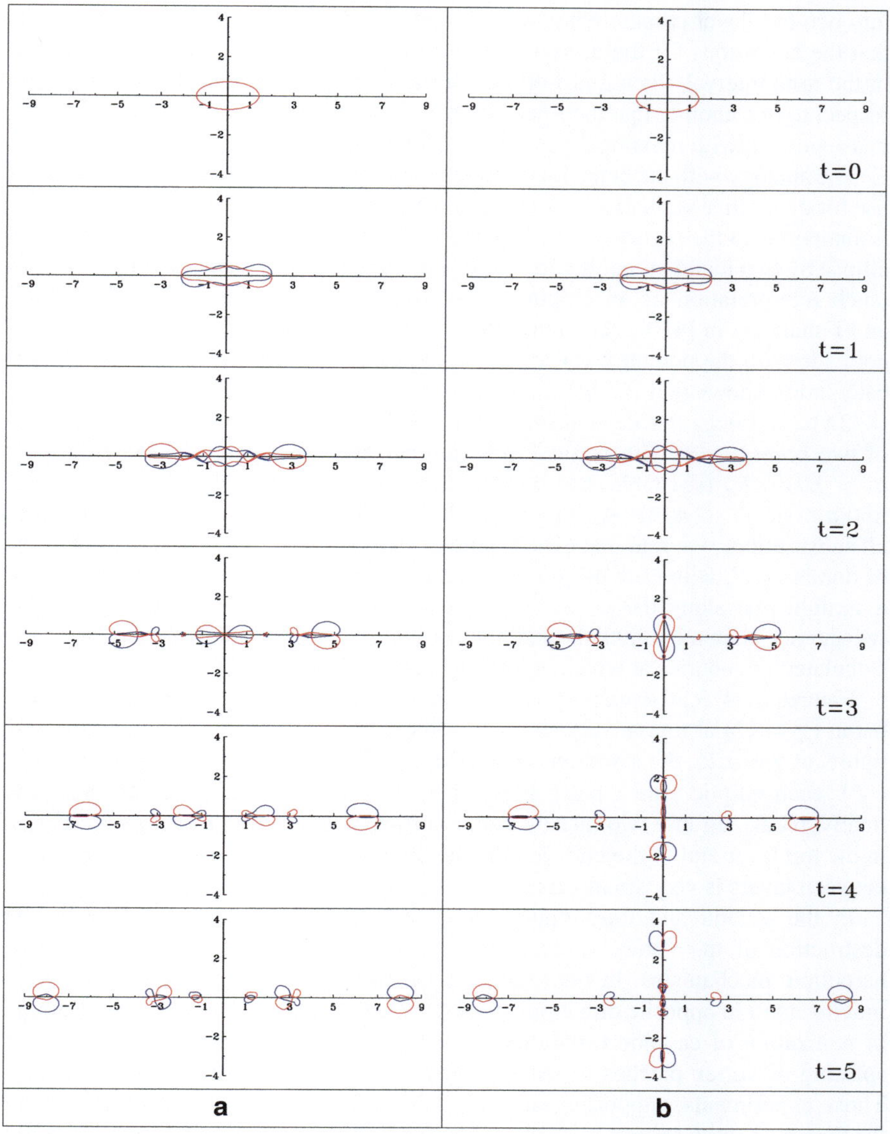

Fig. 3.33 Synchronous configurations of unstable elliptic hetons with vertical axes at $\chi = 2$ and (**a**) $\gamma = 4.55$ (type U_2^{bc}); (**b**) $\gamma = 4.57$ (type U_3^{bc}). Adapted from [859] (Fig. 4) by permission of Springer

self-intersection points (deceleration points). However, when vortex patches finds not enough room within the domain of closed contour lines of the phase plane (an increase in γ causes a decrease of that domain on the plane of variables x and y), the central neck breaks (the beginning of this process can be seen at the fourth time moment in the right part of the figure) and, as the result, the elliptic heton divides into two two-layer pairs moving away in opposite directions. It is important to note that the conditions for the decomposition of initially compact vortex patches form in the time interval when quasi-elliptic vortices are located in such a manner with respect to one another that their head parts lie near hyperbolic singular points of the phase portrait with maximal stress in the field of stream function (Hamiltonian).

The analogy with the behavior of point hetons, which was used before, is again of use for qualitative explanation of the features of diagram 3.30. Thus, the regime U_1^{bc} is analogous to the regime of moving apart of two discrete hetons. We replace each ellipse by two identical point vortices located in the rotation centers of half-ellipses (such interpretation of an elliptic vortex was used, in particular, by Kawakami and Funakoshi in [410]). The intensities of such effective point vortices obviously coincide with the potential vorticity integrated over the area of half ellipse. Direct calculation shows that a_c (the modulus of the x-coordinate of the point vortex) is $(4/3\pi)a$ and does not depend on b. The condition of decomposition of a system of two point hetons with vertical axes into two two-layer pairs with tilted axes is $a_c > 0.717 \cdot R_d$ (see table 2.1). Substituting parameters for effective vortices, we find $a > 0.717 \cdot (3\pi/8) \cdot R_d$. In the variables (γ, χ), we obtain $\gamma\sqrt{\chi} > C$ and after all substitutions, we find that $C \approx 2$. Similar scaling can be applied to the boundary of domain U_2^{bc}, with each ellipse replaced by four identical point vortices lying on a straight line along the a-axis. In this case, we obtain $C \approx 16$. The appropriate two curves, shown by thick dashed lines in Fig. 3.30 are qualitatively similar to the calculated boundaries at which elliptic hetons change their behavior.

Figures 3.34–3.36 exemplify two different types of motion of a highly elliptical heton ($\chi = 12$) at a relatively strong stratification of a two-layer fluid. In the first figure, at $\gamma = 0.25$, the evolution in each layer follows the barotropic scenario (type U_2^{bt}): each elliptic vortex patch first bends, next forms bulges near its ends, and finally, breaks up into two quasi-circular vortices rotating in the cyclonic direction in the top layer and in the anticyclonic direction in the bottom one. The interaction between layers is very small here.

In the second and third figures, at $\gamma = 0.75$ and $\gamma = 2.5$ respectively, the destruction of the vortex structure takes place under the predominant effect of baroclinic mechanisms. In Fig. 3.35, we see two two-layer pairs with tilted axes moving apart in opposite directions (type U_1^{bc}), while in Fig. 3.36 we see an example of realization of cascade instability (type U_2^{bc}). In all three cases, the process of splitting of vortex patches is accompanied by active formation of vortex threads. Those experiments, supporting diagram 3.30, demonstrate a considerable drop in the threshold (with respect to parameter γ) for the formation of baroclinic instability of hetons at finite perturbations of circular shape.

Splitting of vortices has been observed in ocean and atmosphere: decomposition of the "Allen" quasi-elliptic ring of Gulf Stream into two unequal (type U_1^{bt}) circular vortices—the larger "Allen" and the smaller "Arthur" was observed in December 1976 north of the Bermudas [750]; the decomposition of a stratospheric polar vortex in the Southern Hemisphere in September 2002 was recorded in [168, 693, 768]. Laboratory evidence of vortex splitting was observed, for example, by Freymuth et al. [282, 283] and Eisenlohr and Eckelmann [246].

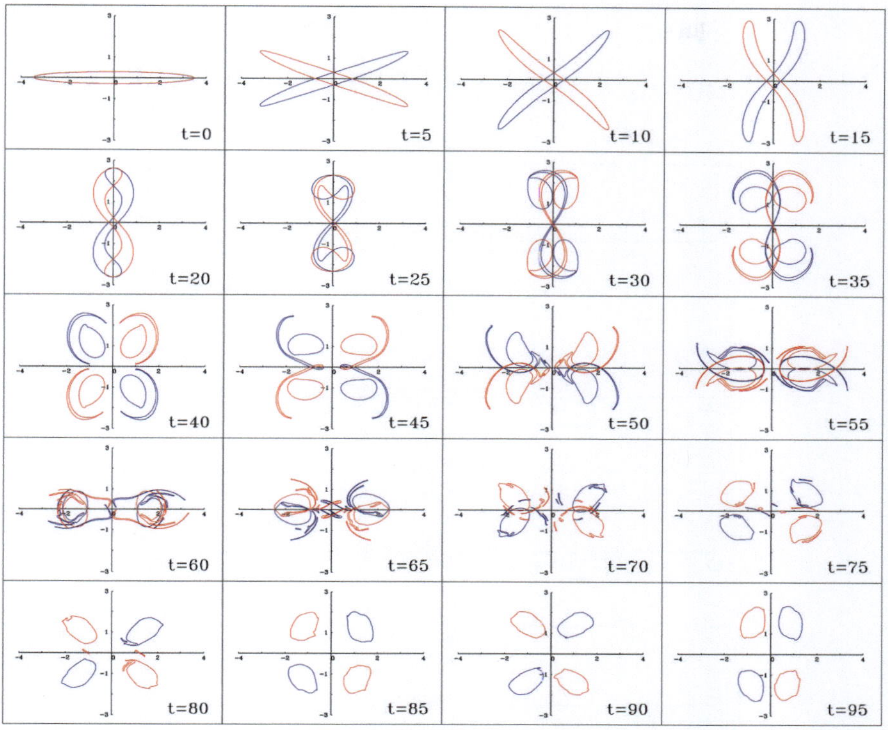

Fig. 3.34 Evolution of an unstable elliptic heton with vertical axis at $\chi = 12$ and $\gamma = 0.25$ (type U_2^{bt})

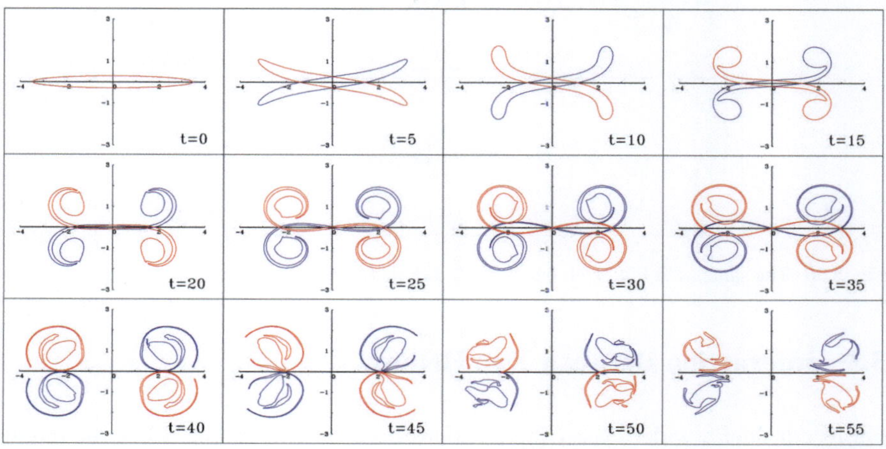

Fig. 3.35 The same as in Fig. 3.34, but at $\gamma = 0.75$ (type U_1^{bc})

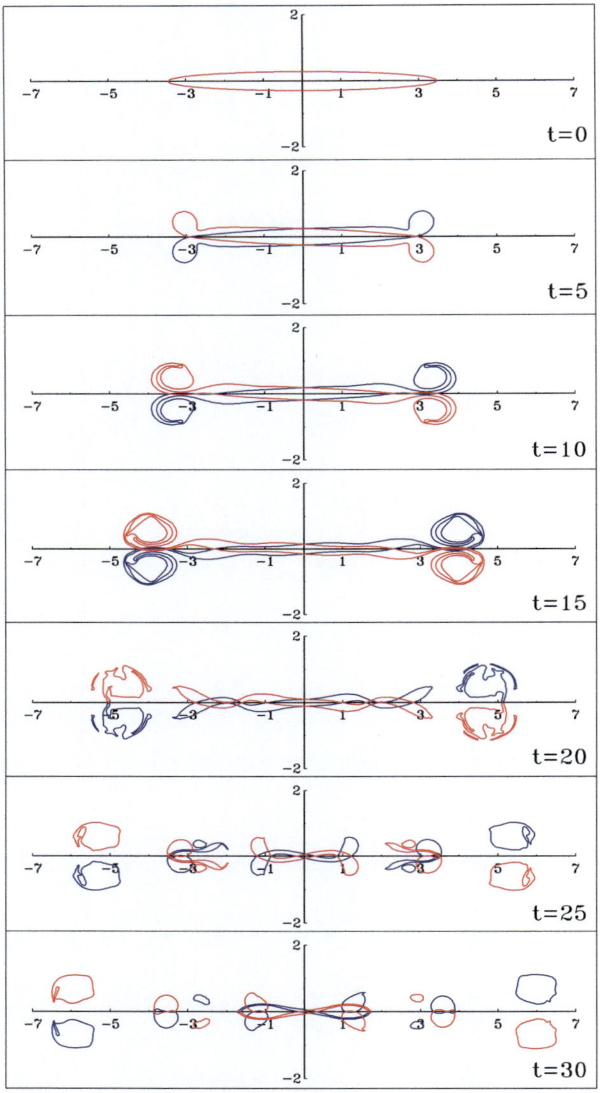

Fig. 3.36 The same as in Fig. 3.34, but at $\gamma = 2.5$ (type U_2^{bc})

3.3 Interaction Between Two Hetons

Let us consider a wider class of interactions between two hetons, when the initial arrangement of four circular vortex patches is determined in the general case by the scheme shown in Fig. 3.37.

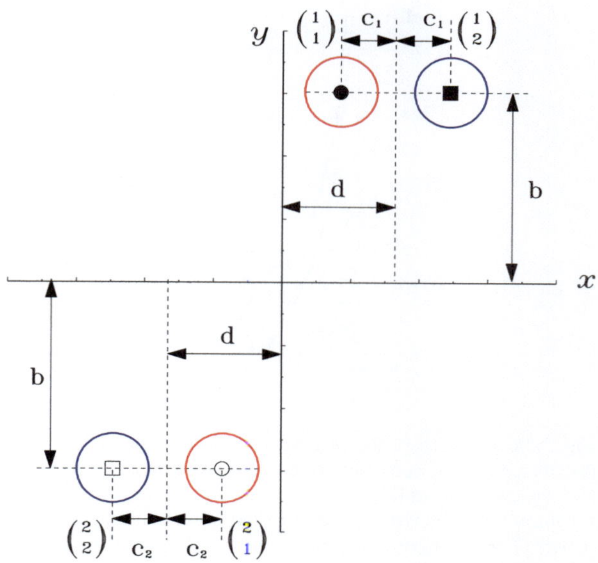

Fig. 3.37 General scheme of the initial arrangement of vortex patches at the interaction between two hetons

3.3.1 Two Hetons with Vertical Axes

Suppose that $c_1 = c_2 = d = 0$. In the simplest case of discrete vortices only two types of motion, $\{3\}$ or $\{2\}$, may take place, depending on parameter B (in phase portraits, the initial positions correspond to points of the axis $x = 0$ in Fig. 2.25a or the axis $t_3 = 0$ in Fig. 2.25b). In the former case, at $B = b\gamma < R^S_{min}$ and $A = 2$ in Table 3.1, within-layer interaction is dominating, and the vortices move along closed coinciding O-shaped trajectories in opposite directions in the top and bottom layers; in the latter case, the axes of each heton tilt and the newly formed two-layer pairs move apart in opposite directions.

The situation becomes much more complicated in the case of distributed vortices because of two important features of finite-size vortices: a tendency toward merging of nearby identical vortex patches and the ability of hetons to decompose (under the conditions corresponding to their instability).

The results of a large series of numerical experiments, aimed at studying the interaction between two vertically aligned hetons, are summarized in Fig. 3.38, showing the diagram of possible states of the system of vortices in the rectangular domain on the plane (b, γ). It can be seen that the dashed line separating the existence domains of solutions of types $\{2\}$ and $\{3\}$ for discrete vortices corresponds well to the case of distributed vortices between domains S_2 and S_3. It is important to note that a similar effect was obtained in [919], where calculations were carried out with the use of a difference pseudospectral code and with dissipation

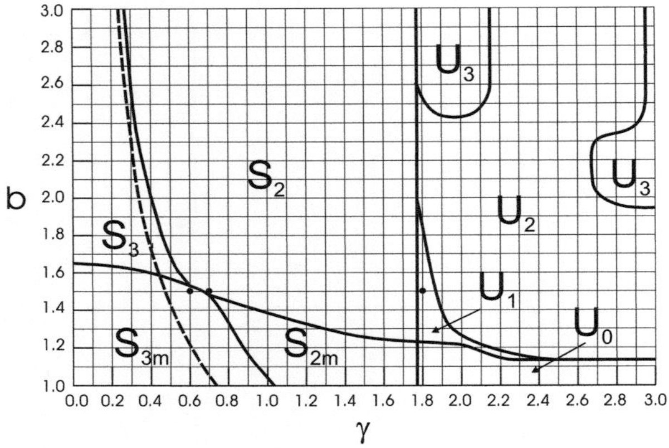

Fig. 3.38 Diagram in the rectangular domain ($\gamma \in (0; 3]$; $b \in [1; 3]$) of possible states of two distributed hetons, consisting of circular unit-radius vortex patches and originally having vertical axes spaced $2b$ apart ($c_1 = c_2 = d = 0$). The *dashed line* is the boundary between the subdomains for solutions of the types $\{2\}$ and $\{3\}$ for discrete vortices. Regions are denoted as follows: S_2 the scatter of hetons in opposite directions along the straight line passing through their centers at the initial moment (an analogue of motions of type $\{2\}$ for discrete vortices), S_{2m} the same as in S_2, but after the temporary merging of vortex patches in the layers and their subsequent separation, S_{3m} the merging of originally circular patches into quasi-elliptic vortex structures rotating in opposite directions, S_3 layer-by-layer rotation of vortex patches in opposite directions (an analog of motions of type $\{3\}$), U_0 the merging of hetons and their subsequent separation into more than two scattering hetons, U_1 the decomposition of two-layer vortices and the formation of four new hetons, U_2 the decomposition of hetons into noncompensated two-layer vortices, followed by the merging of vortex patches and the repeated formation of hetons, U_3 the decomposition of hetons into scattering two-layer vortices without subsequent merging. Markers correspond to the pairs of parameters γ and b used in the numerical experiments shown in Figs. 3.39–3.41. Adapted from [851] (Fig. 8) by permission of Cambridge University Press

(biharmonic friction) taken into account.[4] In this figure, the regimes denoted by letters S and U are realized due to stable or unstable states of two-layer vortices, respectively.

As noted above, the interactions in the cases of discrete and distributed vortices have some specific features. The first series of examples (Figs. 3.39–3.41) refers to motion types S_{3m}, S_2, and U_1. In this case, the distance between the centers of vortex patches is fixed and the stratification parameter γ takes different values.

Figure 3.39 shows the evolution of configuration in the case where vortex patches in both layers merge to form vortices of quasi-elliptic shape and carrying out rotations in opposite directions. Note that within the interval between 2 and 5 dimensionless time units, the system shows almost no evolution, since the center of

[4]Legras and Dritschel [530], who compared the results of calculations by the pseudospectral method and with the use of CDM, note their qualitative agreement. However, the authors of [530] note that CDM is more convenient in studying the formation of vortex threads and the merging and separation of vortices.

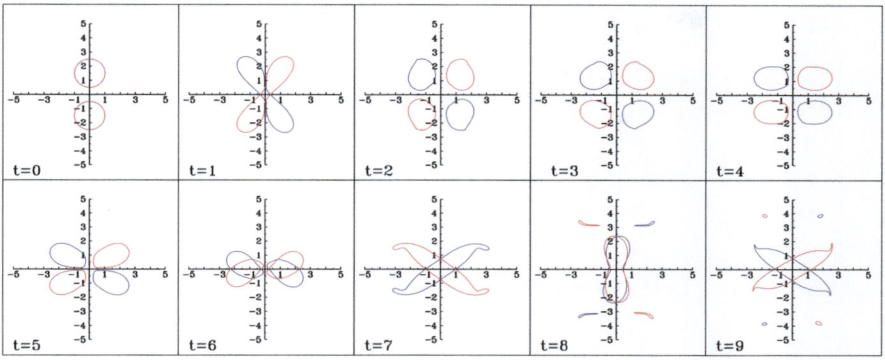

Fig. 3.39 An example of formation of a vortex structure of the type S_{3m} ($\gamma = 0.6; b = 1.5$). Adapted from [851] (Fig. 9a) by permission of Cambridge University Press

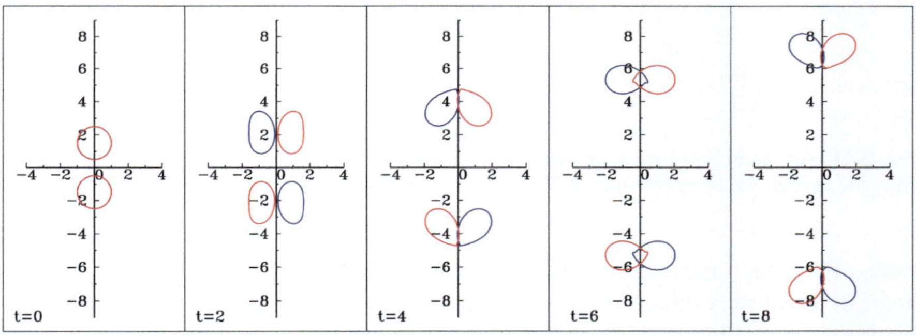

Fig. 3.40 An example of formation of a vortex structure of the type S_2 ($\gamma = 0.7; b = 1.5$). Adapted from [851] (Fig. 9b) by permission of Cambridge University Press

each vortex lies near a fixed saddle point (deceleration point)—the corresponding marker in Fig. 3.38 is situated very close to the boundary between the domains of finite and infinite motions.

Figure 3.40 exemplifies the formation of hetons with tilted axes and their subsequent moving apart in opposite directions. This is a complete analogy with motions of type $\{2\}$ for discrete vortices. The fact that the insignificant difference between the values of γ in this and the previous experiments leads to qualitatively different results suggests the "rigid" character of the solutions of the appropriate system of equations.

An interesting picture of interaction between hetons, when they are unstable, is given in Fig. 3.41 for the case of relatively weak stratification (or small values of deformation radius). At the initial stage, the hetons start moving apart, but, at the same time, each one breaks up into two two-layer pairs, such that one continues moving in the same direction, while the other moves in the opposite direction. Next, the vortex patches of the hetons that took part in the counter motion merge for a short

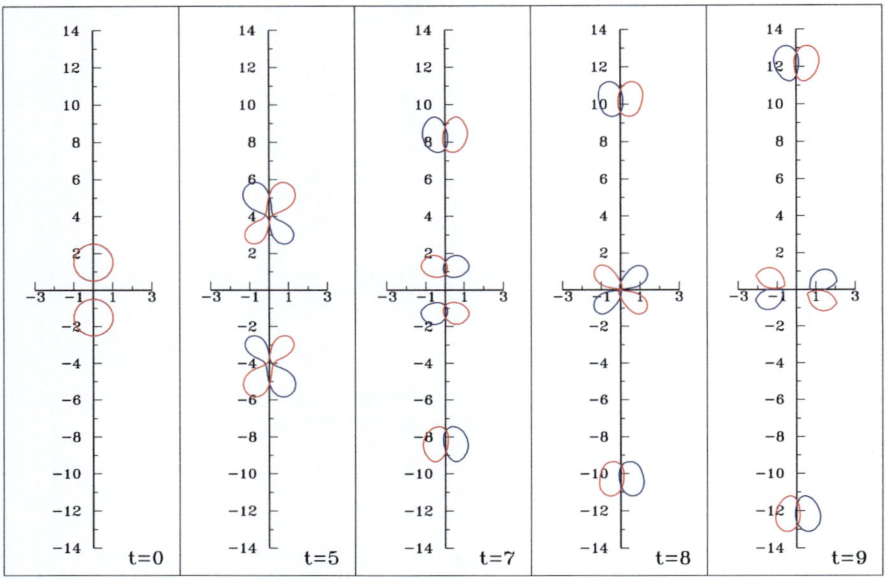

Fig. 3.41 An example of formation of a vortex structure of the type U_1 ($\gamma = 1.8$; $b = 1.5$). Adapted from [851] (Fig. 9c) by permission of Cambridge University Press

time, after which they again separate and the newly formed two-layer pairs move apart in the normal direction. If we compare the resulting effects of the interaction between two hetons at this and the previous figures, without going into details of intermediate processes, we will find that two two-layer pairs moving apart will form in the former case, while four such pairs will form in the latter case.

The series of Figs. 3.42 elucidates the role of parameter b (the initial distance between heton centers) at a fixed value of $\gamma = 1.8$ in the formation of new vortex structures. Note that those experiments were carried out with the same γ as those shown in Fig. 3.41. Compare it with Fig. 3.42a, where the initial distance between the hetons is somewhat increased. It can be seen that qualitatively, all intermediate stages of formation of vortex structures are similar, but at the last stage, the scattering of new two-layer pairs now takes place in the direct, rather than perpendicular, direction. This is due to the weaker effect the originally formed pairs have on them, because the latter pairs in this case have moved much further than in the previous experiment. Clearly, this vortex structure can still be referred to the type U_1.

The evolution of vortices at $\gamma = 1.8$ and $b > 2$ is different. At the initial stage, almost no motion of hetons is induced in the y-axis direction, and the mechanism of destruction via instability is predominant. An interesting class of motions U_2, which is very wide in the space of parameters (γ, b), is shown in Fig. 3.42b. Each heton out of the two hetons with originally vertical axes starts decomposing into two quasi-hetons with very weak non-compensation; however, the vortex patches located in the inner part of the vortex construction experience a considerable impact of the nearest

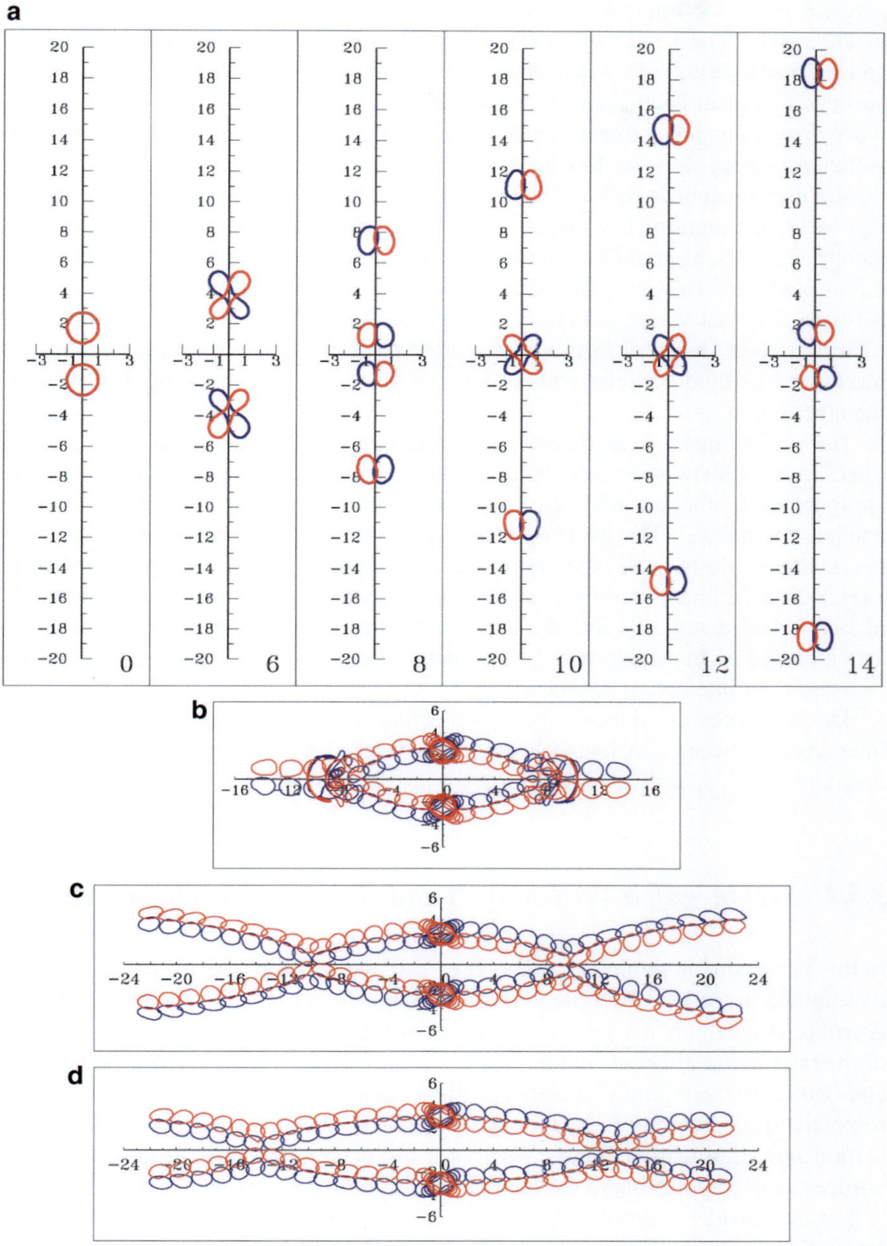

Fig. 3.42 Examples of formation of vortex structures at $\gamma = 1.8$: (**a**) $b = 1.7$, type U_1, (**b**) $b = 2.4$, $t = 0, 1, \ldots, 13$, type U_2, (**c**) $b = 2.7$, $t = 0, 1, \ldots, 19$; (**d**) $b = 3.0$, $t = 0, 1, \ldots, 19$—both belong to type U_3. In the last three fragments, the contours are aligned. Adapted from [851] (Fig. 10) by permission of Cambridge University Press

partners from the opposite layer, which belong to the other half-plane relative to the x-axis. The newly formed two-layer vortices acquire arc-like trajectories, and the quasi-hetons approach one another. After approaching close enough, the vortices merge in each layer with the formation of two new hetons, which now move apart in opposite directions along the x-axis. Those structures, because of the sufficiently large slope of their axes, show a stable behavior (see Sect. 3.2.1). Thus, the dominating contribution of the instability mechanism in this case has resulted in that we see the motion of hetons in perpendicular direction instead of their scattering along the y-axis, as could be initiated by the interaction of hetons at their specified initial positions. However, one should take into account that the actual difference between the final vortex structures in Figs. 3.40 and 3.42b consists not only in a rotation by 90°, but also in that each vortex patch in the former figure contains a mixture of its fluid particles and particles of the vortex that originally belonged to the other heton.

The further increase in parameter b results in that the two-layer vortices with tilted axes, which have formed after the decomposition of the initial hetons, approach each other periodically, though not close enough for the analogous vortex patches to merge.[5] The vortex-vortex repulsion at the moment when they are maximally close to each other takes place because of changes in the tilt of their axes, which facilitate the effect of internal "pairwise" interaction. The comparison of Fig. 3.42c, d suggests that the subsequent increase in the distance b should be accompanied by the straightening of trajectories of scattering hetons because of the weakening of interaction between them.

The examples considered above, reflecting only some particular features of interaction between distributed hetons with vertical axes, demonstrate their wide diversity.

3.3.2 Heton with a Vertical Axis and Heton with a Tilted Axis

In this paragraph, assuming $c_1 = d = 0, c_2 > 0$, we will consider the case similar to that studied in section "Two Hetons with Nonzero Total Linear Momentum and Zero Angular Momentum". A general summary of the resutls of numerous calculations of the interaction of hetons in this case is given in Fig. 3.43. As before, the letters S and U refer to the regimes when the original hetons have stable and unstable states, respectively. Comparing Figs. 3.43 and 2.16, we note that at small γ (small vortices at moderate stratification or large vortices at strong stratification), the model of point vortices is also applicable to the qualitative description of vortex patch motion.

Let us consider examples illustrating the specific features of different types of interaction.

[5] At the second stage of motion, the vortex structures in the left and right parts of this figure are analogs of interactions of the type $\{1a\}$ for discrete vortices (Fig. 2.23).

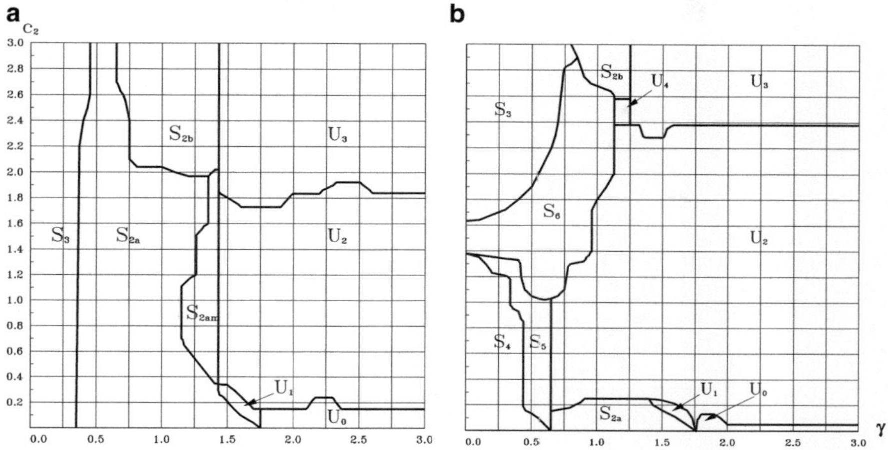

Fig. 3.43 Diagrams in the rectangular domain ($\gamma \in (0; 3]$; $c_2 \in [0; 3]$) of possible states of two distributed hetons consisting of circular unit-radius vortex patches at $c_1 = d = 0$ and (**a**) $b = 2.5$, (**b**) $b = 1.5$. Regions are denoted as follows: S_3 vortex leapfrog, S_{2a} "distant collision" and passage to reverse motion, S_{2am} the same as S_{2a}, but with short-time merging of vortex patches in both layers and subsequent splitting into unequal parts, S_{2b} an analog of motions of the type $\{2b\}$ for point hetons, it is a parallel motion of two two-layer pairs in the positive direction of the y-axis with the distance between them increasing with time, S_4 the formation of a large heton by merging of vortex patches and its translational motion in the direction of original motion of the approaching pair, S_5 the formation of a large heton, which moves along with two small peripheral vortices, belonging to different layers and rotating about the heton in opposite directions, S_6 intermediate merging of vortices with their subsequent separation and passage to the regime of vortex leapfrog U_0 it is a collapse of both hetons and the scatter of the newly formed small vortex structures, U_1 the decomposition of the approaching heton and the subsequent scatter of vortex pairs, U_2 the decomposition of an originally fixed heton into noncompensated two-layer vortices, the subsequent merging of vortex patches, and the passage to motions of the type S_{2am}, U_3 the decomposition of a fixed heton due to the "breakthrough" effect of the approaching heton and the scatter of the newly formed small two-layer vortices at an angle, U_4 the decomposition of a fixed heton into unequal parts, the merging of its larger parts with vortex patches of the approaching heton and the subsequent motion of the newly formed large heton, accompanied by small vortices

- *Stable configurations.* We will assume a configuration stable either when all four vortex patches stay compact or when the configuration has formed after the merging of vortices from a single layer and their subsequent separation.
  - When γ is small (strong stratification), we have the motions of type S_3—a vortex leapfrog [931]. In that case, within-layer interaction dominates, resulting in that the top-layer vortex patches rotate in cyclonic direction, while the vortex patches in the bottom layer rotate in the opposite direction. However, when the centers of vortices from different layers are at the distance of their maximal interaction, we can see the pair effect, facilitating the motion of heton in the positive direction of the y-axis. The combination of those two mechanisms creates the conditions for a baroclinic analog of vortex leapfrog illustrated in Fig. 3.44.

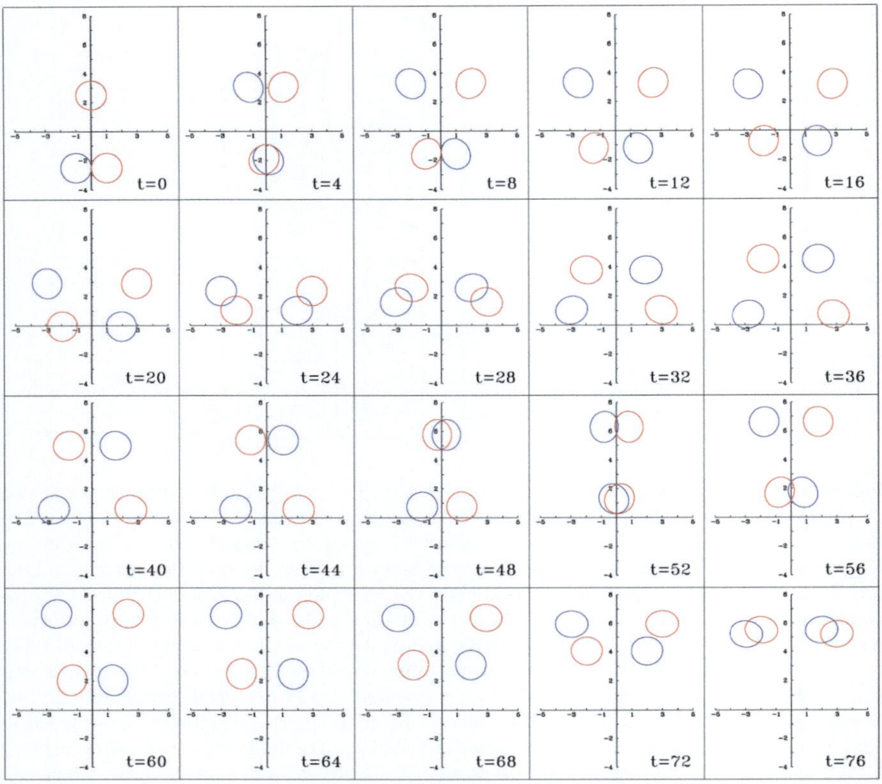

Fig. 3.44 An example of formation of a vortex structure of type S_3 at $\gamma = 0.3$ and $b = 2.5$, $c_1 = 0$, $c_2 = 1$

- When γ increases by 0.1, the contribution of the interlayer interaction becomes quite large for the vortex patches behave as discrete ones in the case of type $\{2\}$ motions. They "collide at a distance" and perform a reverse motion (type S_{2a} in Fig. 3.45).
- On the boundary of domain S_{2a} and in its narrow neighborhood, a zone is situated that corresponds to billiard klapstos type, where the originally fixed pair starts moving away, while the moving one stops, this is the type S_{2c} (an analog of motions of type $\{2c\}$, see Fig. 3.46).
- The further increase in γ is accompanied by the passage into the domain S_{2am}—a qualitative analog of S_{2a} with the only difference that a short-time merge of vortex patches in both layers takes place at an intermediate stage, followed by their separation into unequal parts. As the result, if we ignore small vortices and threads, we have two-layer pairs moving apart in opposite directions: the stronger one moves in the positive direction of the y-axis, while the weaker one moves in the opposite direction (Fig. 3.47). in this case, the area ratio of the vortex patches moving in different directions is 2.16 : 1.

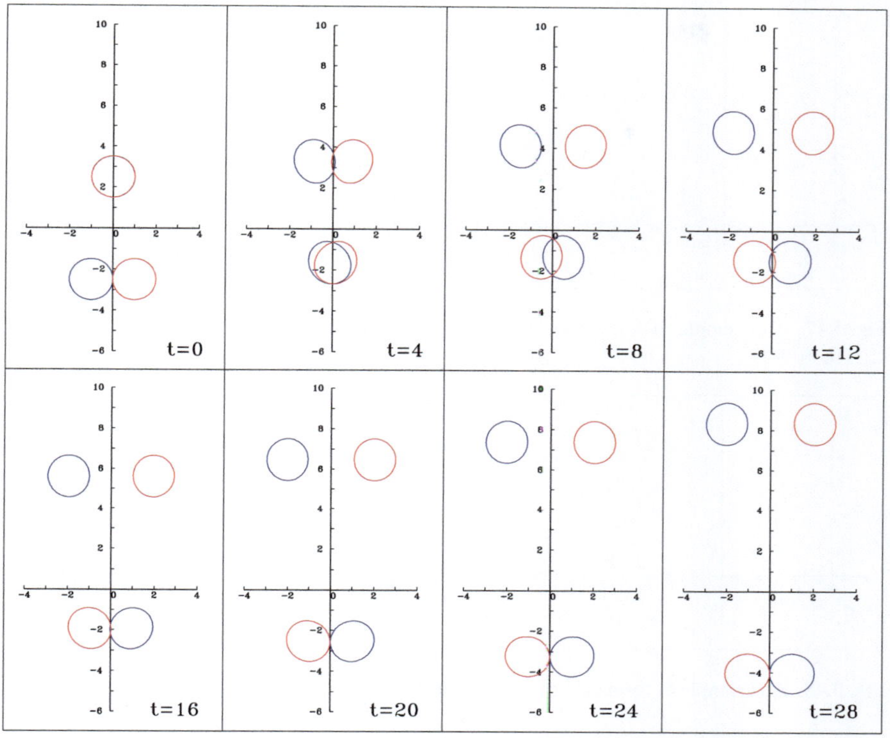

Fig. 3.45 An example of formation of a vortex structure of type S_{2a} at $\gamma = 0.4$ and $b = 2.5$, $c_1 = 0$, $c_2 = 1$

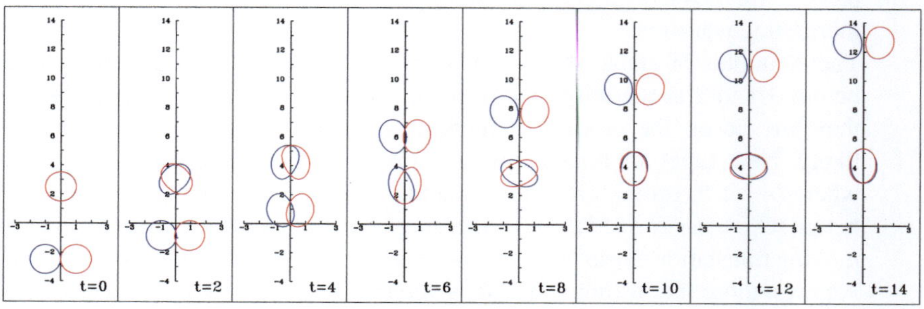

Fig. 3.46 An example of formation of a vortex structure of type S_{2c} at $\gamma = 1.1$ and $b = 2.5$, $c_1 = 0$, $c_2 = 1$

– Domains S_{2a} and S_{2am} occupy only the bottom part of Fig. 3.43a. As the centers of vortices of the heton with number 2 move apart significantly ($C_2 > 1.9$), the passage to the regime S_{2b} takes place in interval $0.8 < \gamma < 1.5$. An example of motions of this type is given in Fig. 3.48—this is an analog of

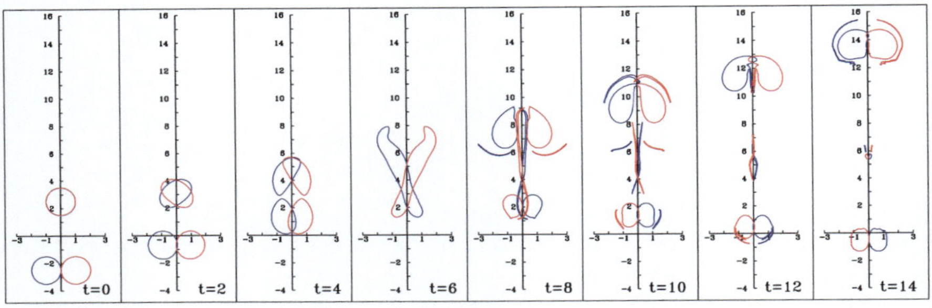

Fig. 3.47 An example of formation of a vortex structure of type S_{2am} at $\gamma = 1.2$ and $b = 2.5$, $c_1 = 0$, $c_2 = 1$

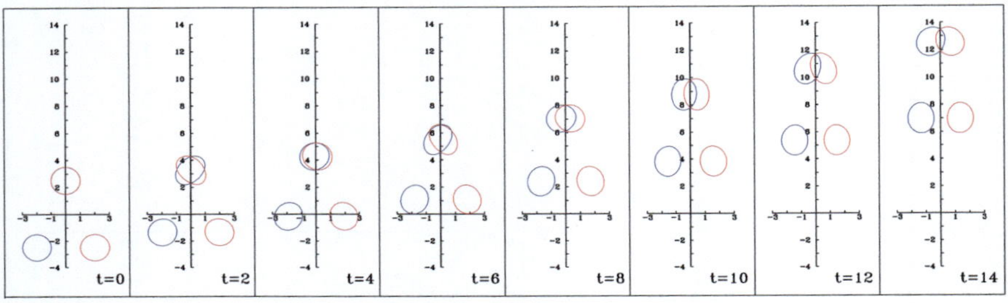

Fig. 3.48 An example of formation of a vortex structure of type S_{2b} at $\gamma = 1.1$ and $b = 2.5$, $c_1 = 0$, $c_2 = 2$

motions of type $\{2b\}$ for point hetons—the parallel motion of two two-layer pairs in the positive direction of the y-axis with the distance between them increasing over time.

- The examples of stable states considered above referred to the case where hetons 1 and 2 at the initial moment were spaced far apart ($b = 2.5$). When they are closer, the process of merging of similar vortex patches is more likely. New types of interaction can appear in this case. Thus, Fig. 3.49, where $b = 1.5$, exemplifies the formation of a structure of type S_4—a large heton, which has formed through merging of vortex patches and which now is moving translationally in the direction of original motion of the incoming pair. An even more interesting situation is given in Fig. 3.50 for the case with the same geometric parameters, but with somewhat weaker stratification: this is a large heton moving with two small peripheral vortices, rotating around it in opposite directions and belonging to different layers (motions of the type S_5). In Figs. 3.49 and 3.50, to show in greater detail the initial stage of evolution, the contours are given at shorter time intervals. In both examples, the long vortex filaments, bounding the vortex constructions, soon collapse, and the vortex patches acquire a quasi-steady shape.

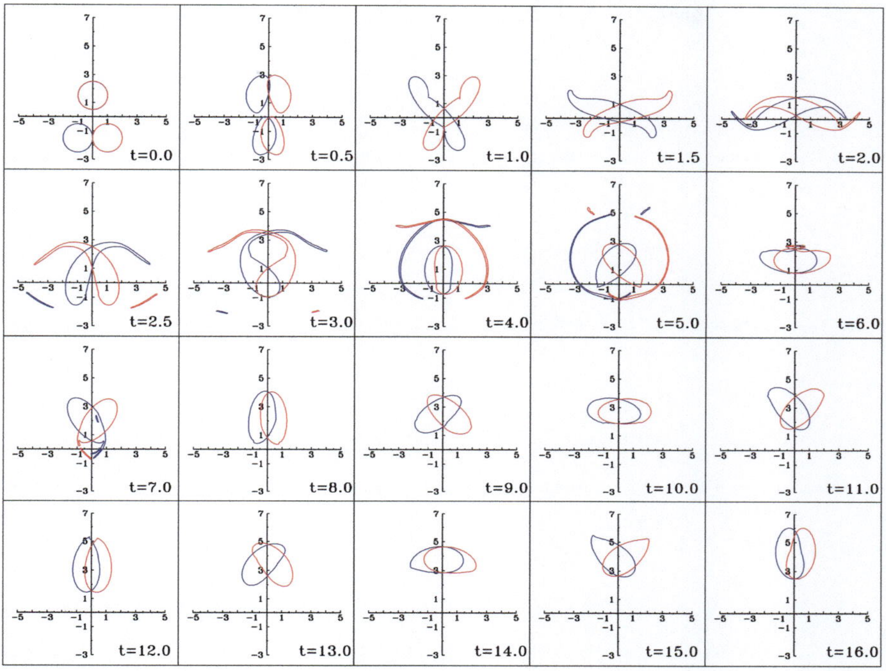

Fig. 3.49 An example of formation of a vortex structure of type S_4 at $\gamma = 0.3$ and $b = 1.5$, $c_1 = 0$, $c_2 = 1$

– With an increase in the initial tilt of the incoming heton (as compared with the experiment in Fig. 3.50), the efficiency of the mechanism of vortex patch merging decreases (Fig. 3.51), resulting in the separation of the merged vortices and the subsequent passage to vortex leapfrog regime—type S_6.

- *Unstable configurations.* We will call unstable the configurations where the decomposition of at least one heton begins before the two-layer pairs approach each other.

  – The characteristic features of motions of the type U_2 (Fig. 3.52) manifest themselves successively in the form of

    (a) decomposition of the initially fixed heton into unequal parts,
    (b) the merging of the larger parts with similar vortex patches of the incoming heton,
    (c) small unsimilar vortices "flowing around" a heton that has formed as the result of merging,
    (d) the formation of a new heton from small vortices,
    (e) scatter of unequal two-layer pairs in opposite directions.

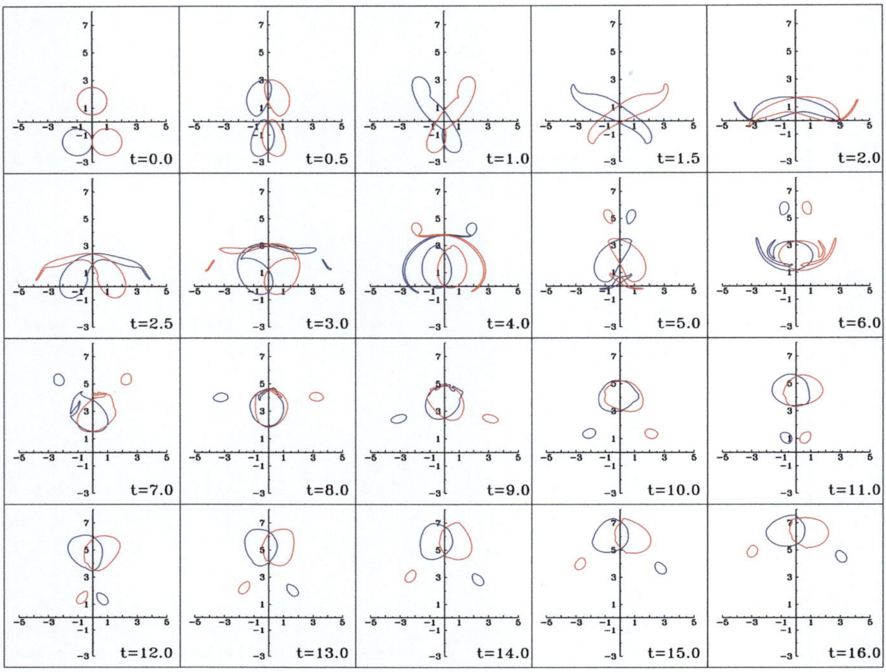

Fig. 3.50 An example of formation of a vortex structure of type S_5 at $\gamma = 0.4$ and $b = 1.5$, $c_1 = 0$, $c_2 = 1$

Essentially, the final vortex structure has the same form as in the case of motions of the type S_{2am}, though their genesis is different. In this case, the area ratio of the vortex patches running apart in opposite direction is 2.96 : 1.

- Fig. 3.53 demonstrates the evolution of a vortex system following the scenario of solutions of the type U_1. The latter are of interest because it is the incoming rather than the fixed heton that decomposes in this case. It breaks up into two hetons. The smaller one follows the originally immobile heton, whose axis is now tilted, while the bigger one moves in the opposite direction.
- The most complex dynamic structure is observed in the domain U_0 at a relatively weak stratification and small tilt of the axis of the incoming heton. Here, both two-layer vortices are unstable (Fig. 3.54). The destruction of heton 2 comprises two stages. At the first stage, it forms a two-layer, U-shaped structure, whose two halves belong to different layers, and two compact vortices, also located in different layers. At the second stage, each half of the U-shaped structure separates into two heton structures: one moves toward heton 1 (which, in its turn, separates into two noncompensated two-layer vortices), while both vortices of the second structure are captured by the remaining two compact vortex patches. As the result, we see the evolution of ten vortex patches, which form

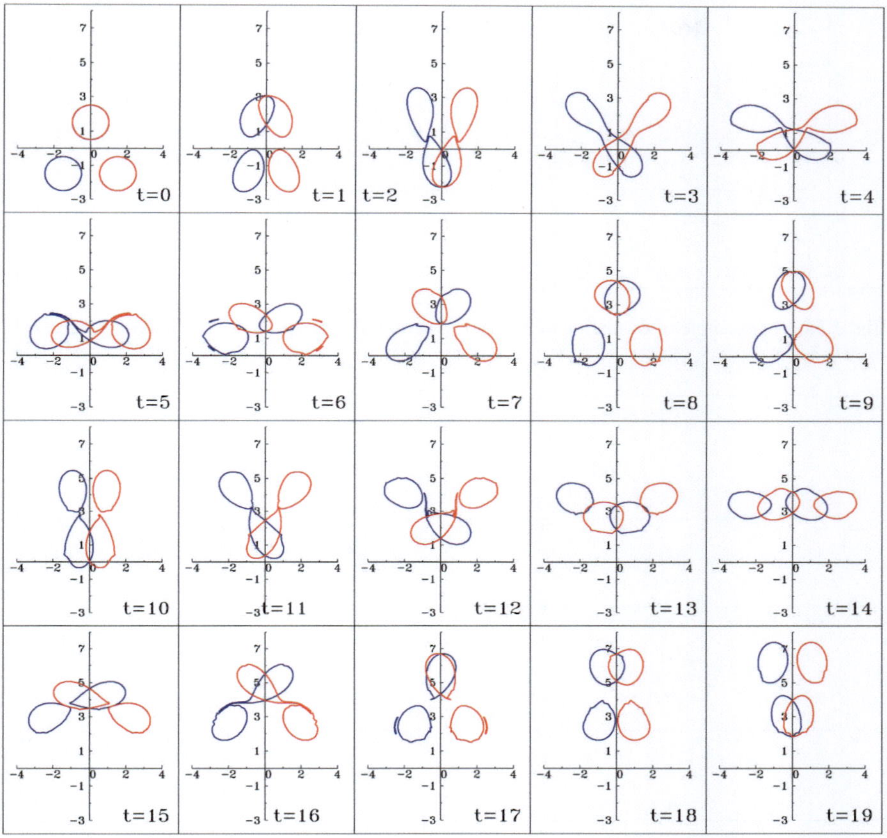

Fig. 3.51 An example of formation of a vortex structure of type S_6 at $\gamma = 0.4$ and $b = 1.5$, $c_1 = 0$, $c_2 = 1.5$

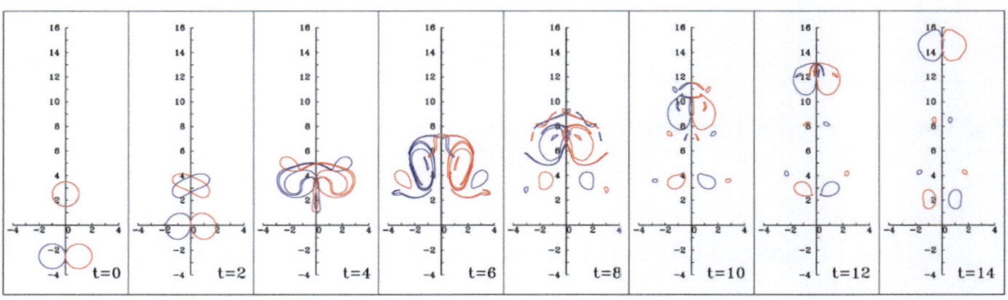

Fig. 3.52 An example of formation of a vortex structure of type U_2 at $\gamma = 1.6$ and $b = 2.5$, $c_1 = 0$, $c_2 = 1$

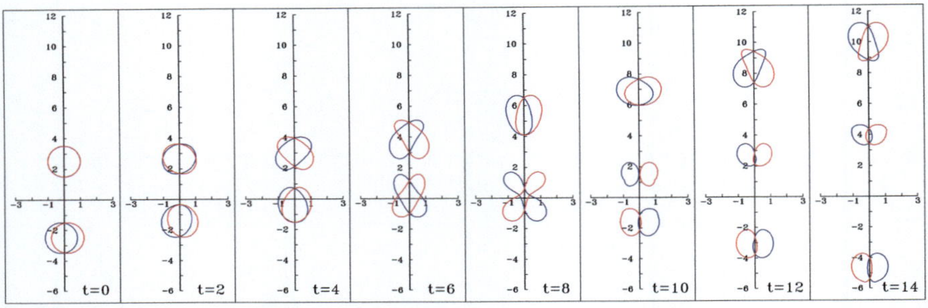

Fig. 3.53 An example of formation of a vortex structure of type U_1 at $\gamma = 1.6$ and $b = 2.5$, $c_1 = 0$, $c_2 = 0.2$

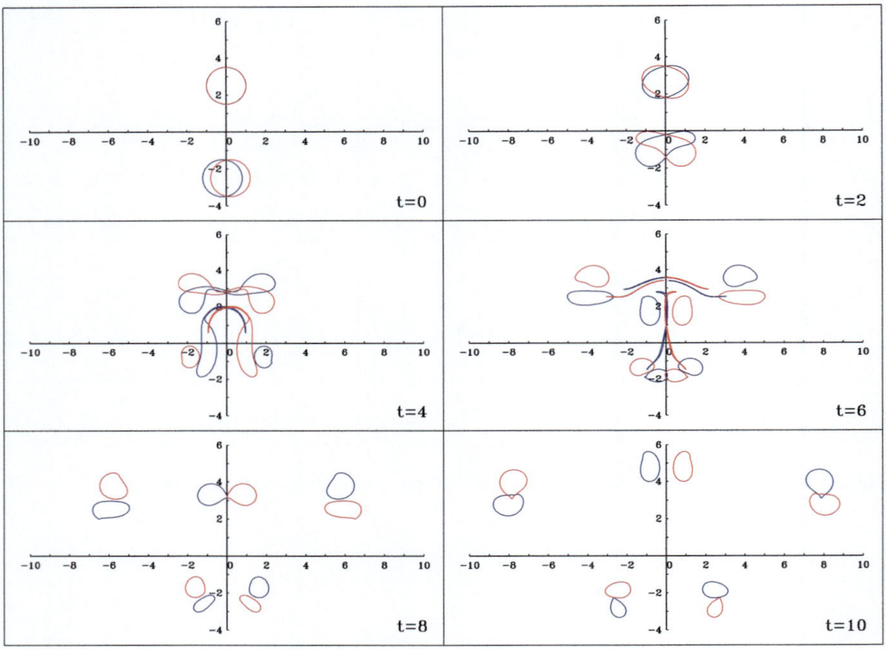

Fig. 3.54 An example of formation of a vortex structure of type U_0 at $\gamma = 2.3$ and $b = 2.5$, $c_1 = 0$, $c_2 = 0.2$

(a) four noncompensated two-layer vortices, rotating about their rotation centers,
(b) a heton moving in the positive direction of the y-axis.
(c) In the implementation of regime U_3 (Fig. 3.55), as well as in the case U_1, it is only the heton with number 1 that decomposes. It divides into two two-layer vortices scattering at an angle to the y-axis. Strictly speaking, those vortices are not hetons (the area ratios of the vortex patches are $1.11 : 1$),

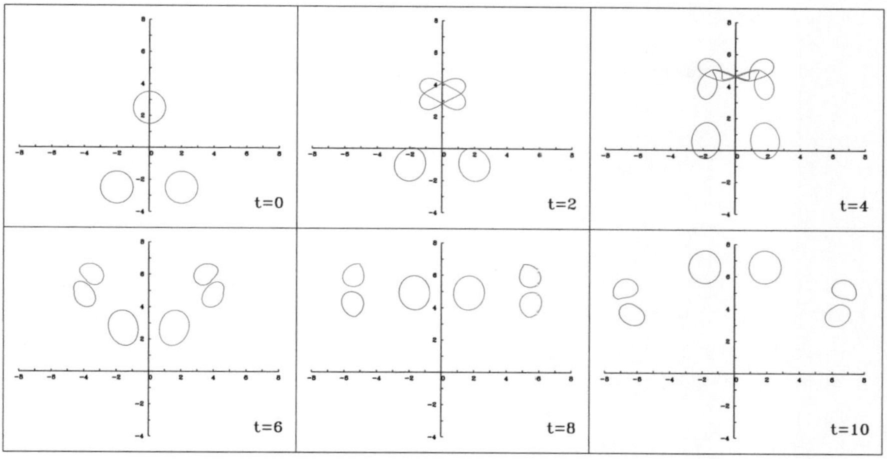

Fig. 3.55 An example of formation of a vortex structure of type U_3 at $\gamma = 1.6$ and $b = 2.5$, $c_1 = 0$, $c_2 = 2$

but their trajectories are nearly rectilinear. In this case, the decomposition is due to both the instability mechanism and the "breakthrough" effect of the incoming heton with inclined axis.

– A small domain in the plane of parameters (γ, c_2) at $c_1 = 1.5$ is occupied by the domain of existence of solutions of the type U_4, whose typical feature is the decomposition of an immobile heton into unequal parts, with the subsequent merging of the larger parts with vortex patches of the incoming heton, and the further motion of the newly formed large heton, accompanied by small vortices (Fig. 3.56). In this example, the ratio of the total areas of the large and small vortices in the configuration at $t = 18$ is $1 : 0.31$.

3.3.3 Two Hetons with Tilted Axes, the Case of Zero Total Momentum

The results of this paragraph allow us to establish the general and distinctive properties of models of distributed and discrete vortices at head-on collision of two hetons, where $d = 0$. Figures 3.57 and 3.58 give diagrams of possible states of discrete and distributed hetons at two different values of parameter b.

The comments to the first of them summarize the main properties of motions of types $\{1\}$, $\{2\}$, $\{3\}$ for discrete vortices, given in section "Two Hetons with Zero Total Linear and Angular Momenta"; the corresponding characteristic trajectories are given in Fig. 2.26. Figure 3.57 shows that at strong stratification (small values of γ), the regime $\{3\}$ takes place, and two vortices of the top layer and two vortices of

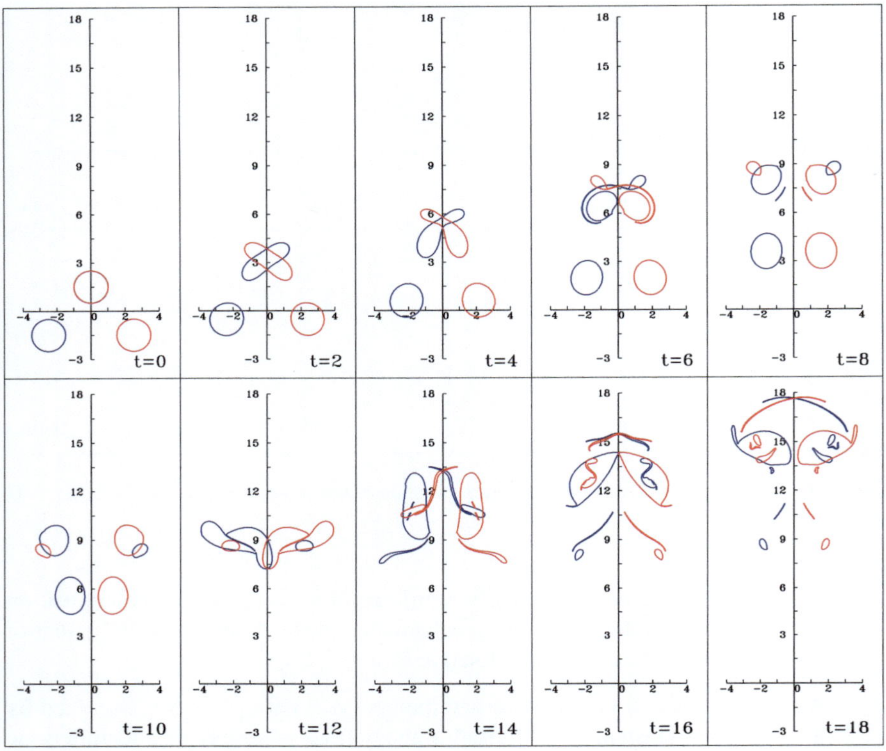

Fig. 3.56 An example of formation of a vortex structure of type U_4 at $\gamma = 1.2$ and $b = 1.5$, $c_1 = 0$, $c_2 = 2.5$

the bottom layer move in opposite directions along O-shaped trajectories lying one under the other. Within-layer interaction dominates in this case, because of which this periodic motion takes place. The result of interaction between layers can be seen only in the effects of deviation of the shape of trajectories from the circular one and in the unevenness of rotation of vortices. With increasing γ, at small and large distances between vortices in the two-layer pairs (parameter c), we pass to motions $\{2\}$ and $\{1\}$, respectively. Note that the presence of a peak in the boundary of domain $\{2\}$ in diagrams 3.57 is caused by the frequently mentioned fundamental property **(!)** of nonmonotonic (in terms of distance) character of the intensity of interaction between two-layer vortex motions.

The structure of head-on collision is much more complicated in the case of distributed hetons (Figs. 3.58–3.65), where, except for the obvious situations, in which vortex patches behave like point vortices (domains S_1, S_2, S_3 in diagrams 3.58), the mechanisms of instability and merging of continual vortices are gaining in importance. Here, as before, the denotations with letter U with subscripts refer to

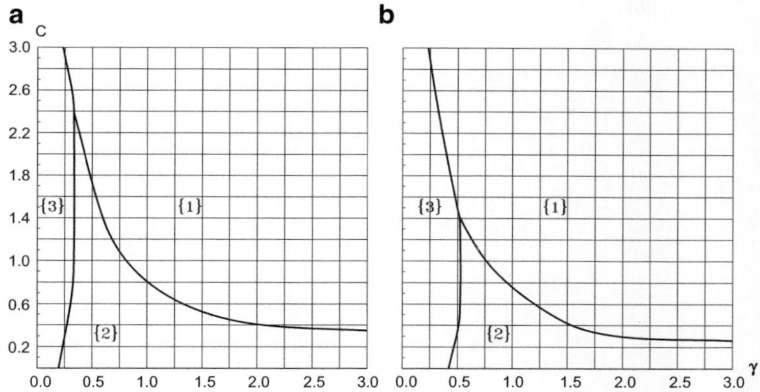

Fig. 3.57 Diagrams in the rectangular domain ($\gamma \in (0; 3]$; $c \in [0; 3]$) of possible states of two discrete hetons at head-on collision ($c \equiv c_1 = c_2$, $d = 0$) and (**a**) $b = 2.5$, (**b**) $b = 1.5$. Denotations for the subdomains: {1} the initial mutual approaching of the hetons with subsequent exchange of partners and moving apart of the newly formed hetons in the normal direction, {2} the initial mutual approaching of the hetons with the subsequent change of the sign of the tilt angles of their axes and the moving apart of hetons in opposite directions, {3} the motion of vortices in different layers in opposite directions along closed *O*-shaped trajectories. Adapted from [851] (Fig. 15a, b) by permission of Cambridge University Press

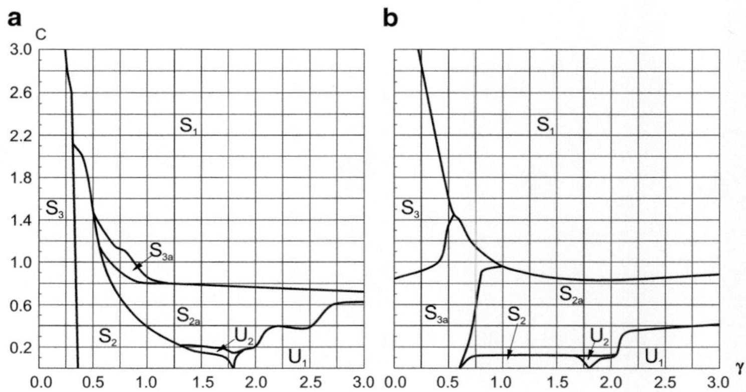

Fig. 3.58 The same as in Fig. 3.57, but for two distributed hetons consisting of circular unit-radius vortex patches. Denotations for the subdomains: S_1, an analog of motions of the type {1} for discrete vortices, S_2 an analog of motions of type {2}, S_{2a} a short-time merging of vortex patches with their subsequent separation into hetons running apart in opposite directions along the *y*-axis; S_3, an analog of motions of type {3}, S_{3a} the same as for S_3, but with the merging of vortex patches in both layers, U_1 the decomposition of hetons into unequal parts with their subsequent merging and regrouping into a system of smaller hetons moving apart, U_2 the decomposition of hetons with their subsequent merging and the formation of a central two-layer structure and two hetons moving apart. Adapted from [851] (Fig. 15c, d) by permission of Cambridge University Press

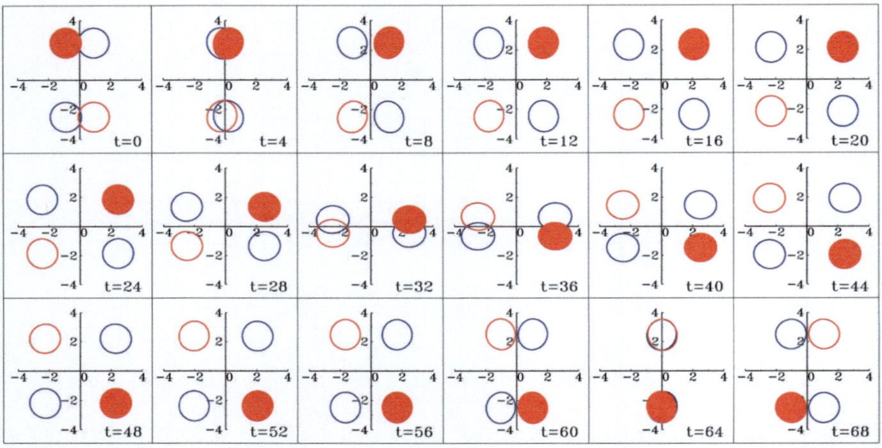

Fig. 3.59 Interaction between two cold hetons at head-on collision: $d = 0, b = 2.5, c = 0.9, \gamma = 0.3$ (type S_3, analog {3}, Fig. 2.26a)

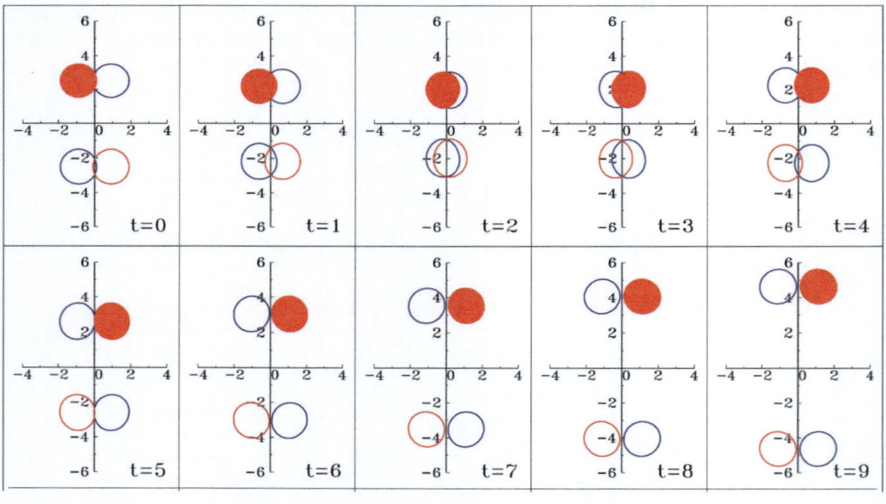

Fig. 3.60 The same as in Fig. 3.59, but at $\gamma = 0.5$ (type S_2, analog {2}, Fig. 2.26b)

the types of motion where the instability process is prevailing and vortex patches start to decompose before hetons can approach each other.

Note that, as well as in the previous paragraph, the correspondence between discrete and distributed vortices is lager at small values of γ, if the vortex patches of the interacting hetons at the initial moment are not too close to each other (as, for example, in the case $b = 1.5$ at small c, see Fig. 3.58b).

Now let us consider in more detail the specific features of different regimes.

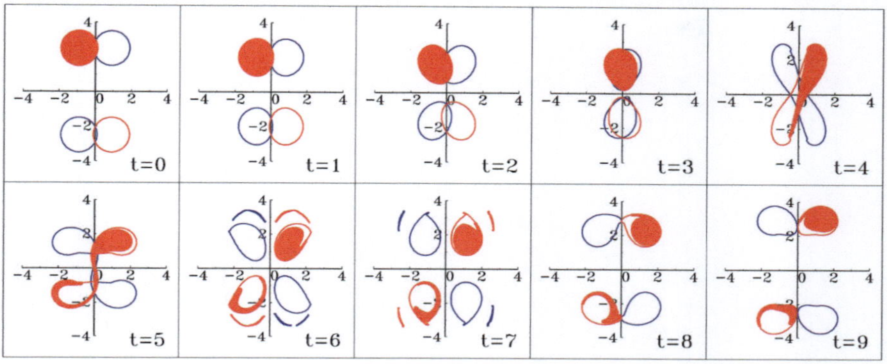

Fig. 3.61 The same, but at $\gamma = 0.7$ (type S_{2a})

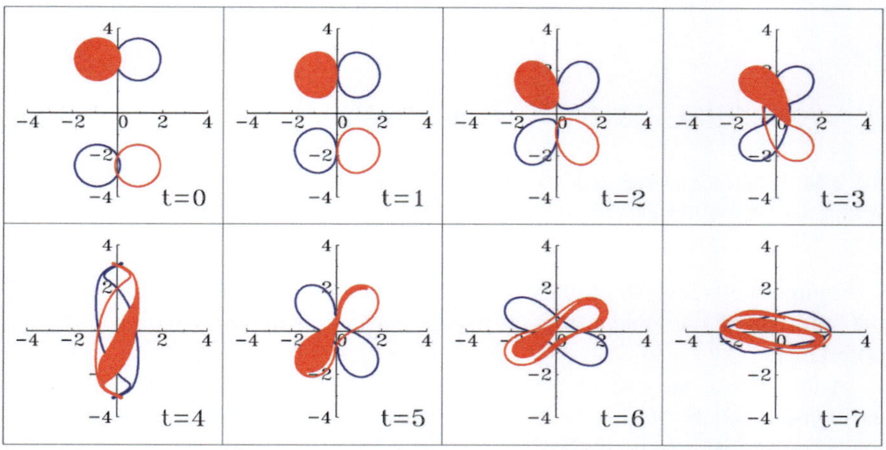

Fig. 3.62 The same, but at $\gamma = 0.9$ (type S_{3a})

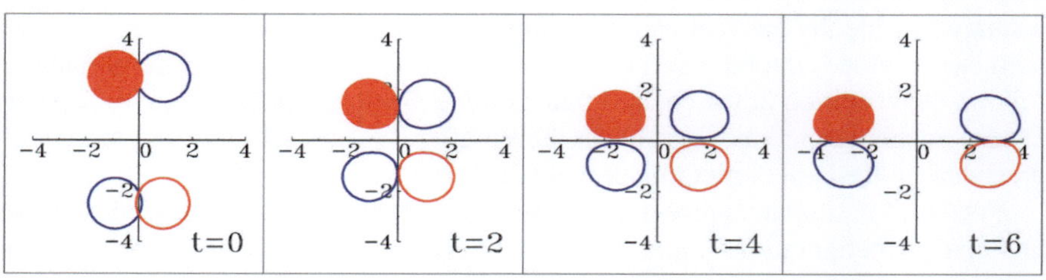

Fig. 3.63 The same, but at $\gamma = 1.1$ (type S_1, analog $\{1\}$, Fig. 2.26c)

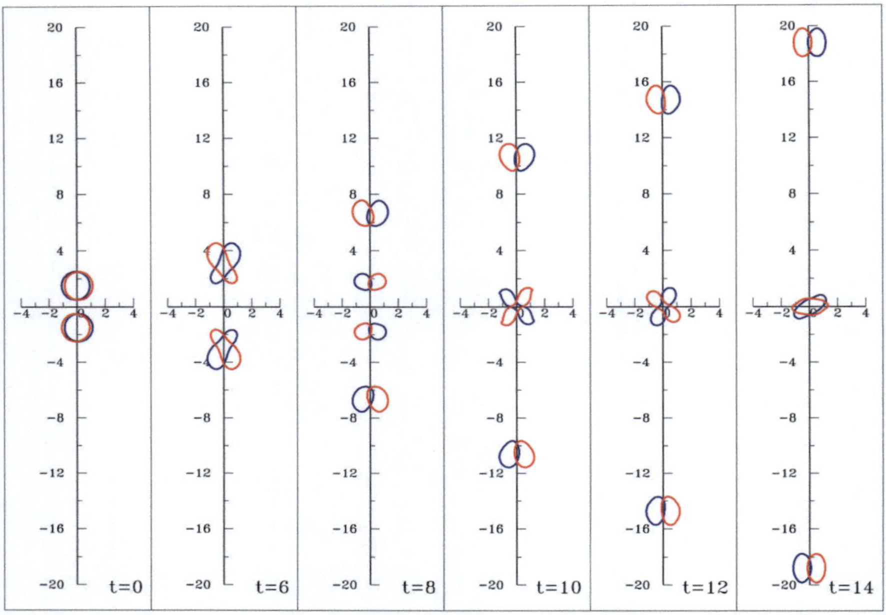

Fig. 3.64 Interaction between two cold hetons at head-on collision: $d = 0, b = 1.5, c = 0.1, \gamma = 1.8$ (type U_2). Adapted from [851] (Fig. 16) by permission of Cambridge University Press

Figures 3.59–3.63 show the results of a series of calculations for modeling the collision of two cold hetons at fixed values of geometric parameters, but at different γ.

In this series, the one of two top-layer vortices is dark colored. This allows us to more reliably trace the motion of the appropriate vortex patch or, more importantly, the dynamic stage of the features of mixing process during the merging and splitting of vortex patches. From the viewpoint of contour dynamics, the *coloration* is equivalent to the introduction of an additional patch with zero potential vorticity, which at $t = 0$ covers the inner part of the vortex. The auxiliary contour, displaced inside the patch at an infinitesimal distance (to avoid singularity), at any moment envelopes the fluid particles located inside the patch, and its deformations follows the evolution of the colored fluid volume. Such approach allows tracing the displacements of the inner part of fluid even after the application of the "surgery" procedure in the case of merging of main contours.

Note that a similar approach was used in sections "Regular Advection Near Stationary Configurations", and 2.3.3.6 to study the advection in a field of point vortices, where fluid particles were simulated by zero-intensity vortices.

Figure 3.59 shows the behavior of vortex patches for the examined class of interaction at relatively strong stratification, where the value of parameter $\gamma = 0.3$ is close to its upper limit in the case of finite motions of the type S_3. Though the

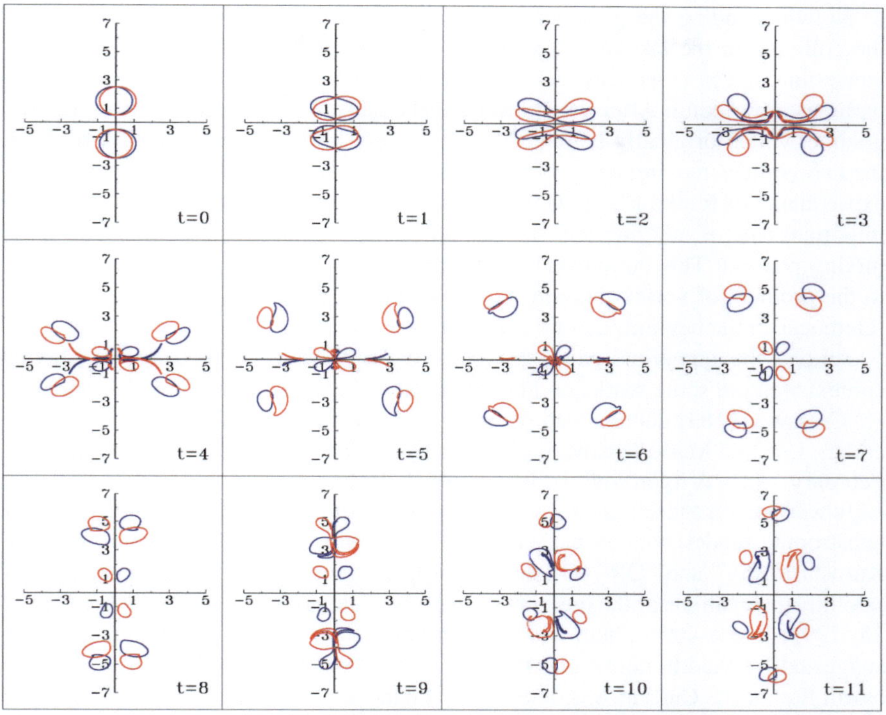

Fig. 3.65 The initial and intermediate stages of the interaction between two cold hetons at head-on collision: $d=0$, $b=1.5$, $c=0.1$, $\gamma=3$ (type U_1)

simulation time of this numerical experiment was rather long, the vortex patches, have performed only a bit more than a half-cycle, and practically kept their circular shape. Here, the effect of deceleration can be clearly seen in time intervals $16 \div 24$ and $44 \div 52$, when vortex centers are near an imaginary square, whose vertices correspond to motionless hyperbolic points of the phase plane for the appropriate problem in discrete formulation (the boundary between S_3, on the one hand, and S_2 or S_1, on the other hand, corresponds to those points).

With a slightly weaker stratification (γ=0.5, see Fig. 3.60), we pass to motions of type S_2, where the approaching of hetons to each other is accompanied by a change in the sign of the incline slope of their axes, and those hetons, similarly to the case of elastic reflection (but without collision), after approaching each other at some limiting distance, pass to reverse motion in opposite directions.

A subsequent decrease in stratification ($\gamma = 0.7$, see Fig. 3.61) yields a new effect: after moving closer, the vortex patches of two hetons merge, but next, the newly formed large vortices separate, yielding, as in the previous case, two two-layer dipoles moving apart in opposite directions. This type of motion was assigned the label S_{2a}.

Notwithstanding the geometric similarity of the vortex structures formed after the collision in the last two experiments, they are different from the Lagrangian viewpoint, since, as we can see, during the merging of vortex patches, their fluid particles are mixing. After the splitting of the quasi-elliptic vortex patch, the fluid volume, which originally belonged to a top-layer vortex, is redistributed between the two vortices moving apart. Note that a similar effect was observed in numerical experiments of Radko [728]. Thus, this and many other examples demonstrate the important role of merging and splitting of vortices as mechanisms of water mass mixing process. This circumstance, in particular, is the cause of the great interest to the problem of vortex merging in the ocean and the atmosphere (see the list of references in the beginning of this chapter).

Among the numerous papers devoted to the analysis of this process, in this context we note some works on field observations of vortex merging phenomenon: (1) Cresswell [183] documented the rapid and complex merging of two warm-core eddies, Leo and Mario (eastward of Australia: 33–36° E, 151–153° W), in January–February 1981. A hydrology survey in April 1981 confirmed this merging. A new eddy was named Mario; (2) Yasuda [997], in the framework of the two-layer quasi-geostrophic model, studied in detail the mechanism of merging of warm rings of Kuroshio "86V" and "OR" (**O**ffshore **R**ing); (3) Lander [512] gives an example of satellite observations of the merging of tropical cyclones Ruth and Pat on September 25, 1994, in the atmosphere above the northwestern Pacific; (4) Vortex merging suggested by the altimetric observing in April–December 1974 at 166° W in the North Equatorial Current was detected by Flament et al. [270]; (5) Kuo et al. [493] recorded the merging of tropical typhoons Alex and Zeb in October 1998 in the Pacific, eastward of the Philippines; (6) Yussef and Marcus [1004] studied the merging of two anticyclones in Jupiter atmosphere, which took place in the same year; (7) Prieto et al. [723] presented a scenario of joint rotation and subsequent merging of atmospheric cyclonic vortices Gil and Henrietta in the northeastern Pacific between September 4 and 10, 2001; (8) Sangrá et al. [801] note the merging of anticyclones A1, A2 and AT topographically generated by Gran Canaria island in June 1998.

Figure 3.62, where $\gamma = 0.9$, illustrates the formation of an associated structure S_{3a}, where the vortex patches of both layers merge and never separate, representing a quasi-elliptic heton, which was studied in Sect. 3.2.2. We can see here that the colored and noncolored fluids mix more and more and the fluid volume tends to become more and more homogeneous [66].

Finally, at $\gamma = 1.1$ (Fig. 3.63), we see a phenomenon of head-on collision of two-layer pairs accompanied by exchange of partners and the moving apart of new hetons in perpendicular direction (S_1 is a prototype of motions of type $\{1\}$ for discrete vortices). Note the obvious analogy with the behavior of a vortex pair incoming perpendicularly to a motionless plane wall. The problem of interaction between a dipole and the wall, which has important applied aspects, has been considered by many authors, using both numerical and experimental approaches, in particular, in [24, 140, 674, 934, 957].

Two examples of collisions with the manifestation of instability properties are given in Figs. 3.64 and 3.65. The former figure gives an exotic case, where, after the separation of each heton (in the case of their instability), the two newly formed two-layer pairs move apart in opposite directions, while the other two pairs move toward one another, after which their vortex patches merge to form a central motionless core-heton (type U_2). A more complex scenario can be seen in the implementation of interactions of type U_1. Such interaction, without its final stage, is illustrated in Fig. 3.65. Here, at the nonlinear stage of instability development, the third mode is excited, and, at $t = 4$, in the central part, we see a structure of two small-scale hetons, which later start moving in opposite directions along the y-axis. At the same time, four noncompensated two-layer vortices form in the peripheral part, and, turning around, they also approach the y-axis. At time $t = 8$, all 12 distributed vortices arrange symmetrically along the y-axis. Next, the vortex patches of noncompensated structures merge for a short time and again separate, now to form two pairs of unequal hetons. The smaller hetons move apart in opposite directions along the y-axis, while the larger ones move toward the two-layer pairs approaching them from the central part and, drawing apart the incoming partners, the hetons squeeze between them. At this stage, the numerical experiment was stopped because of the accumulation of errors caused by the repeated application of contour surgery procedures (those errors manifest themselves in the loss of symmetry). Nevertheless, the direction of the further evolution is quite obvious: the two large hetons, after changing their partners (according to the type S_1) will form two-layer dipoles moving apart along the x-axis, and four small hetons will move along the y-axis, two dipoles moving in each of the opposite directions.

3.3.4 Two Hetons with Tilted Axes, the Case of Nonzero Total Momentum

Suppose that the initial arrangement of vortex patches is such that their centers are situated in the vertices of a parallelogram, when $d > 0$ and $c_1 = c_2 \equiv c$. We will consider only one particular case where the centers of top-layer vortices are originally located on the y-axis (i.e., $d = c$) at fixed $b = 4$. With such arrangement of hetons, their semi-head-on collision becomes possible. A diagram of the states of the system of vortices that can realize on a part of (γ, d) parameter plane is given in Fig. 3.66.

Figure 3.67 gives two examples of the motion of vortex patches for two different types of interaction at which the vortices slightly change their shape, remaining compact. That is why the combining in one figure of vortex contours in discrete moments does not confuse, but rather makes the picture of vortex system evolution more vivid. Thus, Fig 3.67a gives an example of quasi-barotropic motions S_3, when within-layer vortex interaction is dominant. Qualitatively, the behavior of the centers of vortex patches remains the same as in the case of motions of type $\{3\}$ in discrete case: the structure as a whole carries out periodic motions, such that the top-layer

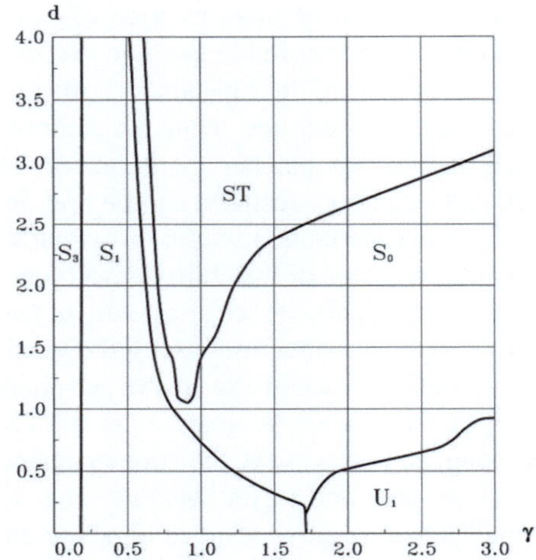

Fig. 3.66 Diagram in rectangular domain ($\gamma \in (0;3]$; $d \in [0;4]$) of the possible states of two distributed hetons consisting of circular unit-radius vortex patches at $d = c = c_1 = c_2$ and $b = 4$. Denotations for the subdomains: S_3 quasi-barotropic motions at dominating intralayer interaction, S_1 hetons scatter in opposite directions, S_0 intermediate merging of top-layer vortex patches with their subsequent separation, partial loss of mass, and passage to the regime of scatter of noncompensated two-layer vortices with tilted axes, *ST* the merging of top-layer vortex patches and the formation of Λ-shaped tripolar structure, U_1 the decomposition of hetons and the subsequent scatter of vortex pairs

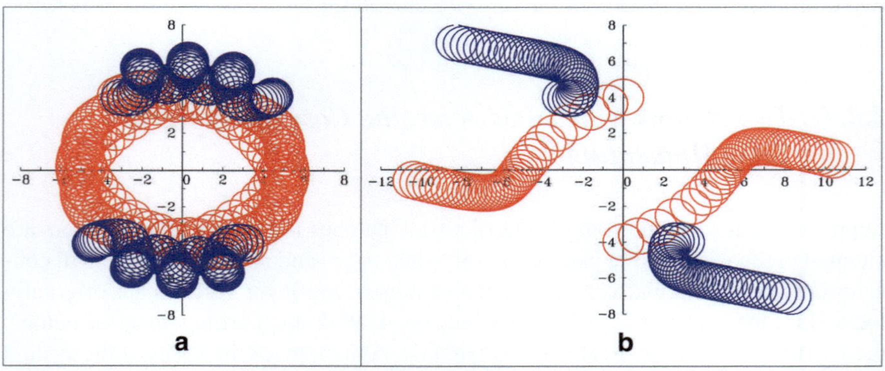

Fig. 3.67 Two series of superposed contours of vortex patches at $b = 4$, $d = 3$ and (**a**) $\gamma = 0.18$, $t = 0, 4 \ldots, 380$ (type S_3, an analog of motions of type $\{3\}$ for discrete vortices, see Fig. 2.10.1.a); (**b**) $\gamma = 0.20$, $t = 0, 4 \ldots, 140$ (type S_2, an analog of motions of type $\{2\}$ for discrete vortices, see Fig. 2.13)

vortices move along a trajectory filling some ring domain, while the bottom-layer vortices move along spiral peripheral curves. In this case, the vortex structures periodically form either collinear or diamond-shaped forms (when the bottom-layer vortices are located in the points of their loops that are closest to the center or maximally remote from it, respectively).

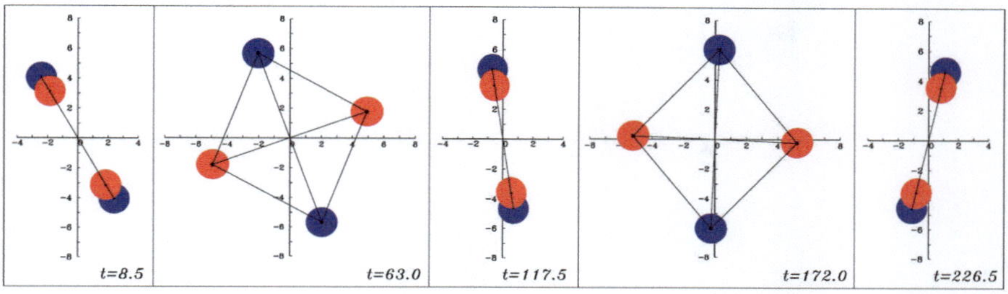

Fig. 3.68 Illustration of the periodic alternation of collinear and *diamond-shaped* constructions for motions of type S_3, given in Fig. 3.67a. The *dashed lines* connect the centers of vortex patches shown by markers and denote the diagonals of rhombs

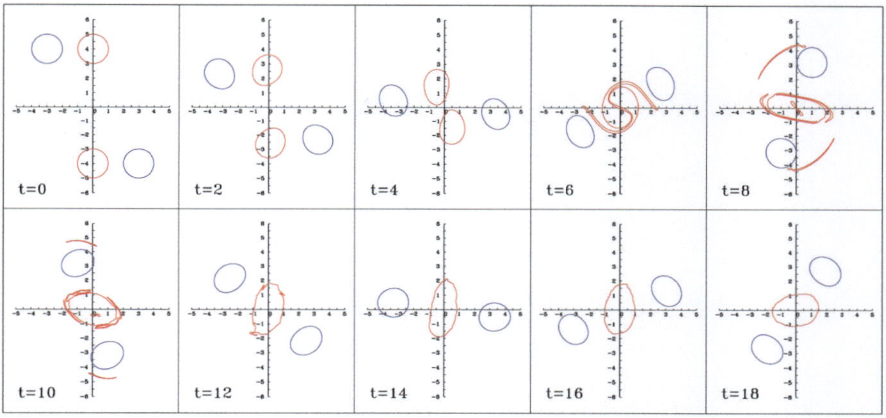

Fig. 3.69 An example of calculation of the evolution of vortex patches at $b = 4$, $d = 3$ and $\gamma = 1.5$ (type ST)

Figure 3.68, where each vortex patch has individual marking, demonstrates this effect and vividly shows that the periods of motion of the top and bottom-layer vortices are in the ratio 2 : 1 (in the calculation period, the upper ones complete one cycle, while the bottom ones—two such cycles). In this case, the period for the bottom layer is 109 units of dimensionless time. A small increase in parameter γ (Fig. 3.67b) is accompanied by the passage into domain S_1. Here, the rotation motion of the top-layer vortices in accordance with scenario S_3, which has just started, is interrupted by the impact of partners from the bottom layer, resulting in the formation of two-layer pairs (S_1), moving apart, by analogy with motions of type $\{1\}$ for discrete vortices.

Figure 3.69 shows the process of formation of a tripolar two-layer structure because of a sidelong attack of two hetons with tilted axes—type ST. The top-layer vortices approach each other and merge to form an indivisible central vortex domain in the form of a quasi-elliptic pulsating figure, while the bottom-layer vortices rotate along a peripheral quasicircular trajectory in the direction induced by the central core. Thus, we can formulate a possible scenario of formation of two-layer

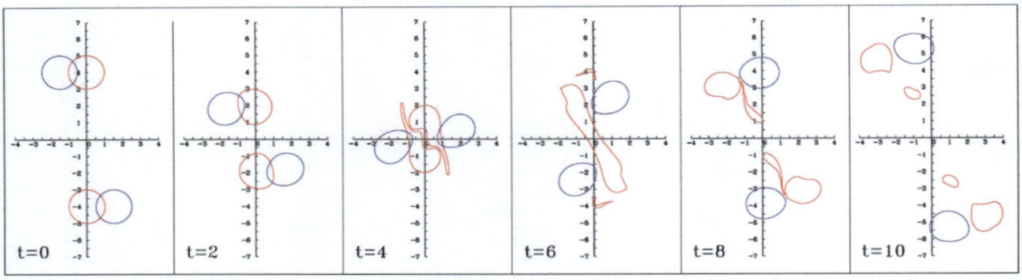

Fig. 3.70 An example of calculation of the evolution of vortex patches at $b = 4$, $d = 1.5$ and $\gamma = 1.5$ (type S_0)

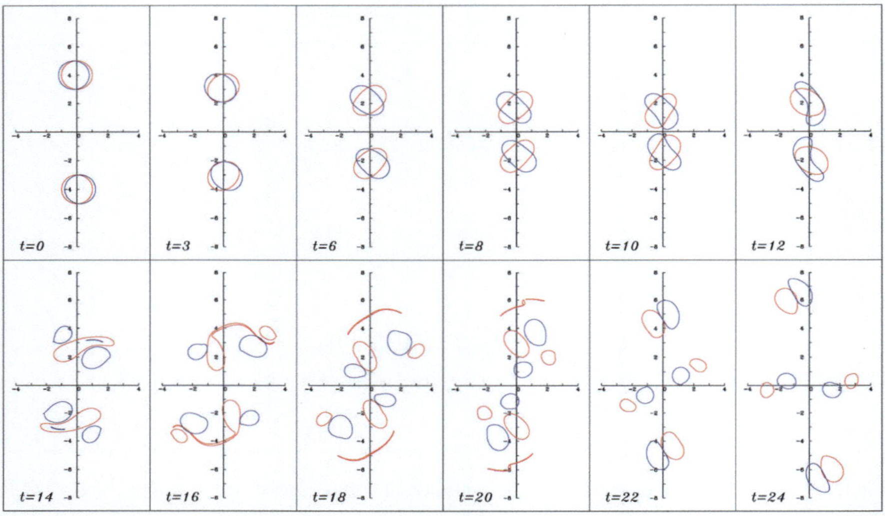

Fig. 3.71 An example of calculation of the evolution of vortex patches at $b = 4$, $d = 0.2$, and $\gamma = 1.75$ (type U_1)

Λ-shaped tripolar structures as a result of an eccentric counter collision of hetons. Note that an analogous numerical solution, but in the barotropic formulation, was obtained by Larichev and Reznik in [519].

According to diagram 3.66, at least at $\gamma > 0.8$, the existence domain of those tripolar structures becomes narrower with decreasing tilt of heton axes d and with increasing γ. As an example, Fig. 3.70 shows that indeed the vertical alignment of the initial arrangement of hetons (other parameters remaining the same), results in that the newly formed central core does not stay continuous (type S_0), and its separation yields two-layer vortex structures moving apart.

Finally, Fig. 3.71 gives an example of a semi-head-on collision of two hetons with slightly tilted axes in the case of realization of instability mechanism. In this case, as the hetons approach each other, first, the bottom-layer vortex patches separate into two parts, after which analogous decomposition takes place in the top layer. Next, each heton forms two two-layer noncompensated vortices, which, after approaching each other, join to form two two-layer pairs. As the result of such interaction, we have a structure of two two-cascade hetons moving apart in opposite directions, where, the interaction between different two-layer pairs will obviously attenuate with distance between them and the motion of each one will asymptotically approach linear motion.

In this paragraph, we considered only a particular case of interaction between two hetons with nonzero total momentum, when $d = c$. The results of a more detail study of this class of motions are given in [852].

3.3.5 Interaction Between a Warm and a Cold Hetons

We discuss the generalization of discrete vortex theory from the Sect. 2.2.2.4 (two vortices with different signs specified in each layer) to the continuous case.

Let us suppose first that, as before, the initial vortex configuration is determined by the scheme shown in Fig. 2.27, the only difference being that the markers in it represent the centers of vortex patches rather than point vortices. Figure 3.72, constructed based on many numerical experiments, shows a diagram of possible states of such vortex system. We chose variables γ and b, where the minimal value for the geometric parameter b in each fragment was chosen such that the initially circular unit-radius vortex patches belonging to the same layer are tangent to each other. Note the general trends: at small γ (strong stratification), horizontal dipoles will form, while at weak stratification, the formation of Z-shaped finite structures is most likely.

The series of Figs. 3.73–3.76 gives examples of formation of different vortex structures. We start from the case $a = 0$, where, in the initial moment, the central two-layer vortex with vertical axis is a cyclonic antiheton (a barotropic vortex), while the two peripheral anticyclonic vortices are located in different layers symmetrically with respect to the central one.

Figure 3.73a gives a scenario of the evolution of HD type configuration. This is an analog of motion of type $\{1\}$ for discrete vortices (compare with Fig. 2.31a) where two moving-apart "horizontal" dipoles $\binom{1}{1}\binom{2}{1}$ and $\binom{1}{2}\binom{2}{2}$), form. In this case, because of strong stratification ($\gamma = 0.1$), the interaction between layers is almost absent, and the distributed vortex pairs in each layer also move apart in opposite directions, as would be the case in the absence of the other layer.

The behavior of vortex patches in the case of moderate or weak stratification (Figs. 3.73b and 3.73c) is radically different compared with the strong stratification. In this case, the central antiheton, which originally had a vertical axis, under the effect of stress from the peripheral anticyclones, shows the instability of the second mode, so its vortex patches in each layer divide into two parts. In Fig. 3.73b (at $\gamma =$ 1.4, i.e., very close to the boundary with HD domain), the smaller cyclones that have formed as the result of decomposition of the central two-layer vortex, have axes with relatively large tilt. Thus, each newly formed two-layer structure is a configuration in the form of a tilted letter L (hence the name of this type of motion); this letter is

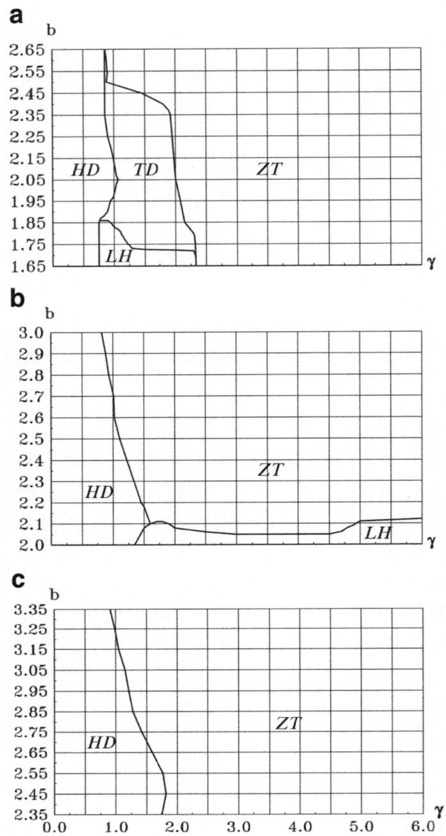

Fig. 3.72 Diagrams in rectangular domain (γ, b) of possible states of two distributed hetons, consisting of unit-radius circular vortex patches, lying on the same straight line, at different values of parameter a: (**a**) $a = 0.35$, (**b**) $a = 0$, and (**c**) $a = -0.35$. Denotations for subdomains: HD horizontal dipoles (an analog of motions of type $\{1\}$), TD tilted dipoles or hetons (an analog of motions of type $\{2\}$), LH L-shaped hetons, ZT Z-shaped tripoles (an analog of motions of type $\{3\}$). Adapted from [847] (Fig. 9) by permission of IOP Publishing

ordinary in the left part of the figure and turned upside-down in its right part. Note that, with an increase in parameter γ, the tilts of the axes of those small two-layer cyclones decrease and, for example, at $\gamma = 4$ (Fig. 3.73c), we see an interesting picture, in which the decomposition of the central antiheton takes place at almost complete preservation of the vertical alignment. This type of motion, accompanied by the decomposition of the central antiheton, has no analogs in the dynamics of discrete vortices. It is worth mentioning, however, that the domain of its existence is rather narrow: $|b-2| \ll 1$ at $|a| \ll 1$ (as in Fig. 3.72b, where $a = 0$). If we neglect the insignificant mass losses of the central vortex at its splitting (they account for slightly more than 1 % in this case), we can say that, essentially, in Fig. 3.73b, c, we see the generation of two triples of vortices with zero total vorticity (whose discrete ancestor is considered in addendum 2 to Sect. 2.3.2), consisting of a cyclonic two-layer vortex and an anticyclone of the bottom (top) layer in the left (right) parts of the figure. Those vortex structures move apart in opposite directions, such that the motions in Fig. 3.73b are analogous to type $\{2b\}$ (compare with Fig. 2.82c), and Fig. 3.73b shows the realization of a variant of moving apart of two asymptotically limiting tritones.

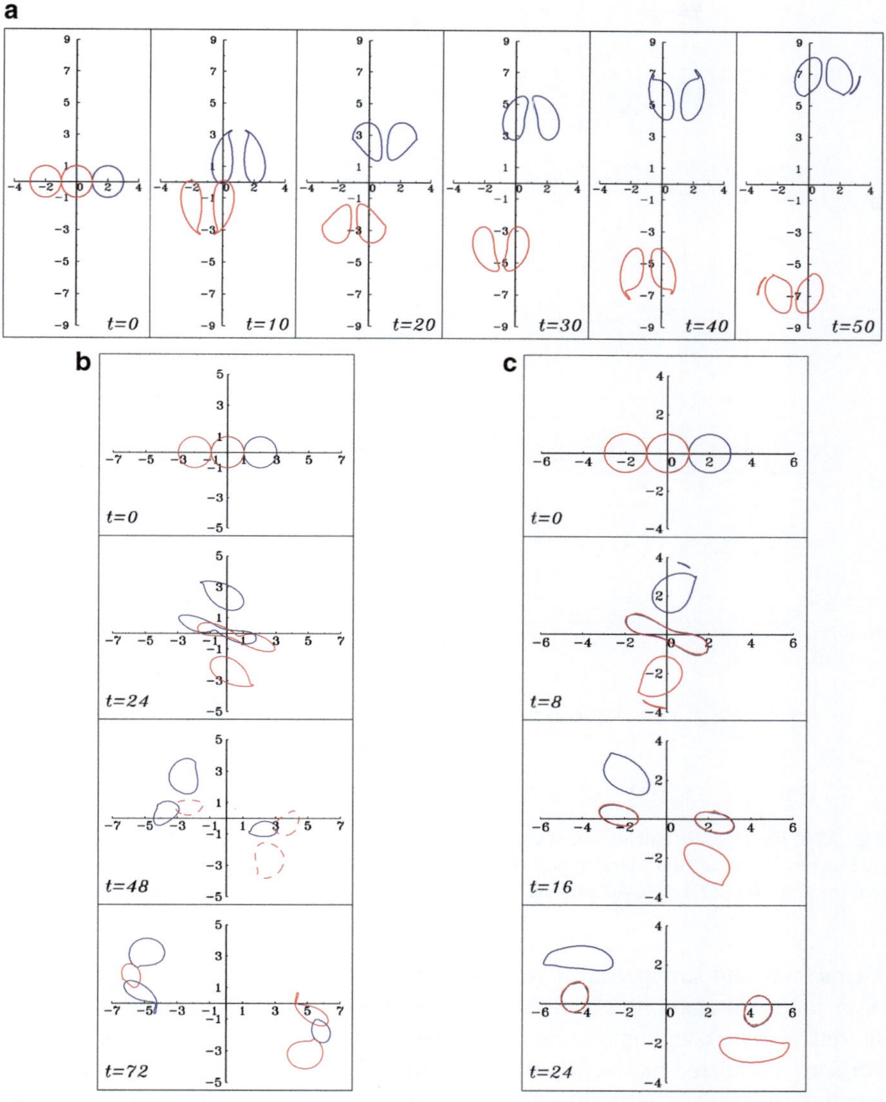

Fig. 3.73 Examples of calculation of vortex patches evolution at $a = 0$, $b = 2$ and (**a**) $\gamma = 0.1$ (*HD* type) and (**b**) $\gamma = 1.4$, (**c**) $\gamma = 4.0$ (both of *LH* type). Adapted from [847] (Fig. 10) by permission of IOP Publishing

Figure 3.74 gives examples of three types of interaction at $a > 0$. Panels (a) and (b) present a scenario of transformation of the both initial *Z*-shaped configurations into a structure of two moving-apart horizontal dipoles (*HD*). The transition between regimes *HD* (Fig. 3.74b) and *TD* (Fig. 3.74c) takes place between $\gamma = 1.035$ and $\gamma = 1.1$. Again it is natural that weaker layer coupling favors horizontal vortex

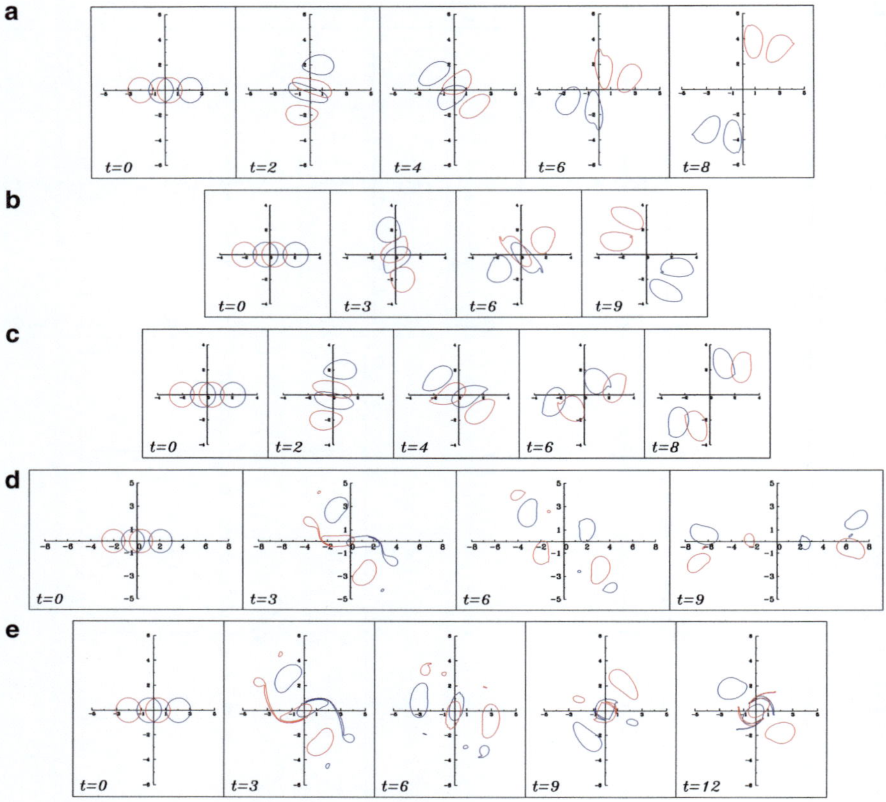

Fig. 3.74 Examples of calculation of vortex patches evolution at $a = 0.35$, $b = 2.05$ and (**a**) $\gamma = 0.1$ (*HD* type), (**b**) $\gamma = 1.035$ (*HD* type), (**c**) $\gamma = 1.1$ (*TD* type), (**d**) $\gamma = 1.95$ (unstable *ZT* type) and (**e**) $\gamma = 2.05$ (*ZT* type). Adapted from [847] (Fig. 11) by permission of IOP Publishing

interactions and stronger layer coupling favors vertical (tilted) vortex interaction. As γ increases again, the boundary between the *TD* and *ZT* regimes is reached. Indeed, as layer coupling increases, the vertical alignment of the like–signed central vortices is rendered more efficient, and *Z*-shaped tripoles are more likely to form. For $\gamma = 1.95$ (Fig. 3.74d), the unstable regime is of type *ZT*. The central cyclones undergo active filamentation and shed small vortex patches (the area loss of the cyclonic vortices is 29 %). For $\gamma = 2.05$, a stable *Z*-shaped tripolar structure is formed (Fig. 3.74e). This structure is the finite-area equivalent of the end-state of type $\{3\}$ for point vortices. In this case, stabilization is preceded by active filamentation and cyclonic vorticity losses are about 34 %.

Figure 3.75 illustrates the effect of variations in the size of the initial collinear configuration (the distance between the rotation center of the entire structure and the center of the peripheral vortex patch at constant values of other parameters) on

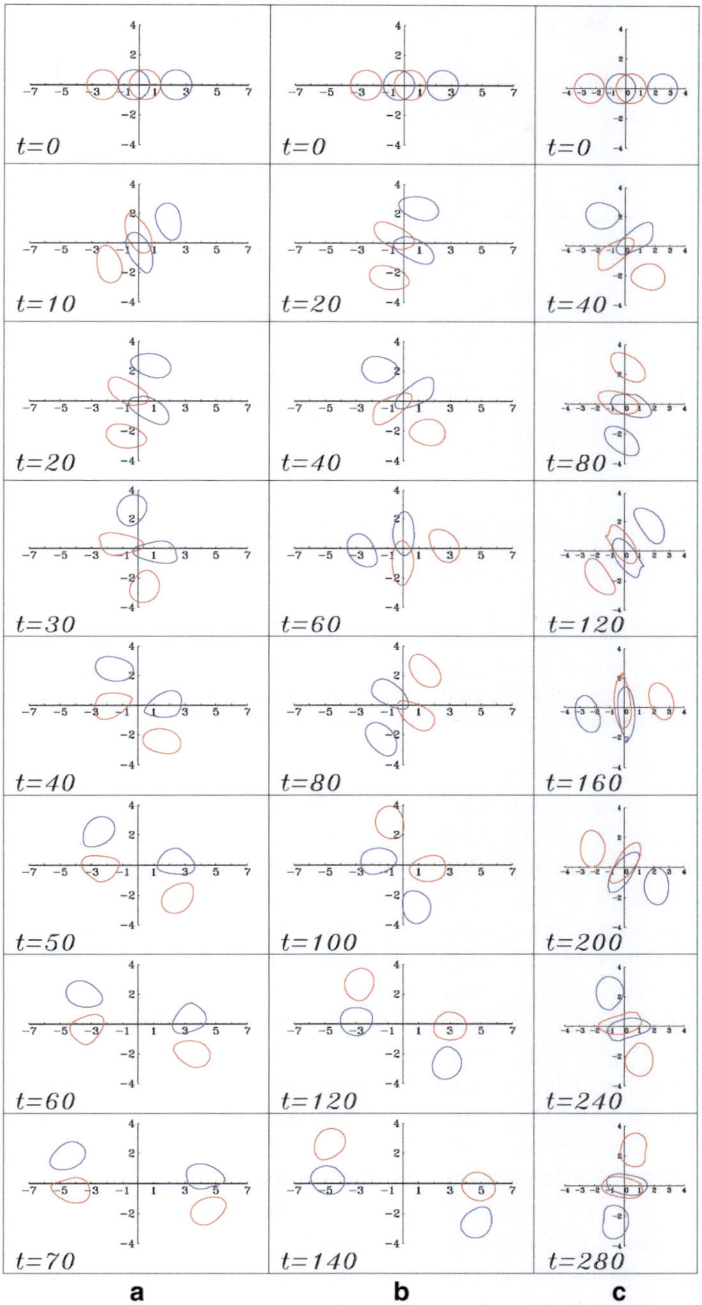

Fig. 3.75 Examples of calculation of vortex patches evolution at $a = 0.35$, $\gamma = 1.5$ and (**a**) $b = 2.35$, (**b**) $b = 2.4435$ (both of TD type) and (**c**) $b = 2.45$ (ZT type). Adapted from [847] (Fig. 13) by permission of IOP Publishing

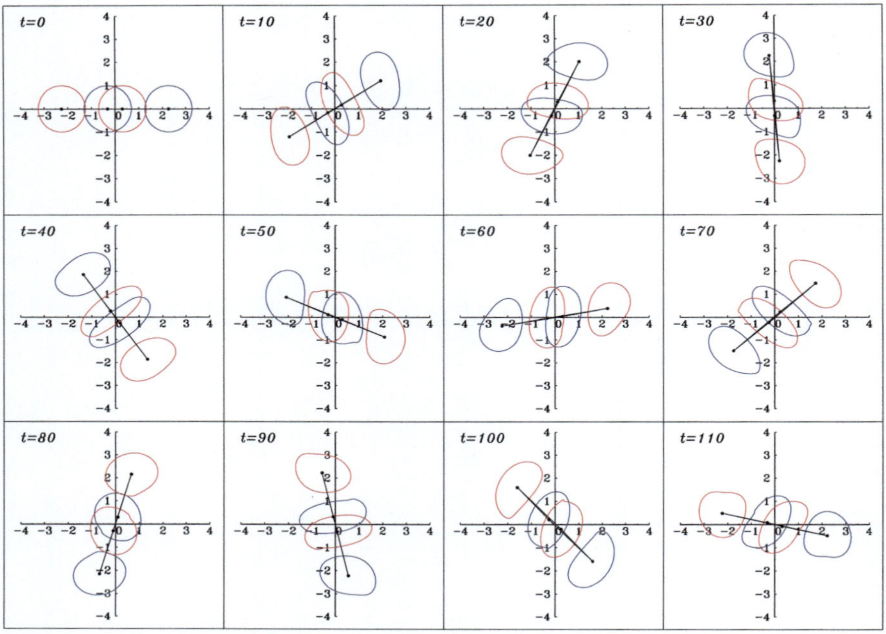

Fig. 3.76 The motion of a collinear vortex structure (finite-core *zigzag-shaped* rotating heton quartet) at $a = 0.311$, $b = 2.2576$, $\gamma = 1$. The centroids of vortex patches are shown by markers and connected by *straight line* segments. Adapted from [847] (Fig. 15) by permission of IOP Publishing

the character of its motion. The panels 3.75a, b show series of instantaneous sets of the contours of vortices performing rotations of TD type. While in the former case, before the warm and cold hetons start moving apart, the vortex construction turned by an angle of about $\pi/2$, in the latter case, where the point corresponding to the initial state of the system in the space of parameters (γ, b) is situated near the boundary between TD and ZT domains, this angle is near $3\pi/2$. So, during the passage from 3.75a to b the warm and cold hetons "exchange" their directions. The subsequent growth of the initial distance b transfers the vortex structure into the ZT state (Fig. 3.75c), in which (unlike 3.74c) no mass loss is observed. The comparison of Fig. 3.74b, c suggests a "rigid" dependence of the state of the system on parameter b. Thus, at $t = 40$, the configurations of vortices are almost the same for both cases, but further, the difference acquires qualitative character.

In terms of the "heton" theory, we can say that during motions of ZT type, in the general case, two antihetons—internal and external—rotate around the system center of rotation with different angular velocities. However, as we know from 2.2.3, for discrete vortices, with dispersion relationship (2.29) satisfied, all four vortices, while lying on a straight line, can rotate with the same constant angular velocity. Numerical experiments show that such quasi-stationary state can take place in the case of distributed vortices as well. Figure 3.76 shows the behavior of a collinear

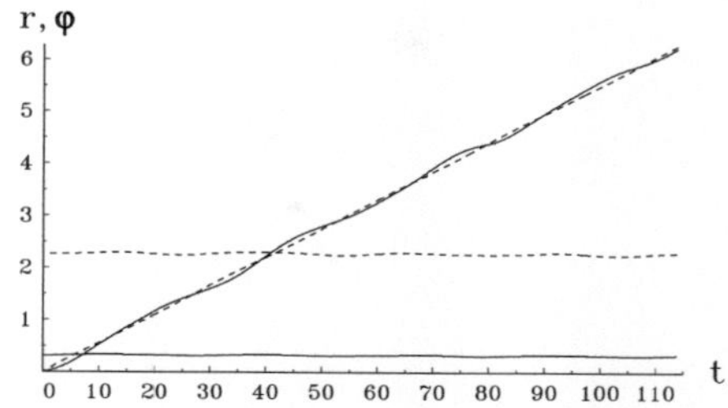

Fig. 3.77 The radius vector $r(t)$ and the angle $\varphi(t)$ between r and x-axis for the centroids of vortex patches $\binom{1}{1}$ (*full line*) and $\binom{2}{2}$ (*dashed line*) for a collinear structure (Fig. 3.76). Adapted from [847] (Fig. 16) by permission of IOP Publishing

structure of four vortices with distances between their centers satisfying the equation of its solid-body rotation with a good accuracy (finite-core zigzag rotating heton quartet). Here, the centers of vortex patches are connected by straight-line segments, which can be seen to lie on the same straight line almost all the time. Figure 3.77, which gives the time dependences of the distances from the rotation center to the nearest and furthers vortices and the angles between the appropriate radius-vectors and the x-axis, strongly suggests that deviations of the distances and angular velocity from constant values vary insignificantly.

It should be noted, that both pulsing shapes of vortex patches in Fig. 3.76 and insignificant deviations of curves from strait lines in Fig. 3.77, can be explained by the fact, that the initial configurations of vortex patches are circular, what is different from the stationary ones. The stationary configurations satisfying the conditions of strict preservation of the shape and radius-vectors of the vector-patch centroids, may be found similarly to the procedure described in Sect. 3.2.2 for the hetons with tilted axes. This problem is not solved yet.

Until now, in this subsection, we considered examples of interactions for the case where the initial configuration is collinear. In fact, as shown in section "Vortex Structures: Warm Heton–Cold Heton, Two Antihetons, Two 'Horizontal' Pairs", the rectilinear state cannot be attained for some possible vortex motions of the class under consideration. In particular, in the case of discrete vortices, this is type $\{4\}$: at the counter motion of horizontal dipoles, a "distant collision" takes place, accompanied by exchange of partners and the subsequent moving apart of now vertical dipoles (Figs. 2.29 and 2.30 give the phase portraits and trajectories of vortices, respectively). The inverse process—the transformation of vertical dipoles into horizontal—is also possible when a warm and a cold heton collide in this manner.

This is the type of interaction we will consider here for distributed hetons only for a particular case (though in great detail), where in the initial moment, the cyclonic vortices of the top (in the top half-plane of the figures) and the bottom (in its bottom half-plane) layers are spaced a fixed distance apart and placed on the y-axis (semi-head-on collision). The diagram in Fig. 3.78 demonstrates the wide diversity of possible regimes. Let us consider examples of various scenarios of vortex motions.

We start from the simplest configuration of a warm and a cold heton with vertical axes, which are originally spaced $2c$ apart (in this case, $c = 4$), but at different values of parameter γ, where the initial states correspond to points of

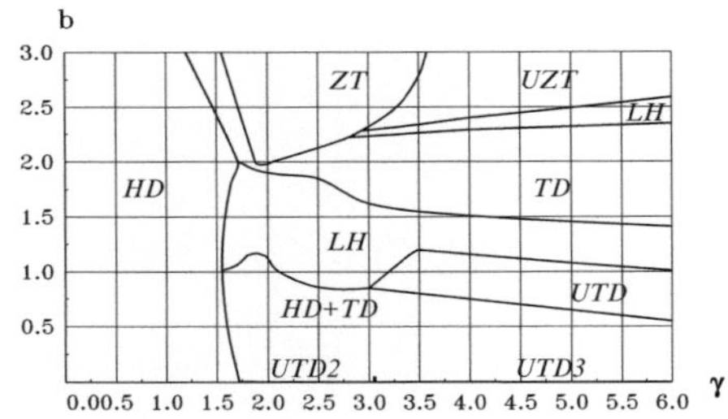

Fig. 3.78 Diagram in plane (γ, b) of possible states of the warm and cold hetons at semi-head-on collision: $a = 0$, $c = 4$. Denotations for subdomains: HD horizontal dipoles, TD tilted dipoles (hetons with tilted axes), $HD + TD$ a combination of two horizontal and two tilted dipoles, $UTD2$ and $UTD3$ combinations of two and three pairs of vertical dipoles forming as a result of decomposition of unstable vortices (this refers to the points of the plane that lie on the γ-axis or at $b \ll 1$); UTD a pair of noncompensated two-layer vortices, ZT Z-shaped tripolar structure; UZT, unstable Z-shaped tripolar structure; LH, L-shaped hetons. Adapted from [847] (Fig. 17) by permission of IOP Publishing

the horizontal axis $b = 0$ in Fig. 3.78. Note that appropriate solutions for point vortices were not considered in section "Vortex Structures: Warm Heton–Cold Heton, Two Antihetons, Two 'Horizontal' Pairs", since they are well known from [350]. According to calculations of Hogg and Stommel (Fig. 3 in [350]), with this type of initial conditions, the final result of interaction is the scatter of horizontal dipoles (HD) in opposite directions along the x-axis. At a strong stratification, i.e., at small γ, this motion begins almost immediately. With increasing γ, the role of interlayer interaction also increases, while that of within-layer decreases. Each heton first acquires a tilt of axis that facilitates their counter motion (the greater γ, the later this takes place), and only after that, when the hetons are close enough to each other, heton pairs that have opposite signs and are moving apart will form in each layer. As can be seen from Figs. 3.79 and 3.80, such situation also takes place for distributed hetons in the interval $\gamma \in (0; 1.705)$. In the second of those figures, where the value of γ is close to its limit, the presented time moments clearly show how slow is the transformation of the system at the first stage and how rapid it is at the second one.

When stratification is weaker ($\gamma > 1.705$), each heton starts demonstrating instability properties and the interaction becomes much more complicated. Thus, at an increase of γ by 0.1 (Fig. 3.81), on the other side of the boundary between the stability and instability domains, each heton separates into two two-layer vortices with tilted axes moving apart in opposite directions. The comparison of time intervals here and in Fig. 3.80 shows that the process of the patches' destruction begins much earlier than the visible relative displacement of their centers in the previous case. Next, the two external, newly formed, vertical dipoles of smaller size move outward, while the two internal ones, having moved closely to each other, continue their translational motion in the same directions. As the result, we have two combinations of a warm and a cold heton moving apart in opposite directions. This is a typical picture in the realization of the mode with $m = 2$.

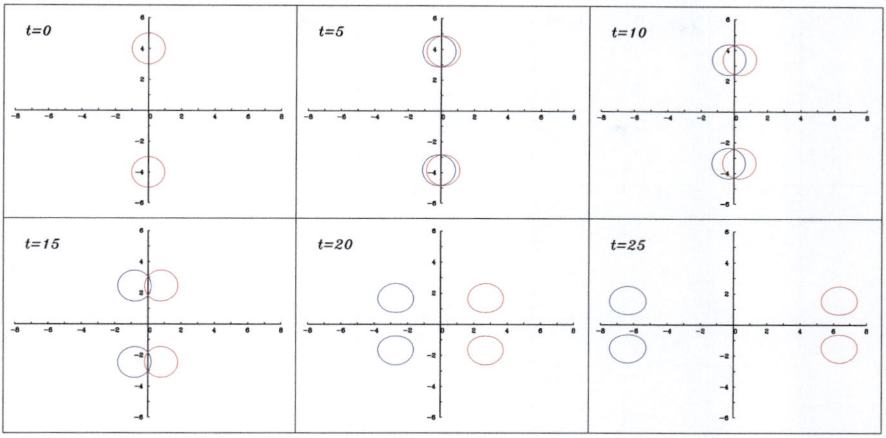

Fig. 3.79 The interaction between a warm and a cold heton at $a = 0$, $c = 4$, $\gamma = 0.5$ and $b = 0$. *HD* regime

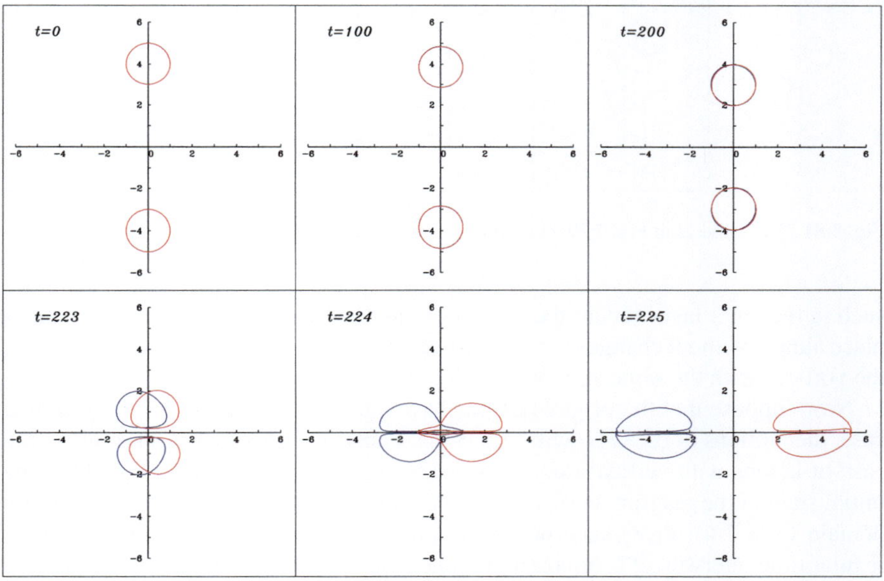

Fig. 3.80 The same as in Fig. 3.79, but with $\gamma = 1.7$

At $m = 3$, the structure of interaction is of the same qualitative character (Fig. 3.82), except that, as it is commonly the case at $m > 2$, after the destruction of original hetons because of instability, two-layer cores persist in their central part. Note that here, in the time interval between $t = 9$ and $t = 10$, as well as in the previous example at $t \approx 26$, the passage of one cyclonic vortex in one layer over

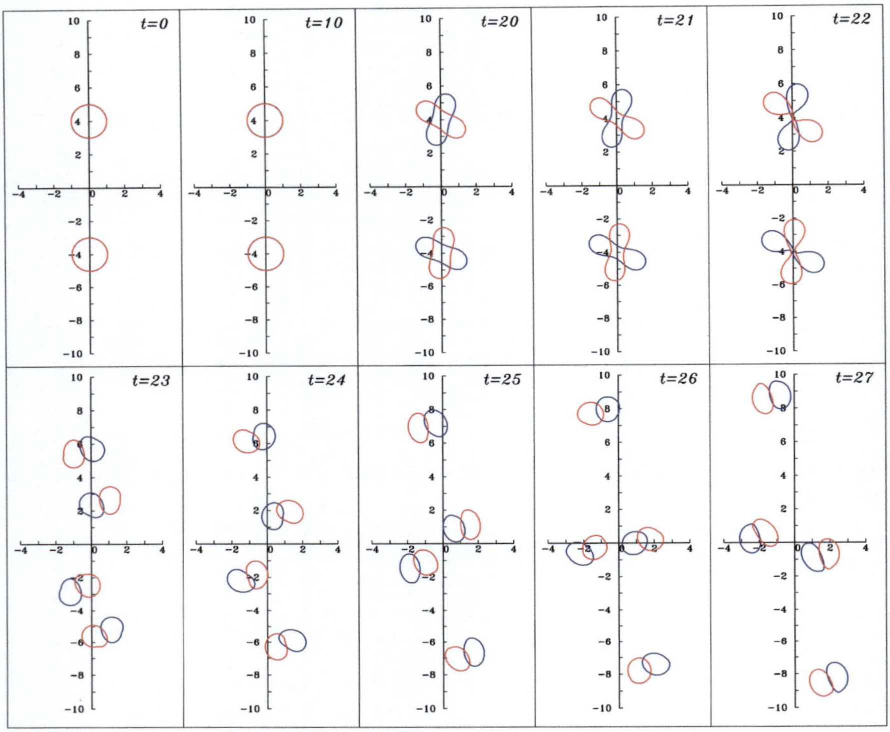

Fig. 3.81 The same as in Fig. 3.79, but with $\gamma = 1.8$. Regime $UTD2$

such in the other layer (under the impact of their partners and at a small turn) takes place almost without changes in the shape of their contours. In a homogeneous fluid, the vortices with the same sign would, clearly, merge.

Now suppose that the original hetons have slightly tilted axes ($b = 0.1$). In this case, the motions of type HD will follow the scenario similar to that observed in the case of hetons with vertical axes. Figure 3.83a, where, as well as in Fig. 3.79, the initial state in the diagram corresponds to a point near the boundary of the stability domain ($\gamma = 1.6$), gives superposed instantaneous contours of vortex patches at regular time intervals. It is important to note that, though, as before, the initial stage of counter motion proceeds in a slowered regime, in this case, owing to the initially specified tilt of axes, this stage is much shorter in time.

In the same manner, we give the results of calculations for $\gamma = 2.1$ (Figs. 3.83b), where the case of unstable vertical dipoles is realized. In the presence of a tilt in heton axes and the accompanying asymmetry of the initial configuration, the formation of the second unstable mode is accompanied by the splitting of vortex patches into unequal parts and, thus, each heton breaks up into two vertically noncompensated two-layer vortices with tilted axes. Clearly, those vortex structures move not rectilinearly, but along arc-shaped trajectories and in such a manner that

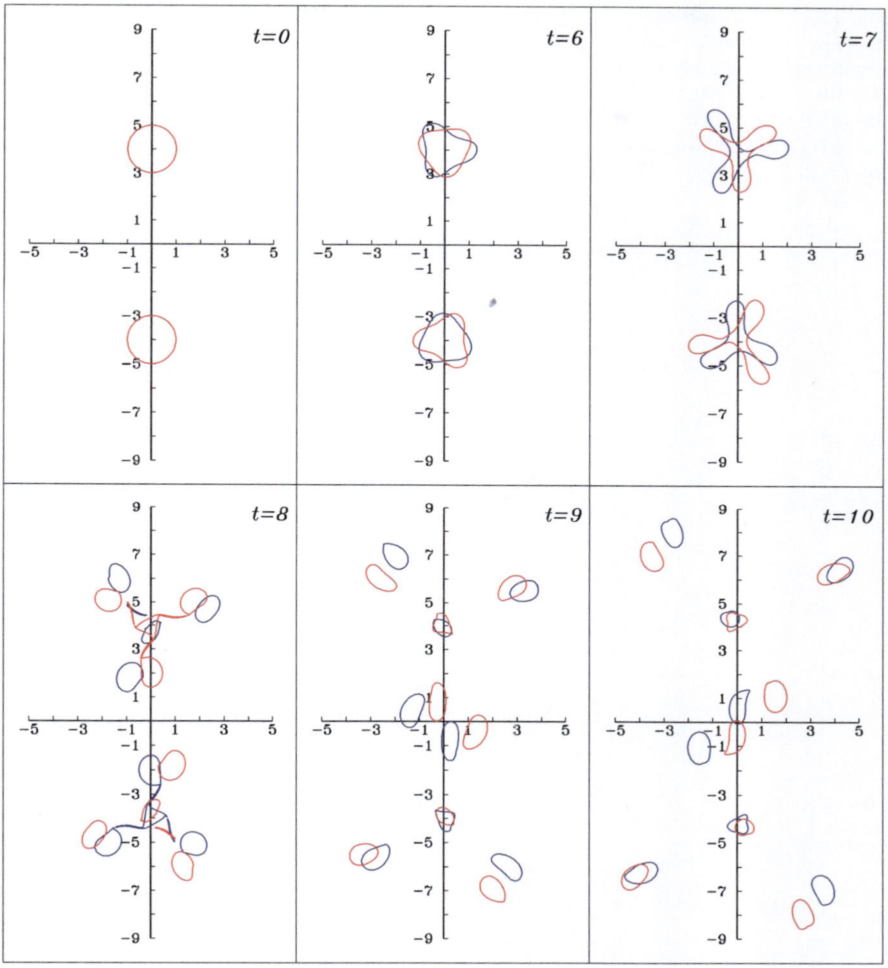

Fig. 3.82 The same as in Fig. 3.79, but with $\gamma = 3.6$. Regime $UTD3$

after passing a half of the imaginary closed curve, the vortex patches again meet, merge, and next decompose into equal parts. The result of such restructuring is the formation of a configuration similar to that observed in Fig. 3.81 just after the decomposition of original hetons (for example, at $t = 22$), but at larger distance between the vortex structures in the top and bottom parts of the figure.

Figure 3.84, constructed at the same value of b, but at much weaker stratification, where the formation of vortices at the initial stage takes place mostly due to the instability of hetons, exemplifies the formation of a deformed third mode. This includes two small-size hetons with tilted axes, which start moving toward each other, and four noncompensated two-layer vortices, which, moving in pairs along

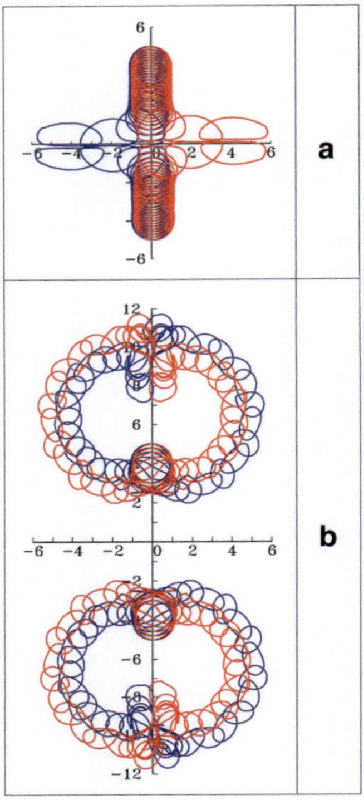

Fig. 3.83 The interaction between a warm and a cold hetons (superposed contours) at $a = 0$, $c = 4$, $b = 0.1$ and (**a**) $\gamma = 1.6$, $t = 0, 1, \ldots, 23$ (*HD* regime); (**b**) $\gamma = 2.1$, $t = 0, 1, \ldots, 20$ (*UTD*2 regime)

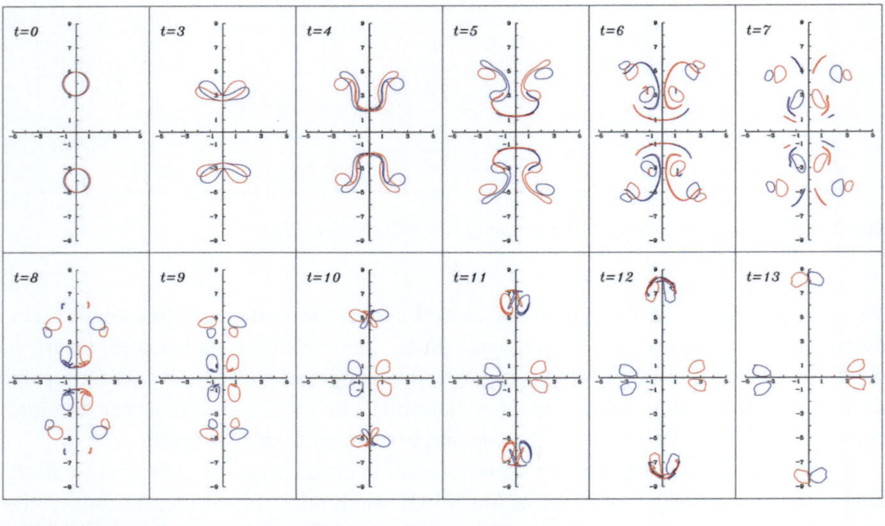

Fig. 3.84 The interaction between a warm and a cold heton at $a = 0$, $c = 4$, $b = 0.1$ and $\gamma = 3$. Regime $HD + TD$

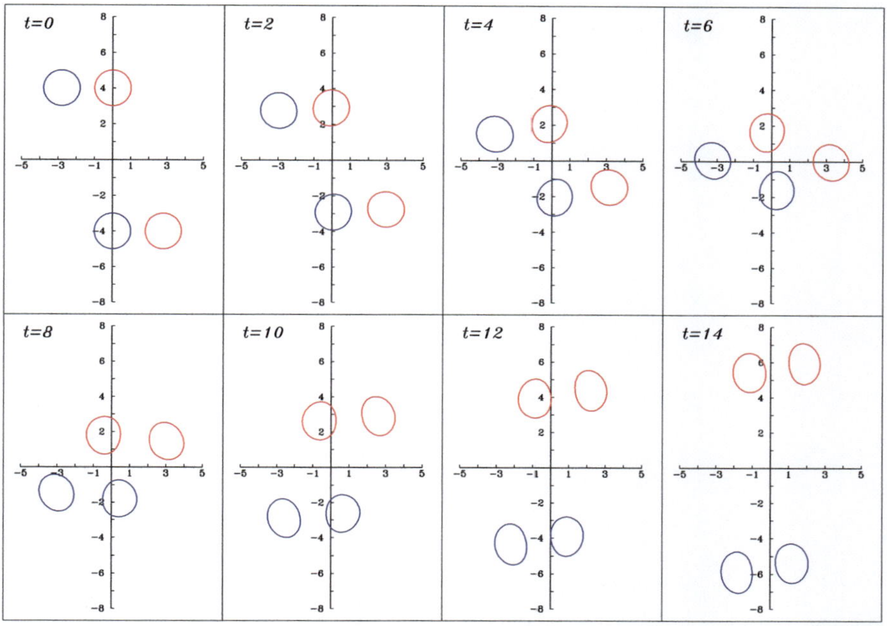

Fig. 3.85 The same as in Fig. 3.84, but with $\gamma = 1$ and $b = 2.8$. *HD* regime

arc-shaped trajectories, meet and, after the merging of vortex patches in both layers, form two hetons moving apart. The former two hetons, after their meeting, group to form two horizontal hetons moving apart. The result of this multistage interaction process is the formation of two horizontal and two vertical vortex pairs moving apart perpendicularly ($HD + TD$ type).

The series of experiments given in Figs. 3.85–3.88, has been carried out at a fixed (relatively large) tilt of axes and at different values of stratification parameter γ. The first figure gives an example of the formation of *HD* structure, where, unlike Figs. 3.79, 3.80 and 3.83a, the asymmetry is strong. The exchange of partners proceeds in such a manner that the cyclones that originally lay on the *y*-axis, after approaching each other, fall under the dominating within-layer impact of their anticyclonic partners, causing them to turn, and the two newly formed horizontal dipoles move apart in opposite directions.

As parameter γ increases (Fig. 3.86), the interaction between the central cyclones, which have approached each other, becomes stronger. Therefore, at the stage when the hetons are maximally close to each other the entire vortex construction turns in the cyclonic direction, but each two-layer dipole still remains an indivisible configuration. The hetons move apart in opposite directions, but the line of their motion is now different. Note that in the interval $T = 4 \div 8$, the vortex structure shows the properties of a *Z*-shaped tripole, which is unstable.

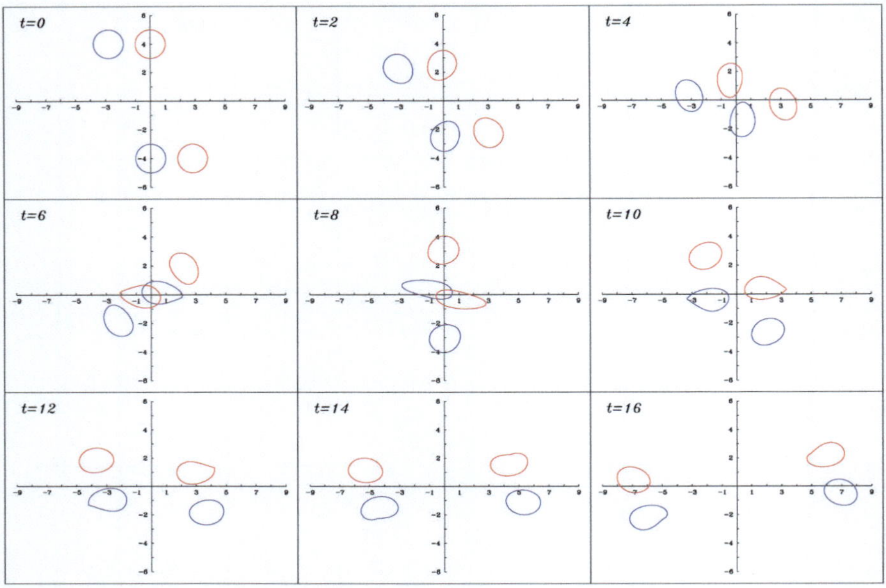

Fig. 3.86 The same as in Fig. 3.84, but with $\gamma = 1.5$ and $b = 2.8$. TD regime

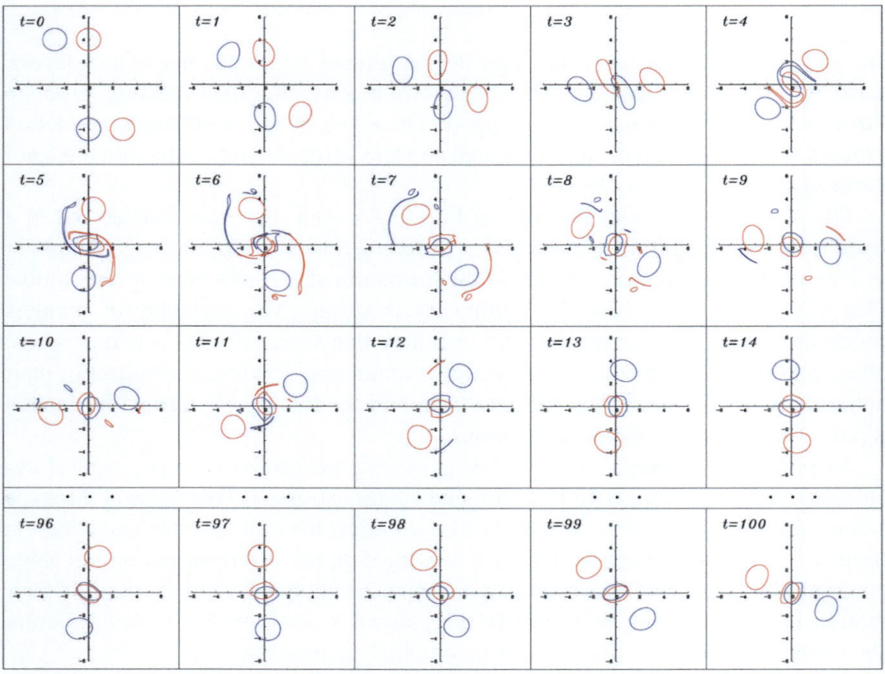

Fig. 3.87 The same as in Fig. 3.84, but with $\gamma = 2$ and $b = 2.8$. ZT regime

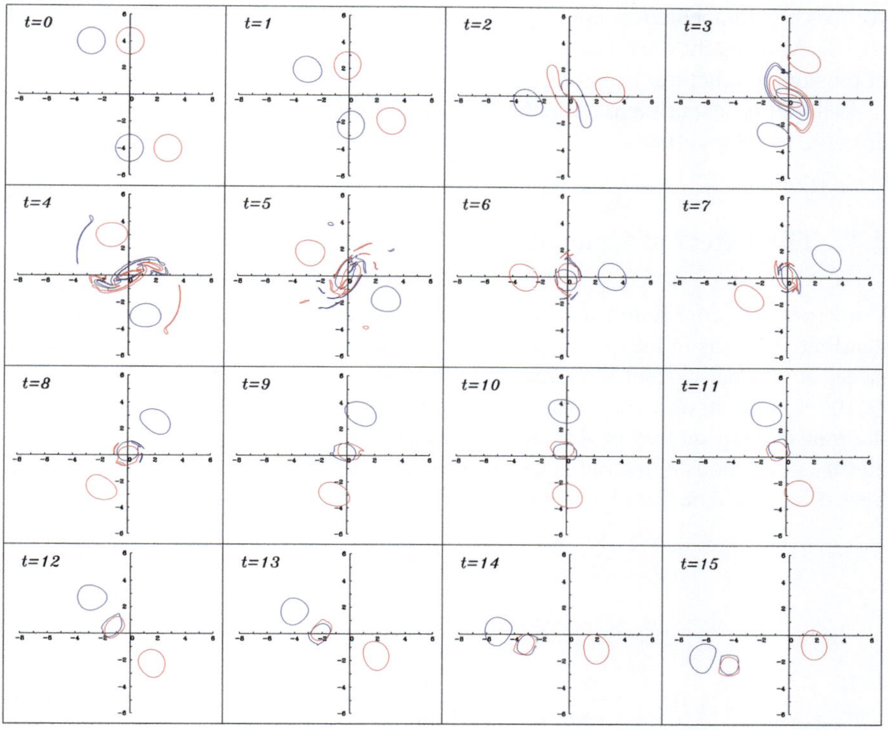

Fig. 3.88 The same as in Fig. 3.84, but with $\gamma = 4$ and $b = 2.8$. *UZT* regime. Adapted from [847] (Fig. 19) by permission of IOP Publishing

The formation of captured (associated) vortex structures is an important property of non-head-on collision of hetons. Such regime is illustrated by Fig. 3.87. The transitional stage shows an intense release of vortex threads and small vortices (the mass loss of central vortices here is 32.6 %), but next, a steady state is reached at about $t = 13$, and we see a stable *Z*-shaped configuration in the form of a central cyclonic antiheton with tilted axis and peripheral anticyclones, rotating about it in opposite layers. The figure does not show the interval covering about 4.5 rotation periods of the entire construction. Note that, given the value of b, the existence domain of stable *Z*-shaped structures is determined by the interval $\gamma \in [1.60; 3.55]$ (Fig. 3.78). Beyond this domain, the bound constructions that have formed at the collision become unstable.

An example of unstable *Z*-shaped configuration is given in Fig. 3.88 for the case of weaker stratification, the values of other parameters remaining unchanged. Here, after the formation of a tripolar structure, the central antiheton is captured by a peripheral anticyclone, thus determining the passage to an *L*-shaped configuration, which moves away from the center, and an isolated anticyclone, which some time later must acquire a quasi-static position. Note that in this case, the losses of central vortices for filamentation are even greater than in the previous case—they reach 40.1 %. Most likely, the relative "weakness" of the core is the cause of the instability of the vortex structure.

Laboratory experiments carried out in [280, 436] confirm the existence of unstable tripolar vortices.

3.4 The Effect of External Field on Heton Motion

Until now, we considered the problems of self-action or the interaction between distributed hetons in the assumed absence of external field. Here, we will study the effect of a zonal current with velocity $U(t)$ and an isolated small ($\sigma = O(Ro) = O(10^{-2})$) bottom obstacle in the form of a straight round cylinder with radius a on the translational motion of a heton with a tilted axis. In this case, the functions p_0 and η_0, continuous in the radial coordinate r and determining the external field (see Sect. 1.2.6), will be found as the solution of problem (1.73):

$$p_0 = -U(t)y - \frac{\sigma a^2}{4}\begin{cases} (r/a)^2, & r \le a, \\ 1 + 2\ln(r/a), & r > a; \end{cases} \tag{3.17}$$

$$\eta_0 = -\frac{\sigma F}{h_2\gamma^2}\begin{cases} 1 - \gamma a \mathrm{K}_1(\gamma a)\mathrm{I}_0(\gamma r), & r \le a, \\ \gamma a \mathrm{I}_1(\gamma a)\mathrm{K}_0(\gamma r), & r > a. \end{cases} \tag{3.18}$$

From (1.73), it follows, in particular, that an axisymmetric obstacle generates an anticyclonic vortex with vorticity $-\sigma/h_2$ localized above the obstacle and concentrated in the bottom layer.

Consider the problem of the motion of a heton with a tilted axis above the obstacle, in the case where vortex patches are symmetrically shifted relative the x-axis, along which the external barotropic flow is also directed. As the initial condition, we assume the resting condition, which requires us to introduce a free cyclonic vortex with vorticity σ/h_2 in the bottom layer directly above the obstacle to neutralize the effect of topography (see Fig. 3.89). Thus, in parallel, we will solve the problem of initialization of a topographic vortex near the obstacle.

Having in mind the possible oceanographical applications of the obtained numerical results, we also assume more realistic relationships between the thicknesses of the top and bottom layers. We will use the procedure for constructing calibration functionals, proposed by Flierl in [272], to ensure that the parameters (h_1 and $\Delta\rho$) of the two-layer model optimally correspond to its continuous analog. In [272], the bottom boundary of the top layer, i.e., the mean position of the density jump horizon, is associated with the position of the peak in the vertical distribution of Brunt–Väisälä frequency. This enables us to obtain simple functional relationships

$$h_1 = \int_0^1 zN^2(z)dz \Big/ \int_0^1 N^2(z)dz, \quad \Delta\rho = \frac{\rho_0}{g}\int_0^1 N^2(z)dz. \tag{3.19}$$

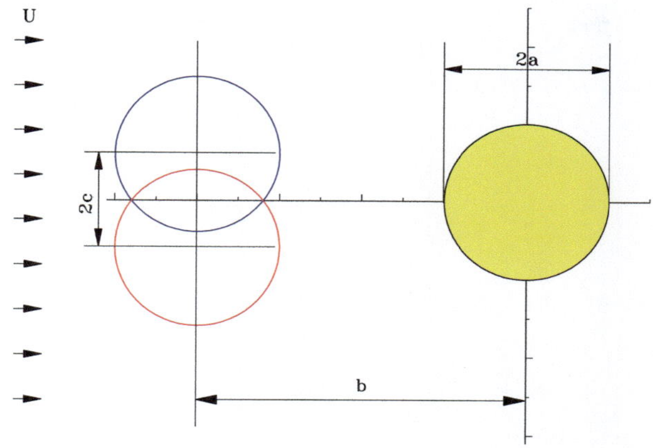

Fig. 3.89 Scheme of the initial state of the problem of heton motion over an obstacle. Here $2c$ is the distance between the centers of vortex patches in the *top layer* (*red line*) and *bottom layer* (*blues line*), B is the distance between heton "axis" and the center of the obstacle. The *yellow domain* is a cyclonic vortex in the *bottom layer*, which originally covers the cylindrical circular obstacle (*green*) with the center in the coordinate origin

The averaged profiles of vertical density distribution, given in Table 1.2.1 in [404], according to formulas (3.19) give approximately

$$h_1 = 0.104,\ \Delta\rho = 3.47\cdot 10^{-3}\Delta\rho \quad \text{and} \quad h_1 = 0.095,\ \Delta\rho = 2.63\cdot 10^{-3}\Delta\rho$$

for the Pacific and Atlantic, respectively. Thus, assuming $h_1 = 0.1$, $\rho_0 = 1$, $\Delta\rho = 3\cdot 10^{-3}$ in the calculations, we can consider the model used here quite adequate for typical oceanic conditions.

In the context of this model, we consider several examples of heton motion over an obstacle.

Figure 3.90 gives the results of calculations at $C = 1$, $B = 4$, $a = 1$, $U = 0$, $\sigma = 0.05$, and $\gamma = 1$, i.e., when there is no external flow; the tilt of heton axis is such that, in the vertical projection, the vortices of the top and bottom layers are tangent in the initial moment; the radius of the obstacle is equal to the radius of each vortex patch; the height of the obstacle in dimensional units is 250 m (the depth of the ocean being 5,000 m); and the fluid is moderately stratified. As the heton approaches the obstacle, it increases the tilt of its axis. In this case, the top-layer vortex, which, obviously, experiences the least effect of the obstacle, deviates to the right and almost does not change its shape. The bottom-layer vortex is partially captured by the topography and, simultaneously, experiences considerable deformations. Before its destruction, it carries out about 3/4 of full revolution around the obstacle. The bottom-layer cyclonic vortex, originally localized above the obstacle (it is dyed), is also strongly destroyed. In the period of numerical experiment, part of each cyclone in the bottom layer was captured by the obstacle, and a second part, after intense filamentation, merged with the top-layer anticyclone, thus forming a self-moving bipolar structure, moving away from the obstacle neighborhood.

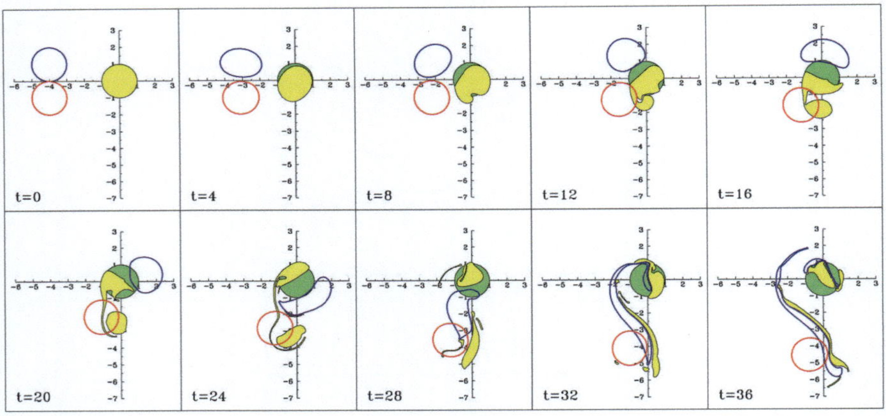

Fig. 3.90 Heton motion over an obstacle at $c = 1, B = 4, a = 1, U = 0, \sigma = 0.05, \gamma = 1$

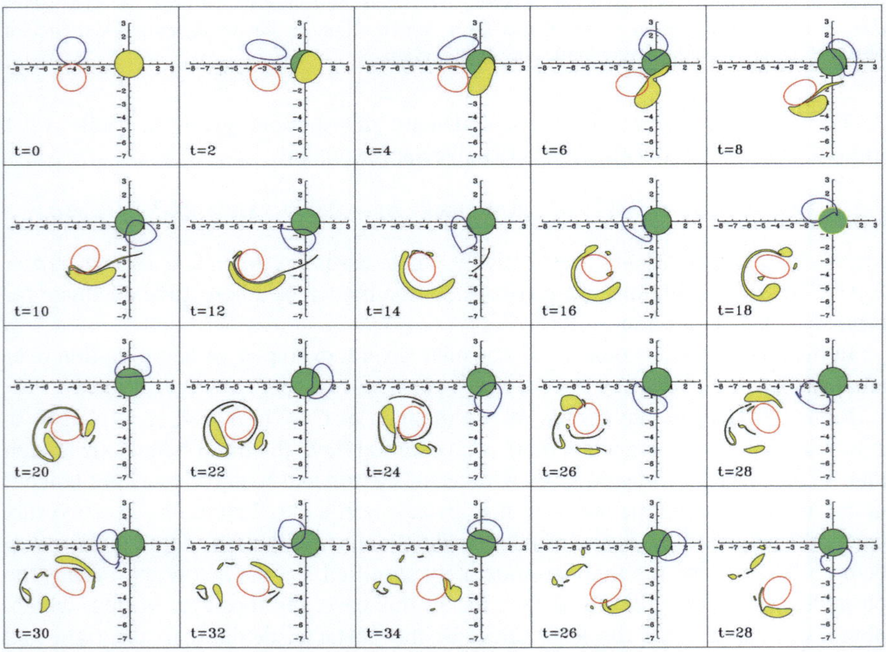

Fig. 3.91 Heton motion over an obstacle at $c = 1, B = 4, a = 1, U = 0, \sigma = 0.05, \gamma = 2.6$

The next example in Fig. 3.91 corresponds to the same values of external parameters except for γ: now, $\gamma = 2.6$, which corresponds to weaker stratification, when a heton with *vertical axis* loses its stability and the only unstable mode is the second one (see Fig. 3.1). We remind, however, that in Sect. 3.2.1, where the stability of heton with *tilted axis* was studied, it was shown that at $c > 0.6$,

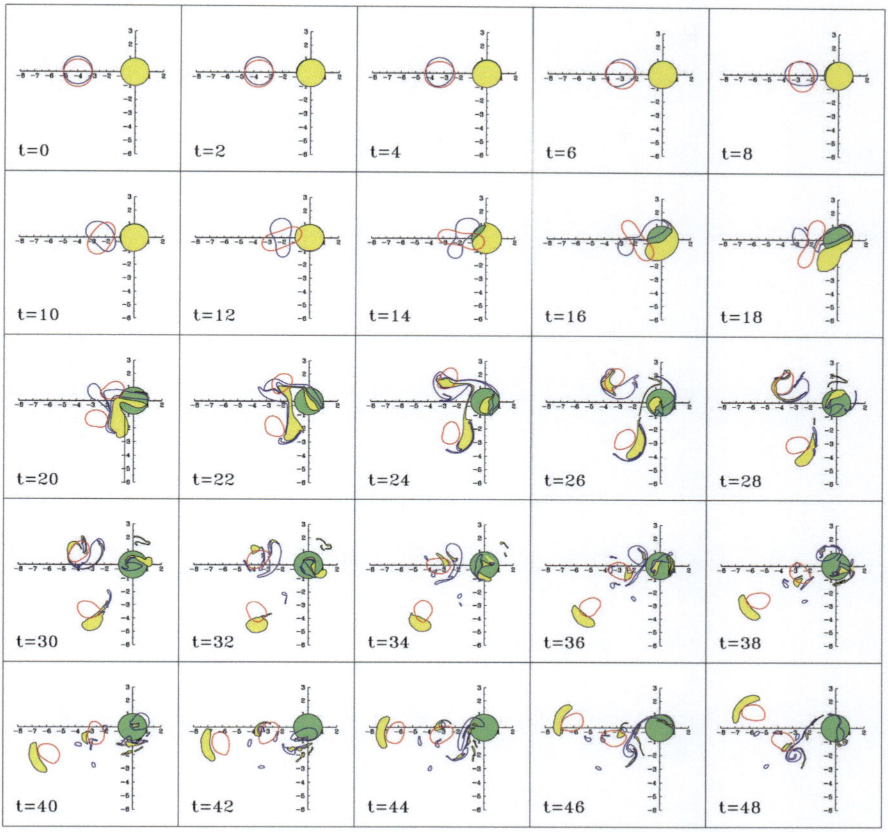

Fig. 3.92 Heton motion over an obstacle at $c = 0.1$, $B = 4$, $a = 1$, $U = 0$, $\sigma = 0.05$, $\gamma = 2.6$

two-layer vortex is stable at any γ. Thus, the instability is impossible at $c = 1$, and we can examine the effect of parameter γ on the resulting pattern of *per se* vortex motion. The weakening of stratification leads to the following effects: the bottom-layer cyclone, a component of heton, remains quite compact and is fast trapped by bottom topography (during the calculation time, it carried out nearly three revolutions around the mountain); the second cyclonic vortex, which originally was situated above the mountain, completely leaves it, falling under the influence of the rotating effect from the top-layer anticyclone. This is accompanied by its separation into many small parts and two relatively large ones. The latter, along with the top-layer vortex, forms a practically motionless quasi-tripolar, two-layer Λ-shaped structure.

In the next experiment (Fig. 3.92), the tilt of heton axis is much less ($c = 0.1$), and, when approaching the obstacle, the two-layer pair shows unstable behavior—the patches divide into two parts, first in the bottom and next in the top layer. The two newly formed top-layer patches remain compact, while the bottom layer

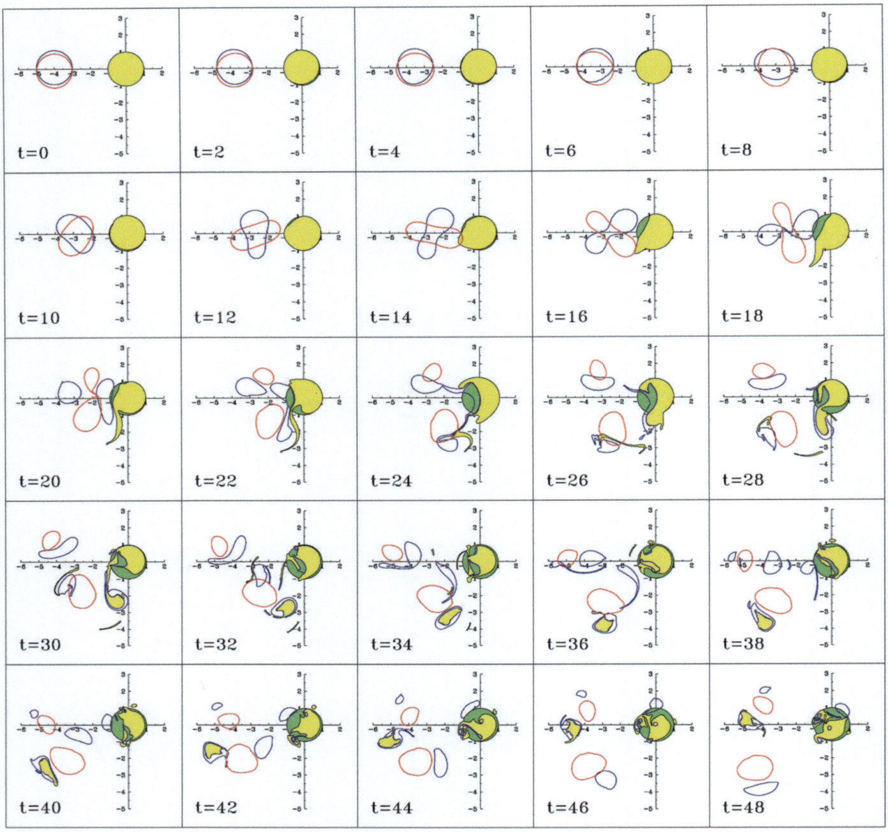

Fig. 3.93 Heton motion over an obstacle at $c = 0.1$, $B = 4$, $a = 1$, $U = -0.0015$, $\sigma = 0.05$, $\gamma = 2.6$

shows considerable deformations of contours with the formation of vortex threads and small vortices. The topographic vortex is also subject to destruction; practically all its particles will leave the obstacle neighborhood by the end of calculation period. Two-layer dipole structures can be clearly seen in time interval $t = 24 \div 48$ in such a manner that in one structure in the bottom layer, the leading role belongs to the heton vortex patch, and in the other—to the free topographic vortex. The dynamics here is mostly determined by the rotation of such structures about a common center; in this case, a small part of the heton cyclonic patch is captured by the obstacle.

Above, it was assumed that, in the initial moment, the heton starts moving only due to the self-motion mechanism, caused by a vertical shift of its axis. Now we consider the effect of the external zonal current. Let a weak homogeneous current, hampering the motion of heton is specified (Fig. 3.93). The comparison with Fig. 3.92 shows that the motion of heton toward the obstacle is slower, but its decomposition at the initial stage proceeds similarly. The topographic vortex is more conservative in this case—its major portion remains captured by the obstacle. Note an interesting feature of this case—a series of mass exchanges between individual vortex structures.

The following stages can be conventionally identified: (1) in the interval $t = 30 \div 34$, we see a tendency toward the formation of a bipolar two-layer structure in the upper half-plane of the figure and a tripolar one in its lower part; (2) at $t = 36$, the bottom-layer vortex in the former structure decomposes, and the bipolar structure thus becomes tripolar; while in the latter structure, one bottom-layer vortex is captured by bottom topography, and this structure becomes an asymmetric quasi-heton (in which the bottom vortex now contains both a part of the topographic vortex and a part of the originally heton structure); this state still persists at $t = 38$; (3) at $t = 40$, a bottom-layer vortex, belonging to the tripolar structure, is captured by the bipolar structure (they "exchange" by their orders); (4) at $t = 42 \div 44$, reverse exchange takes place (but now another bottom-layer vortex is involved in the process); (5) at $t = 46 \div 48$, we have fully formed structures, which again qualitatively resemble those observed in the interval $t = 36 \div 38$. Note that the part of the bottom-layer heton cyclone captured within the time period from 40 to 48 units has made about half-revolution around the mountain.

Suppose that the velocity of the incoming flow is still directed against heton motion, but now it counterbalances its position. Under such conditions, a stable heton, in the absence of topographic perturbation, must be immobile. As can be seen from Fig. 3.94, the two-layer vortex practically does not change its position and shape for a long time. However, at $t \geq 10$, the second mode becomes unstable and the vortex structure breaks up into two moving apart hetons with tilted axes. The heton that moves mostly in the direction of the external flow, obviously moves away fast enough, but the pair that has a large positive zonal velocity component because of the formed large tilt of its axis can also overcome the decelerating effect of the flow. In this case, its interaction with the topographic vortex and the obstacle is relatively weak. The topographic vortex experiences small deformations, and its major portion stays captured by the mountain.

In the next experiment (Fig. 3.95), the external-flow direction coincides with that of the motion of heton itself, and flow velocity is five times greater than in the previous case. Obviously, the heton more rapidly approaches the obstacle, stays captured by it for a short time ($t = 6 \div 10$), acquires above it an anticyclonic rotation, and starts to decompose. After that, it breaks downstream and its vortex patches in each layer mostly decompose into two parts. Although the interaction of the heton with the topographic vortex is not too active here (because of their relative remoteness), the joint impact of this cyclonic vortex and the obstacle causes the two-layer vortex to deviate to the right from the general direction. It is important to note that, at the specified velocity, not large as it may be, the topographic vortex will be completely removed from the obstacle.

The situation will radically change if we consider a submarine obstacle, which is two times higher (Fig. 3.96). Now, only part of the topographic vortex is carried downstream, the vortex patch belonging to the heton forms long tongues near the obstacle, and a part of the top-layer vortex patch is captured by the obstacle.

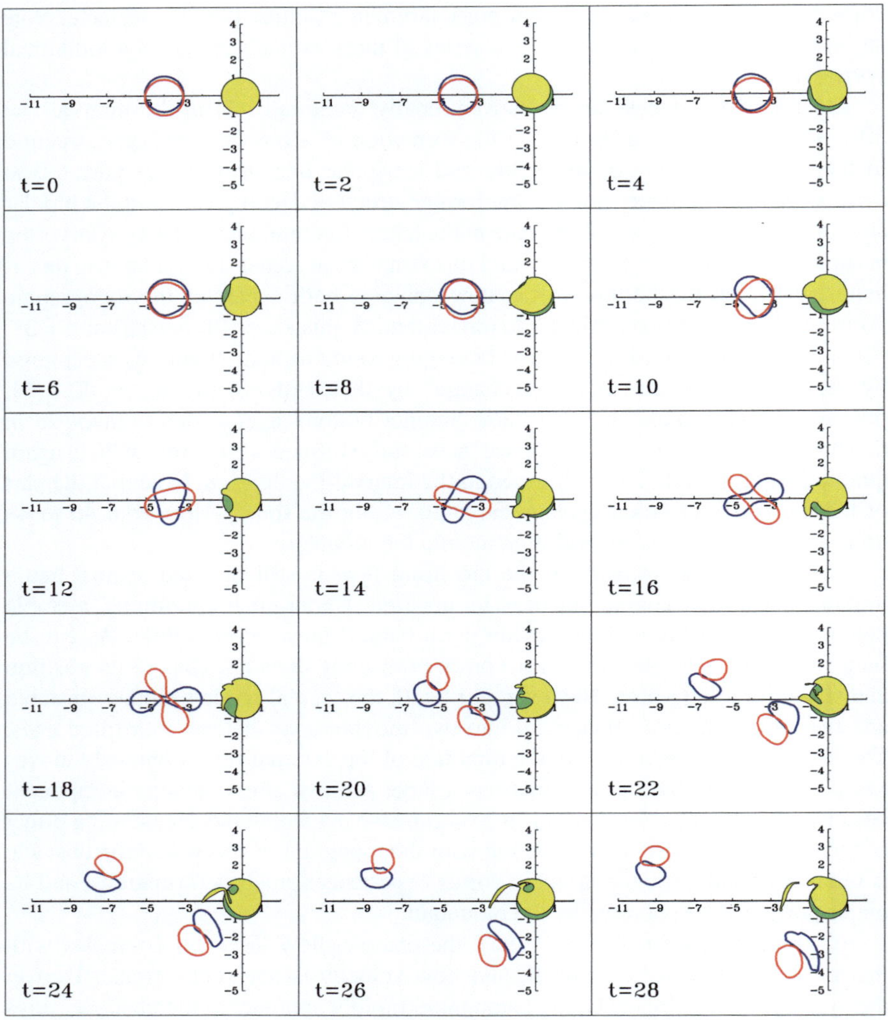

Fig. 3.94 Heton motion over an obstacle at $c = 0.1$, $B = 4$, $a = 1$, $U = -0.004$, $\sigma = 0.05$, $\gamma = 2.6$

In the next two experiments, we vary the horizontal size of the obstacle, its height remaining unchanged. Thus, if the radius of the round cylindrical mountain is increased twofold (Fig. 3.97), it captures the topographic vortex very fast. Its center carries out small-amplitude quasi-circular motions about obstacle center, while its contour deforms, protruding long tongues. The bottom-layer vortex patch rounds the mountain; during the calculation period, it makes slightly more than half rotation. The top-layer vortex experiences considerable destruction, and its largest parts accompany the bottom-layer vortex patch.

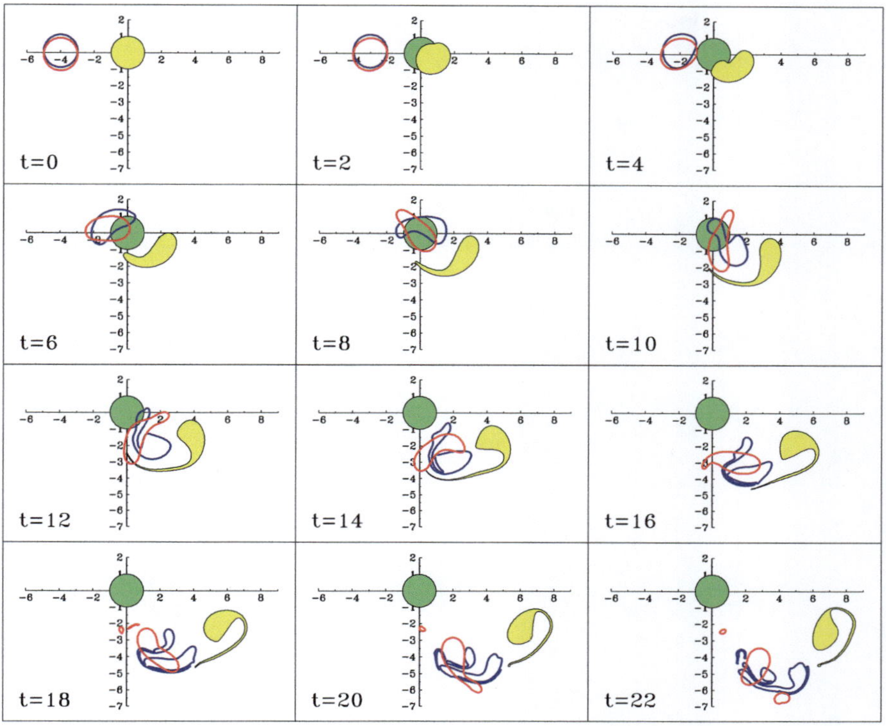

Fig. 3.95 Heton motion over an obstacle at $c = 0.1$, $B = 4$, $a = 1$, $U = 0.02$, $\sigma = 0.05$, $\gamma = 2.6$

Figure 3.98, where the submarine obstacle is, conversely, two times less than heton vortex patches, the topographic vortex very rapidly moves from the obstacle and leaves its neighborhood, the vortex shape remaining practically unchanged. Under the effect of the topographic vortex, the vortex pair deviates right from rectilinear motion. The top-layer vortex takes a quasi-elliptic shape and remains indivisible during the calculation period, while the vortex patch experiences strong stretching and practically decomposes, and only its very small part is captured by the obstacle.

3.5 Vortex Patch Dynamics in a Three-Layer Model

In this section we will consider some problems of vortex dynamics in the formulation given in Sect. 1.2.9.

In many cases, the real structure of vertical density distribution can be adequately approximated by a three-layer model. For example, in [648], the processing of velocity measurements at 63 anchored buoy stations operated in 1984–1989 in

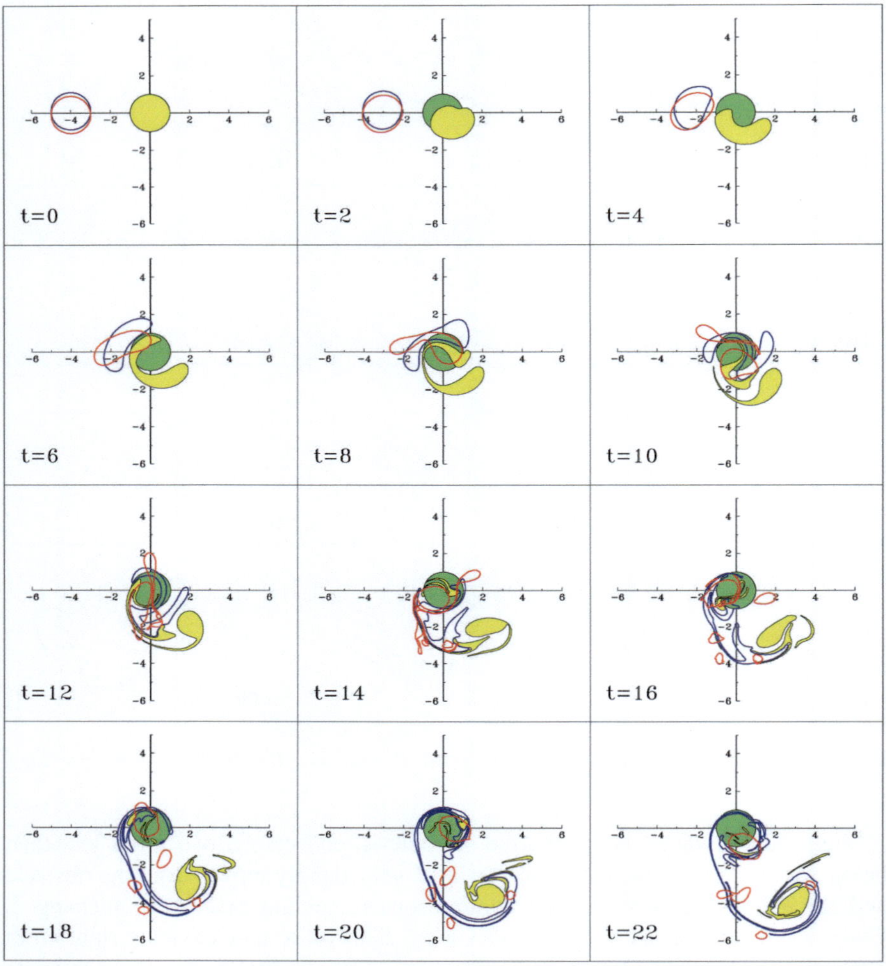

Fig. 3.96 Heton motion over an obstacle at $c = 0.1$, $B = 4$, $a = 1$, $U = 0.02$, $\sigma = 0.1$, $\gamma = 2.6$

the abyssal tropic zone of the Pacific ("Abyssal" project), revealed *a three-layer structure* of low-frequency synoptic currents, in [831], the *three-layer character* of convective oceanic boundary layer is mentioned. In [373] *a three-layer model* of the ocean is used to analyze the stability of the current along bottom slope; in [171], it is used to study the exchange between subtropic and subpolar vortices in the North Atlantic; in [360, 361], to explain the existence of undercritical, postcritical, and hypercritical states of drift circulation in the ocean; in [638], to prove the existence of the passage from the antisymmetrical viscous regime to the axisymmetric regime by pitchfork' type bifurcation; Mirabel and Monin [613] investigated instability of

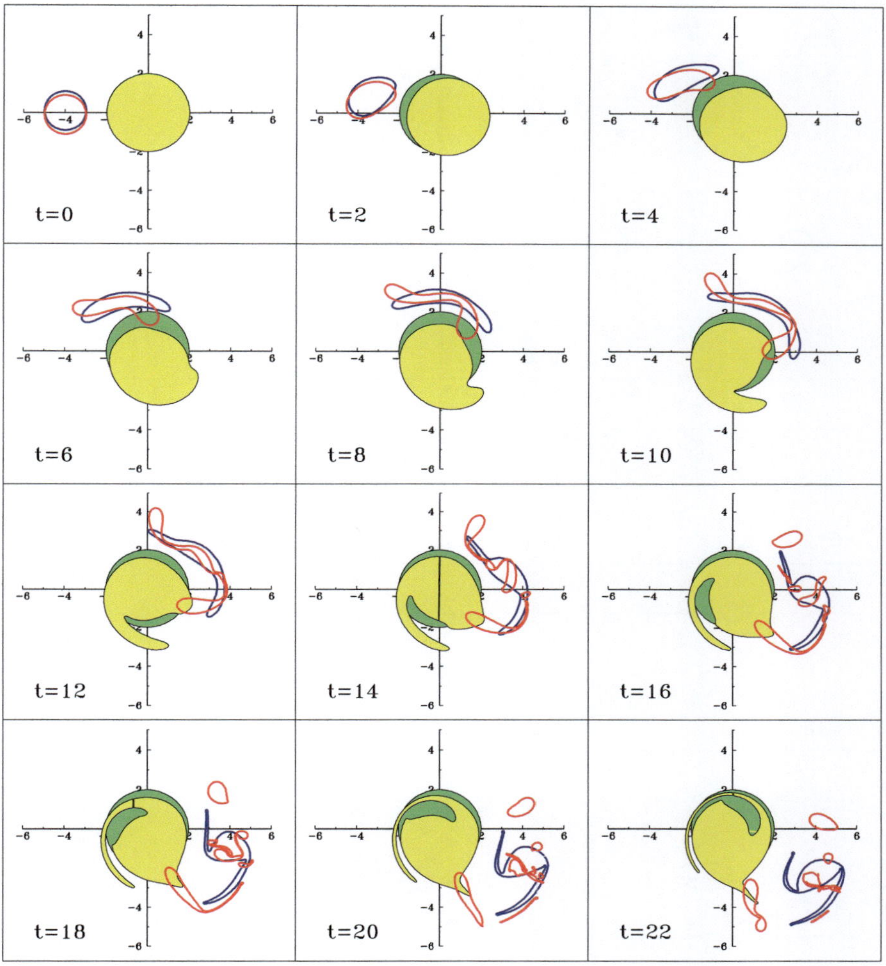

Fig. 3.97 Heton motion over an obstacle at $c = 0.1$, $B = 4$, $a = 2$, $U = 0.02$, $\sigma = 0.1$, $\gamma = 2.6$

ocean circulation patterns in the *three-layer ocean model*; in [305] the sensitivity of a large scale coupled ice ocean model to the parameterization of vertical mixing in the upper ocean has been investigated; a *three-layer numerical model* is used to evaluate the asymmetric evolution of hurricane [818] and tropical cyclone [819]. Ikeda [373] note that at least *three layers* should be used to study the instability of current flowing along a slope.

Thus, there is reason to believe that this model is informative and can be used to study a wide range of processes in the ocean.

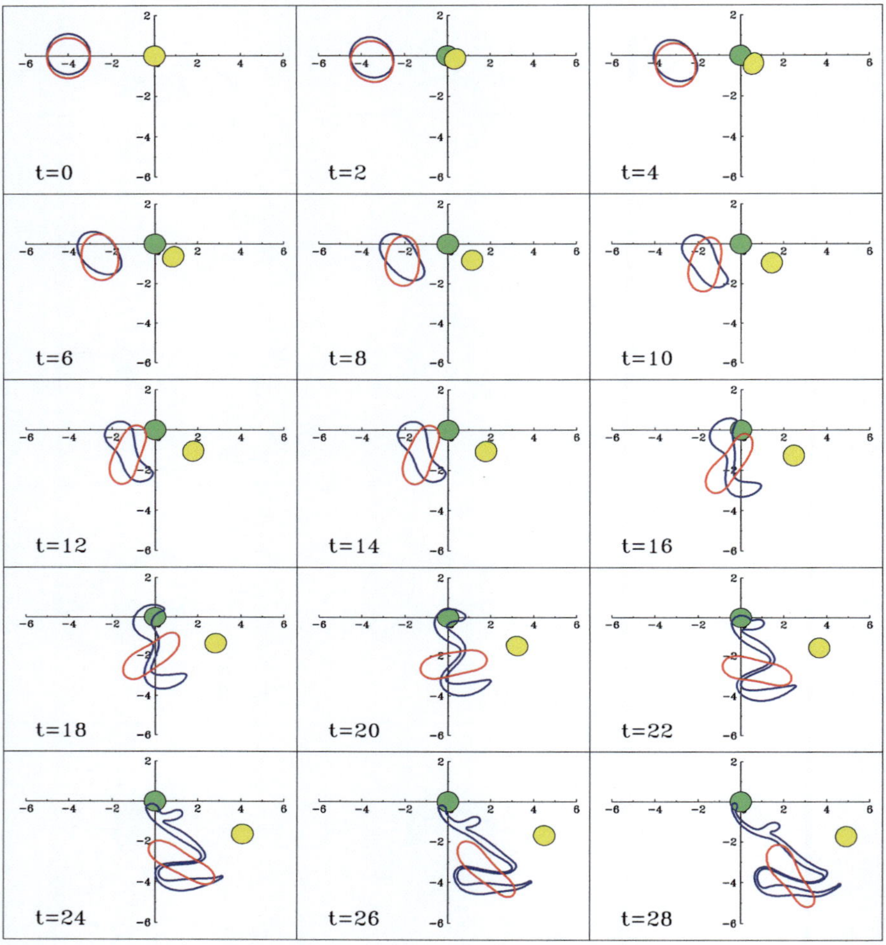

Fig. 3.98 Heton motion over an obstacle at $c = 0.1$, $B = 4$, $a = 0.5$, $U = 0.02$, $\sigma = 0.1$, $\gamma = 2.6$

3.5.1 Stability Study of a Three-Layer Vortex

The stability of an axisymmetric three-layer 2-modal heton (S-vortex), which induces velocity field (1.99), for the general case of different radii ρ_j $(j = 1, 2, 3)$ of vortices in layers was studied in sufficient detail in [845, 846]. The general scheme of stability analysis is given in the Sect. 1.2.9 of this book.

Figures 3.99 and 3.100 show the existence of a distinct boundary between the stability and instability domains. Thus, in the former figure, where $F = 0.37$, vortex patches, while pulsating appreciably, still preserve a compact quasi-elliptic shape all over the very long calculation time.

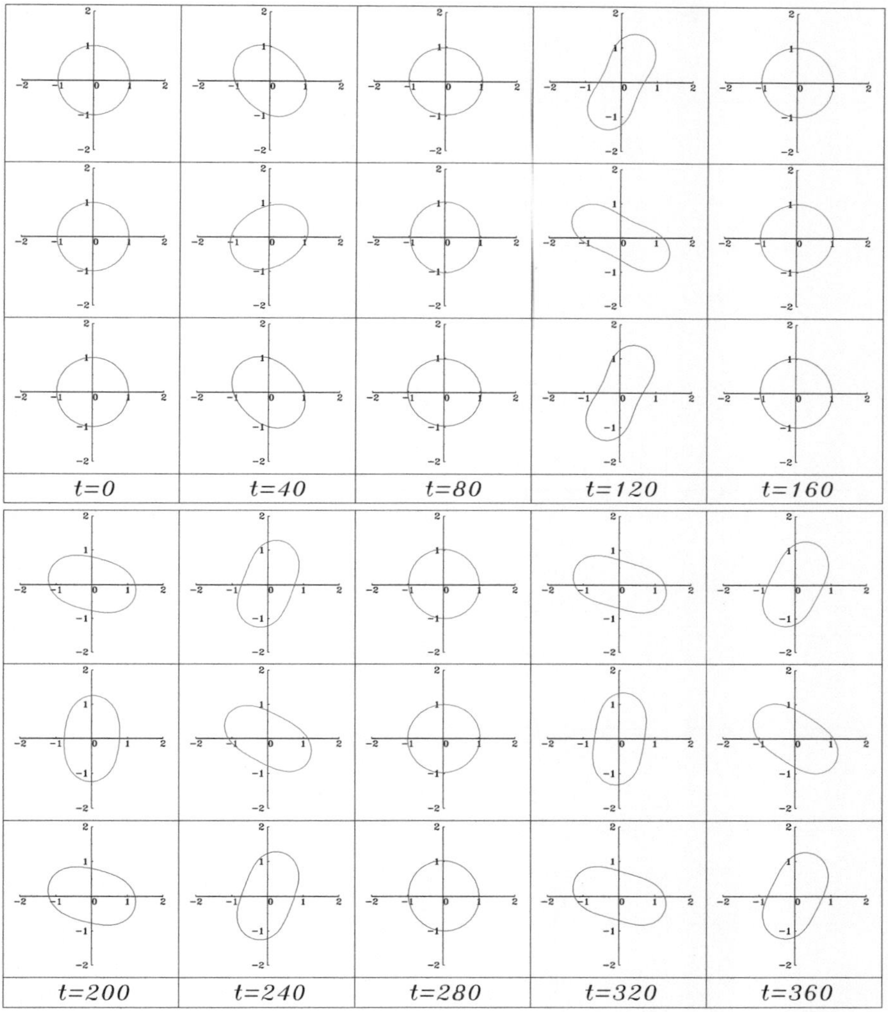

Fig. 3.99 Synchronous configurations of vortex patches contours of the *top* (*red line*), *middle* (*magenta line*), and *bottom* (*blue line*) *layers*, from *top* to *bottom*, of a stable three-layer cylindrical ($\rho_1 = \rho_2 = \rho_3 = 1$) vortex with zero total vorticity at $h_1 = h_2 = h_3 = 1/3$, $F_1 = F_2 = 0.370$ ($\gamma_1 = 1.054, \gamma_2 = 1.825$), $\Pi_1 = \Pi_3 = -0.5, \Pi_2 = 1$

With a small change in the Froude number ($F = 0.38$, Fig. 3.100), we pass into the instability domain of the second mode, where the heton breaks up into two three-layer vortices, having arc-shape axes and moving in opposite directions. It is worth mentioning that the initial, though rather long, stage of evolution in both cases follows similar scenarios (compare Figs. 3.99 and 3.100 at $t = 0$ and $t = 40$), but next, in the presence of instability, i.e., in the latter case, the loss of the simple connectedness of vortex patches is very rapid.

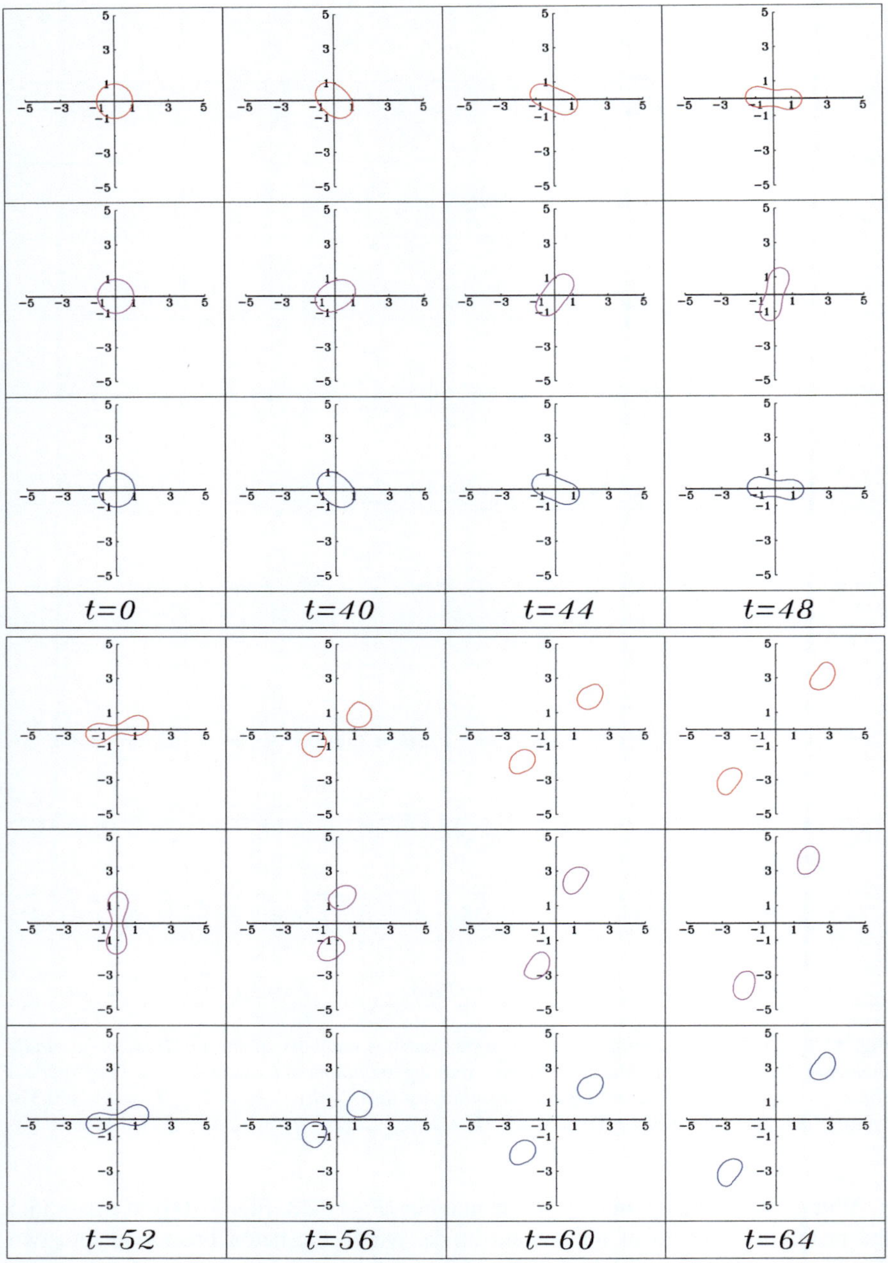

Fig. 3.100 The same as in Fig. 3.99 for an unstable three-layer vortex at $F_1 = F_2 = 0.380$ ($\gamma_1 = 1.068, \gamma_2 = 1.849$), when the only unstable mode is the second one

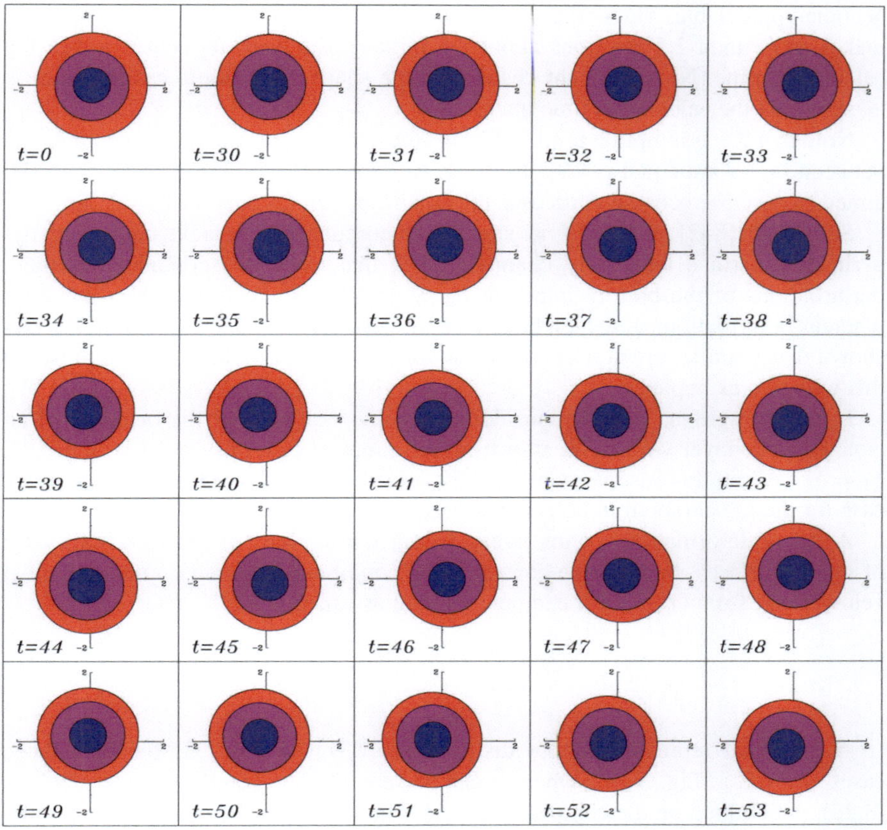

Fig. 3.101 The realization of the first unstable mode. Combined instantaneous configurations of vortex patches of a *cone-shaped* three-layer heton at $\rho_1 = 1.5$, $\rho_2 = 1$, $\rho_3 = 0.5$, $h_1 = h_2 = h_3 = 1/3$, $F_1 = F_2 = 1$, $W_1 = W_3 = -0.5, W_2 = 1$, where $W_j = \Pi_j/(\rho_j)^2$. Each vortex patch has its own color: *red* for the *top layer*, *magenta* for the *middle layer*, and *blue* for the *bottom layer*

An important distinction between the three-layer and the two-layer situations is the fact that the three-layer vortex can be unstable with respect to perturbations of *the first* mode, at which the vortex structure has to move as a whole. In the modeling of the instability process of a "cylindrical" three-layer vortex, the first mode is very difficult to identify, since the increment of its growth is generally much less than that of higher modes, and the instability domains (in the space of external parameters) of the first and second modes have considerable intersections. It is more convenient to carry out the calculations for a "cone-shaped" vortex, where the instability domain of the first mode can be identified explicitly [845, 846].

An example of evolution of such three-layer heton at $\rho_1 = 1.5$, $\rho_2 = 1$, and $\rho_3 = 0.5$ is given in Fig. 3.101. The given geometry of the vortex allows all three vortex patches to be shown in each panel. The heton preserves its configuration and position unchanged for a long time, but at $t \geq 30$, the displacements of vortex patch centers become appreciable. The entire vortex formation, as a whole, starts nutations, and makes more than 2.5 rotations about the initial position of its center within the calculation time. No significant changes in the shape of contours can be seen (at least, within the calculation time interval).

Note that the assumption used so far that the density jumps at the two boundaries between layers are equal is very artificial. A more realistic for a three-layer model aimed to describe oceanic vortices is the condition $\Delta\rho_1 \gg \Delta\rho_2$, i.e., $F_1 \ll F_2$.

Smeed in [832] pioneered in showing theoretically and next confirmed experimentally (in a laboratory channel) [833] that, with this condition satisfied, perturbations of the bottom interface η_2 generate small-scale waves in the flow, whereas perturbations of the top boundary η_1 cause large-scale waves. In [844], it is shown that a similar effect takes place for vortices as well. Below, we will illustrate this with two examples.

In this paragraph, we will consider only some examples of the behavior of an isolated three-layer vortex. We start from the simplest case of $h_1 = h_2 = h_3 = 1/3$, $F_1 = F_2 \equiv F$; in this case it is obvious that a full geometric symmetry exists in this case for the top and bottom layers of a three-layer heton.

As local deformations of interfaces appear due to velocity shift on both sides of such interfaces, they can be controlled by conditions imposed on the azimuthal velocities V_j from (1.99). For example, we can assume

$$V_2\Big|_{r=1} = V_3\Big|_{r=1} \quad \text{or} \quad V_1\Big|_{r=1} = V_2\Big|_{r=1}. \tag{3.20}$$

A graphical representation of distributions $V_j(r)$ $(j = 1,2,3)$ for both those cases is given in Fig. 3.102, where it can be seen that the conditions for velocities (3.20), which hold on the contours of circular unit-radius vortex patches, imply the closeness of functions V_j, uniform over r, on both sides of the appropriate interface between layers. From this it follows that the inequality $\eta_1 \gg \eta_2$ will hold in the former case, and $\eta_1 \ll \eta_2$ will hold in the latter case. Note also that the three values of potential vorticity in the layers (coefficients in formulas (1.99)) for the construction of velocity profiles were chosen with Eqs (3.20) taken into account and with the assumption of vertical compensation of vorticity $\Pi_1 + \Pi_2 + \Pi_3 = 0$ at $\Pi_2 = 1$.

The values of potential vorticity, satisfying the first and second conditions (3.20), in axisymmetric vortices in the initial time moment were chosen in numerical experiments, illustrated in Figs. 3.103 and 3.104, respectively. In the former (latter) case, when the interface between the second and third (the first and second) layers is almost not perturbed, the second (third) mode is maximally unstable. Thus, those calculations support the conclusion of work [832] that at $\Delta\rho_1 \gg \Delta\rho_2$, the perturbations of the bottom interface η_2 induce wave (vortex) structures of smaller scale, than perturbations in the top interface η_1 do.

Given the vertical distribution of potential vorticity, in a baroclinic vortex, we can, owing to the above regularity, state that if a change in its sign is observed in surface layers, the vortex tends to separate into larger parts than those in the case where the rotation sign changes in the bottom layers.

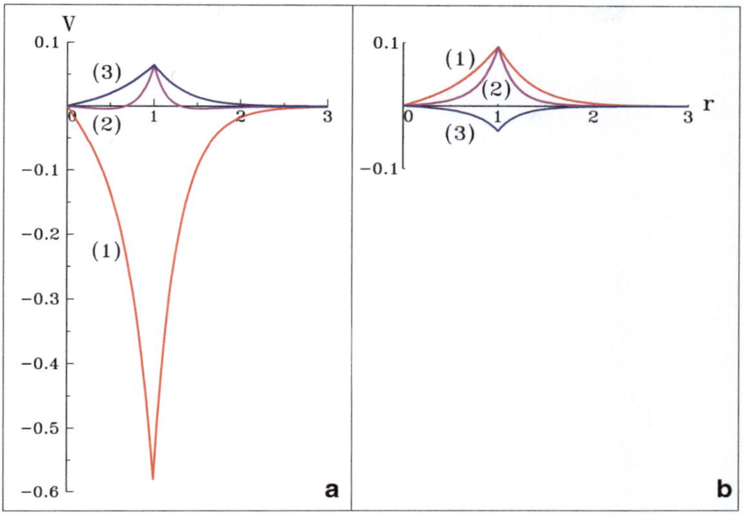

Fig. 3.102 Azimuthal velocity profiles induced by a three-layer cylindrical ($\rho_1 = \rho_2 = \rho_3 = 1$) vortex in (*1*) the top, (*2*) middle and (*3*) bottom layers at $F_2 = 4F_1$: (**a**) the first case in conditions (3.20) and (**b**) the second case in (3.20)

3.5.2 Modeling the Motion of Meddies

It is known from oceanographic observations [379] that intrathermocline eddies (meddies) are usually identified by the anomalous values of temperature and salinity and almost do not differ in density from the surrounding water masses in the horizontal direction. Mediterranean Sea water, flowing from the Gibraltar as a bottom current, occupy intermediate depths in the eastern North Atlantic at depths from 500 to 1,500 m [259]. This fact is a motivation for using a three-layer quasi-geostrophic model with vortex patches located in the middle layer to construct an adequate model of oceanic eddies. In this case, owing to the geostrophic balance, anticyclonic (cyclonic) eddies in the middle layer have a form of double-convex (double concave) lenses. Note that the idea of applying a three-layer model to the description of motions of intrathermocline lenses seems to have been first proposed by Hogg and Stommel in [352], after which it was successfully used in the recent work of Carton et al. [147].

The specific values of geophysical parameters of the model will be chosen taking into account the actual distribution of many-year mean density profiles in the Atlantic [404] (the full smooth line in the left part of Fig. 3.105) and mean horizontal scales of the eddies [1001].

The vertical density distribution is approximated by a two-step piecewise-constant function with density jumps at interfaces between layers, such that $\Delta\rho_1/\Delta\rho_2 = 4$ and the dimensionless thicknesses of the layers are $h_1 = 0.1$, $h_2 = 0.2$

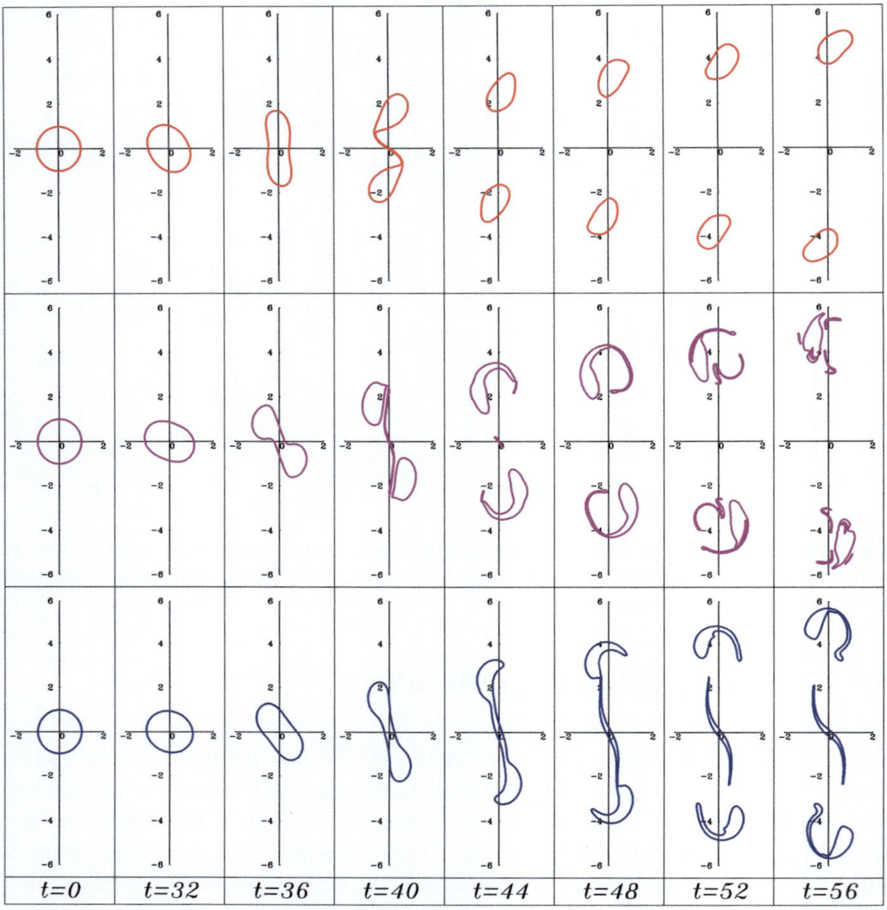

Fig. 3.103 Evolution of an unstable three-layer vortex at $F_1 = 1$ and $F_2 = 4$ in the case $\eta_2 \ll \eta_1$ (excitation of the second mode)

and $h_3 = 0.7$. This choice of parameters is based on the following considerations. In Sect. 3.4, with reference to [404], we obtained $h_1 = 0.1$; further, the value $h_2 = 0.2$ was accepted to ensure that at the depth of the ocean of 5,000 m, the middle layer occupies horizons from 500 to 1,500 m. In this case, the condition $F_2 = 4F_1 \equiv 4F$ will hold with a sufficient accuracy.

In the context of this model, we will consider two problems: (1) the interaction of two circular vortex patches and (2) the evolution of an elliptic vortex patch. In both cases, we suppose that the vortices are located only in the middle layer. Hence, they are prototypes of the meddies.

Indeed, Fig. 3.106 shows the presence of high concentration of lenses in the North Atlantic, especially near the Azores frontal zone, which enables their effective

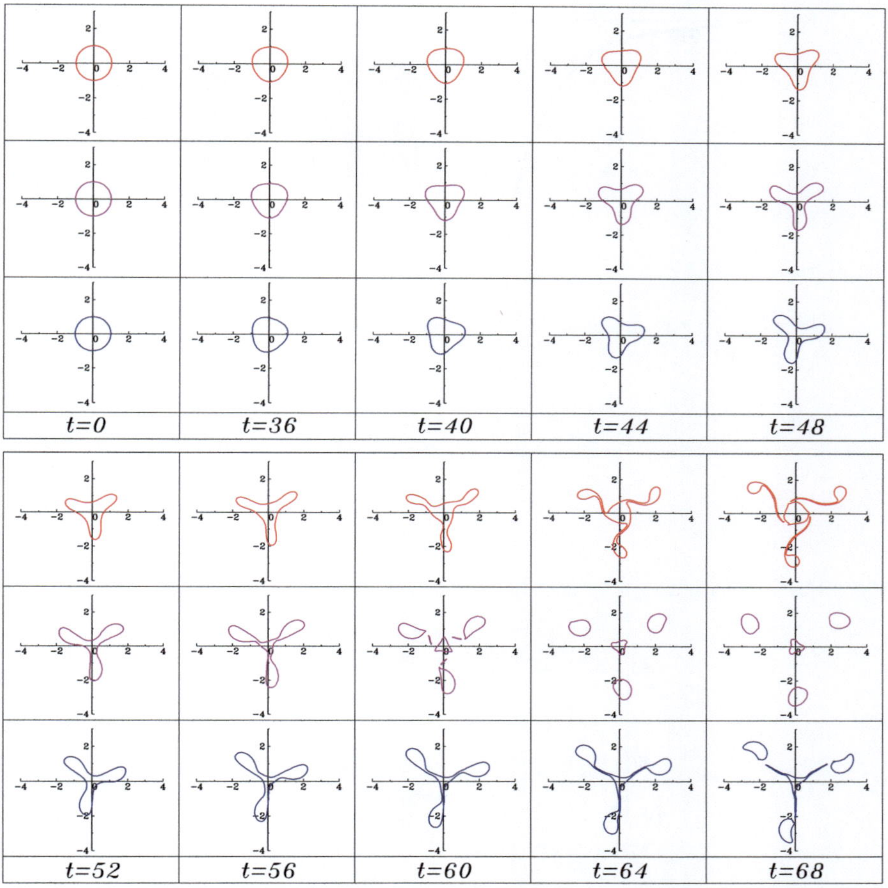

Fig. 3.104 Evolution of an unstable three-layer vortex at $F_1 = 1$ and $F_2 = 4$ in the case $\eta_1 \ll \eta_2$ (excitation of the third mode)

interaction and suggests the high probability of merging of neighboring vortices and the subsequent formation of elongated (as a first approximation, quasi-elliptic) vortex patches. Therefore, the two problems mentioned above have quite real applications.

Merging of Two Initially Circular Vortex Patches

It is known [510] that two point vortices equal in intensities and signs, rotate with a constant angular velocity relative to the center of the segment connecting these vortices in the sense determined by the sign of the vorticity. Two identical distributed circular vortex patches behave similarly if they are located sufficiently far from each

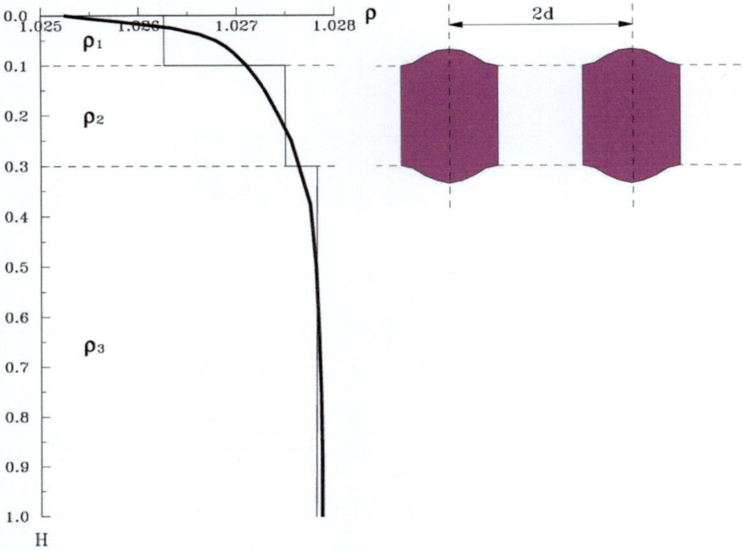

Fig. 3.105 A schematic representation of the three-layer model and two anticyclonic lenses

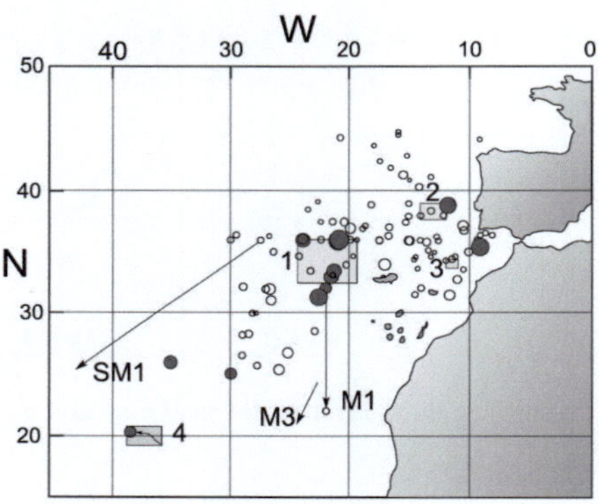

Fig. 3.106 The distribution of Mediterranean lenses (meddies) in the North Atlantic by their volume (the size of the circle is proportional to lense volume). The shadowed markers show the largest lenses: *1* the domain of Azores frontal zone [408, 751, 914], *2* the domain of detected merging of two lenses [806]; *3* the domain of formation of lenses with dipole structure [257]; *4* the domain where the large-scale Mesopolygon experiment was conducted [1001]; *SM1* the trajectory of a lens in the zone of SEMAPHORE experiment; *M1*, *M3* the trajectory of Sharon lense [752]

other [357], but they can merge when they are sufficiently close to each other. The problem of merger of vortices of the same sign in a homogeneous or two-layer fluid (plasma) is one of the fundamental problems in hydrodynamics and many publications are dedicated to this problem (see the list of citations in the beginning of Sect. 3).

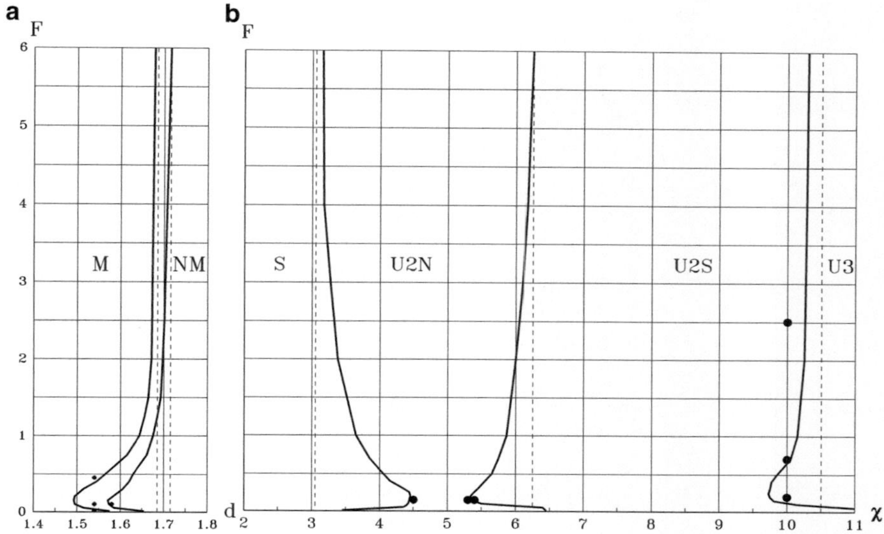

Fig. 3.107 Diagram of different states: (**a**) two circular vortex patches on the plane of parameters (d,F) in a barotropic fluid (the *dashed lines*) and in the *middle layer* of a three-layer fluid (the *full lines*); (**b**) the same for a single elliptic vortex on the plane of parameters (χ,F). The positions of round markers correspond to the coordinates on the plane of those parameters used in the numerical experiments shown in Figs. 3.108–3.111a and Figs. 3.112–3.114b

The key problem is the estimation of the critical distance between the centers of the vortex patches, at which the merging process occurs. It was found that two cylindrical circular eddies of a unit radii merge if $d < d^* \sim 1.6 \div 1.7$ where is half distance between the eddy centers (the right part of Fig. 3.105) [357, 529]. We note that these results for the critical value d^* were obtained for vortices in a barotropic fluid. In the case considered here, the vortex patches of unit radius in the middle layer are located between two "free" interface surfaces, which play a critical role according to the calculations.

In Fig. 3.107, the interval for $d^* \in [1.69; 1.71]$ for a barotropic case, obtained by CDM-calculations, is limited by two dashed lines. To the left of the left line (*M* domain), the vortex patches always merge, while to the right of the right line (*NM* domain), they do not merge. Within the narrow interval between them, we have either a single merging with subsequent separation, or the multiple ones.

For meddies, the appropriate interfaces between the regimes (full lines) behave in a different manner: at large F, we essentially have a barotropic limit, while with parameter F varying from 0 to about 1.5, the distribution is nonmonotonic with respect to parameter F with a minimum at $F \approx 0.15$. Assuming, for example, $f = 10^{-4}\ \mathrm{s}^{-1}$, $g = 10^3\ \mathrm{g}\cdot\mathrm{cm}^3\cdot\mathrm{s}^{-2}$, $H = 5\cdot 10^5$ cm, $\Delta\rho_1 = \frac{4}{5}\Delta\rho = 2.4\cdot 10^{-3}\ \mathrm{g}\cdot\mathrm{cm}^{-3}$ and $F = 0.15$, we obtain that for the vortices that are most prone to merging, the characteristic radius is $L = 3.96\cdot 10^6$ cm $= 39.6$ km.

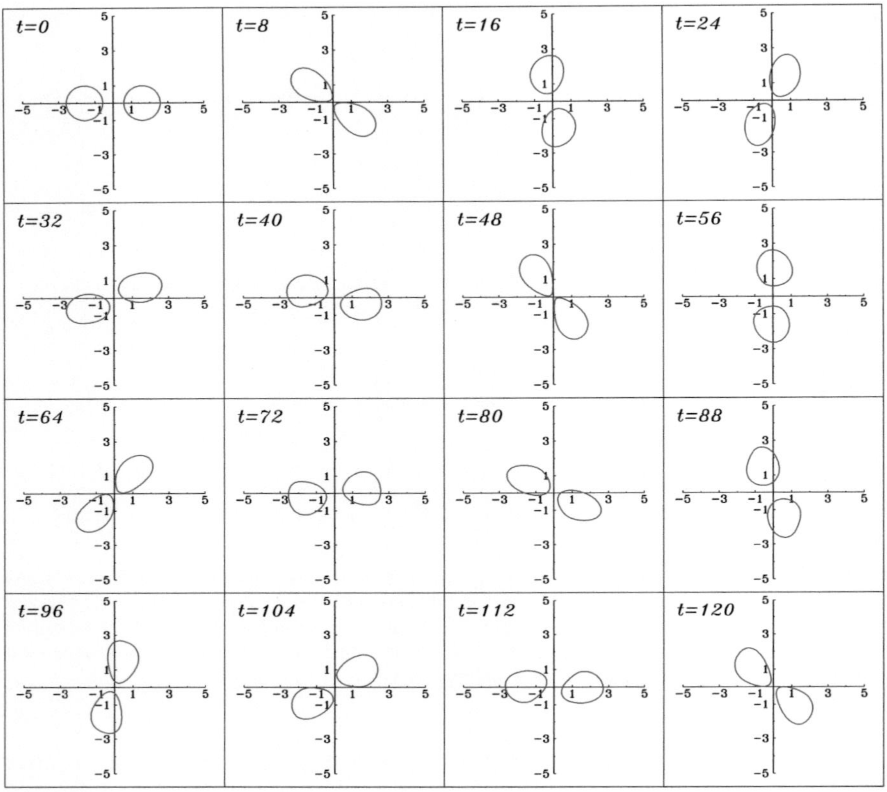

Fig. 3.108 The motion of originally circular vortex patches of the *middle layer* without merging: $F = 0.15$, $d = 1.58$

Figures 3.108–3.111 give the results of calculations for modeling the evolution of two identical circular unit-radius vortex patches in the middle layer at the same initial distance between their centers but at different values of parameter F (three dark markers in Fig. 3.107a).

Figure 3.108 gives an example of the behavior of anomalously close vortex patches ($d = 1.58$) at $F = 0.15$ (the bottom marker in Fig. 3.107a), where the vortices rotate about a common vorticity center without merging. In this case, as can be seen from the figure, the vortex patches periodically approach each other and move apart as was the case in the interaction of Aska meddies, B1 and B2 [806].

Figure 3.109, where $F = 0.40$ (the middle marker in Fig. 3.107a) illustrates the process of repeated merging and subsequent separation of vortex patches. This example, in particular, provides a possible explanation to the chaoticity that can often be seen in the behavior of drifters [120]: such drifter, originally localized in a circular vortex patch, can later (after merging–separation) appear either in the same or in another patch, and such process can occur repeatedly.

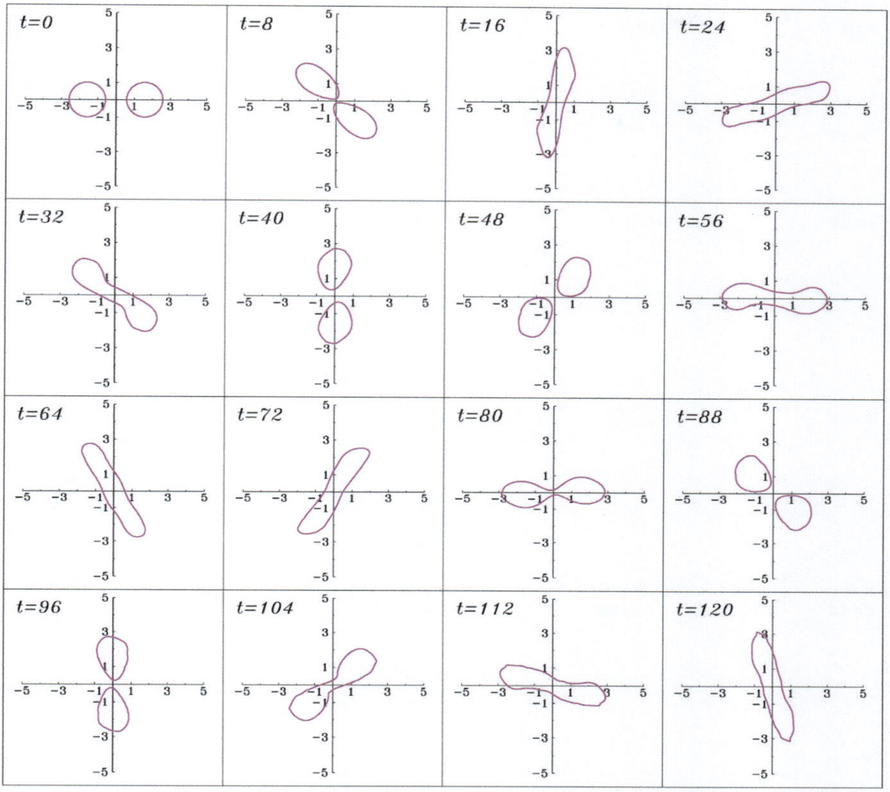

Fig. 3.109 The motion of originally circular vortex patches of the *middle layer* with periodical merging and subsequent separation: $F = 0.4$, $d = 1.58$

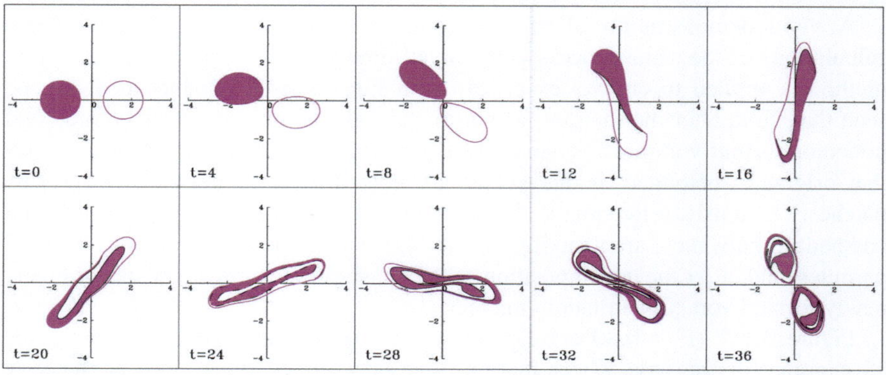

Fig. 3.110 Vizualization of the mixing process during merging and splitting of the vortex patches under the conditions of the experiment in Fig. 3.109, i.e. at $F = 0.4$, $d = 1.58$ (the calculation time-step is reduced by 2)

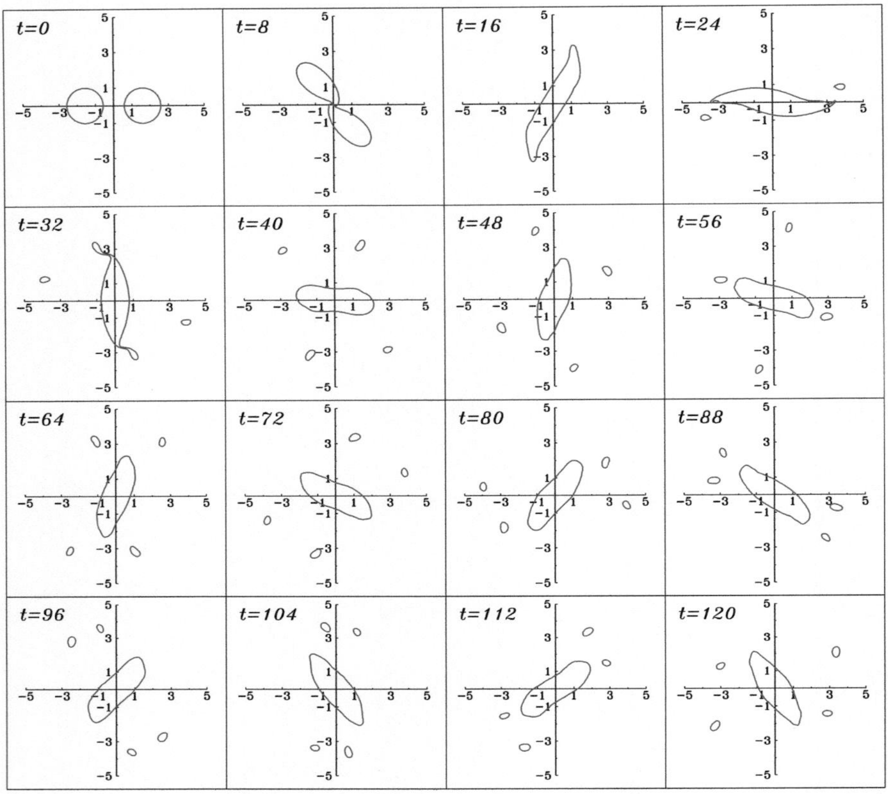

Fig. 3.111 The motion of originally circular vortex patches of the middle layer with merging: $F = 0.8$, $d = 1.58$

A vivid demonstration of this phenomenon is given in Fig. 3.110, where calculations of the same process with halved time step are given and a coloration method is applied to one vortex patch (as in Figs. 3.59–3.62). Indeed, it can be seen that, first, after the merging of vortex patches, the mixing of the colored and noncolored fluid volumes is chaotic, and, second, after mixing, the fluid particles that originally belonged to one vortex are now distributed between two vortex patches. Obviously, a hypothetic drifter, which is located in one of the lenses and has neutral buoyancy, after mixing, has to take part in the chaotic motion of fluid particles, and, after the decomposition of the quasi-elliptic vortex can appear in any newly formed vortex in a random manner.

Figure 3.111 ($F = 0.80$ corresponds to the top marker in Fig. 3.107a) gives an example of the case where merging of two vortices takes place at the very first stage of evolution. The vortex structure that formed as the result of merging of the initially circular patches has the form of a pulsating quasi-elliptic lense rotating in the anticyclonic direction and surrounded by small-scale vortices that have separated at the intermediate stage. After the separation of vortex filaments[6] and small vortices, the remaining core takes a compact form with the ratio between half-axes not exceeding 4. Thus, it is stable (see Fig. 3.107b).

Everywhere, the configurations of vortex patches are shown in dimensionless times. The scaling in such that one dimensionless time unit corresponds to the rotational period, with the fluid velocity equal to 30 cm/s at the vortex radius [447]. Thus, one unit of dimensionless time corresponds approximately to 10, 16, and 24 days in the first, second, and third cases, respectively.

Evolution of Elliptic Vortex Patch

In Sect. 3.2.2, we considered the problem of the behavior of elliptic hetons. Here, we investigate another problem—the possible scenarios of the evolution of an elliptic vortex patch located in the middle layer of a three-layer fluid, depending on the fluid stratification. Keeping in mind that in the previous series of calculations the quasi-elliptical vortices could be generated as a results of merging of two circular vortices, we consider initial vortex patches with the area equal to the area of two unit circular vortices, i.e., assume $ab = 2$, where a and b are semiaxes of the ellipse.

Results of numerical experiments are summarized in the diagram shown in Fig. 3.107b. The dashed vertical lines represent the boundaries of stability domains in the barotropic case: in the domain S the elliptical patch is stable, in $U2N$, $U2S$, and $U3$ domains it is instable with splitting of the vortex patch into two nonequal parts,[7] into two equal parts, and into three parts, respectively.

The thick full lines show the appropriate boundaries for elliptic lenses. A specific feature of the baroclinic case, as well as of the problem of vortex merging, is the nonmonotonic behavior of regime boundaries with respect to parameter F. It is important to note that the extreme properties are observed at the same critical value $F \approx 0.15$, as in the case of merging of circular lenses. On the one hand, it corresponds to the conditions of maximal stability of elliptic vortex patch with semiaxes ratio up to $\chi = 4.4$ against $\chi = 3$ in the barotropic case. On the other hand, at $F \approx 0.15$, we have a minimal interval in terms of χ for the domain of asymmetric splitting of elliptic vortex into two parts ($4.5 < \chi < 5.3$ against $3.3 < \chi < 6.5$), as well as the most favorable conditions for the formation of the unstable symmetrical second ($5.3 < \chi < 9.7$ against $6.5 < \chi < 10.5$) and third ($\chi > 9.7$ against $\chi > 10.5$) modes.

Figures 3.112 and 3.113, where $\chi = 4.5$ and $\chi = 5.4$ give examples of vortex patch behavior at $F = 0.15$ and at extreme values of ellipse eccentricity. In the

[6]The role of filamentation in the merger of anticyclonic lenses is investigated in detail by Cushman-Roisin [193].

[7]The asymmetry effect during the destruction of elliptic vortex patches was found for the first time by Kozlov and Makarov in [472], where the authors also presented their consideration about inevitability of asymmetry at $3.3 < \chi < 6.5$; see also [117,958].

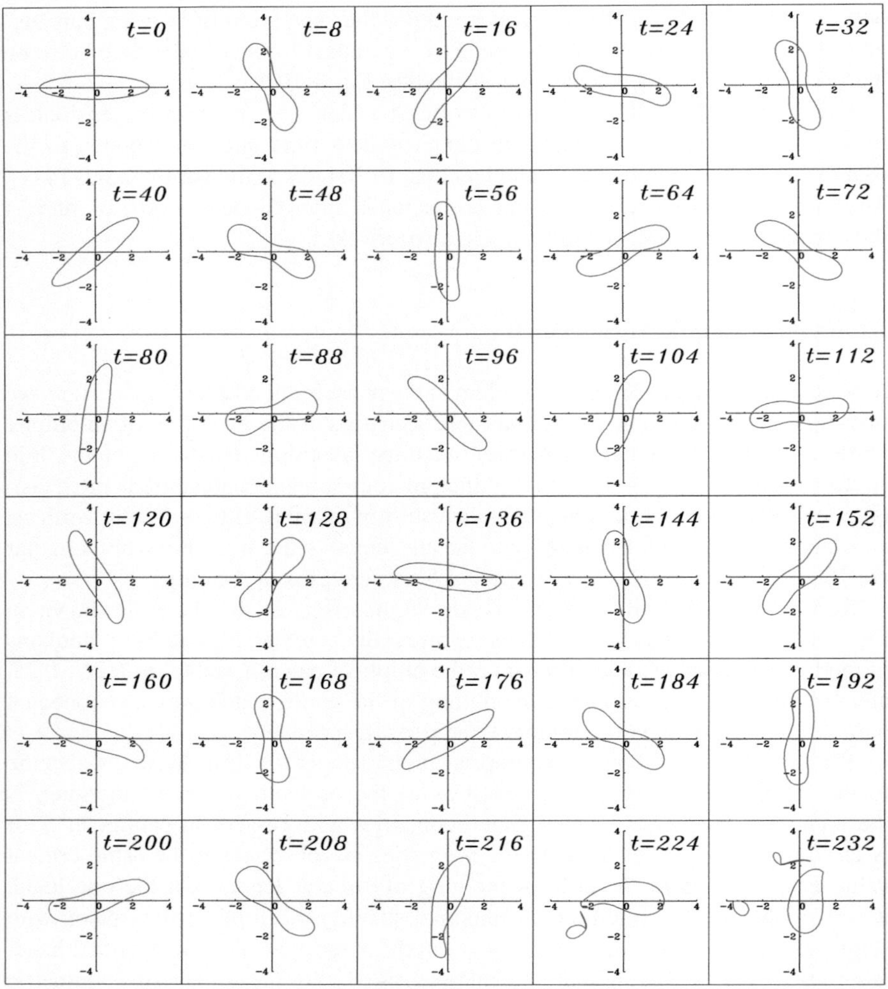

Fig. 3.112 Evolution of an elliptic meddy at $\chi = 4.5$, $F = 0.15$ (splitting into asymmetric parts, i.e. separation of small vortices)

former case, the vortex patch behavior is stable for a long time, and it is only at $t > 200$ that the vortex patch becomes asymmetric and next breaks up into two unequal parts.[8]

[8]The asymmetry effect during the destruction of elliptic eddy patches was found for the first time in [472], where the authors also presented their considerations about inevitability of asymmetry at $3.3 < \chi < 6.5$; see also [117].

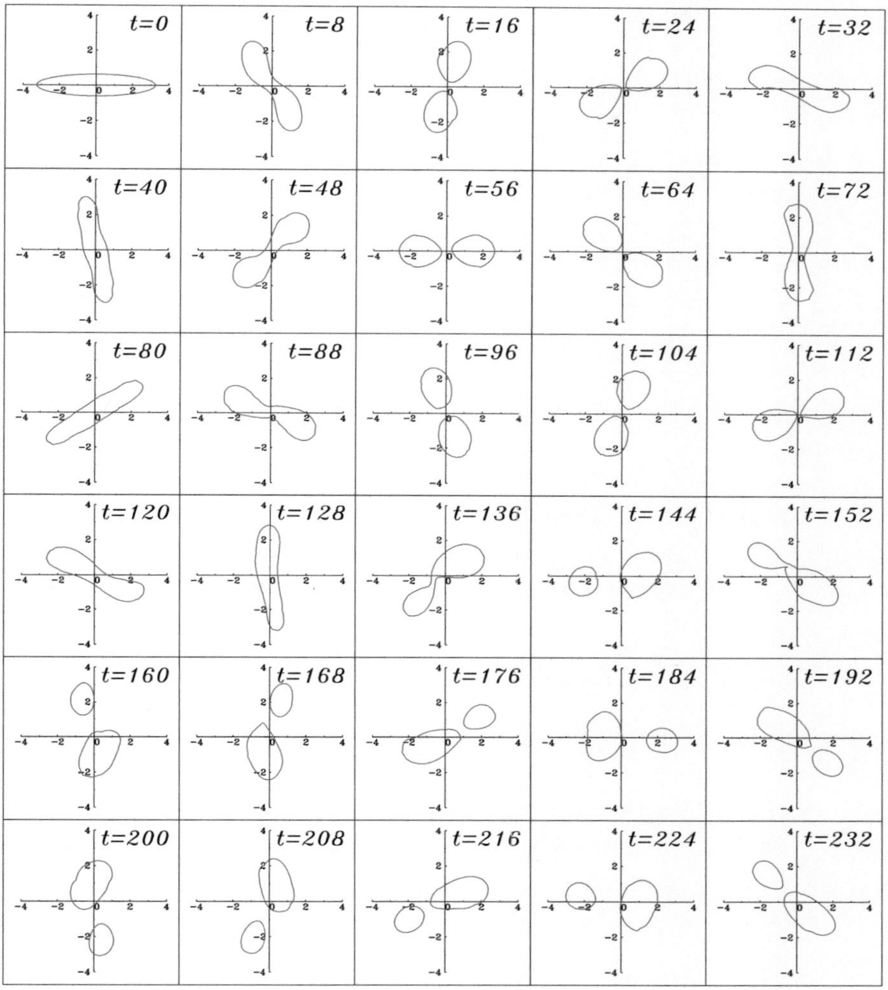

Fig. 3.113 Evolution of an elliptic meddy at $\chi = 5.3$, $F = 0.15$ (splitting into asymmetric parts)

In the latter case, before its final decomposition into a doubly connected figure, comprising two patches with different areas, the vortex patch three times divides into practically equal parts, which repeatedly merge to form a single structure.

The experiment in which the value of parameter χ was increased by as little as 0.1 (Fig. 3.114), shows the already irreversible decomposition into two identical quasi-circular parts. This calculation gives an additional confirmation to the result that is shown in Fig. 3.108 and demonstrates the possibility of two circular vortices anomalously approaching each other at $F = 0.15$ without merging.

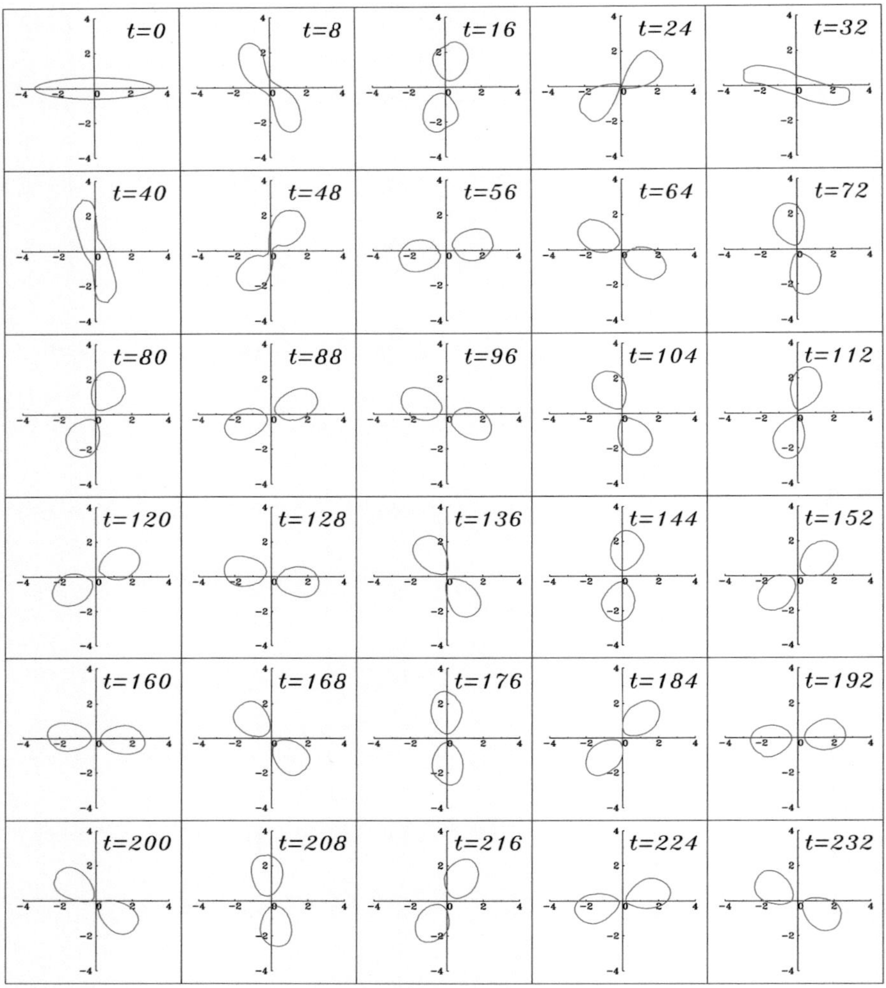

Fig. 3.114 Evolution of an elliptic meddy at $\chi = 5.4$, $F = 0.15$ (splitting into symmetric parts)

The following three figures demonstrate the behavior of a strongly elongated ($\chi = 10$) elliptic vortex at different values of F. Such a strongly elongated configuration can serve, for example, as a model of the Mediterranean salt tongue investigated in [865]. In Fig. 3.115, where $F = 2.5$, i.e., at relatively weak stratification, when the marker in Fig. 3.107b lies within domain $U2S$, at the intermediate stage $t \approx 12 \div 24$, we see a tendency toward the formation of a third mode, but next the central bulge extends to become a thread, and a second symmetrical mode forms as a result. Note that the behavior of barotropic elliptic patches is similar [472].

In Fig. 3.116, where $F = 0.7$ (the corresponding marker in Fig. 3.107b is already very close to the boundary with the domain $U3$), the effect of the third mode is much

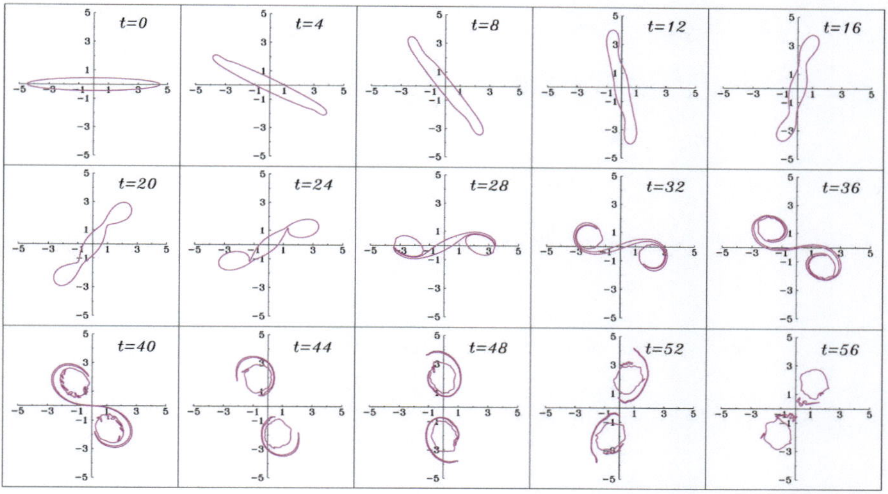

Fig. 3.115 Evolution of an elliptic meddy at $\chi = 10$, $F = 2.5$ (formation of the second mode)

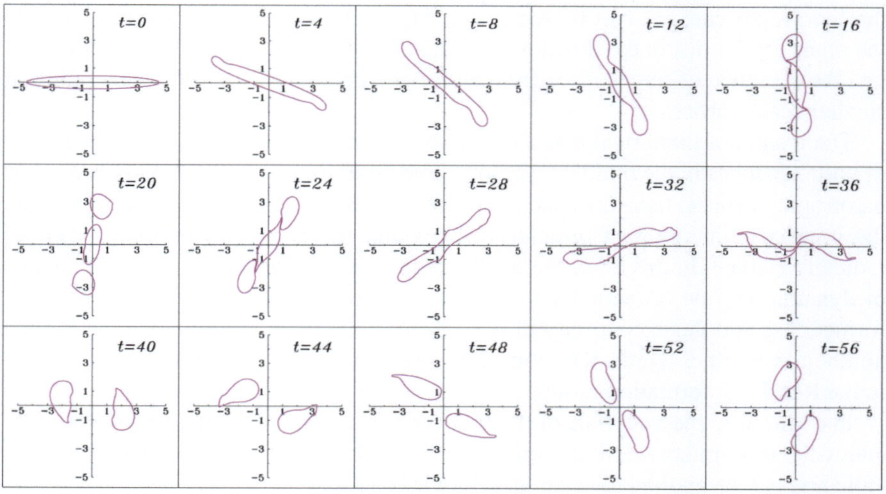

Fig. 3.116 Evolution of an elliptic meddy at $\chi = 10$, $F = 0.7$ (successive formation of the third mode, merging and splitting into two parts)

stronger. Therefore, at $t = 20$, we see *de facto* the decomposition of the elliptical vortex into three parts. Later merging takes place, and, finally, the mode $m = 2$ appears again.

Figure 3.117 shows a scenario of explicit formation of the third mode at $F = 0.2$. The behavior of this tripolar structure within the calculation period is quite stable, but, because of instability, it will decay with time. It is worth mentioning that, in

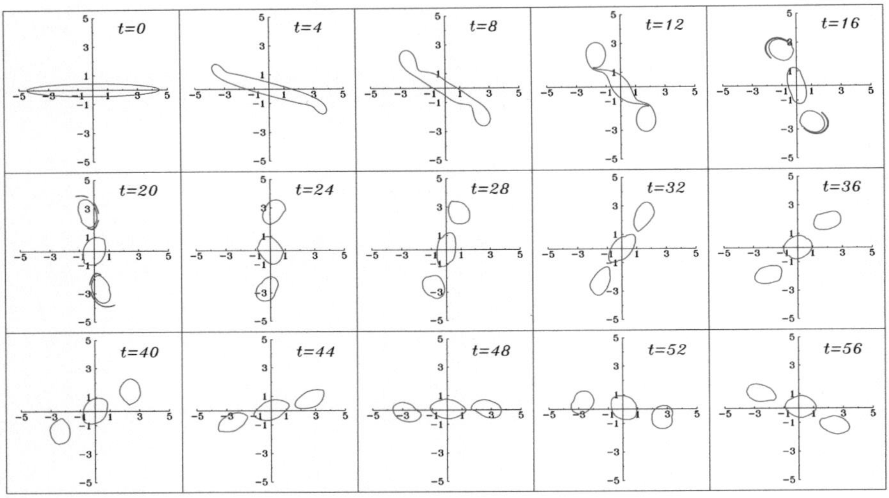

Fig. 3.117 Evolution of an meddy at $\chi = 10$, $F = 0.2$ (formation of the third mode)

the barotropic case, as can be seen from Fig. 3.107b, the corresponding point in the plane (χ, F) lies in the domain of realization of the second unstable mode $U2S$, i.e., the presence of free surface from above and below facilitates the generation of smaller scale vortices.

The results of numerical experiments show that meddies, simulated in the form of vortex patches in the middle layer in a three-layer rotating fluid, largely behave as barotropic vortices, but under the conditions of moderate and strong stratification, the criteria of merging of circular vortices and the behavior of elliptic vortices radically change. In this case, the major parameters, characterizing the bifurcations of dynamic regime, in addition to the relative closeness of vortex patches for the former case and the excentricity of the ellipse in the latter, also include the Froude number or, in other words, the ratio of the characteristic size of the vortex structure to the Rossby deformation radius.

In addition to the materials of the two last paragraphs regarding the evolution of anticyclonic intrathermocline vortices, we note that all results obtained above are valid and can be applied to ocean cyclonic lenses.

This statement is based on many calculations, in particular, on the numerical experiment shown in Fig. 3.118. This figure compares synchronous configurations of vortex patches at eccentric counter interaction between two pairs of lenses for two situations, where the head-on collision involves cyclonic vortices in one case and anticyclonic vortices in the other case. In both those examples, the collision results in a temporary formation of tripolar structures, which next decompose to form two pairs moving apart, resembling the motions of type S_0 (Fig. 3.71), with the only difference that now all vortices lie in the same layer.

The figure shows that in each time moment, all configurations have an obvious symmetry with respect to rotation by 180°; the insignificant deviations from it

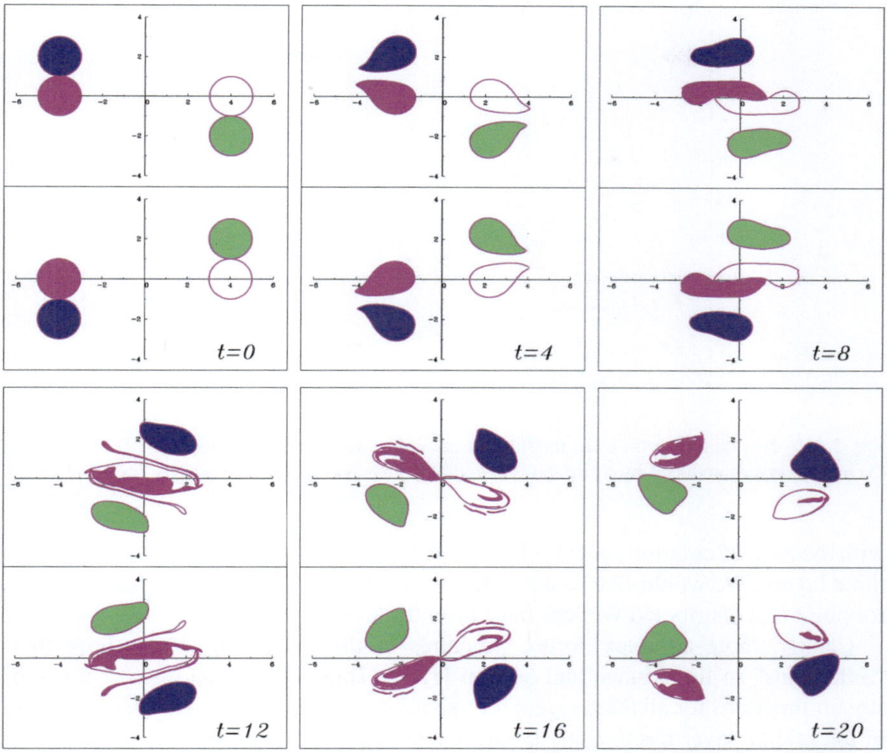

Fig. 3.118 Synchronous configurations of vortex patches at eccentric counter interaction of two lense pairs, when a head-on collision involves cyclonic vortices (in the *bottom part*) and anticyclonic vortices (in the *top part*). In both cases, we have $F = 0.2$. For the identification of vortices, each vortex is given its own color

at later stages of interaction are due to the accumulated computation errors. The cyclone-anticyclone identity mentioned above now takes place for all processes—the self-motion of vortices, their merging, mixing, and subsequent separation.

On Detecting Lenses on Oceanic Surface

Notwithstanding the fact that the intrathermocline vortices are localized in intermediate levels, their effect extends throughout the ocean depth. Indeed, using the formula (1.99) and assigning in it $\Pi_1 = \Pi_3 = 0$, we can construct the profiles of azimuthal velocities, induced in all three layers by a circular vortex patch in the middle layer.

Figure 3.119 shows that, though the *potential* vorticity in the top and bottom layers is zero, the deformations of the interfaces between layers under the effect of the lense of the middle layer induce *a relative* vorticity, in the top and bottom layers,

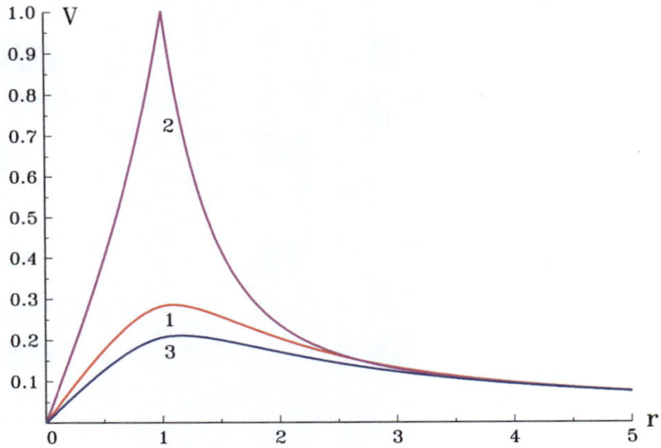

Fig. 3.119 Normalized profiles of modules of azimuthal velocities in (1) the top, (2) middle, and (3) bottom layers, induced by a circular unit-radius vortex with potential vorticity $\Pi_2 = -1$

with the radii of circumferences with maximal velocities practically coinciding in all three layers. We would like to note, that analogous profiles of azimuthal velocities for point and distributed vortices have been discussed in [352].

Clearly, more complex vortex structures of the middle layer also have their "reflections" in the surface and bottom layers. Thus, Fig. 3.120 gives isolines of stream functions for all three layers for two moments, when the vortex configuration of the middle layer has a singly connected *(a)* and doubly connected *(b)* structure. It can be seen that the kinematic field of the middle layer is clearly reproduced at the qualitative level in the top layer, and partially, in the bottom one. Although the calculations were carried out under the "rigid lid" assumption, it is obvious that by nature, anticyclonic vortex structures of the top layer cause deformations of the ocean surface above them in the form of depressions. Thus, it becomes clear that the positions and, partly, the configuration of intrathermocline lenses can be visualized with the help of remote sensing altimetric observations [147, 869]. Certainly, confirmation of these theoretical results requires the simultaneous analysis of satellite observations and direct observation from research ships using CTD–surveys, clusters of moorings, and autonomous floats located in the body of the eddy [147].

Recently, Bashmachnikov and Carton [64] and Emelianov et al. [250] have shown reliably the possibility of lens detection by remote sensing.

On the Effect of Bottom Topography on the Motion of Lenses

Let us briefly consider some aspects of the effect of an isolated circular cylindrical submarine obstacle of unit-radius on the motion of a circular lense of the same size, brought by a relatively weak zonal current with velocity $U = 0.2$. According to estimate (1.107), with such flow velocity, stagnant zones will appear in all three layers already at $\sigma > 0.4$.

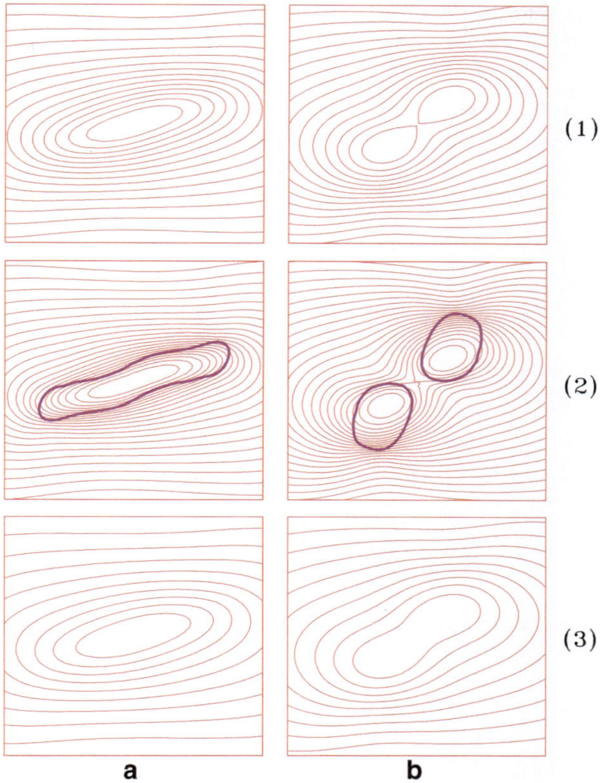

Fig. 3.120 *Thin curves* are isolines of the stream function of horizontal motion in (*1*) the *top*, (*2*) *middle*, and (*3*) *bottom layers* at times (**a**) $t = 24$ and (**b**) $t = 48$ in Fig. 3.109. The *thick lines* are the contours of vortex patches in the *middle layer*

An example of isolines of a steady streamflow function (in the absence of a meddy) in the middle layer at $\sigma = 1.2$ is given in Fig. 3.121. A captured vortex (the inner part of the separatrix loop), surrounded by the external flow, forms above the southern periphery of the obstacle. Clearly, in the presence of a intrathermocline vortex, moving with the flow, an unsteady perturbation will be superimposed in this phase portrait.

In Fig. 3.122, where $\sigma = 0.8$, the originally circular lense flows around the obstacle along its northern part, acquires quasi-elliptic shape, and this quite stable, rotating, and pulsating vortex structure moves downstream.

The behavior of the meddy in Fig. 3.123 at $\sigma = 1.0$ is different. Because of the stronger effect of bottom topography, the vortex patch becomes more elongated and, as the result of instability, it divides into two smaller scale vortices at the downstream periphery of the obstacle. The latter vortices, rotating about a common center, are also carried away by the flow.

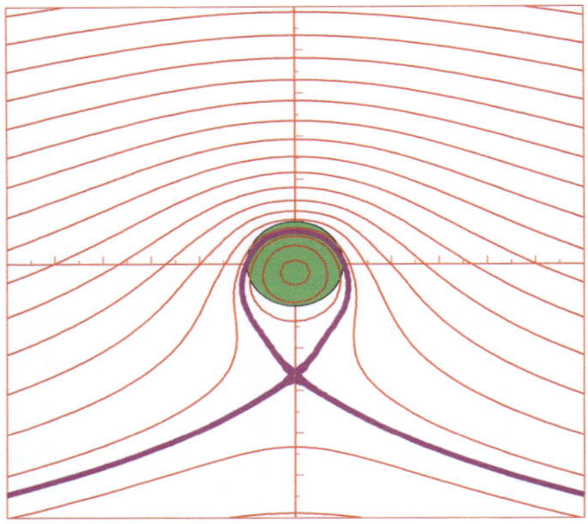

Fig. 3.121 Isolines of stream function for the *middle layer* in a zonal flow over a circular cylindrical obstacle (marked by color) at $U = 0.2$, $F = 0.2$, $\sigma = 1.2$. The *thick line* is the separatrix separating the domains of infinite and finite motions

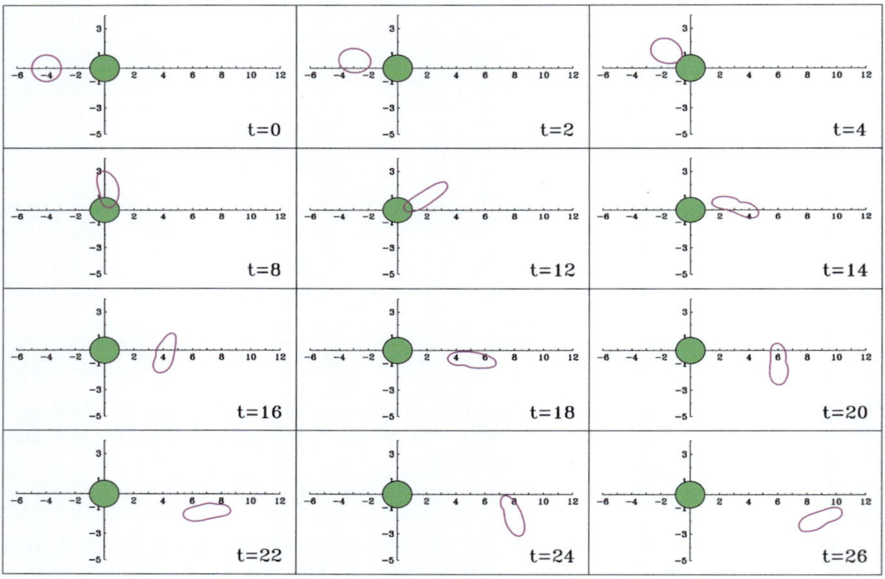

Fig. 3.122 The effect of a cylindrical bottom obstacle on the motion of an anticyclonic meddy in a zonal flow at $U = 0.2$, $F = 0.2$, $\sigma = 0.8$

The further increase in obstacle height (σ= 1.2, see Fig. 3.124) results in that one of the vortices that have originated from the decomposition of the lense is captured by the obstacle, and only the other vortex is carried away by the flow.

Calculations show that a changeover exists between the latter two regimes. This is shown in Fig. 3.125, which gives simultaneously the trajectories of the centers of rotation of vortex patches for two cases with the values of topographic parameter σ, differing only in the fifth place after decimal point. The upper arc-shaped thick continuous line represents practically merged trajectories of the centers of vortex patches (thick magenta and thin yellow lines) before their separation. Next, after

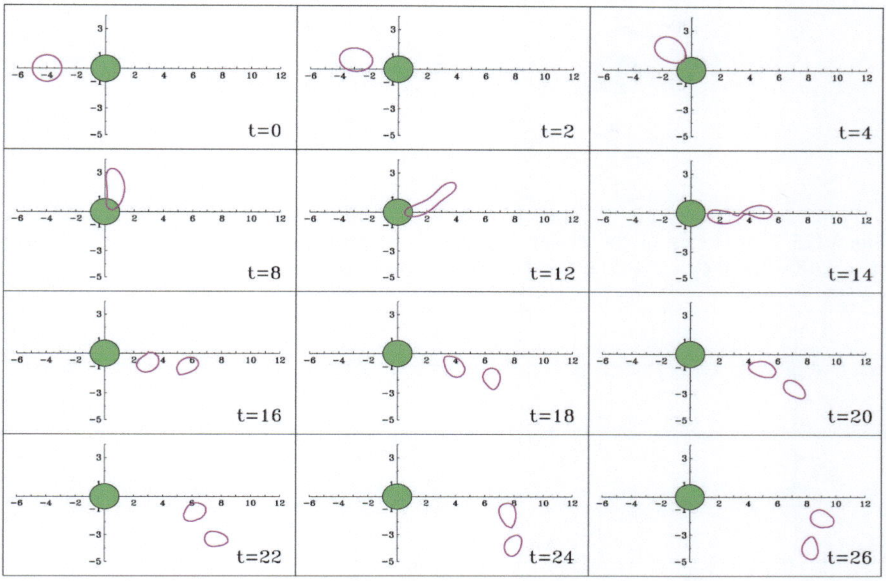

Fig. 3.123 The same as in Fig. 3.122, but at $\sigma = 1.0$

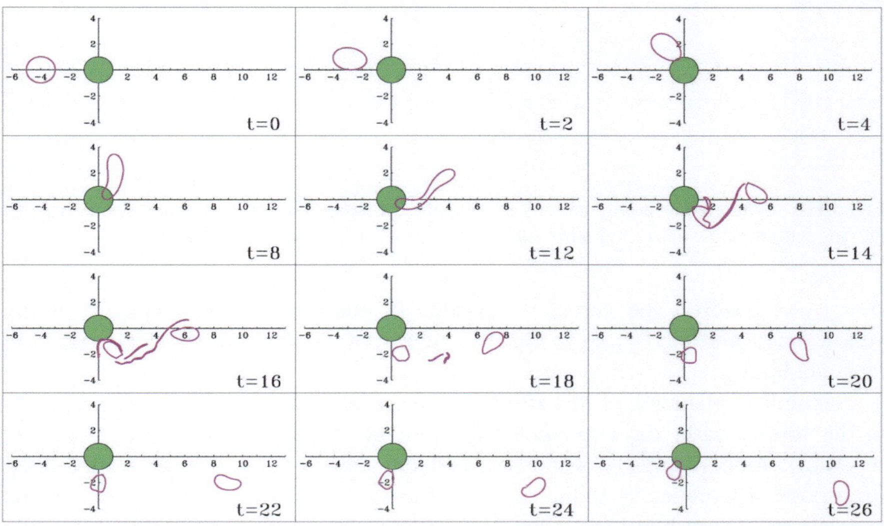

Fig. 3.124 The same as in Fig. 3.122, but at $\sigma = 1.2$

the decomposition, both trajectories divide to form two pairs of merged trajectories along which the centers of the newly formed smaller scale vortices move. One of these pairs (the right one) remains undivided all the time (the vortices in both cases are carried away by the flow), while the second pair soon divides into two—at

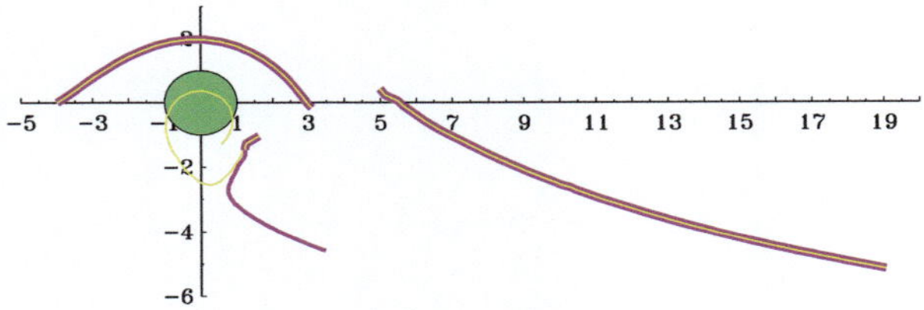

Fig. 3.125 Trajectories of the centers of rotation of vortex patches at $U = 0.2$, $F = 0.2$, $\sigma_1 = 1.18512$, and $\sigma_2 = 1.18511$. See the text for explanations

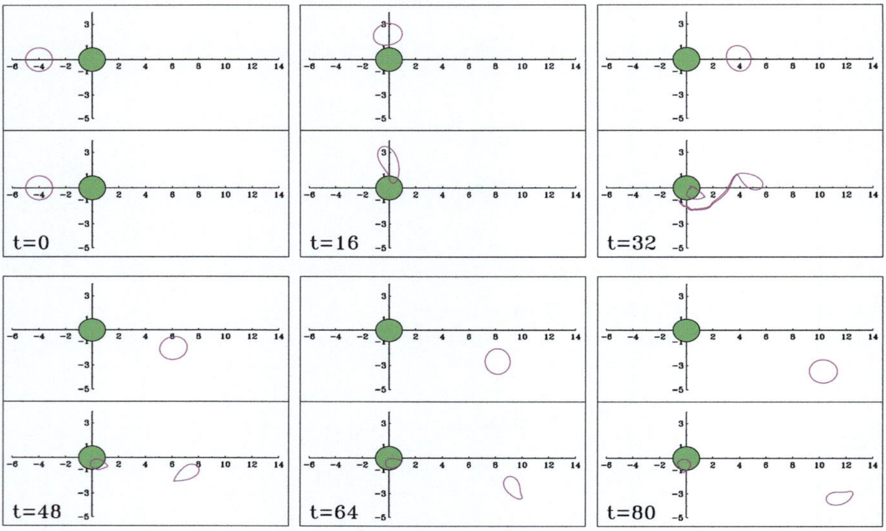

Fig. 3.126 Synchronous configurations of (*top*) the cyclonic and (*bottom*) anticyclonic meddies moving with the flow onto a cylindric submarine obstacle at $F = 0.2$, $U = 0.2$, $\sigma = 1.4$

$\sigma = \sigma_1 = 1.18512$, the vortex is captured by the obstacle (its trajectory is the yellow line), and at $\sigma = \sigma_2 = 1.18511$, it leaves its neighborhood (the semithin magenta line).

A simple explanation of this effect can be derived from the phase portrait 3.121: in the former case, the corresponding point of the phase plane lies within the separatrix loop (capture), while in the latter case, it lies outside of the loop (transport). Theoretically, in the case of a steady flow, the captured vortex remains near the mountain; however, if the flow velocity increases due to some cause, the vortex can be carried away.

It should be noted that the cyclone-anticyclone symmetry mentioned above takes place only for free vortices. In the presence of bottom topography, the cyclonic and anticyclonic lenses behave differently. This is obvious, as an additional anticyclonic vortex always forms above the obstacle, and this vortex interacts with the lenses. Figure 3.126 gives an example of synchronous configurations of cyclonic (top panels) and anticyclonic (bottom panels) vortices, carried by the flow onto an isolated obstacle. The cyclone, practically without changes in its shape, moves about the obstacle, after which it is carried away by the flow, while the anticyclone divides into two parts, one of which is fast captured by the mountain.

The analysis of experimental observations has demonstrated that meddies propagation in the northeastern part of the Atlantic is an important element of hydrological structure of the ocean. Taking into account that the total number can be as large as 100 [261], it becomes obvious that the considered mechanisms of anticyclonic lenses merger can be helpful in explaining the regularities of the appearance of large-volume intrathermocline eddies in different regions of the ocean. In particular, the appearance of such large lenses in the region of the Azores Frontal Zone allows us to explain the existence of long-living meddies at great distances from this region.

Transport of meddies at intermediate depths by individual eddies over large distances during many years can be comparable with the advective transport of heat and salt [45, 752]. Estimates by different researchers indicate that the meddies' contribution to the total flux of the salinity anomaly can be substantial. For example, Richardson et al. [748, 752] estimate this contribution as 25 %, Arhan et al. [42] as more than 50 %, and Mazé et al. [577] as almost 100 %. So, we can speak about the climatic role of the long-living eddy transport to maintain of the Mediterranean salt tongue as prominent hydrographic features of the mid-depth North Atlantic [123, 488, 989].

Dynamics of Medies in the Flow Over Submerged Hills

In this subsection, we use the CDM and discuss the results of modeling the interaction of cyclonic and anticyclonic lenses captured by the mean currents of large-scale cyclonic gyres, during their motion over submerged obstacles. The conditions of the model experiment are close to the real one resembling the four submerged banks of the Horseshoe mountain system on the continental slope of the Iberian Peninsula and mean velocity values of the dominant current field in this area.

For this region, model parameters are the following: the mean ocean depth is $H = 4,000$ m, we set the thicknesses of the upper, intermediate and bottom layers as $H_1 = 600$ m, $H_2 = 1,000$ m and $H_3 = 2,400$ m (thus, the dimensionless thicknesses are $h_1 = 0.15$, $h_2 = 0.25$, $h_3 = 0.6$); the first and the second deformation radii are $Rd_1 = 32$ km, $Rd_2 = 15$ km. From 1.96, we have $\lambda_1 = 0$, $\lambda_{2,3} = -(D/Rd_{1,2})$, and assuming $D = Rd_1$, we obtain $F_1 = 0.14, F_2 = 0.7378$.

Let us consider the meddies' evolution, which are located in the intermediate layer and transported by the barotropic current of the type

$$\Psi = \frac{A}{2}\left[Bx^2 + (1-B)y^2\right], \tag{3.21}$$

and are affected by four cylindrical submarine elevations located only in the bottom layer and model the bottom topography in the area of the Horse Seamounts (see Fig. 3.127).

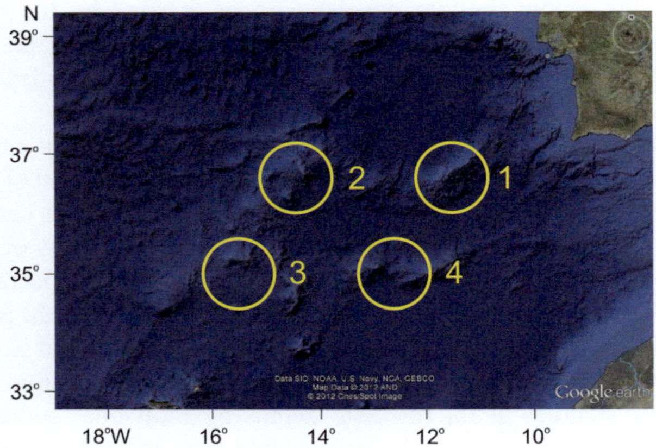

Fig. 3.127 Schematic drawing of four circular cylindrical elevations (their boundaries are shown with a *solid yellow line*) used in this work for modeling the real bottom topography in the vicinity of the Horseshoe Seamounts. The map of the bottom topography is taken from Google Earth

At $B = 0$ or $B = 1$ the formula (3.21) sets zonal or meridional shear currents, correspondingly; at $0 < B < 1$ we have elliptical, and at $B > 1$ hyperbolic singular points in the origin of coordinates $(0;0)$; the parameter A defines the velocity scale: $A = 1$ corresponds to the velocity of 10 cm/s.

Of course, such model of bottom topography is a very rough representation of the real one, but it allows us to perform the general study of capture and partial destruction processes of the intrathermocline vortices. It is shown in the work by Kozlov [456] that when analyzing the influence of the bottom topography on currents in the ocean, the volume but not the shape of the obstacle plays the main role. We shall identify the dimensionless heights of the elevations by σ_i, where $i = 1,2,3,4$ correspond to the elevation denominations in the figure. In the coordinate system selected in such way, the centers of the circles, that match the boundaries of cylindrical mounts, have the following dimensionless values:

$$(x_1;y_1) = (-4;6), (x_2;y_2) = (-13.2;6), (x_3;y_3) = (-17.2;), (x_4;y_4) = (-8;0).$$

Value εH, where $\varepsilon = 0.1$ is the vertical scale of bottom topography. Thus, $\sigma = 1$ corresponds to the height of 400 m. It should be noted, that similar three-layer model was used by Wang and Dewar [968] for studying lens dynamics in the vicinity of islands, which were modeled by cylindrical columns, located in all three layers.

As before, the horizontal scale is taken as Rd_1, i.e. 32 km; so the circles in the figure with radii $R_i = 2$ represent the boundaries of circular mountains with diameters equal to 128 km. The time scale is selected so that one-dimensionless unit of time corresponds to half of the period of fluid particle rotation; the particle is located on the vortex contour of a unit radius. A particle having the velocity of 10 cm/s, the time scale is equal approximately to 10 days.

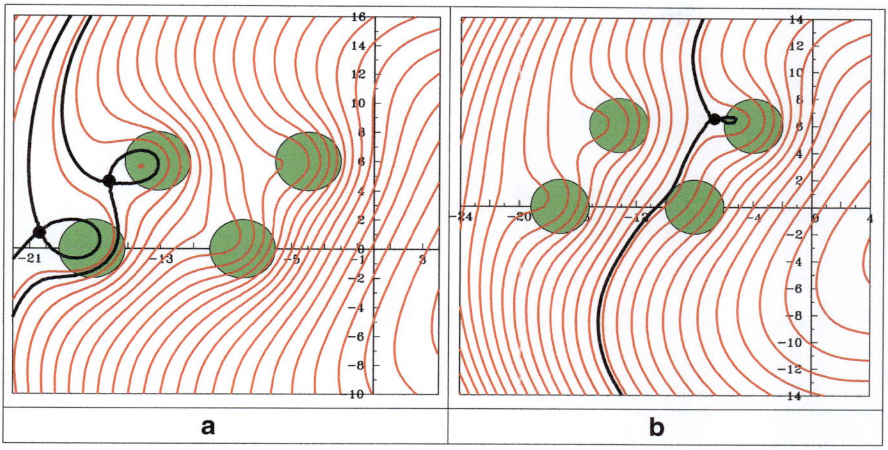

Fig. 3.128 Isolines of stream functions in the intermediate layer: (**a**) $\sigma_1 = \sigma_2 = \sigma_3 = \sigma_4 = 2$, (**b**) $\sigma_1 = 2.5, \sigma_2 = \sigma_3 = 2, \sigma_4 = 1, A = 0.2, B = 0.5$

In Fig. 3.128, the typical pattern of isolines of the stream function in the intermediate layer is shown with the given above external parameters when the background barotropic current represents a large-scale cyclonic vortex with the center in the coordinate origin. In Fig. 3.128a, where the current of 2 cm/s corresponds to the characteristic scale (as $A = 0.2$) and all the elevations have the same heights of 800 m, the flow deviates weekly over the seamounts 1 and 4, and closed circulations appear over the seamounts 2 and 3. In the situation when the mount 1 has the larger height, namely, 1,000 m, and the mount 4, inversely, the smaller, i.e. 400 m (what is closer to real conditions), a small area with a local vortex over the periphery of the mount 1 was obtained only after slightly increasing the current velocity (Fig. 3.128b).

Below, we give the results of numerical experiments for studying movement of the meddies placed into the external field shaped as a cyclonic flow, crossing a topographic ridge. We discuss different scenarios of the vortex patch evolution in the intermediate layer. Calculations have been made with a three-layer version of CDM [844].

In the first series of calculations we examine the behavior of an originally circular lens twice smaller in horizontal size than the cross section of each cylindrical elevation. Initial position of the lens center is defined by coordinates (x_0, y_0). The external field is shown in Fig. 3.128a.

In Fig. 3.129, at $t = 0$ the center of the anticyclonic lens is located in the same meridian as the center of the mount 1. As the picture shows, the vortex patch rounds the submerged ridge along its Eastern periphery, lingers on in the vicinity of the elevations 1 and 4, warps significantly, forming a small quasi-circular core and a long vortex line (*a streamer*, according to the terminology of Bashmachnikov et al. [65]). Then the vortex structure is brought by the flow in the south direction.

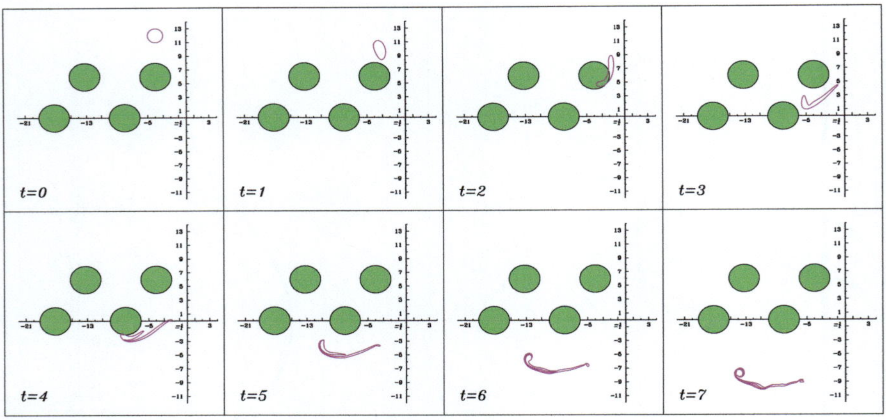

Fig. 3.129 Successive positions of the lens contour (*magenta lines*) in the given moments of dimensionless time at $(x_0; y_0) = (-4; 12)$. Filled *green circles* represent elevations in *top* projection

In Fig. 3.130 the original lens position is much closer to the mount 2. Brought by the flow, it passes between the elevations of the eastern and western seamounts, losses its compact shape and decays practically totally. It should be noted that in the time interval $[4 \div 6]$, it is observed some tendency to capturing a part of the vortex by the elevation 4, but later, the residual part of the lens overcomes all the area of the bottom disturbance, and only independent streamers move downstream.

Figure 3.131 demonstrates the evolution of an intrathermocline anticyclonic vortex for the case when the zonal coordinate of its center position in the initial time moment is outside the area of the bottom elevation, but the vortex motion is affected by topography. Moreover, a part of the meddy is strongly captured by the mount 3 and remains in the mount vicinity. Though the meddy presence in the intermediate layer changes significantly the topological properties of the stream-function field (Fig. 3.129a), the capture of the vorticity may be explained by the fact, that the significant part of the lens found itself in the internal part of the separatrix loop at the periphery of the mount 3.

Till now in all experiments the lens represented an anticyclonic intrathermocline vortex. In Fig. 3.132, where all the external parameters are the same as in the previous case, the vortex rotates in the cyclonic direction. Here, as before, the main impact is produced by the mount 3, but the capture is weaker. As a result, the vortex patch divides; one part of the vortex remains over the elevation, and the second one, connected with it by the streamer, is brought downstream.

The second series of the experiments was carried out taking into account the external field given in Fig. 3.128b. Figure 3.133 gives a set of the vortex-patch contours; this patch was initially a circle of the same radius as the cylindrical elevations. The center of the vortex was located above the center of seamount number 1. It is obvious, that the mount 2 strongly affects the lens; the effect of the mount 4 is smaller. However, after the lens passage over the ridge only a small part

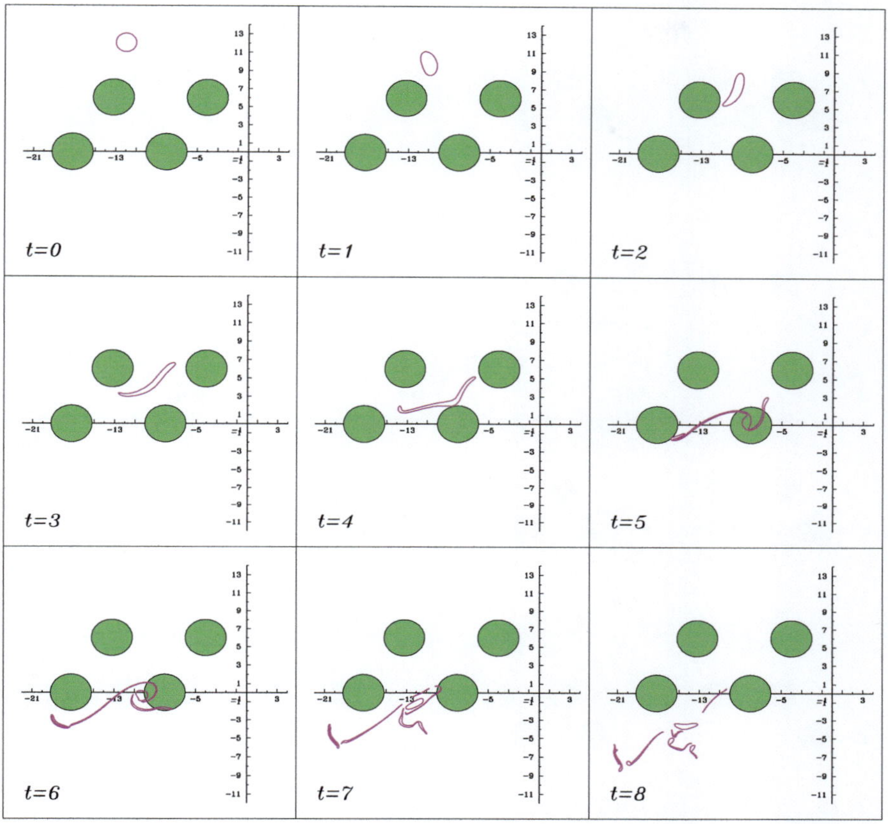

Fig. 3.130 The same as in Fig. 3.129 but at $(x_0; y_0) = (-12; 12)$

of the vorticity in filaments form remains over it. The main part of the patch is brought by the flow outside the mountain system.

In the last example (Fig. 3.134) we examine the problem of the vortex pair having non-zero total momentum and being influenced additionally by the same barotropic flow. The pair consists of an anticyclone similar to that one from the previous experiment, and a cyclone of twice smaller size. However, their potential vorticities are compensated in such a way that, being not affected by the external current and bottom topography, they would move rectilinearly. Calculations show, that the pair passes easily in the vicinity of the seamount, and affected by the elevation, the anticyclonic vortex patch divides into two parts, one of them winding up around the cyclonic vortex, and the second one being connected with the first part by a streamer. In whole, the vortex construction has a quasi-tripolar structure. The results of this experiment suggest that a similar structure, composed of two meddies and of a cyclone, and observed in the region [145], could originate from the interaction of a vortex pair with bottom topography.

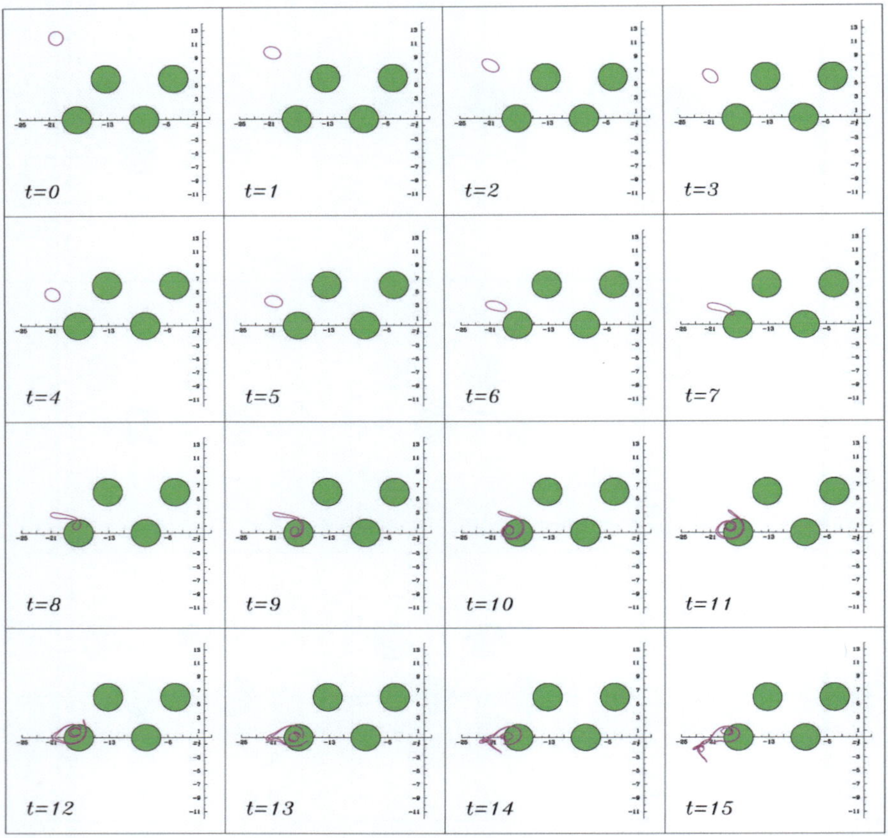

Fig. 3.131 The same as in Fig. 3.129 but at $(x_0; y_0) = (-20; 12)$

So, in the framework of a three-layer quasi-geostrophic model and using the simple bottom topography and external background current, we carried out numerical modeling of the vortex-patches' evolution in the intermediate layer that we consider as meddies. It was shown that depending on the initial lens position, the meddies, being influenced by the submerged elevation, may deform significantly up to the total decay. In the real ocean, evolution and destruction of lenses provide additional mechanisms of water mixing at intermediate depths. When lenses drift in the area with complex bottom topography, a set of lens-like vortices of different scales and narrow elongated jets (streamers, spikes) can appear. This may cause a complex structure of temperature and salinity distribution at hydrological sections.

Complete and detailed observations of the meddy interaction with Seine seamount, located south of the Horseshoe Seamounts were carried out in the OASIS experiment (**O**ce **A**nic **S**eamounts: an **I**ntegrated **S**tudy) in 2003–2004 [65]. A close

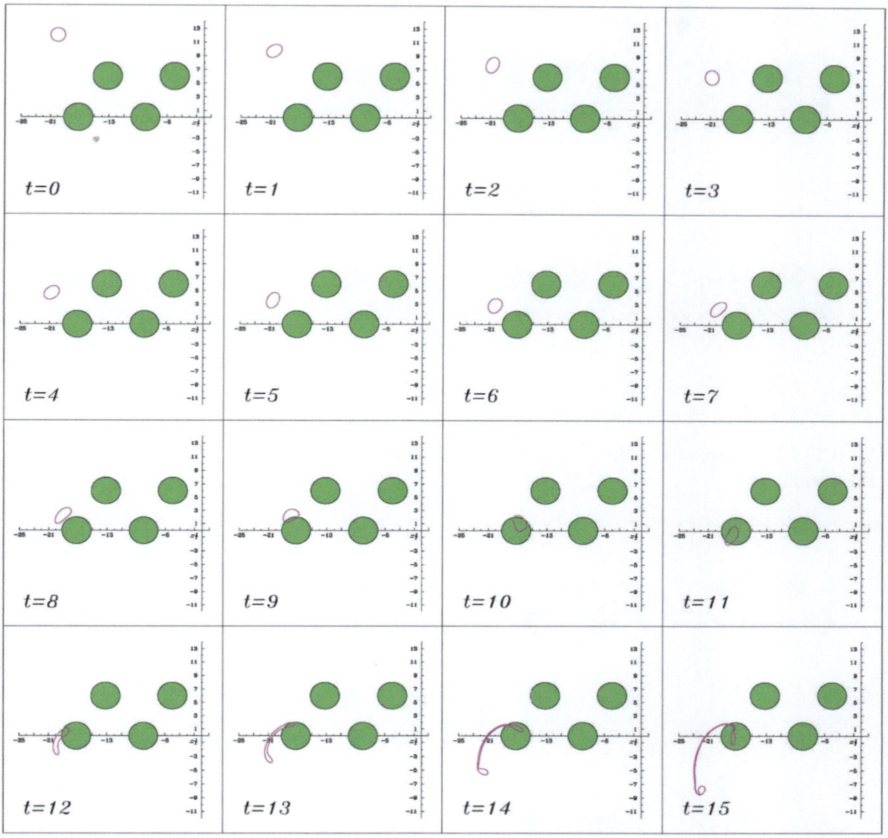

Fig. 3.132 The same as in Fig. 3.131, but the vortex being cyclonic

contact of the anticyclonic vortex with Seine mount took place in December 2003. This meddy with unique salinity core at depths of 1,100–1,200 m drifted near this seamount. Then it was captured by a jet of the Azores Current and transported out of the study area. Filament-like elongated structures with anomalously high salinity (streamers) were found, in the area adjacent to the mount. Admittedly, these streamers (or independent meddy's fragments) were captured by the seamount and formed an anticyclonic vortex with a rotation period of about 2 months. After a few rotations around the seamount, this vortex was expelled to the east by the Azores Current after June 2004.

Due to a long and close contact with seamount, the meddy left it loosing a part of its core. Streamers formation from the waters of the meddy core and their consequent destruction seem to provide a mechanism for MW mixing at intermediate depth. At the same time, the observations show that the possibility of new streamers' merging with the main core is excluded. This means that meddies,

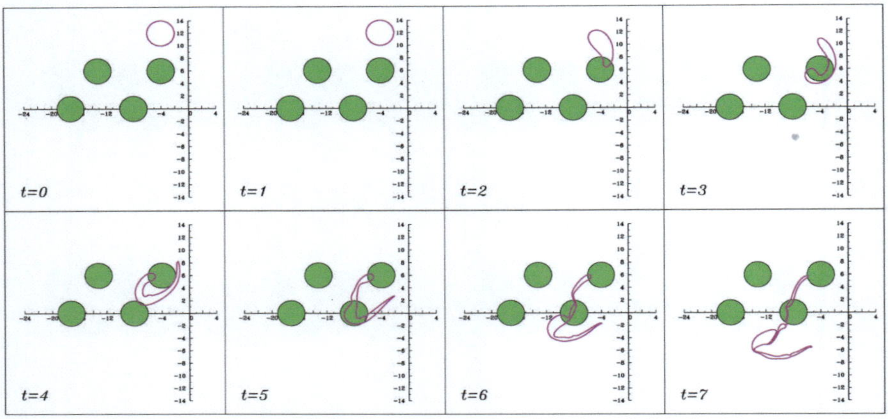

Fig. 3.133 The same as in Fig. 3.129, for the external field of the type 2b and for the lens with radius equal to 2

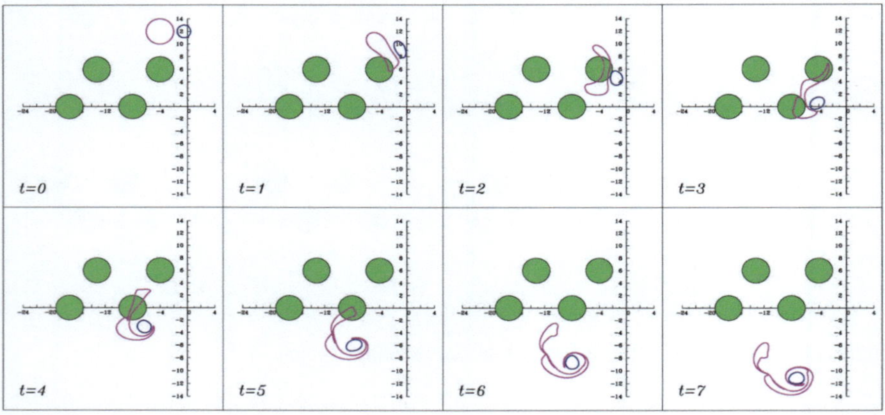

Fig. 3.134 The same as in Fig. 3.133, but in the presence of the cyclonic vortex (its contour is shown by the *blue line*)

drifting through the system of submerged mounts, lose a part of their core water volume. A qualitative confirmation of this is based on a volumic distribution of meddies in the Eastern North Atlantic (Fig. 3.106). This map shows that the meddies with the largest volumes drift to the southwest along the Moroccan coast and to the west of the Horseshoe Seamount system. Only meddies of small volumes were observed to the south of this mount system. It was shown that available observations satisfactorily confirm the results of our model experiments. Such calculations allow us to forecast the main scenarios of the evolution of the intrathermocline vortices, which cannot be observed so far in the conditions of complex bottom topography, using a simple hydrodynamic model.

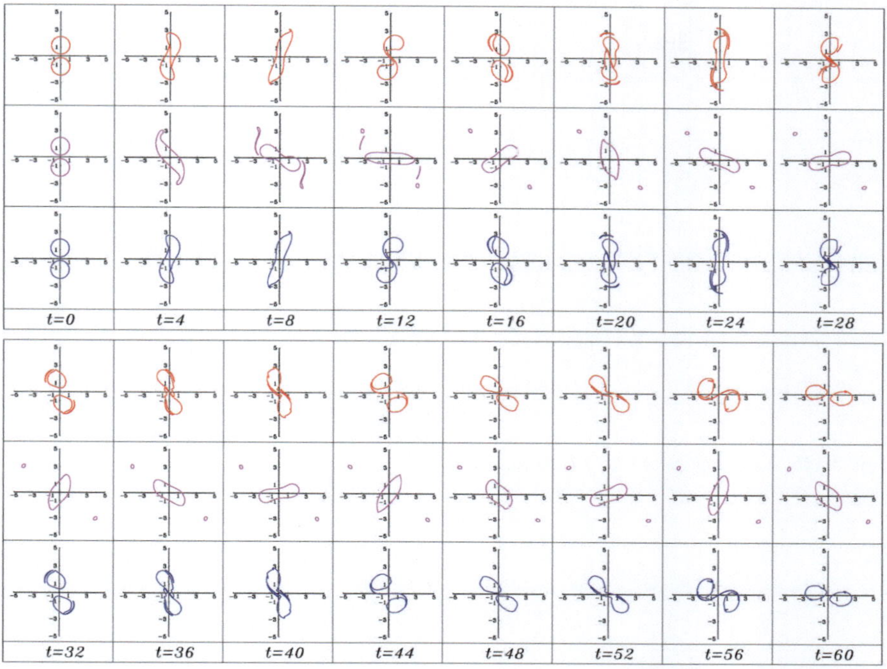

Fig. 3.135 Interaction between two three-layer vortices with zero total vorticity ($\Pi_1^1 = \Pi_1^2 = \Pi_3^1 = \Pi_3^2 = -0.5$, $\Pi_2^1 = \Pi_2^2 = 1$) at $h_1 = h_2 = h_3 = 1/3$, $F = 0.1$ ($\gamma_1 = 0.548$, $\gamma_2 = 0.949$), $d = 1.2$

3.5.3 Examples of Interaction Between Three-Layer Vortices

The possible scenarios of interaction between three-layer vortices can be studied, for example, with the use of a series of numerical CDM-calculations and the subsequent construction of the state diagram, analogous to those shown in Figs. 3.43, 3.58, 3.66, 3.72, 3.78. Here, we will give only several examples, which have no complete analogies in two-layer model and demonstrate the diversity of the types of interaction.

We will start with the simplest case of two identical three-layer hetons with originally vertical axes, located on the y-axis and spaced $2d$ apart. In the case of stronger stratification, when $F_1 = F_2 \equiv F$, each heton in itself is stable. Obviously, if they are close enough (and calculations at $1 \leq d < 1.2$ confirm that), the vortices of the same sign will merge in all three layers and, after thin vortex threads are filtered out, a single heton with a vertical axis will form.

However, at $d = 1.2$ (Fig. 3.135), the situation becomes more complicated: the vortices in the middle layer merge completely, while in the top and the bottom layers, periodic merging–separation takes place with three such cycles observed during the calculation period.

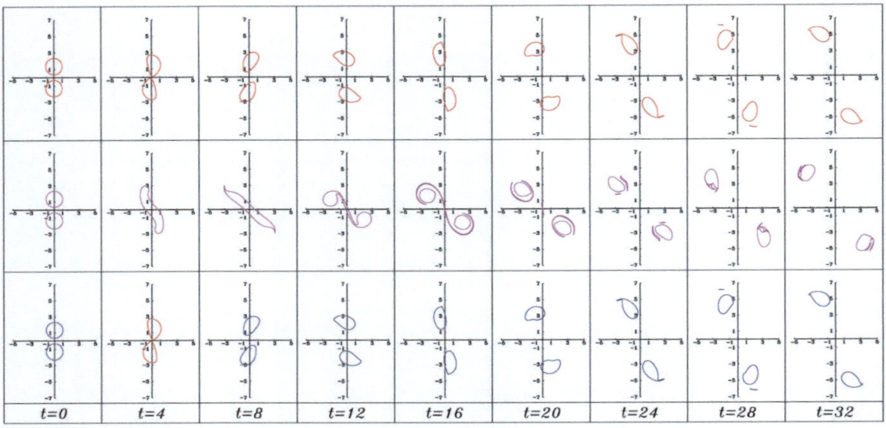

Fig. 3.136 The same as in Fig. 3.135, but at $d = 1.34$

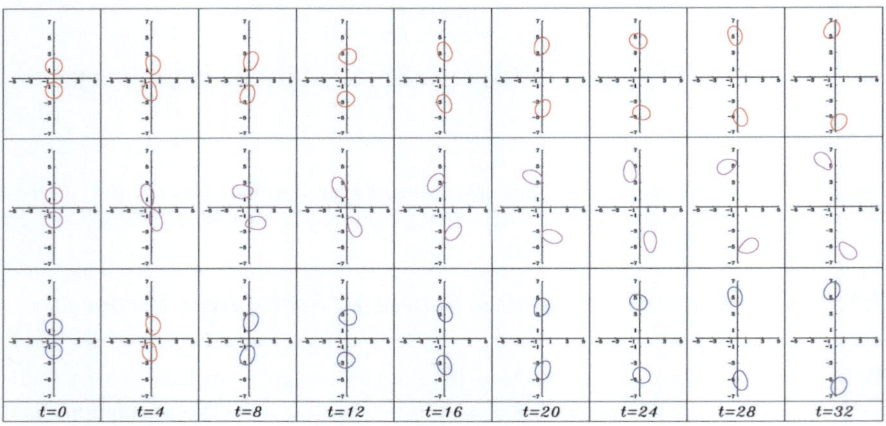

Fig. 3.137 The same as in Fig. 3.135, but at $d = 1.48$

Thus, the original configuration of two hetons periodically takes the form of a single three-layer heton with a vertical axis or the form of a structure consisting of a quasi-elliptic cyclonic vortex in the middle layer, accompanied by two vortices in the top layer and two vortices in the bottom layer. This type of motion corresponds to a very small neighborhood of the value $d = 1.2$.

With the further increase in parameter d, the motion acquires infinite character. However, in the interval $1.21 \leq d < 1.47$, the process has an intermediate stage, in which short-time merging takes place either in all three layers or only in the middle layer (it is this situation that is given in Fig. 3.136), after which the merged vortices separate again to form two moving apart hetons with arc-shape axes.

At $d \geq 1.48$, as can be seen from Fig. 3.137, the moving apart begins immediately after the bending of the axes of both hetons because of within-layer interaction.

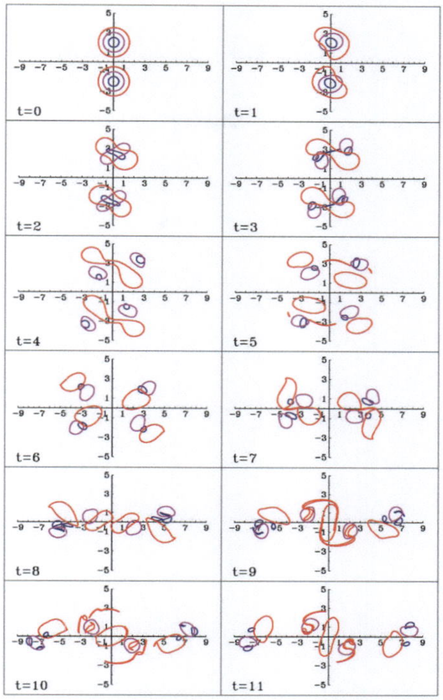

Fig. 3.138 Interaction between two three-layer *cone-shaped* vortices with zero total vorticity at $\rho_1 = 1.5, \rho_2 = 1, \rho_3 = 0.5$, $W_1^1 = W_1^2 = W_3^1 = W_3^2 = -0.5$, $W_2^1 = W_2^2 = 1$, $h_1 = 0.1, h_2 = 0.2, h_3 = 0.7$, $F_1 = 0.2, F_2 = 0.8$ ($\gamma_1 = 1.343, \gamma_2 = 2.5180$), $d = 2$. The configurations of vortex contours of all three layers are aligned (*top view*)

The case of weaker stratification, when the hetons involved in the interaction can be unstable, will be considered here in the particular case of more complex baroclinic structures—cone-shaped vortices, narrowing from top to bottom ($\rho_1 = 1.5$, $\rho_2 = 1$, $\rho_3 = 0.5$)—under the conditions $F_2/F_1 = 4$ and $h_1 = 0.1$, $h_2 = 0.2$, $h_3 = 0.7$ (see Sect. 3.5.2). Under the assumptions introduced above, each heton has a most unstable mode with $m = 2$.

Figure 3.138, where $F_1 = 0.2$, $F_2 = 0.8$, $d = 2$ and the conditions of zero total intensity

$$W = \sum_{j=1}^{3} \sum_{\alpha=1}^{2} W_j^\alpha = \sum_{j=1}^{3} h_j \sum_{\alpha=1}^{2} \Pi_j^\alpha (\rho_j^\alpha)^2,$$

are satisfied, demonstrates the process of interaction of six such vortex patches. Almost immediately, vortex contours start deforming significantly, and by moment $t = 5$, each vortex patch is completely divided into two parts. The complex interaction, including repeated merging in the top layer, results in that an actually tripolar two-layer structure, comprising a core in the top layer and satellites in the middle layer, forms in the center, and two three-layer complexes, each containing a vortex from the top and bottom layer and two smaller scale vortices, form in the bottom layer.

Chapter 4
The Concluding Chapter

Abstract This chapter summarizes the main results obtained in the book, defines possible generalizations and evokes perspectives for future investigation as well as related problems.

4.1 Concluding Remarks

In this book, we mainly spoke about the results obtained in our joint works (of course, with the participation of many co-authors as mentioned in the Introduction). Following are what we consider as some of the most salient points:

- We applied the point vortex theory in two- and three-layer quasi-geostrophic models to examine the possible relative and absolute vortex motions, especially the motion of heton type vortices.
- For the investigation of the simplest vortex structures' behavior we used trilinear coordinates that allowed us to make a quite complete classification of relative vortex motions. In particular, we obtained a series of new stationary solutions for three and four vortices in a two-layer rotating fluid. Tables 4.1 and 4.2 synthesize the main stationary states of heton structures discussed in Chaps. 2 and 3.
- The notion of "choreography" was a useful tool for studying closed stationary vortex trajectories. We first introduced the concept of "generalized absolute choreographies", which represent matching (in the horizontal) of closed trajectories of vortices belonging to different layers.
- Regular and chaotic advection of fluid particles in the vicinity of quasi-stationary symmetric (with affecting background barotropic flow) and asymmetric states of three vortices in a two-layer fluid have been examined. We have shown that chaotic domains, non-linear resonances, transport corridors and other effects are possible depending on external parameters.
- We examined the stability of two-layer axisymmetric distributed vortex structures with respect to small perturbations of their shapes. And we studied the effect of

M.A. Sokolovskiy and J. Verron, *Dynamics of Vortex Structures in a Stratified Rotating Fluid*, Atmospheric and Oceanographic Sciences Library 47,
DOI 10.1007/978-3-319-00789-2_4,

Table 4.1 Main stationary states of vortex structures in a two-layer fluid

Vortex structure type	Formula and explanation or comment	Figure
Two-vortex discrete structures:		
Two-tier top	(2.4)—angular velocity relative to the centre with coordinates (2.6)	
Heton with tilted axis	(2.8)—translational velocity	2.1
Four-vortex discrete structures:		
Regular splitting		2.13, 2.17a, 2.20, 2.22b, d 2.23h, i 2.26a, b
O-shaped vortex structure	Generalized absolute choreography	2.26c
Zigzag-shaped translating heton quartet	(2.26)—dispersion equation (2.25)—translational velocity	2.23e, 2.24
Zigzag-shaped rotating heton quartet	Generalized absolute choreography (2.29)—dispersion equation (2.30)—angular velocity	2.28, 2.34a
Unstable rhomb-shaped structure	(2.21)–(2.23)—dispersion equations (2.22)—angular velocity relative	2.9
Tripole-shaped vortex structure (double capture)	Absolute choreographies	2.10b, 2.11, 2.12a
A-heton discrete structures:		
Regular splitting	A=3	2.2a
Regular splitting	A=6	2.2b, 2.3
Three-vortex discrete structures:		
Ordinary roundabout	(2.39)—angular velocity relative to the central vortex ($R > R_0$)	2.38b (left)
Static roundabout	the same formula ($R = R_0$)	2.38b (center)
Inverse roundabout	the same formula ($R < R_0$)	2.38b (right)
Symmetric eccentric roundabout	(2.41)—angular velocity relative to the centre with coordinates (2.40)	2.39b
	Generalized absolute choreography	2.39b (center)
Eccentric roundabout	(2.73)—dispersion equation (2.77)—angular velocity relative to the centre with coordinates (2.76)	2.65, 2.72a, 2.73a
Triton	(2.78)—dispersion equation (2.79)—translational velocity	2.74a
Unstable triangular configuration	(2.81)—dispersion equation (2.84)—angular velocity relative to the centre with coordinates (2.77) (bottom line) (2.86)—translational velocity ($\mu = -2$)	2.66–2.68

(continued)

Table 4.1 (continued)

Vortex structure type	Formula and explanation or comment	Figure
Finite-core vortex with	General scheme of heton	1.1a
a vertical axis	Possible scheme of heton formation	1.2
Stable noncompensated vortex		3.11b
Stable quasi-elliptic heton		3.32a
Finite-core heton with	General scheme	1.1b
a tilted axis	Possible scheme of heton formation	1.3
Double symmetric heton		3.25
Zigzag-shaped rotating quartet	The approximate numerical solution	3.76

Table 4.2 Main stationary states of vortex structures in a three-layer fluid

Vortex structure type	Formula and explanation	Figure
Three-layer point hetons:	Translational velocity	
1-modal heton with tilted axis,	(2.128)	1.5a
2-modal heton with an arc-like axis	(2.129)	1.5b
Rotating point vortex structures:	Angular velocity	
Three-tier roundabout	(2.130)	
Tripole in the upper or lower layer	(2.131)	
Tripole in the middle layer	(2.132)	
Roundabout comprising a central vortex of the upper (lower) layer and satellite vortices in the middle layer	(2.133)	
Roundabout comprising a central vortex in the middle layer and satellite vortex in the upper (lower) layer	(2.134)	
Roundabout comprising a central vortex in the upper (lower) layer and satellite vortices in the lower (upper) layer	(2.135)	
Top comprising vortices in the upper and lower layers	(2.136)	
Top comprising two upper or two lower layers	(2.138)	
Nonmodal point heton in the two upper or two lower layers	Translational velocity (2.137)	
Three-layer finite-core 2-modal heton (S-vortex)		3.99

the vortex geometry and the fluid stratification on their stability. We have pointed out the possibility of appearance of the so called "cascade" instability when the stratification is relatively weak.

- The Method of Contour Dynamics allowed us to look into the peculiarities of non-linear stage of unstable vortex instability. In particular, we determined an

asymptotic “ballistic”; law for propagation of the temperature anomaly patch boundary. We also explained the mechanism for the formation of mushroom-shaped structures and of a system of peripheral small-scale vortices—the ringlets. We stated a qualitative analogy between the finite-core heton and the symmetrical structure of discrete hetons.
- We numerically studied the stability of two-layer vortices affected by finite perturbations (the hetons with tilted axes and the quasi-elliptical ones). The classification of regimes, at which unstable vortex structures evolve, has been performed.
- A two-parameter family of translating and static doubly symmetric finite-core heton equilibria was presented. We examined the stability and bifurcations of the solutions found, as well as the properties of the hetons in the limiting cases.[1]
- The mechanisms of the two-heton interaction for different initial position of vortex patches have been investigated. We charted their possible resulting states in dependence on stratification and kinematic parameters. Domains of existence for new stationary states, both stable and unstable, finite and infinite ones, have been found. In particular, we gave explanation of the mechanism of formation for the tripolar baroclinic vortex Λ-, L-, and Z-shaped structures.
- We obtained criteria of merging and destruction of baroclinic vortices, including their possible re-merging with the subsequent collapse, and we proposed a mechanism for mixing process of water masses in the ocean.
- During our simulations, we obtained interesting results on the behavior of intrathermocline vortices (lenses, meddies), that are assumed to be vortex patches in the intermediate layer of the three-layer ocean model, such as: (1) merging criteria for two initially circular vortex patches, (2) the stability of elliptical lenses, (3) the possibility of detecting lenses by remote sensing of the ocean surface was pointed out, and (4) specific effect of bottom topography on the lens motion, carried by an external barotropic flow.

4.2 Outlook to Heton Problems

The review of publications and the original results given in this book show that, within a little less than 40 years of its appearance, the theory of localized baroclinic vortices has evolved considerably and contributed to many, sometimes surprising, results. At the same time, this theory has found fields for application in both meteorology and oceanology.

The concept of hetons—baroclinic vortices with zero total intensity—has proved to be useful and has enabled the identification of a range of important problems that can be treated completely and coherently within the context of the approximation of

[1] In the near future, we hope to carry out similar studies for few other stationary structures of the heton type: those ones having both translation and rotation motions, and tripolar and quadropolar ones.

quasi-geostrophic discrete and finite-core vortices. The two-layer geostrophic heton model has shown itself to be a basic minimal model that takes into account the baroclinic nature of vortices in a stratified uniformly rotating fluid in its simplest form.

The heton theory adequately describes the essentially nonlinear stages in the development of purely baroclinic flows (with a zero barotropic component), which are naturally generated by thermal sources under real geophysical conditions and in laboratory experiments. The model adequately reproduces:

- the vertical and horizontal structure of geophysical fields (temperature and velocity);
- the typical scales of both individual stable vortex structures and vortex clusters (vortex ensembles);
- decay mechanisms of large vortex formations;
- the types of possible circulation regimes and the principles of their classification;
- asymptotic non-diffusion (ballistic); heat transfer laws.

Promising directions for research into the generalization of the heton theory involve studying the interaction between vortices and large-scale flows [156, 226–228, 324, 328, 527, 978], bottom topography [156, 157, 217, 218, 222, 266, 345, 502, 503, 813, 860], shore relief [217], and their interaction in closed basins, as well as problems of solute transport [227, 323]. To apply to studying the effects of differential rotation (even in the case of a homogeneous gradient—beta-plane approximation), however, would require some essential modifications being made to it, mostly because of the need to allow for the radiation of Rossby waves [320, 497, 736]. In the cases where Rossby wave radiation is suppressed (eastward heton spreading), some exact solutions can be obtained for individual hetons [273, 317, 324, 579] and von Kármán heton streets [96, 321, 322].

The effects of equivalent two-dimensional fluid compressibility on heton dynamics at scales $D \geq \lambda$ can be taken into account by rejecting the rigid-lid approximation on the top boundary, which was used in deriving Eqs. (1.11) and (1.12), and replacing it by the free surface condition [315]. In this limit, the barotropic mode is screened and the logarithmic Green function in (1.31) must be changed to $(-1/2\pi)\mathrm{K}_0(\gamma^* r)$.

4.3 Discussion

Naturally, in this book, we only touched upon many of the problems of the vortex-structure dynamics in a rotating fluid, or they remained out of our attention. Here, are some of them.

4.3.1 On the Role of Baroclinic Vortices in the Formation of Thermohaline Structure of the Ocean

As mentioned above, the motion of a baroclinic vortex, whose most known model is heton, is always accompanied by transfer of available potential energy and hence, of heat and salt.

A single distributed vortex, whose horizontal size is much greater than the Rossby deformation radius, is found to be unstable, and in this case, it splits to form smaller vortex structures, which can propagate in different directions. During the decomposition of a large baroclinic vortex, the domain of anomalous concentration of thermohaline characteristics, associated with this vortex, can extend significantly. On the other hand, the merging of vortices with the same signs is accompanied by intense mixing, resulting in significant horizontal redistribution of water mass properties. During the possible subsequent division of a large vortex, the newly formed thermohaline structures can be again transported over large distances. Such processes can be seen in the behavior of vortices with different structure, in particular, intrathermocline lenses. Thus, even the enumeration of those simplest elements of vortex interaction demonstrates the key role of vortices in the formation of the thermohaline structure of the ocean.

In addition, we note that the role of mesoscale vortices in the processes of horizontal and vertical mixing in the ocean, which appears to have been first mentioned by Rossby in 1936 [769], was studied in [14, 15, 43, 50, 106, 161, 195, 198, 268, 281, 291, 292, 306, 307, 355, 362, 366, 370, 376, 378, 398, 409, 417, 443, 448, 487, 498, 502, 515, 537, 538, 543, 594, 628, 636, 670, 676–678, 712, 764, 776, 786, 800, 821, 837, 838, 868, 870, 874, 923, 924, 937, 938, 940, 951, 952, 979, 982, 1024]; the regional aspects of the problem were studied in [1, 69, 106, 109, 245, 284, 293, 294, 296, 297, 305, 377, 434, 499, 520, 533, 539, 616, 721, 758, 761, 762, 778, 814, 835, 863, 865, 973, 985, 988, 1008, 1029] in particular.

A significant breakthrough in the understanding of the mixing process took place due to the appearance of such research domain as chaotic mixing of advection trajectories of fluid particles in Lagrangian coordinates.[2]

Notwithstanding the fact that the property of mixing of trajectories that initially were close to one another, which is known now as *chaotic advection* [27, 28], in unsteady regular geophysical flows was mentioned by Eckart [241] more then 60 years ago, and later demonstrated by Welander [975]. Systematic studies of chaotic advection in the atmosphere, natural estuaries and oceans began only four decades later. Significant results have been obtained in this field in the recent 20 years. A more detail description of these problems can be found, for example, in the books [223, 442, 676, 800, 1007], reviews [441, 566, 677, 981, 1018] and papers [2, 16, 30, 33–

[2]We have touched upon this problem in the sections "Origination of Chaos at A = 2" and "Chaotic Advection Near Stationary Configurations" in the context of the chaotic behavior of liquid particles in the field of quasi-stationary states of systems consistent of three vortices in a two-layer rotating fluid.

35, 49, 55, 74, 101, 110, 116, 124–128, 213, 214, 226–228, 299, 323, 380–382, 394, 440, 443–446, 467–471, 498, 500, 501, 534–536, 569, 596, 609, 620, 641, 645–647, 662, 678, 683, 694, 695, 697–699, 703, 708, 709, 717, 718, 722, 753, 754, 765, 779–784, 795, 799, 826, 862, 875, 916–918, 938–940, 949, 964, 970, 974, 990–996, 1000, 1005, 1026].

4.3.2 Bottom Topography and Vortices

We only mentioned the problem of bottom topography influence on vortex dynamics (Sects. 3.4 and "Dynamics of Medies in the Flow Over Submerged Hills"). This problem was of an inexhaustible interest in the last 60 years. It was caused in the end of 1950th by first records of deep-water currents made by neutral buoyancy floats [456]. A comprehensive analysis of the problems associated with the effect of bottom topography on circulation in the ocean can be found in monographs [52, 201, 442, 456, 689, 1010, 1032, 1034], reviews [143, 357, 441, 605, 932, 933], and papers [5–7, 16, 22, 53, 54, 62, 72, 73, 78, 80, 92, 112, 114, 114, 115, 124, 126, 135, 137, 138, 140, 153–155, 157, 176, 177, 203, 204, 206, 208, 211, 212, 251, 252, 266, 314, 335, 345, 346, 349, 364, 365, 368, 369, 373, 375, 381–383, 385, 389–392, 429, 446, 450–455, 464, 465, 469–471, 481–483, 491, 494, 556, 563, 578, 582, 637, 649, 663, 666, 720, 727, 734, 735, 739–741, 779–784, 796, 804, 805, 811, 813, 815–817, 839, 844, 849, 860, 861, 873, 878, 881, 882, 887, 889, 891, 893, 894, 898, 901–903, 906, 928–930, 939, 941–943, 946, 947, 962, 968, 976, 1011–1016, 1020–1023, 1031, 1033, 1035–1038] in particular.

4.3.3 More on Lenses

We considered some particular aspects of the dynamics of the intrathermocline vortices (lenses, meddies) in Sect. 3.5.2 in the context of the evolution of the intermediate-layer vortex patches in the three-layer quasi-geostrophic ocean model.

Intrathermocline vortices having core with maximum orbital velocities located at intermediate depths, were first discovered in the 1930s years from the analysis of the measurements made at the drifting station "North Pole–1" during the expedition led by P.P. Shirshov [824]. Later, similar vortex structures have been found repeatedly in the records of both direct and indirect observations in different areas of the World Ocean [75, 76]. The peak of interest to the investigation of the intrathermocline vortices sprang up in the mid-1980s. At this moment a collection of papers [253], prepared under the editorship of K.N. Fedorov appeared. It was written by researchers of P.P. Shirshov Institute of Oceanology AS USSR, Moscow, and was completely devoted to the study of lenses. Continued interest to the study of the lens dynamics characterizes this rather incomplete list of papers: [7, 10, 13, 18–20, 41–46, 60, 64, 65, 67, 95, 107–109, 111, 123, 132, 145, 147, 154, 155, 175, 176, 178, 207, 212, 215, 216, 230, 250, 253–255, 257–266, 341, 345, 346, 348, 352, 359, 378, 379, 393,

397, 408, 415, 422, 447, 521, 555, 556, 577, 588, 600, 630, 650, 681, 682, 719, 748, 749, 751, 752, 775–778, 806, 807, 812, 813, 815–817, 825, 848, 863, 865, 868, 869, 883, 888, 891, 914, 915, 923–926, 968, 989, 1001, 1002, 1019].

4.3.4 On Modons

We didn't touch an important dynamics problem *modons* [872] (or *solitary Rossby wave* [516, 517], or *submesoscale coherent vortex (SCV)* [582]) which is worthy of a special book. A modon is a vortex structure represented by a vortex pair, located inside a circular area, and which is an exact stationary solution of the equation of the absolute vorticity conservation on the β-plane.

Solutions obtained by Stern [872] and Larichev and Reznik [516, 517], gave rise to numerous investigations, and in particular, to works [83–86, 139, 276, 277, 418–423, 427–429, 431–433, 518, 519, 582, 585, 606, 642, 657–661, 797, 879, 884, 885]; they gave generalizations to the case of baroclinic fluid and also examined the questions of stability and interaction of modons with each other and with bottom topography.

Interesting effects of the emergence of cyclonic-anticyclonic asymmetry were discussed by many authors, [e. g. 170, 337, 338, 459–461, 492, 559, 643, 644, 774] and, in particular, within the two-layer model [51, 143, 871, 920]. It is worth mentioning in conclusion that, although the quasi-geostrophic motions constitute a considerable part of flows of geophysical nature, the requirement for the Rossby number to be small ($Ro \ll 1$) is not satisfied universally. In the case of intense vortices $Ro \sim 1$, a cyclostrophic rather than geostrophic balance can take place. Solutions for localized cyclostrophic vortices are shown in [143, 209, 300–302, 318, 367, 386, 399–402, 742, 836, 927, 953].

We hope, this book should provide insights into solving these and other problems of geophysical hydrodynamics and oceanography.

Appendix A
E.J. Hopfinger. Experimental Study of Hetons

This Appendix is a summary of the experiments with hetons, conducted by Griffiths and Hopfinger [309, 310], aimed at illustrating heton generation in the laboratory. These experiments were motivated by the work of Hogg and Stommel [350, 351] who considered discrete, baroclinic geostrophic vortices that have the capacity of transporting heat. Real vortices have a finite core size and it was of interest to compare experiments with the theoretical, idealized point vortex solutions. Numerical solutions of, finite core size geostrophic vortices have followed and are presented in this book.

Griffiths and Hopfinger [309, 310] considered the simplest situation of a rotating, two layer stratified fluid in a tank 100 cm in diameter and 45 cm deep with the fluid layers being of equal depth, H = 20 cm. The rotating tank was first filled with fresh water to 20 cm depth and then a sugar solution of desired density was slowly injected through a tube placed near the sidewall at the bottom of the tank. When the lower, denser layer reached 20 cm the system was left to spin-up to solid body rotation. It needs to be mentioned that the two layer system never reached complete solid body rotation. A weak azimuthal drift, generally $<3\%$ of the tank rotation, remained that switched from being axisymmetric to non-axisymmetric and reverse. Griffiths and Linden [312] interpreted this drift as being the result of a meridional Eddington-Sweet circulation, driven by the diffusion of solute across the isopycnals. The use of sugar solution instead of salt solution reduced this diffusion, hence the importance of this circulation. Lower diffusion resulted also in a thinner interface thickness, which was close to 2 cm at the time when the heton experiments were stared. The tank rotation was Ω = 1.0 rad s^{-1} $(f = 2s^{-1})$ and the reduced gravity $g' = g\Delta\rho/\rho$ was chosen such that the internal Rossby radius of deformation $\lambda' = (g'H)^{1/2}/f$ = 5, 10 and 15 cm. Note that the Rossby radius λ used in this book is defined as $\lambda = [g'h_1h_2/(h_1+h_2)]^{1/2}/f$, hence, when $h_1 = h_2 = H$, $\lambda = \lambda'/\sqrt{2}$.

The vortices in this two layer stratified, rotating system were generated by sources (anticyclones) and sinks (cyclones) placed at the free surface or the bottom. These vortices have a core radius R' of about 4 cm at which the azimuthal velocity is maximum. The vortex strength is a function of the flow rate. With sinks strong

M.A. Sokolovskiy and J. Verron, *Dynamics of Vortex Structures in a Stratified Rotating Fluid*, Atmospheric and Oceanographic Sciences Library 47,
DOI 10.1007/978-3-319-00789-2, © Springer International Publishing Switzerland 2014

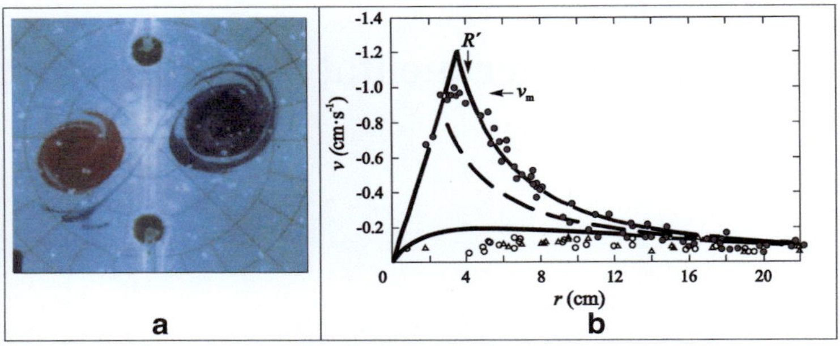

Fig. A.1 (**a**) Image of two anticyclones in the *top layer* taken at six tank rotation periods $(6T_\Omega)$ after vortex generation (after stop of the source flow). The source separation seen in (a) is 18 cm, $\lambda' = 14.1$ cm and $R' \approx 4$ cm. (**b**) Azimuthal velocity in a two-layer, $\lambda' = 7$ cm, anticyclonic vortex. •, upper layer vortex velocity at $9T_\Omega$ after vortex generation. *Open symbols* are for lower layer velocity. *Solid line* is solution of inviscid, quasi-geostrophic potential vorticity equations (see [310]). $R' \approx 4$ cm is intersection of theoretical solution with $v = v_m$. *Dashed line* is $v = s/2r$ of barotropic vortex of same strength

vorticity concentration can be achieved with the absolute value of angular velocity of the cyclonic vortices reaching $0.5 - 1.0\Omega$, whereas with a source (anticyclones) the angular velocity was limited to $0.3 - 0.5\Omega$. Note also, that, in order to prevent large perturbations, the source flow velocity needs to be kept low (large source flow area). A source or sink flow duration of 30 s (5 rotation periods) was sufficient to establish a well developed, depth independent vortex. During this time the vortex displacement remained negligible. The spin-down time of the vortices in the bottom layer (influence of bottom boundary) is $H/(f\nu)^{1/2} \approx 140$ s and is about twice as long in the top layer. The vortices were visualised by dye injection and the azimuthal velocity was measured by the streak line optical method. A typical image of two anticyclones generated in the top layer is shown in Fig. A.1a and the azimuthal velocity of a typical anticyclonic vortex is shown in Fig. A.1b.

In the Arctic oceans, a fresh source flow by ice melting can generate an anticyclone and ice freezing can give rise to a dense source flow, generating a cyclonic vortex in the top layer (see Fig. 1.2 in this book). In the laboratory it would also be possible to generate cyclonic vortices in the top layer by a dense source flow instead a sink flow. A tilted heton can emerge from baroclinic instability (Fig. 1.3 in this book).

It is of interest to compare experimental observations with numerical simulations presented in this book. The simplest case is a pair of vertical axis hetons corresponding to the situations shown in Figs. 3.46 and 3.47.

Figure A.2 illustrates an attracting heton pair in the laboratory tank. The anticyclones are generated in the top layer (red dye) by two sources at the free surface, visible in the images, and the cyclones are generated in the lower layer by two sinks at the bottom (blue and green dyes), placed at the same locations as the sources above; the vortices are initially of equal strength. The two (hot) hetons

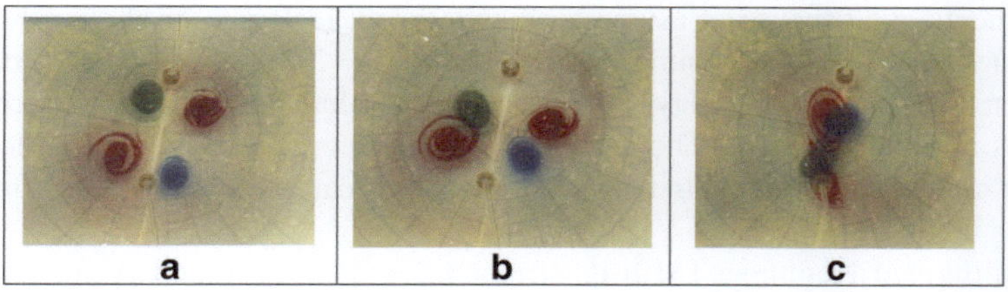

Fig. A.2 Photographs of an unstable, coalescing heton pair, initially separated by $2b^* = 20\,\mathrm{cm}$ (separation of injection tubes seen in the images). The suction tubes at the *bottom* are at the same location as the sources on *top* and initially the anticyclones (*red dye*) are of the same strength as to cyclones in the *bottom layer* (*green* and *blue dyes*). Initially, the hetons are axial but are torn apart by the vortex interaction in each layer. The vortex radii are $R' \approx 4\,\mathrm{cm}$, $\lambda' = 15\,\mathrm{cm}$ and images (a) to (c) are at 6, 15 and 22 T_Ω after vortex generation

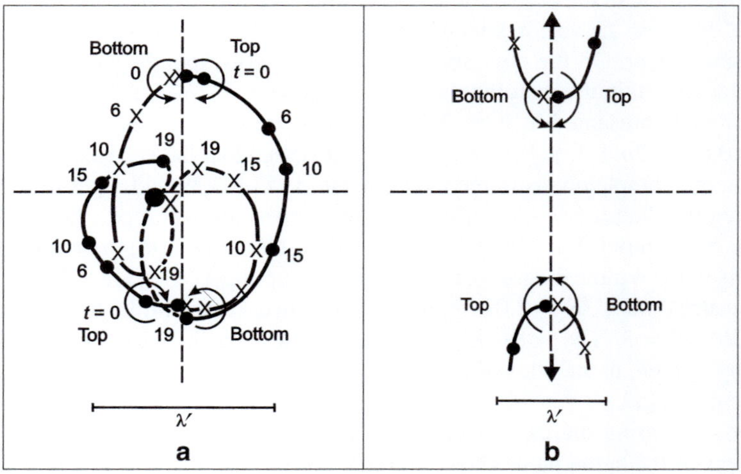

Fig. A.3 Paths of the centres of anticyclones in the *top layer* (*solid circles*) and cyclones in the *lower layer* (*crosses*). (**a**) Conditions corresponding to those of Fig. A.2. (**b**) initial heton separation $2b^* = 20\,\mathrm{cm}$ and $\lambda' = 10\,\mathrm{cm}$

generated in this way are initially separated by $2b^*/R' \approx 5$ and the dimensionless Rossby radius is $\lambda'/R' \approx 3.75$. In this case the interaction of vortices in the same layer override the interaction between vortices in different layers. The two anticyclones in the top layer orbit in the anticyclonic sense and the cyclones in the lower layer orbit in the cyclonic sense and eventually merge. The initially distance between the vortices in the same layer is $2b^* \approx 1.3\lambda'$, which is close to the critical value of $1.4\lambda'$ of point vortices Hogg and Stommel [350].

Figure A.3 shows the paths of the centres of the baroclinic vortices in the top and bottom layers, shown in Fig. A.2 ($2b^*/\lambda' = 1.3$), from the start, time $t = 0$ after generation to $t = 19T_\Omega$ and, finally, to complete merging (dashed traces). The paths of the vortices is not symmetric because of various imperfections inherent in experiments and because of asymmetric spin down. In Fig. A.3b the behaviour of the same heton pair is shown, but now for $2b^*/\lambda' = 2$. In this case the interaction between the top and bottom layer vortices is greater than the mutual influence of the vortices in the same layer. The hetons propagate in opposite directions. However, for this movement tilting is first necessary, which is caused by the interaction of the vortices in the same layer.

Griffiths and Hopfinger [310] determined the stability boundary for two anticyclones generated in the top layer. For barotropic vortices the critical distance is $2b^*/R = 3.3$; below this value vortices merge and remain stable, in orbital motion, above. Experiments showed that the stability boundary of baroclinic vortices depends strongly on the internal Rossby radius. For the conditions in Fig. A.3, that is $2b^*/R' \approx 5$, the stability boundary obtained by Griffiths and Hopfinger [310] is $\lambda'/R' \approx 2.4$. The heton repulsion shown in Fig. A.3b is close to this stability boundary, which, in the case of two hetons, likely to shift to somewhat larger values of λ'/R' because the vortex interaction in opposite layers opposes the interactions in the same layer. On the contrary, the conditions corresponding to Fig. A.3a are situated well within the merging domain $(2b^*/R' \approx 5,\ \lambda'/R' \approx 3.75)$.

The results presented in Figs. 3.39 and 3.40, are for $2b = 2b^*/D = 3$ and $\gamma = D/\lambda = 0.6$ and $2b = 3$, $\gamma = 0.7$ respectively. It is seen that the vortices coalesce (but never merge completely as in the experiments) in Fig. 3.39 and repel in Fig. 3.40. In the regime diagram $b - \gamma$ of Fig. 3.38, the coalescing heton pair lies in domain S_{3m} and heton repulsion (Fig. 3.40) in S_2. When taking $D = R'$, the corresponding positions in the regime diagram of Griffiths and Hopfinger [310] are at $\lambda'/R' = 2.35$ (coalescence) and $\lambda'/R' = 2.01$ (repulsion), in both cases the initial heton separation being $2b^*/R' = 3$. The value $\lambda'/R' = 2.01$ is close to the experimental stability boundary, which, as mentioned above, is likely to shift to slightly larger values in the case of hetons.

Now, comparing the experiments with the numerical results, the experimental conditions of the heton paths shown in Figs. A.3a, b are, respectively, $b \approx 2.5$, $\gamma \approx 0.37$ and $b \approx 2.5$, $\gamma \approx 0.56$. The latter lies well within the heton scatter domain S_2 whereas the former (heton merging) is close to the boundary of S_2 and S_{3m} but still in S_2. Discrepancies between experiments and numerical simulations are to be expected because the vortex radius is less well defined, the interface thickness is finite, the vorticity distribution in the vortex core varies and spin-down occurs.

Appendix B
M.A. Sokolovskiy. In Memory of My Teacher

Vadim Fedorovich Kozlov
(1933–2005)

Vadim Fedorovich Kozlov was born on March 17, 1933 in Vladivostok into a family of seamen. His father, Fedor Kirillovich Kozlov, was a sea captain; his mother, Valentina Iosifovna, née Gein, was a seaman's wife and kept the house. In her youth, she also studied in a marine technical school, but after the birth of her son, took charge of his upbringing. Later, she also contributed much to the upbringing of her grandson Kirill. These were marine families. His paternel elder brother, Matvey Kirillovich Kozlov, was also a sea captain, and the younger brother, Ivan Kirillovich Kozlov, was a chief engineer of Sakhalin Shipping Company. The maternal grandfather, Iosif Antonovich Gein, worked as a chief engineer on ships in the Far East and Black Sea Shipping Companies since the 1910s. The uncle, Anatoliy Iosifovich Gein, also worked as a chief engineer on ships of Murmansk Shipping Company, while another uncle, Gleb Iosifovich Gein, served in the navy.

M.A. Sokolovskiy and J. Verron, *Dynamics of Vortex Structures in a Stratified Rotating Fluid*, Atmospheric and Oceanographic Sciences Library 47,
DOI 10.1007/978-3-319-00789-2, © Springer International Publishing Switzerland 2014

Vadim Fedorovich spent his early childhood and schooldays in his hometown. After graduating from school with a Silver Medal, VF enrolled without examination (according to the rules of that time, the persons who had graduated from school with a Silver Medal could be enrolled after an interview) in the Division of Mechanics, the Faculty of Mechanics and Mathematics, Moscow M.V. Lomonosov State University (MSU).

In that time, the faculty was situated in its historic building in Mokhovaya Street, but in 1953 it moved into the new high-rise building in Leninskie Gory. The students, including VF, settle in a comfortable dormitory in MSU Main Building. After their second year, students were to choose a department for their area of expertise. VF chose the Department of Gas and Wave Dynamics, which was then headed by Academician of Uzbek Academy of Sciences, Kh.A. Rakhmatulin. After graduating from the University, VF was invited to continue his study as a postgraduate student.

A curious thing that happened with VF during his postgraduate study taught him a valuable lesson: to keep up with the scientific literature. It was like this: the research adviser formulated a problem for him about the reflection of an acoustic wave from a deformable plane. The young postgraduate student solved the problem and reported the obtained result at a conference. Next, he wrote a paper, which he was going to publish in the Vestnik (Herald) of Moscow State University. The reviewer of his paper was firm but fair: he informed the author that this result was obtained by Rayleigh as long ago as 1878. Reviewing the past many years later, VF joked: "Lord Rayleigh and I have obtained an elegant solution", and continued: "Though I did these 80 years later, but still independently!" This lesson was as severe as useful. As a result, the future professor finished his post-graduate studies without defending his thesis. (Scientists in Russia have two scientific degrees, i.e., Candidate of Science and Doctor of Science; the successful completion of post-graduate study is commonly a defended Candidate thesis).

Immediately after his graduation from the university, VF married his classmate Tatyana Leonidovna Gavrilova. She, also graduated from the Faculty of Mechanics and Mathematics, was doing her postgraduate studies in the Department of Applied Mathematics, Steklov Mathematical Institute, Russian Academy of Sciences, under scientific supervision of A.A. Lyapunov.

After the completion of their postgraduate studies in 1960, the newly married couple completed the Moscow period of their life and moved to Vladivostok. In Vladivostok, they immediately started their teaching activity in the Far Eastern State University (FESU). VF started learning the fundamentals of lectureship at the Faculty of Physics and Mathematics, where he lectured on the basic courses of *Theoretical Mechanics* (for students in mathematics) and *Methods of Mathematical Physics* (for students in physics) and carried out practical exercises in those courses; he also was a scientific supervisor of term and (later) graduation works in both lines.

In 1964, after the organization of the Faculty of Geophysics in FESU, he transferred to the Department of Oceanography of this Faculty, where he lectured on basic and special courses, such as *Higher Mathematics*, *Hydromechanics*, *Dynamics of Currents*, *Nonlinear Problems of Current Dynamics*, *Theory of Geostrophic Flows*, *Theory of Gravitational Waves*, *Nonstationary Acoustic Waves*. He also led seminars on wave dynamics and dynamics of flows.

It will be observed that VF was a masterly lecturer. His direct or indirect disciples will always remember his splendid lectures. Although the theory of oceanic flows was absolutely new for VF, he, due to his extraordinary capacity for work and the ability to see the essence of problems, very quickly learned the basic works in this field and prepared a course of lectures, forming the basis for teaching this discipline. Several years later, he published an excellent textbook [449], in which he generalized all results of the theory of stationary flows in the ocean available by that time. All books went like hot cakes. Not only students and postgraduates, but also many experts in oceanography appreciated the basic character of this work and the exhaustive list of references in the field. Until his last years, VF received requests from his colleagues to send them the book, but, even in his private library, he had only one copy of it. The former disciples of VF, who became professors in universities, used this book as a basis for their lectures to the following generations of students in oceanography. Large teaching activity in the first years of his work in the university was not an obstacle for VF in his research. From the first days he joined the researches in Marine Physics.

In the early 1960s, a weekly seminar on Marine Physics started working at the Department. For two decades, this seminar was essentially the center of oceanographic studies in Vladivostok. This seminar was attended, in addition to members of the Department, by professors and researchers from other educational and research institutions. VF was a permanent member of this seminar, and he set there a high level of reports and discussions. Even his first publications on the parameterization of vertical turbulent exchange in the ocean and studies of the specific features of subequatorial circulation placed him among the leading theoretical oceanographers in the country.

These studies, supplemented by results in the theory of thermocline, formed the basis for his candidate thesis. He prepared the thesis himself, without a scientific advisor, and, according to the existing requirements, he needed in a special permission of the State Commission for Academic Degrees and Titles to defend it. The Commission gave its approval as an exception, provided the defense should take place with two doctors of science as official opponents (generally, one of the opponents is a candidate of science). The official opponents were the well-known oceanographers A.I. Fel'zenbaum and A.S. Sarkisyan, who appreciated the merits of VF very high.

Soon after the defense of his thesis, VF transferred into the Pacific Oceanological Institute, Far East Branch, USSR Academy of Sciences (POI FEB RAS), as a senior researcher in the Laboratory of Thermal Processes and Marine Dynamics. From 1970 to 1974, VF was working at this institute, keeping close contacts with the university, where he lectured several courses and supervised the graduation works of students from the Departments of Oceanography and Computational Mathematics.

The main scientific interests of those years are represented by works aimed to develop mathematically grounded diagnostic calculations of oceanic flows and a concept of integral quasi-geostrophic models reflecting the important properties of three-dimensional models. The paper [450] on the mutual adaptation of the mass and current fields to the bottom topography, in my opinion, is among the brilliant works of VF. The classes of partial analytical solutions of simplified nonlinear equations of ocean dynamics are still most complete. He suggested in this work carrying out a detailed typification of the World Ocean areas by analyzing the contribution of individual terms of the considered equations. Unfortunately, the evaluation of these contributions based on real oceanographic data was never implemented. No doubt, this is an omission of practical oceanographers.

In 1974, the primary employment of VF again became FESU, though now it was the Department of Computational Mathematics, Faculty of Mathematics. At this department, VF prepared an ingenious author's course *Computing Techniques* for students specializing in *Applied Mathematics* and introduced a new area of expertise *Application of Numerical Methods in Oceanography*. He delivered lectures on several special courses for students in this field: *Perturbation Methods in Applied Mathematics*, *Numerical Solution of Problems in Ocean Dynamics*, *Methods for Fluid Oscillations Calculation*, etc. During the 5 years of his work with this department, he "set on their feet" many skilled experts in the application of numerical methods and computation engineering.

Following the principle: "the researcher should change something in his interest every five years", VF again accepted an invitation to transfer to POI FEB RAS. This decision was partially due to the appearance in POI of Valeriy Isaakovich Klyatskin, who transferred from Moscow to Vladivostok for permanent work. VF very much disliked any administrative activity, so he readily agreed to take a position of senior researcher in the Laboratory of Wave Processes, created by V.I. Klyatskin. Later on, when, during POI restructuring, the Laboratory of Wave Processes became a division with the same name, VF had to agree to lead the Laboratory of Geophysical Hydrodynamics created in the division and consisting of his disciples.

A unique feature of VF as a scientist is that, since his graduation from MSU, he never belonged to any school and formulated himself all the problems for his research work. No doubt, he can be regarded as a founder of the Far Eastern Russian School of Geophysical Hydrodynamics. Nowadays, this school is supported by K.V. Koshel and S.V. Prants.

In the 1970s, VF focused on the problem of the effect of bottom topography on oceanic flows. He solved several difficult and elegant problems of vortex structures' generation by irregularities of the bottom relief and shoreline, including stationary Taylor vortices and leeward Rossby waves [451–454, 481, 483, in particular]. A generalization of these basic results is given in book [456] and in his Doctorate thesis defended in 1984.

Having the "doctoral standing" in fact, VF for a long time avoided starting a routine work of preparing a thesis. As in the case with the position of the Head of Laboratory, he again was compelled to yield to the "situation": in the course of a cognac-drinking session during the 2nd All-USSR Oceanological Congress (Yalta, 1982), A.S. Sarkisyan and V.I. Klyatskin, succeeded in making VF to give a word of honor to put hand to this task. The thesis was written within 3 months and, after the completion of appropriate procedures, brilliantly defended at the Scientific Council of the Marine Hydrophysical Institute in Sevastopol.

The history of science knows many real-world examples when an interesting idea comes to the minds of several people practically at the same time. VF knew quite well that the 2D movement of fluid particles in a continuous motionless depends only on the boundary configuration of areas with nonzero relative vorticity. He developed a numerical algorithm for calculating flows under such formulation. It was only to charge a student or a postgraduate with writing a computer program and to solve a simple model problem. Several young persons tried to do that but all failed. However, in 1979, the American mathematicians Zabusky, Hughes and Roberts published their famous work on contour dynamics for two-dimensional Euler equations [1006], and implemented the same idea. VF proposed his disciple and then colleague, V.G. Makarov, to put aside his work and to put hand to this problem. Viacheslav wrote the required software package. This software was more efficient and user friendly than the American one (moreover, it enabled a very wide class of generalizations), but, there is nothing to be done, the Zabusky group merited to be pioneers in the creation of the Contour Dynamics Method (CDM).

In 1983, VF published his first work on CDM [455], which was focused on the application of this method to the problems of geophysical hydrodynamics. Two years later, VF gave the first generalization of the method to the case of a continuously stratified medium [457]. In those years, almost all researchers from Vadim Fedorovich's laboratory joined the work aimed to solve numerous problems with the use of this approach [334, 458, 462, 464, 465, 472, 473, 477–480, 484, in particular, and many other papers written after the death of VF]. Anyway, the more than 10-year period of work, associated with the application of CDM and its modifications [459–461] for numerous problems, was extremely fruitful for both VF and his colleagues.

Works [458] and [459–461] are worth special mentioning. In the first paper, VF suggested a general algorithm for constructing asymptotic expansions, what allowed determining of the boundaries of vortex patches for stationary states of translational and rotational types of various vortex structures up to the fourth degree of a small parameter. This paper specifies appropriate expansions for already known states, gives some new solutions, and describes an exhaustive procedure for their construction. The second series of papers gives an expansion of the well-known problem of elliptic Kirchhoff vortex to the case with the involvement effect taken into account. Both papers are examples of qualitatively new generalizations of classical results.

In the next 5 years of his scientific life, VF was interested in the problem of construction of background flows. The concept, he formulated in [463], was developed up to its implementation both for model cases [466], and for natural water bodies (the seas of Okhotsk and Japan) [474–476].[1] The background flows determine the base state of horizontally uniform and steady-state distribution of potential vorticity, which provides a global minimum of mechanical energy for some

[1]Currently, research in this direction is developed by V.V. Novotryasov and D.V. Stepanov in POI FEB RAS.

area with given geometry of its boundary and bottom topography. The possible mass exchange at the boundary is taken into account. Thus, the specified background state of the system allows one to uniquely determine a flow field that is dynamically consistent with the field of relative or potential vorticity.

This concept turned out to be surprisingly fruitful in studies of many phenomena taking place in the ocean and associated with chaotic mixing of fluid particles, e.g., the diffusion of a conservative admixture. The interest VF showed to chaotic dynamics was due to his desire to understand the nature of diffusion of vortex filaments, separating from vortices and clearly traceable in CDM-based calculations. These filaments form every time when an initially compact vortex moves either in a shear flow, or it decays because of its own instability, or, finally, when two (or more) vortices with the same sign are close enough to one another. Problems of chaotic dynamics of flows in different areas [380–382, 467, 468, 470, 471] were the last object of VF's scientific interest.

For publication, VF preferred the journal "Izvestiya, Atmosphere and Ocean Physics"; 37 his papers got out there. The sympathy was mutual: the editorial board of the journal published their tributes on the occasions of VF's 60th and 70th anniversaries.

Since 1997 until his last days, VF in cooperation with S.V. Prants guided a permanent seminar on nonlinear dynamics in POI. As in the case with oceanographic seminars in FESU, researchers from many allied institutions gathered at its meetings, and they still do this. At this seminar, all the results obtained by VF in the field of chaotic dynamics were also discussed.

VF's devotion to books should also be mentioned. Since his studentship, he started forming his scientific library. Later, because of the evident lack of place in the bookshelves in his home library, VF introduced a rule, according to which his bookshelves were to contain no more than 2,000 the most necessary books. Other books (after thorough choice) would be transferred to his office in the institute. The library of POI still contains a splendid Vadim Fedorovich's contribution, including copies of papers and books on different fields of oceanology. He paid special attention to journal publications In abstract journals and, recently, in the Internet he traced new studies in the fields of interest for him.

During the years of his active work, VF collected several tens of thousands of preprints and copies of papers. Many colleagues addressed him with bibliographic questions and immediately received the required paper from the bowels of his famous bookcases. His collection is unique. Now it is in full disposal of the personnel of Geophysical Hydrodynamic Laboratory, where VF was working until his last days.

VF lived in the central part of Vladivostok and took suburban electric train to reach the institute, situated in the city's suburbs. On his way to the institute, he looked through new journals and preprints of papers delivered by mail. One day in the mid-1980, when we were working on the problem of stability of two-layer vortex, he entered the Laboratory and asked me whether I knew the word "heton". I confessed that this was the first time I heard this word. With a mischievous glint in his eyes, he took from his briefcase a preprint of the new paper by Hogg and Stommel [350], in which this term was introduced. Since then, the "heton" has become an important component of my life, a fact of which this book is a confirmation.

On January 9, 2005, Vadim Fedorovich departed from this life, what decreased the amount of intellect in the world. One can hardly say how many new and fruitful ideas could have been generated by the inquisitive mind of our Teacher, but, unfortunately, the fate gave him too short term. VF was an open man; he always shared his knowledge and experience with all who needed in them.

I was lucky to have had a privilege of practically daily converse with VF during the 20 years of my work in Vladivostok. In one of our numerous informal conversations, I mentioned some of his works that, in my opinion, were of greatest interest (see above) and asked him whether he agreed with my estimate. VF approved my choice and, with his intrinsic sense of humor, he added: "It is brief, but quite sufficient for my epitaph!" All who knew him will remember this wise, life-loving, and witty person.

Acknowledgements I am sincerely grateful to Tatyana Leonidovna Gavrilova for information about Vadim Fedorovich's relatives.

References

1. Abernathey R, Marshall J, Mazloff M, Shuckburgh E (2010) Enhancement of mesoscale eddy stirring at steering levels in the Southern Ocean. J Phys Oceanogr 40(1):170–184
2. Abraham ER, Bowen MM (2002) Chaotic stirring by a mesoscale surface-ocean flow. Chaos 12(2):373–381
3. Abrashkin AA (1987) Theory of interaction between two plane vortices in a perfect fluid. Fluid Dyn 22(1):53–59
4. Abrashkin AA, Yakubovich EI (1984) Planar rotational flows of an ideal fluid. Sov Phys Dokl 29(5):370–371
5. Adcock ST, Marshall DP (2000) Interactions between geostrophic eddies and the mean circulation over large-scale bottom topography. J Phys Oceanogr 30(12):3223–3238
6. Adcroft A, Hill C, Marshal J (1997) Representation of topography by shaved cells in a height coordinate ocean mode. Mon Weather Rev 125(9):2293–2315
7. Adduce C, Cenedese C (2004) An experimental study of a mesoscale vortex colliding with topography of varying geometry in a rotating fluid. J Mar Res 62(5):611–638
8. Afanasyev YaD, Peltier WD (1998) Three-dimensional instability of anticyclonic swirling flow in rotating fluid: laboratory experiments and related theoretical predictions. Phys Fluids 10(12):3194–3202
9. Aguiar ACB, Read PL, Wordsworth RD, Salter T, Yamazaki YH (2009) A laboratory model of Saturn's North Polar Hexagon. Icarus 206(2):755–763
10. Aiki H, Yamagata T (2000) Successive formation of planetary lenses in an intermediate layer. Geophys Astrophys Fluid Dyn 92(1):1–29
11. Akhmetov DG (2009) Vortex rings. Springer, Berlin/Heidelberg, 150 p
12. Alekseenko SV, Kuibin PA, Okulov VL (2007) Theory of concentrated vortices. Springer, Berlin/Heidelberg, 494 pp
13. Aleynik DL (1998) The structure an evolution of a meddy and Azores frontal zone in autumn 1993. Oceanology 38:312–322
14. Alford MH, Gregg MC, D'Asaro EA (2005) Mixing, 3D mapping, and Lagrangian evolution of a thermohaline intrusion. J Phys Oceanogr 35(9):1689–1711
15. Alford MH, Pinkel R (2000) Observations of overturning in the thermocline: the context of ocean mixing. J Phys Oceanogr 30(5):805–832
16. Allen JS, Samelson RM, Newberger PA (1991) Chaos in a model of quasigeostrophic flow over topography: an application of Melnikov's method. J Fluid Mech 226:511–547
17. Allison M, Godfrey DA, Beebe RF (1990) A wave dynamical interpretation of Saturn's Polar Hexagon. Science 247(4946):1061–1063
18. Alves JMR, Carton X, Ambar I (2011) Hydrological structure, circulation and water mass transport in the Gulf of Cadiz. Int J Geosci 2(4):432–456

M.A. Sokolovskiy and J. Verron, *Dynamics of Vortex Structures in a Stratified Rotating Fluid*, Atmospheric and Oceanographic Sciences Library 47,
DOI 10.1007/978-3-319-00789-2, © Springer International Publishing Switzerland 2014

19. Ambar I (1983) A shallow core of Mediterranean water off western Portugal. Deep-Sea Res 30(6A):677–680
20. Ambar I, Serra N, Neves F, Ferreira T (2008) Observations of the Mediterranean undercurrent and eddies in the Gulf of Cadiz during 2001. J Mar Syst 71:195–220
21. Amoretti M, Dukin D, Fajans J, Pozzoli R, Romé M (2001) Asymmetric vortex merger: experiments and simulations. Phys Plasmas 8(9):3865–3868
22. An BW, McDonald NR (2005) Coastal currents and eddies and their interaction with topography. Dyn Atmos Oceans 40(4):237–253
23. Antipov SV, Nezlin MV, Snezhkin EN, Trubnikov AS (1985) Rossby auto-soliton and laboratory model of Jupiter's Great Red Spot. Sov Phys JETP 62:1097–1107
24. Antonova RA, Zhvania BI, Lominadze DG, Nanobashvili DI, Chagelishvili GD, Yan'kov VV (1996) Dynamics of dipole vortices in the interaction with a solid boundary. Plasma Phys Rep 22(9):775–782
25. Aref H (1979) Motion of three vortices. Phys Fluids 22(3):393–400
26. Aref H (1982) Point vortex motions with a center of symmetry. Phys Fluids 25(12): 2183–2187
27. Aref H (1983) Integrable, chaos and turbulent vortex motion in two-dimensional flows. Annu Rev Fluid Mech 15:345–389
28. Aref H (1984) Stirring by chaotic advection. J Fluid Mech 143:1–21
29. Aref H (1989) Three-vortex motion with zero total circulation: addendum. J Appl Math Phys (ZAMP) 40(4):495–500
30. Aref H (2002) The development of chaotic advection. Phys Fluids 14(4):1315–1325
31. Aref H (2009) Stability of relative equilibria of three vortices. Phys Fluids 21:094101. doi:10.1063/1.3216063
32. Aref H (2010) Self-similar motion of three point vortices. Phys Fluids 22:057104. doi:10.1063/1.3425649
33. Aref H, Balachandar S (1986) Chaotic advection in a Stokes flow. Phys Fluids 29(11):3515–3521
34. Aref H, Brøns M (1998) On stagnation points and streamline topology in vortex flows. J Fluid Mech 370:1–27
35. Aref H, Jones SW, Mofina S, Zawadski I (1989) Vortices, kinematics and chaos. Phys D 37(1–3):423–440
36. Aref H, Pomphrey N (1982) Integrable and chaotic motions of four vortices. I. The case of identical vortices. Proc R Soc Lond A 380(1779):359–387
37. Aref H, Rott N, Thomann H (1992) Gröbli's solution of the three-vortex problem. Annu Rev Fluid Mech 24:1–20
38. Aref H, Stremler MA (1999) Four-vortex motion with zero total circulation and impulse. Phys Fluids 11(12):3704–3715
39. Arendt SC (1995) Steadily translating vortices in a stratified fluid. Phys Fluids 7(2):384–388
40. Arendt SC (1996) Two-dimensional vortex dynamics in a stratified barotropic fluid. J Fluid Mech 314:139–161
41. Arhan M, Carton X, Piola A, Zenk W (2002) Deep lenses of circumpolar water in the Argentine Basin. J Geophys Res 107(C1):3007. doi:10.1029/2001JC000963
42. Arhan M, Colin de Verdiére A, Mémery L (1994) The eastern boundary of the subtropical North Atlantic. J Phys Oceanogr 24(6):1295–1316
43. Armi L (1978) Some evidence for boundary mixing in the deep ocean. J Geophys Res 83(C4):1971–1979
44. Armi L, Hebert D, Oakey N, Price JF, Richardson PL, Rossby HT, Ruddick B (1989) Two year in the life of a Mediterranean salt lens. J Phys Oceanogr 19(3):354–370
45. Armi L, Stommel H (1983) Four views of a portion of the North Atlantic subtropical gyre. J Phys Oceanogr 13(5):828–857
46. Armi L, Zenk W (1984) Large lenses of highly saline Mediterranean water. J Phys Oceanogr 14(10):1560–1576

47. Arnold VI (1965) Conditions for nonlinear stability of stationary plane curvilinear flows of an ideal fluid. Sov Math Dokl 6:773–777
48. Arnold VI (1969) On an a priori estimate in the theory of hydrodynamical stability. Am Math Soc Transl 79:267–269
49. Babiano A, Boffetta G, Provenzale A, Vulpiani A (1994) Chaotic advection in point vortex models and two-dimensional turbulence. Phys Fluids 6(7):2465–2474
50. Badin G, Tandon A, Mahadevan A (2011) Lateral mixing in the pycnocline by baroclinic mixed layer eddies. J Phys Oceanogr 41(11):2080–2101
51. Baey J-M, Carton X (2002) Vortex multipoles in two-layer rotating shallow-water flow. J Fluid Mech 460:151–175
52. Baines PG (1997) Topographic effects in stratified flows. Cambridge University Press, Cambridge, 500 pp
53. Baines PG, Boyer DL, Xie B (2005) Laboratory simulations of coastally trapped waves with rotation, topography and stratification. Dyn Atmos Oceans 39(3–4):153–173
54. Baines PG, Smith RB (1993) Upstream stagnation points in stratified flow past obstacles. Dyn Atmos Oceans 18(1–2):105–113
55. Balasuriya S, Jones CKRT (2001) Diffusive draining and growth of eddies. Nonlinear Process Geophysics 8(4/5):241–251
56. Baker GR (1990) A study of the numerical stability of the method of contour dynamic. Phil Trans R Soc 333(1931):391–400
57. Baker-Yeboah S, Flierl GR, Sutyrin GG, Zhang Y (2010) Transformation of an Agulhas eddy near the continental slope. Ocean Sci 6:143–159. http://dx.doi.org/10.5194/os-6-143-2010
58. Barba LA, Leonard A (2007) Emergence and evolution of tripole vortices from net-circulation initial conditions. Phys Fluids 19(1):017101. doi:10.1063/1.2409734
59. Barbosa JP, Métais O (2000) Large-eddy simulations of deep-ocean convection: analysis of the vorticity dynamics. J Turbul 1(009):1–31
60. Baringer MO, Price JF (1997) Mixing and spreading of the Mediterranean outflow. J Phys Oceanogr 27(8):1654–1677
61. Barker SJ, Crow SC (1977) The motion of two-dimensional vortex pairs in a ground effect. J Fluid Mech 82(4):659–671
62. Barnier B, Le Provost C (1993) Influence of bottom topography roughness on the jet and inertial recirculation of a mid-latitude gyre. Dyn Atmos Oceans 18(1–2):29–65
63. Basdevant C, Couder Y, Sadourny R (1984) Vortices and vortex-couples in two-dimensional turbulence, or long-lived couples are Batchelor's couples. Lect Notes Phys 230:327–346
64. Bashmachnikov I, Carton X (2012) Surface signature of Mediterranean water eddies in the North-East Atlantic: effect of the upper ocean stratification. Ocean Sci 8:931–943. doi:10.5194/os-8-931-2012. doi:10.5194/osd-9-2457-2012
65. Bashmachnikov I, Mohn C, Pelegrí JL, Martíns A, Jose F, Machí F, White M (2009) Interaction of Mediterranean water eddies with Sedlo and Seine Seamounts, subtropical Northeast Atlantic. Deep-Sea Res II 56(25):2593–2605
66. Batchelor GK (1967) An introduction to fluid mechanics. Cambridge University Press, Cambridge
67. Batteen ML, Martinho AS, Miller HA, McClean JL (2007) A process-oriented modelling study of the coastal Canary and Iberian Current system. Ocean Model 18(1):1–36
68. Bauer L, Morikawa GK (1976) Stability of rectilinear geostrophic vortices in stationary equilibrium. Phys Fluids 19(7):929–942
69. Beal LM, Chereskin TK, Lenn YD, Elipot S (2006) The sources and mixing characteristics of the Agulhas Current. J Phys Oceanogr 36(11):2060–2074
70. Beckers M, Clercx HJH, van Heijsts GJF, Verzicco R (2002) Evolution and instability of monopolar vortices in a stratified fluid. Phys Fluids 15(4):1033–1045
71. Beckers M, van Heijst GJF (1998) The observation of a triangular vortex in a rotating fluid. Fluid Dyn Res 22(5):265–279

72. Beckmann A, Haidvogel DB (1993) Numerical simulation of flow around a tall isolated seamount. Part 1. Problem formulation and model accuracy. J Phys Oceanogr 23(8):1736–1753
73. Beckmann A, Haidvogel DB (1997) A numerical simulation of flow at Fieberling Guyot. J Geophys Res 102(C3):5595–5613
74. Beerens SP, Ridderinkhof H, Zimmerman JFE (1994) An analytical study of chaotic stirring in tidal areas. Chaos Solit Fract 4(6):1011–1029
75. Belkin IM, Emelyanov MV, Kostyanoy AG, Fedorov KN (1986) Thermohaline structure of intermediate waters of the ocean and intrathermocline eddies. In: Fedorov KN (ed) Intrathermocline eddies in the ocean. P.P. Shirshov Institute of Oceanology, Moscow, pp 8–34 (in Russian)
76. Belkin IM, Kostyanoy AG (1992) Intrathermocline eddies in the Word ocean and their regional peculiarities. In: Barenblatt GI, Seidov DG, Sutyrin GG (eds) Coherent structures and self-organisation of currents in the ocean. Nauka, Moscow, pp 112–127 (in Russian)
77. Benilov ES (2000) The dynamics of a near-surface vortex in a two-layer ocean on the beta-plane. J Fluid Mech 420:277–299
78. Benilov ES (2001) Baroclinic instability of two-layer flows over one-dimensional bottom topography. J Phys Oceanogr 31(8):2019–2025
79. Benilov ES (2003) Instability of quasi-geostrophic vortices in a two-layer ocean with a thin upper layer. J Fluid Mech 475:303–331
80. Benilov ES (2005) Stability of a two-layer quasigeostrophic vortex over axisymmetric localized topography. J Phys Oceanogr 35(1):2019–2025
81. Benilov ES (2005) On the stability of oceanic vortices: a solution to the problem? Dyn Atmos Oceans 40(3):133–149
82. Bennett A (2006) Lagrangian fluid dynamics. Cambridge University Press, Cambridge, 310 pp
83. Berestov AL (1979) Solitary Rossby waves. Izv Atmos Ocean Phys 15:443–447
84. Berestov AL (1981) Some new solutions for the Rossby solitons. Izv Atmos Ocean Phys 17:82–87
85. Berestov AL (1985) Dispersion relationships for the Rossby solitons. Izv Atmos Ocean Phys 21:332–334
86. Berestov AL, Monin AS (1980) Solitary Rossby waves. Adv Mech 3:3–34
87. Birkhoff G (1960) Hydrodynamics. A study in logic, fact, and similitude. 2nd edn. revised and enlarged. Princeton University Press, Princeton, 184 pp
88. Billant P, Chomaz J-M (2000) Experimental evidence for a new instability of a vertical columnar vortex pair in a strongly stratified fluid. J Fluid Mech 418:167–188
89. Blackmore D, Ting L, Knio O (2007) Studies of perturbed three vortex dynamics. J Math Phys. doi:10.1063/1.2428272
90. Bogomolov VA (1977) Vorticity dynamics on a sphere. Fluid Dyn 6(6):863–870
91. Bogomolov VA (1985) On motion on a rotating sphere. Izv Atmos Ocean Phys 21:391–396
92. Bograd SJ, Rabinovich AB, LeBlond PH, Shore JA (1997) Observations of seamount-attached eddies in the North Pacific. J Geophys Res 102(C6):12441–12456
93. Bolsinov AV, Fomenko AT (1995) Orbital equivalence of integrable Hamiltonian systems with two degrees of freedom. A classification theorem. I. Sbornik: Mathematics 81(2):421–466
94. Bord EG (2006) On the nonlinear disturbances of vortex polygon. Russ Sci J Nonlinear Dyn 2(3):353–360 (in Russian)
95. Borenäs KM, Wåhlin AK, Ambar I, Serra N (2002) The Mediterranean outflow splitting—a comparison between theoretical models and CANIGO data. Deep-Sea Res II 49(19):4195–4205
96. Borth H (1999) Von Kármánsche Wirbelstraßen und barokline Jetströme in einem 2-Schichten Kanal auf der beta-Ebene. PhD thesis, University of Bremen, Germany, 1999 AWI Rep. 96, 147 pp

97. Borisov AV, Bolsinov AV, Mamaev IS (1999) Lie algebras in vortex dynamics and celestial mechanics: 4. Regul Chaotic Dyn 4(1):23–50
98. Borisov AV, Lebedev VG (1998) Dynamics of three vortices on a plane and a sphere: 2. General compact case. Regul Chaotic Dyn 3(2):99–114
99. Borisov AV, Lebedev VG (1998) Dynamics of three vortices on a plane and a sphere: 3. Noncompact case. Problem of collapse and scattering. Regul Chaotic Dyn 3(4):74–86
100. Borisov AV, Kilin AA, Mamaev IS (2005) Absolute and relative choreographies in the problem of the motion of point vortices in a plane. Dokl Math 71(1):139–144
101. Borisov AV, Kilin AA, Mamaev IS (2006) Transition to chaos in dynamics of four point vortices on a plane. Dokl Phys 51(5):262–267
102. Borisov AV, Mamaev IS (1999) Poisson structures and Lie algebras in the Hamiltonian mechanics. Publisher House Udmurt University, Izhevsk, 464 pp (in Russian)
103. Borisov AV, Mamaev IS (2005) Mathematical methods in the dynamics of vortex structures. Institute of Computer Sciences, Moscow–Izhevsk, 368 pp (in Russian)
104. Borisov AV, Mamaev IS, Kilin AA (2004) Absolute and relative choreographies in the problem of point vortices moving on a plane. Regul Chaotic Dyn 9(2):101–111
105. Borisov AV, Pavlov AE (1998) Dynamics and statics of vortices on a plane and a sphere: 1. Regul Chaotic Dyn 3(1):28–39
106. Bower AS (1991) A simple kinematic mechanism for mixing fluid parcels across a meandering jet. J Phys Oceanogr 21(1):173–182
107. Bower AS, Armi L, Ambar I (1995) Direct evidence of the meddy formation off southwestern coast of Portugal. Deep-Sea Res 42:1621–1630
108. Bower AS, Armi L, Ambar I (1997) Lagrangian observations of meddy formation during a Mediterranean Undercurrent Seeding Experiment. J Phys Oceanogr 27(12):2545–2575
109. Bower AS, Rossby T (1989) Evidence of cross-frontal exchange processes in the Gulf Stream based on isopycnal RAFOS float data. J Phys Oceanogr 19(6):1177–1190
110. Bower AS, Rossby HT, Lillibridge JL (1985) The Gulf Stream—barrier or blender? J Phys Oceanogr 15(1):24–32
111. Bower AS, Serra N, Ambar I (2002) Structure of the Mediterranean undercurrent and Mediterranean water spreading around the southwestern Iberian Peninsula. J Geophys Res 107(C10):3161. doi:10.1029/2001JC001007
112. Boyer DL, Davies PA (2000) Laboratory studies of orographic effects in rotating and stratified flows. Annu Rev Fluid Mech 32:165–202
113. Boyland P, Stremler M, Aref H (2003) Topological fluid mechanics of point vortex motions. Phys D 175(1–2):69–95
114. Bracco A, Pedlosky J (2003) Vortex generation by topography in locally unstable baroclinic flows. J Phys Oceanogr 33(1):207–219
115. Bracco A, Pedlosky J, Pickart RS (2008) Eddy formation near the West Coast of Greenland. J Phys Oceanogr 38(9):1992–2002
116. Bracco A, Provenzali A, Scheuring I (2000) Mesoscale vortices and the padadox of the plancton. Proc R Soc Lond B 267:1795–1800
117. Brandt LK, Cichocki TK, Nomura KK (2010) Asymmetric vortex merger: mechanism and criterion. Theor Comput Fluid Dyn 24:163–167
118. Brandt LK, Nomura KK (2006) The physics of vortex merger: further insight. Phys Fluids 18:051701. doi:10.1063/1.2201474
119. Brickman D (1995) Heat flux partitioning in deep-ocean convection. J Phys Oceanogr 25(11), part 1:2609–2623
120. Brown MG (1990) Are SOFAR float trajectories chaotic? J Phys Oceanogr 20(1):139–149
121. Brunner-Suzuki A-MEG, Sundermeyer MA, Lelong M-P (2012) Vortex stability in a large-scale internal wave shear. J Phys Oceanogr 42(10):1668–1683
122. Brutyan MA, Krapivskii PL (1988) Hamiltonian formulation and fundamental conservation laws for a model of small elliptical vortices. J Appl Math Mech 52(1):133–136
123. Bubnov VA (1971) Structure and dynamics of the Mediterranean waters in the Atlantic Ocean. Ocean Res 22:220–286 (in Russian)

124. Budyansky MV, Uleysky MYu, Prants SV (2002) Fractals and dynamic traps in the simplest model of chaotic advection with a topographic vortex. Dokl Earth Sci 387(8):929–932
125. Budyansky MV, Uleysky MYu, Prants SV (2004) Chaotic scattering, transport, and fractals in a simple hydrodynamic flow. J Exp Theor Phys 99(5):1018–1027
126. Budyansky MV, Uleysky MYu, Prants SV (2004) Hamiltonian fractals and chaotic scattering of passive particles by a topographical vortex and an alternating current. Phys D 195(3–4):369–378
127. Budyansky MV, Uleysky MYu, Prants SV (2007) Lagrangian coherent structures, transport and chaotic mixing in simple kinematic ocean models. Commun Nonlinear Sci Numer Simul 12(1):31–44
128. Budyansky MV, Uleysky MYu, Prants SV (2009) Detecting barriers to cross-jet Lagrangian transport and its destruction in a meandering flow. Phys Rev E 79(5):056215. doi:10.1103/PhysRevE.79.056215
129. Burbea J (1982) On patches of uniform vorticity in a plane of irrotational flow. Arch Ration Mech Anal 77:349–358
130. Burbea J (1982) Motions of vortex patches. Lett Math Phys 6:1–16
131. Burbea J, Landau M (1982) The Kelvin waves in vortex dynamics and their stability. J Comput Phys 45(1):127–156
132. Byshev VI (1992) Properties of an intra-thermocline lens on a subpolar front in the North Atlantic. Oceanology 32(4):701–707
133. Cabral HE, Schmidt DS (2000) Stability of relative equilibria in the problem on N+1 vortices. SIAM J Math Anal 31(2):231–250
134. Capéran P, Verron J (1988) Numerical simulation of a physical experiment on two-dimensional vortex merger. Fluid Dyn Res 3(1–4):87–92
135. Carmack EC, Kulikov EA (1998) Wind-forced upwelling and internal Kelvin wave generation in Mackenzie Canyon, Beaufort Sea. J Geophys Res 103(C9):18447–18458
136. Carnevale GF, Kloosterziel RC (1994) Emergence and evolution of triangular vortices. J Fluid Mech 259:305–331
137. Carnevale GF, Kloosterziel RC, van Heijst GJF (1991) Propagation of barotropic vortices over topography in a rotating tank. J Fluid Mech 233:119–139
138. Carnevale GF, Purini R, Orlandi P, Cavazza P (1995) Barotropic quasi-geostrophic f-plane flow over anisotropic topography. J Fluid Mech 285:329–347
139. Carnevale GF, Vallis GK, Purini R, Briscolini M (1988) The role of initial conditions in flow stability with an application to modons. Phys Fluids 31(9):2567–2572
140. Carnevale GF, Velasco Fuentes OU, Orlandi P (1997) Inviscid dipole-vortex rebound from a wall or coast. J Fluid Mech 351:75–103
141. Carr LE III, Williams RT (1989) Barotropic vortex stability to perturbations from axisymmetry. J Atmos Sci 46(20):3177–3191
142. Carton XJ (1992) On the merger of shielded vortices. Europhys Lett 18(8):697–703
143. Carton XJ (2001) Hydrodynamical modelling of oceanic vortices. Surv Geophys 22(3):179–263
144. Carton XJ (2009) Instability of surface quasigeostrophic vortices. J Atmos Sci 66(4):1051–1062
145. Carton X, Chérubin L, Paillet J, Morel Y, Serpette A, Le Cann B (2002) Meddy coupling with a deep cyclone in the Gulf of Cadiz. J Mar Syst 32(1–3):13–42
146. Carton XJ, Corréard SM (1999) Baroclinic tripolar vortices: formation and subsequent evolution. In: Sorensen JN, Hopfinger EJ, N Aubry (eds) Simulation and identification of organized structures in flows: IUTAM/SIMFLOW symposium. Dordrecht, Kluwer, pp 181–190
147. Carton XJ, Daniault N, Alves J, Chérubin L, Ambar I (2010) Meddy dynamics and interaction with neighboring eddies southwest of Portugal: observations and modeling. J Geophys Res. doi:10.1029/2009JC005646

148. Carton XJ, Flierl GR, Perrot X, Meunier T, Sokolovskiy MA (2010) Explosive instability of geostrophic vortices. Part 1: baroclinic instability. Theor Comput Fluid Dyn 24(1–4):125–130
149. Carton XJ, Flierl GR, Polvani LM (1989) The generation of tripoles from unstable axisymmetric isolated vortex structures. Europhys Lett 9(4):339–344
150. Carton XJ, Legras B (1994) The life-cycle of tripoles in two-dimensional incompressible flows. J Fluid Mech 267:51–82
151. Carton XJ, Meunier T, Flierl GR, Perrot X, Sokolovskiy MA (2010) Explosive instability of geostrophic vortices. Part 2: parametric instability. Theor Comput Fluid Dyn 24(1–4):131–135
152. Carton XJ, McWilliams JC (1989) Barotropic and baroclinic instabilities of axisymmetric vortices in a quasigeostrophic model. In: Nihoul JCJ, Jamart BM (eds) Mesoscale/synoptic coherent structures in geophysical turbulence. Amsterdam/Oxford/New York/Tokyo, Elsevier, pp 225–244
153. Cartwright JHE, Feingold M, Piro O (1996) Chaotic advection in three-dimensional unsteady incompressible laminar flow. J Fluid Mech 366:259–284
154. Cenedese C (2002) Laboratory experiments on mesoscale vortices colliding with a seamount. J Geophys Res 107(C6). doi:10.1029/2000JC000599
155. Cenedese C, Linden PF (1999) Cyclone and anticyclone formation in a rotating stratified fluid over a sloping bottom. J Fluid Mech 381:199–223
156. Cenedese C, Marshall J, Whitehead JA (2004) A laboratory model of thermocline depth and exchange fluxes across circumpolar fronts. J Phys Oceanogr 34(3):656–667
157. Cenedese C, Whitehead JA (2000) Eddy-shedding from a boundary current around a cape over a sloping bottom. J Phys Oceanogr 30(7):1514–1531
158. Cerretelli C, Williamson CHK (2003) The physical mechanism for vortex merging. J Fluid Mech 475:41–77
159. Cerretelli C, Williamson CHK (2003) A new family of uniform vortices relating to vortex configurations before merging. J Fluid Mech 493:219–229
160. Cessi P, Ierley G, Young W (1987) A model of the inertial recirculation driven by potential vorticity anomalies. J Phys Oceanogr 17(10):1640–1652
161. Cessi P, Young WR, Polton JA (2006) Control of large-scale heat transport by small-scale mixing. J Phys Oceanogr 36(10):1877–1896
162. Chao S-Y, Shaw P-T (1999) Close interactions between two pairs of heton-like vortices under sea ice. J Geophys Res 104(C10):23591–23605
163. Chao S-Y, Shaw P-T (1999) Fission of heton-like vortices under sea ice. J Oceanogr 55(1):65–78
164. Chao S-Y, Shaw P-T (2000) Slope-enhanced fission of salty hetons under sea ice. J Phys Oceanogr 30(11):2866–2882
165. Chao S-Y, Shaw P-T (2003) Heton shedding from submarine-canyon plumes in an Arctic boundary current system: sensitivity to the undercurrent. J Phys Oceanogr 33(9):2032–2044
166. Chaplygin SA (1899) On a pulsating cylindrical vortex. Trans Phys Sect Imp Mosc Soc Frends Nat Sci 10:13–22 (in Russian). English translation in: Regul Chaotic Dyn 2007, 12(1):101–116
167. Chaplygin SA (1903) One case of vortex motion in fluid. Trans Phys Sect Imp Mosc Soc Frends Nat Sci 11(2):11–12 (in Russian)
168. Charlton AJ, O'Neil A, Lahoz WA, Berrisford P (2005) The splitting of the stratospheric polar vortex in the Southern Hemisphere, September 2002: dynamical evolution. J Atmos Sci 62(3):590–602
169. Charney JG (1963) Numerical experiments in atmospheric hydrodynamics. In: Metropolis NC, Taub AH, Todd J, Tompkins CB (eds) Proceedings of symposia in applied mathematics "Experimental arithmetic high speed computing and mathematics". Am Math Soc 15:289–310
170. Chefranov SG (2001) Centrifugal-dissipative instability of Rossby vortices and their cyclonic-anticyclonic asymmetry. J Exp Theor Phys Lett 73(6):274–278

171. Chen LG, Dewar WK (1993) Intergure communication in a three-layer model. J Phys Oceanogr 23(5):855–878
172. Chenciner A, Gerver J, Montgomery R, Simó C (2002) Simple choreographic motions of N bodies: a preliminary study. In: Newton P, Holmes P, Weinstein A (eds) Geometry, mechanics, and dynamics: volume in honor of the 60th birthday of J.E. Marsden, part III. Springer, New York, pp 287–308
173. Chernyshenko SI (1988) The asymptotic form of the stationary separated circumfluence of a body at hight Reynolds numbers. J Appl Math Mech 52(6):746–753
174. Chernyshenko S (1993) Stratified Sadovskii flow in a channel. J Fluid Mech 250:423–431
175. Chérubin LM, Carton X, Dritschel DG (2007) Vortex dipole formation by baroclinic instability of boundary currents. J Phys Oceanogr 37(6):1661–1667
176. Chérubin LM, Carton X, Paillet J, Morel Y, Serpette A (2000) Instability of the Mediterranean water undercurrents southwest of Portugal: effects of baroclinicity and topography. Oceanol Acta 23(5):551–573
177. Chérubin LM, Morel Y, Chassignet EP (2006) Loop current ring shedding: the formation of cyclones and the effect of topography. J Phys Oceanogr 36(4):569–591
178. Chérubin L, Serpette A, Carton X, Paillet J, Connan O, Morin P, Rousselet R, Le Cann B, Le Corre P, Labasque T, Corman D, Poete N (1977) Descriptive analysis of the hydrology and currents on the Iberian shelf from Gibraltar to cape Finisterre: preliminary results of the INTERAFOS and SEMANE experiments. Ann Hydrogr 21(768):5–81
179. Chirikov BV (1979) A universal instability of many-dimensional oscillator systems. Phys Rep 52(5):263–379
180. Christiansen JP, Zabusky NJ (1973) Instability, coalescence and fission of finite-area vortex structures. J Fluid Mech 61(part 2):219–243
181. Corréard SM, Carton XJ (1999) Formation and stability of tripolar vortices in stratified geostrophic flows. Il Nuovo Cim 22C(6):767–777
182. Couder Y, Basdevant C (1986) Experimental and numerical study of vortex couples in two-dimensional flows. J Fluid Mech 173:225–251
183. Cresswell GR (1982) The coalescence of two East Australian Current warm-core eddies. Science 215(4529):161–164
184. Crow SC (1970) Stability theory for a pair of trailing vortices. AIAA J 8(12):2172–2179
185. Crowdy DG (1999) A class of exact multipolar vortices. Phys Fluids 11(9):2556–2564
186. Crowdy DG (2002) The construction of exact multipolar equilibria of the two-dimensional Euler equations. Phys Fluids 14(1):257–267
187. Crowdy D (2002) Exact solutions for rotating vortex arrays with finite-area cores. J Fluid Mech 469:209–235
188. Crowdy D (2004) Explicit solutions for a steady vortex-wave interaction. J Fluid Mech 513:161–170
189. Crowdy D, Duchemin L (2005) The effect of solid boundaries on pore shrinkage in Stokes flow. J Fluid Mech 531:359–379
190. Crowdy D, Marshall J (2005) Analytical solutions for rotating vortex arrays involving multiple vortex patches. J Fluid Mech 523:307–337
191. Crowdy D, Sunara A (2007) Contour dynamics in complex domains. J Fluid Mech 593:235–254
192. Crowdy D, Tanveer S, Vasconcelos G (2005) On a pair of interacting bubbles in planar Stokes flow. J Fluid Mech 541:231–261
193. Cushman-Roisin B (1989) On the role of filamentation in the merging of enticyclonic lenses. J Phys Oceanogr 19(2):253–258
194. Cushman-Roisin B (1994) Introduction to geophysical fluid dynamics. Prentice Hall, New York, 320 pp
195. Cushman-Roisin B (1995) Effects of horizontal advection on upper ocean mixing: a case of frontogenesis. J Phys Oceanogr 11(10):1345–1356

196. Cushman-Roisin B, Beckers J-M (2011) Introduction to geophysical fluid dynamics. Physical and numerical aspects, 2nd edn. Academic Press, Elsevier Inc., Amstardam/Boston/Heidelberg/London/New York/Oxford/Paris/San Diego/San Francisco/Singapore/Sydney/Tokyo, 828 pp
197. Cushman-Roisin B, Sutyrin GG, Tang B (1992) Two-layer geostrophic dynamics: 1. Governing equations. J Phys Oceanogr 22(2):117–127
198. Danabasoglu G, McWilliams JC, Gent PR (1994) The role of mesoscale tracer transports in the global ocean circulation. Science 264(5162):1123–1126
199. Danilov S, Gryanik V, Olbers D (1998) Equilibration and lateral spreading of a strip-shaped convection region. In: Alfred-Wegener-Institut für Polar- und Meeresforschung, Report 86, 66 pp
200. Danilov S, Gryanik V, Olbers D (2001) Equilibration and lateral spreading of a strip-shaped convection region. J Phys Oceanogr 31(4):1075–1087
201. Darnitskiy VB (2010) Oceanological processes near underwater mountains and ridges of open ocean. Vladivostok, FSUE "TINRO-Center", 200 pp (in Russian)
202. Davey MK (1977) Baroclinic instability in a fluid with three layers. J Atmos Sci 34(8):1224–1234
203. Davey MK (1978) Recycling flow over bottom topography in a rotating annulus. J Fluid Mech 87(3):497–520
204. Davey MK, Hurst RGA, Johnson ER (1993) Topographic eddies in multilayer flow. Dyn Atmos Oceans 18(1–2):1–27
205. Davies I, Truman A, Williams D (1983) Classical periodic solutions of the equal-mass $2n$-body problem, b-ion problem and the n-electron problem. Phys Lett 99A(1):15–18
206. Davies PA (1972) Experiments on Taylor columns in rotating stratified fluids. J Fluid Mech 54(4):691–717
207. Davies PA, Guo Y, Rotenberg E (2001) Laboratory model studies of Mediterranean outflow adjustment in the Gulf of Cadiz. Deep-Sea Res Part II 49(19):4207–4223
208. Davies PA, Rahm L (1982) The interaction between topography and a nonlinearly stratified rotating fluid. Phys Fluids 25(11):1931–1934
209. Davies TV (1948) Rotatory flow on the surface of the earth. Part I. Cyclostrophic motion. Philos Mag Ser 7 39(293):482–491
210. Deem GS, Zabusky NJ (1978) Vortex waves: stationary 'V-states', intarctions, recurrence, and breaking. Phys Rev Lett 40(13):859–862
211. De-Hai L (1994) Quasi-resonant interactions among barotropic Rossby waves with two-wave topography and low frequency dynamics. Geophys Astrophys Fluid Dyn 76(1–4):145–163
212. De-Hai L, Yan L (2001) Dynamics of meddies interacted with a seamount in a 1.5 layer model. J Hydrodyn Ser B 3:93–99
213. del-Castillo-Negrete D, Greene JM, Morrison PJ (1996) Area preserving nontwist maps: periodic orbits and transition to chaos. Phys D 91(1–2):1–23
214. del-Castillo-Negrete D, Morrison PJ (1993) Chaotic transport by Rossby waves in shear flow. Phys Fluids A 5(4):948–965
215. de Verdiere AC (1992) On the southward motion of Mediterranean salt lenses. J Phys Oceanogr 22(4):413–420
216. Dewar WK (1988) Ventilating beta plane lenses. J Phys Oceanogr 18(8):1193–1201
217. Dewar WK (2002) Convection in small basins. J Phys Oceanogr 32(10):2766–2788
218. Dewar WK (2002) Baroclinic eddy interaction with isolated topography. J Phys Oceanogr 32(10):2789–2805
219. Dhanak MR, Marshall MP (1993) Motion of an elliptic vortex under applied periodic strain. Phys Fluids A5(5):1224–1230
220. DiBattista MT, Majda AJ (2000) An equilibrium statistical theory for large-scale features of open-ocean deep convection. J Phys Oceanogr 30(6):1325–1353
221. DiBattista MT, Majda AJ (2001) Equilibrium statistical predictions for baroclinic vortices: the role of angular momentum. Theor Comput Fluid Dyn 14(5):293–322

222. DiBattista MT, Majda AJ, Marshall A (2002) A statistical theory for the "pathiness" of open-ocean deep convection: the effect of preconditioning. J Phys Oceanogr 32(2):599–626
223. Dijkstra HA (2005) Nonlinear physical oceanography: a dynamical systems approach to the large scale ocean circulation and El Niño, 2nd reversed and enlarged edn. Springer Science+Business media, New York, 532 p
224. Dikarev SN (1992) Deep convection—the process of deep water formation in the open sea. In: Barenblatt GI, Seidov DG, Sutyrin GG (eds) Coherent structures and self-organisation of currents in the ocean. Nauka, Moscow, pp 145–155 (in Russian)
225. Dolzhanskii FV, Krymov VA, Manin DYu (1990) Stability and vortex structures of quasi-two-dimensional shear flows. Phys Usp (Adv Phys Sci) 33(7):495–520
226. Doronina TN (1994) On the structure of an intense baroclinic vortex in three-dimensional shearing motions. Dokl (Trans) RAS Earth Sci Sect 339(4):528–532
227. Doronina TN (1995) The structure of circulation cells in intense baroclinic vortices in a current with a velocity shift and the advective transport of a solute. Izv Atmos Ocean Phys 31(2):223–232
228. Doronina TN (1997) Interaction of baroclinic point vortices in quasi-geostrophic barotropic and baroclinic shearing flows. Oceanology 37(4):454–460
229. Doronina T, Gryanik V, Olbers D, Warncke T (1998) A 3D heton mechanism of lateral spreading in lacalized convection in a rotating stratified fluid. Alfred-Wegener-Institut für Polar- und Meeresforschung, Report 87, 84 pp
230. Drillet Y, Bourdallé-Badie R, Siefridt L, Le Provost C (2005) Meddies in the Mercator North Atlantic and Mediterranean Sea eddy-resolving model. J Geophys Res 110:C03016. doi:10.1029/2003JC002170
231. Dritschel DG (1985) The stability and energetics of corotating uniform vortices. J Fluid Mech 157:95–134
232. Dritschel DG (1986) The nonlinear evolution of rotating configurations of uniform vorticty. J Fluid Mech 172:157–182
233. Dritschel DG (1988) Contour surgery: a topological reconnection scheme for extended integrations using contour dynamics. J Comput Phys 77(1):240–266
234. Dritschel DG (1988) The repeated filamentation of two-dimensional vorticity interfaces. J Fluid Mech 194:511–547
235. Dritschel DG (1989) On the stabilization of a two-dimensional vortex strip by adverse shear. J Fluid Mech 206:193–221
236. Dritschel DG (1990) The stability of elliptical vortices in an external straining flow. J Fluid Mech 210:223–261
237. Dritschel DG (1995) A general theory for two-dimensional vortex interactions. J Fluid Mech 293:269–303
238. Dritschel DG (1998) On the persistence of non-axisymmetric vortices in inviscid two-dimensional flows. J Fluid Mech 371:141–155
239. Dritschel D, Legras B (1998) Modeling oceanic and atmospheric vortices. Phys Today 46(3):44–51
240. Dritschel DG, Waugh DW (1992) Quantification of the inelastic interaction of inequal vortices in two-dimensional vortex dynamics. Phys Fluids A4(8):1737–1744
241. Eckart C (1948) An analysis of the stirring and mixing processes in incompressible fluid. J Mar Res 7(3):265–275
242. Eckart C (1960) Hydrodynamics of oceans and atmospheres. Pergamon Press, New York, 290 pp
243. Eckhardt B (1988) Integrable four vortex motion. Phys Fluids 31(10):2796–2801
244. Eckhardt B, Aref H (1988) Integrable and chaotic motions of four vortices. 2. Collision dynamics of vortex pairs. Phil Trans R Soc Lond A 326(1593):655–696
245. Eden C, Böning C (2002) Sources of eddy kinetic energy in the Labrador Sea. J Phys Oceanogr 32(12):3346–3363
246. Eisenlohr H, Eckelmann H (1989) Vortex splitting and its consequences in the vortex street wake of cylinders at low Reynolds number. Phys Fluids A1(2):189–192

247. Elcrat AR, Fornberg B, Horn M, Miller K (2000) Some steady vortex flows past a circular cylinder. J Fluid Mech 409:13–27
248. Elcrat AR, Fornberg B, Miller K (2005) Stability of vortices in equilibrium with a cylinder. J Fluid Mech 544:53–68
249. Elcrat AR, Miller KG (1989) Computation of vortex flows past obstacles with circulation. Phys D 37(1–3):441–452
250. Emelianov M, Claret M, Fraile-Nuez E, Pastor M, Laiz I, Salvador J, Pelegrí JL, Turiel A (2012) Detection of a weak meddy-like anomaly from high-resolution satellite SST maps. In: Espino M, Font J, Pelegrí JL, Sánchez-Arcilla A (eds) Advances in Spanish Physical Oceanography. Sci Mar 76S1:229–234. doi:10.3989/scimar.03619.19I
251. Estrade P, Middleton JH (2010) A numerical study of island wake generated by an elliptical tidal flow. Cont Shelf Res 30(9):1120–1135
252. Fang F, Morrow R (2003) Evolution, movement and decay of warm-core Leeuwin Current eddies. Deep Sea Res Part II 50(12–13):2245–2261
253. Fedorov KN (ed) (1986) Intrathermocline eddies in the ocean. PP. Shirshov Institute of Oceanology, Moscow 142 pp (in Russian)
254. Fedorov KN (1986) Intrathermocline eddies—a specific type of oceanic eddies with a core. In: Intrathermocline eddies in the ocean. P.P. Shirshov Institute of Oceanology, Moscow, pp 5–7 (in Russian)
255. Fedorov KN, Ginzburg AI (1992) The near-surface layer of the ocean. VSP. III, Utrecht/Tokyo, 259 pp
256. Fedorov KN, Ginzburg AI (1989) Mushroom-like currents (vortex dipoles): one of the most widespread forms of stationary coherent motions in the ocean. In: Nihoul JCJ, Jamart BM (eds) Mesoscale/synoptic coherent structures in geophysical turbulence. Elsevier, Amsterdam/Oxford/New York/Tokyo, pp 1–14
257. Filyushkin BN (1989) Investigation of intrathermocline lenses of Mediterranean origin (Cruise 16 of R/V "Vityaz", June 3–September 16, 1988). Oceanology 29(4):535–536
258. Filyushkin BN, Aleynik DL, Demidov AN, Sarafanov AA, Kozhelupova NG (2007) The peculiarity of the formation and spreading Mediterranean water mass at intermediate depths of the Atlantic Ocean. In: Waters masses of the oceans and sea. MAX Press, Moscow, pp 92–129 (in Russian)
259. Filyushkin BN, Aleynik DL, Gruzinov VM, Kozhelupova NG (2002) The thermohaline water structure at the region dynamic degradation of the Mediterranean lenses. In: Proceeding of the State Oceanographic Institution, "Ocean and sea research", S-Ptb. No 208, pp 15–32 (in Russian)
260. Filyushkin BN, Aleynik DL, Gruzinov VM, Kozhelupova NG (2002) Dynamic degradation of the Mediterranean lenses in the Atlantic Ocean. Dokl Earth Sci 387A(9):1079–1082
261. Filyushkin BN, Aleynik DL, Kozhelupova NG, Moshonkin SN (2009) Horizontal transport peculiarities of the Mediterranean water in the Atlantic. In: Komchatov VF (ed) Proceeding of the State Oceanographic Institution, "Ocean and sea research", vol 212. Moscow, 76–88 (In Russian)
262. Filyushkin BN, Plakhin EA (1995) Experimental study of the first stage of Mediterranean water lens formation. Oceanology 35:797–804
263. Filyushkin BN, Sokolovskiy MA (2011) Modeling the evolution of intrathermocline lenses in the Atlantic Ocean. J Mar Res 69(2–3):191–220
264. Filyushkin BN, Sokolovskiy MA, Kozhelupova NG, Vagina IM (2010) Dynamics of intrathermocline lenses. Dokl Earth Sci 434(part 2):1377–1380
265. Filyushkin BN, Sokolovskiy MA, Kozhelupova NG, Vagina IM (2011) Reflection of intrathermocline eddies on the ocean surface. Dokl Earth Sci 439(part 1):986–989
266. Filyushkin BN, Sokolovskiy MA, Kozhelupova NG, Vagina IM (2011) Evolution of intrathermocline eddies moving over a submarine hill. Dokl Earth Sci 441(part 2):1757–1760
267. Fine KS, Driscoll CF, Molmberg JH, Mitchell TB (1991) Measurements of symmetric vortex merger. Phys Rev Lett 67(5):588–591

268. Finnigan TD, Luther DL, Lukas R (2002) Observations of enhanced diapycnal mixing near the Hawaiian Ridge. J Phys Oceanogr 32(11):2988–3002
269. Fjortoft R (1950) Application of integral theorems in deriving criteria of stability for laminar flows and for the baroclinic circular vortex. Geophys Publ 17(6):1–52
270. Flament P, Lumpkin R, Tourmadre J, Armi L (2001) Vortex pairing in an unstable anticyclonic shear flow: discrete subharmonics of one pendulum day. J Fluid Mech 440:401–409
271. Flatau M, Schubert WH, Stovens DE (1994) The role of baroclinic processes in tropical cyclone motion: the influence of vertical tilt. J Atmos Sci 51(1):2589–2601
272. Flierl GR (1978) Models of vertical structure and the calibration of two-layer models. Dyn Atmos Oceans 2(4):341–381
273. Flierl GR (1987) Isolated eddy models in geophysics. Annu Rev Fluid Mech 19:493–530
274. Flierl GR (1988) On the instability of geostrophic vortices. J Fluid Mech 197:349–388
275. Flierl GR, Carton X, Messager C (1999) Vortex formation by unstable oceanic jets. Eur Ser Appl Ind Math 7:137–150
276. Flierl GR, Larichev VD, McWilliams JC, Reznik GM (1980) The dynamics of baroclinic and barotropic solitary eddies. Dyn Atmos Oceans 5(1):1–41
277. Flierl GR, Stern ME, Whitehead JA Jr (1983) The physical significance of modons: laboratory experiments and general integral constraints. Dyn Atmos Oceans 7(4):233–264
278. Flór J-B, Govers WSS, van Heijst GJF, van Sluis R (1993) Formation of a tripolar vortex in a stratified fluid. Appl Sci Res 51(1–2):405–409
279. Flór J-B, van Heijst GJF (1994) An experimental study of dipolar vortex structures in a stratified fluid. J Fluid Mech 279:101–133
280. Flór JB, van Heijst GJF (1996) Stable and unstable monopolar vortices in a stratified fluid. J Fluid Mech 311:257–287
281. Flór J-B, Hopfinger EJ, Guyez E (2010) Contribution of coherent vortices such as Langmuir cells to wind-driven surface layer mixing. J Geophys Res 115:C10031. doi:10.1029/2009JC005900
282. Freymuth P, Bank W, Palmer M (1984) First experimental evidence of vortex splitting. Phys Fluids 27(5):1045–1046
283. Freymuth P, Bank W, Palmer M (1985) Futher experimental evidence of vortex splitting. J Fluid Mech 152:289–299
284. Freeland HJ, Rhines P, Rossby HT (1975) Statistical observations of trajectories of neutrally buoyant floats in the North Atlantic. J Mar Res 33:383–404
285. Friedrich KO (1953) Special topics in fluid dynamics. New York University Press, New York
286. Fujiwhara S (1921) The mutual tendency towards symmetry of motion and its application as a principle in meteorology. Q J R Meteorol Soc 47(200):287–293
287. Fujiwhara S (1923) On the growth and decay of vortical systems. Q J R Meteorol Soc 49(206):75–104
288. Fujiwhara S (1931) Short note on the behavior of two vortices. Proc Phys Math Soc Jpn III Ser 13:106–110
289. Fukumoto Y (2003) The three-dimensional instability of a strained vortex tube revisited. J Fluid Mech 493:287–318
290. Gallaire F, Chomaz J-M (2003) Three-dimensional instability of isolated vortices. Phys Fluids 15(8):2113–2126
291. Garrett C, MacCready P, Rhines P (1993) Boundary mixing and arrested Ekman layers: rotating stratified flow near a sloping boundary. Annu Rev Fluid Mech 25:291–323
292. Garrett C, Munk W (1972) Oceanic mixing by breaking internal waves. Deep-Sea Res 19(12):823–832
293. Gent P, McWilliams J (1986) The instability of barotropic circular vortices. Geophys Astrophys Fluid Dyn 35(1–4):209–233
294. Gent P, McWilliams J (1990) Isopycnal mixing in ocean circulation models. J Phys Oceanogr 20(1):150–155
295. Gill AE (1982) Atmosphere-ocean dynamics. Academic, London, 662 pp

296. Ginzburg AI, Kostianoy AG, Soloviev DM, Stanichny SV (2000) Remotely sensed coastal/deep-basin water exchange processes in the Black Sea surface. In: Halpern D (ed) Satellite, oceanography and society, Chapter 15. Elsevier, Amsterdam/Lausanne/New York/Oxford/Shannon/Singapore/Tokyo, pp 273–287
297. Ginzburg AI, Kostianoy AG, Nezlin NP, Soloviev DM, Stanichny SV (2002) Anticyclonic eddies in the northwestern Black Sea. J Mar Syst 32(1–3):91–106
298. Gledzer AE (1999) Mass entrainment and release in ocean eddy structures. Izv Atmos Ocean Phys 35(6):759–766
299. Gluhovsky AB, Klyatskin VI (1977) On dynamics of flipover phenomena in simple hydrodynamic models. Dokl Earth Sci Sec 237:18–20
300. Goldshtik M, Hussain F (1998) Analysis of inviscid vortex breakdown in a semi-infinite pipe. Fluid Dyn Res 23(4):189–234
301. Goncharov VP, Gryanik VM, Pavlov VI (2002) Venusian 'hot spots': physical phenomenon and its quantification. Phys Rev E66. doi:10.1103/PhysRevE.66.066304
302. Goncharov VP, Pavlov VI (2001) Cyclostrophic vortices in polar regions of rotating planets. Nonlinear Process Geophys 8(4/5):301–311
303. Goncharov VP, Pavlov VI (2003) Hamitonian contour dynamics. In: Borisov AV, Mamaev IS, Sokolovskiy MA (eds) Fundamental and applied problems of the vortex theory. Institute of Computer Science, Moscow–Izhevsk, pp 179–237 (in Russian)
304. Goncharov VP, Pavlov VI (2008) Hamitonian vortex and wave dynamics. GEOS, Moscow, 432 pp (in Russian)
305. Goosse H, Deleersnijder E, Fichefet T, England MH (1999) Sensitivity of a global coupled ocean-sea ice model to the parameterization of verical mixing. J Geophys Res 104(C6):13681–13685
306. Grant ALM, Belcher SE (2011) Wind-driven mixing below the oceanic mixed layer. J Phys Oceanogr 41(8):1556–1575
307. Greenslade MD, Haynes PH (2008) Vertical transition in transport and mixing in baroclinic flows. J Atmos Sci 65(4):1137–1157
308. Greenspan HP (1990) The theory of rotating fluids. Breukelen Press, Brookline, 352 pp
309. Griffiths RW, Hopfinger EJ (1986) Experiments with baroclinic vortex pairs in a rotating fluid. J Fluid Mech 173:501–518
310. Griffiths RW, Hopfinger EJ (1987) Coalescing of geostrophic vortices. J Fluid Mech 178:73–97
311. Griffiths RW, Linden PF (1981) The stability of vortices in a rotating, stratified fluid. J Fluid Mech 105:283–316
312. Griffiths RW, Linden PF (1985) Intermittent baroclinic instability and fluctuations in geophysical circulations. Nature 316(6031):801–803
313. Griffiths RW, Pearce AF (1985) Instability and eddy pairs on the Leeuwin Current south of Australia. Deep-Sea Res 32:1511–1534
314. Grimshaw R, Tanga Y, Broutman D (1994) The effect of vortex stretching on the evolution of barotropic eddies over a topographic slope. Geophys Astrophys Fluid Dyn 76(1–4):43–71
315. Gryanik VM (1983) Dynamics of singular geostrophic vortices in a two-level model of atmosphere (ocean). Izv Atmos Ocean Phys 19(3):171–179
316. Gryanik VM (1983) Dynamics of localized vortex perturbations "vortex charges" in a baroclinic fluid. Izv Atmos Ocean Phys 19(5):347–352
317. Gryanik VM (1988) Localized vortices—"vortex charges" and "vortex filaments" in a baroclinic differentially rotating fluid. Izv Atmos Ocean Phys 24(12):919–926
318. Gryanik VM (1990) About theoretical models of the localized quasi-geostrophic eddies in the atmosphere and ocean. In: Nikiforov EG, Romanov VF (eds) The investigations of vortex dynamics and energetics of the atmosphere, and the problems of climate. Gydrometeoizdat, Leningrad, pp 31–60 (in Russian)
319. Gryanik VM (1991) Dynamics of singular geostrophic vortices near critical points of currents in a N–layer model of the atmosphere (ocean). Izv Atmos Ocean Phys 27:517–526

320. Gryanik VM (1992) Radiation of Rossby waves and adaptation of potential vorticity fields in the atmosphere (ocean). (Dokl) Trans RAS Earth Sci Sect 326(1):976–979
321. Gryanik VM, Borth H, Olbers D (2001) The theory of quasigeostrophic von Kármán vortex streets in two-layer fluids on beta-plane and intermittent turbulent jets. Alfred-Wegener-Institut für Polar- und Meeresforschung, Preprint 106, 59 pp
322. Gryanik VM, Borth H, Olbers D (2004) The theory of quasigeostrophic von Kármán vortex streets in two-layer fluids on beta-plane. J Fluid Mech 505:23–57
323. Gryanik VM, Doronina TN (1990) Advective transport of a conservative solute by baroclinic singular quasigeostrophic vortices in the atmosphere (ocean). Izv Atmos Ocean Phys 26(10):1011–1026
324. Gryanik VM, Doronina TN, Olbers D, Warncke TH (2000) The theory of three-dimensional hetons and vortex-dominated spreading in localized turbulent convection in a fast rotating stratified fluid. J Fluid Mech 423:71–125
325. Gryanik VM, Sokolovskiy MA, Verron J (2003) Dynamics of barocline vortices with zero total intensity (hetons). In: Borisov AV, Mamaev IS, Sokolovskiy MA (eds) Fundamental and applied problems of the vortex theory. Institute of Computer Science, Moscow–Izhevsk, pp 547–622 (in Russian)
326. Gryanik VM, Sokolovskiy MA, Verron J (2006) Dynamics of heton-like vortices. Regul Chaotic Dyn 11(3):417–438
327. Gryanik VM, Tevs MV (1989) Dynamics of singular geostrophic vortices in an N-layer model of atmosphere (ocean). Izv Atmos Ocean Phys 25(3):179–188
328. Gryanik VM, Tevs MV (1991) Dynamics of singular geostrophic vortices near critical points of currents in a N-layer model of the atmosphere (ocean). Izv Atmos Ocean Phys 27(7):517–526
329. Gryanik VM, Tevs MV (1997) Dynamics and energetics of heton interacting in linearly and exponentially stratifed media. Izv Atmos Ocean Phys 33(4):385–398
330. Gudimenko AI (2007) Dynamics of perturbed equilateral and collinear configurations of three point vortices. Russ Sci J Nonlinear Dyn 3(4):379–391 (in Russian)
331. Gudimenko AI (2008) Dynamics of perturbed singular confguration of three point vortices. Russ Sci J Nonlinear Dyn 4(2):429–441 (in Russian)
332. Gudimenko AI (2008) Dynamics of perturbed equilateral and collinear configurations of three point vortices. Regul Chaot Dyn 13(2):85–95
333. Gudimenko AI, Zakharenko AD (2010) Qualitative analysis of relative motion of three vortices. Russ Sci J Nonlinear Dyn 6(2):307–326 (in Russian)
334. Gurulev AYu, Kozlov VF (1988) Numerical modeling of structure changes on a potential vorticity front. Izv Atmos Ocean Phys 34(3):395–403
335. Haidvogel DB, Beckmann A, Chapman DC, Lin R-Q (1993) Numerical simulation of flow around a tall isolated seamount. Part II. Resonant generation of trapped waves. J Phys Oceanogr 23(11):2373–2391
336. Hairer E, Nørsett SR, Wanner G (2008) Solving ordinary differential equations. I: Nonstiff Problems. Springer, Berlin, 528 pp
337. Hakim GJ, Snyder C, Muraki DJ (2002) A new surface model for cyclone-anticyclone asymmetry. J Atmos Sci 59(16):2405–2420
338. Hart JE, Adler B, Leben R (1988) Cyclonic/anticyclonic gyre asymmetries: laboratory and intermediate-model experiments. Dyn Atmos Oceans 27(1–4):219–232
339. Harvey BJ, Ambaum MHP, Carton XJ (2011) Instability of shielded surface temperature vortices. J Atmos Sci 68(5):964–971
340. Hattori Y, Fukumoto Y (2003) Short-wavelength stability analysis of thin vortex rings. Phys Fluids 15(10):3151–3163
341. Hebert D, Oakey N, Ruddick B (1990) Evolution of a Mediterranean salt lens: scalar properties. J Phys Oceanogr 20(9):1468–1483
342. Helfrich KR, Battisti TM (1991) Experiments on baroclinic vortex shelding from hydrothermal plumes. J Geophys Res 96(C12):12511–12518

343. Helfrich KR, Send U (1988) Finite-amplitude evolution of two-layer geostrophic vortices. J Fluid Mech 197:331–348
344. Hénon A (1976) Family of periodic solutions of the planar three-body problem and their stability. Celest Mech Dyn Astron 13:267–285
345. Herbette S, Morel Y, Arhan M (2003) Erosion of a surface vortex by a seamount. J Phys Oceanogr 33(8):1664–1679
346. Herbette S, Morel Y, Arhan M (2005) Erosion of a surface vortex by a seamount on the β plane. J Phys Oceanogr 35(11):2012–2030
347. Hesthaven JS, Lynov JP, Rasmussen JJ, Sutyrin GG (1993) Generation of tripolar vortical structures on the beta plane. Phys Fluids A5(7):1674–1678
348. Hogan PJ, Hubert HE (2006) Why do intrathermocline eddies form in the Japan/East Sea? A modeling perspective. Oceanography 19(3):134–143
349. Hogg NG (1973) On the stratifed Taylor column. J Fluid Mech 58:517–537
350. Hogg NG, Stommel HM (1985) The heton, an elementary interaction between discrete baroclinic geostrophic vortices, and its implications concerning eddy heat-flow. Proc R Soc Lond A 397:1–20
351. Hogg NG, Stommel HM (1985) Hetonic explosions: the breakup and spread of warm pools as explained by baroclinic point vortices. J Atmos Sci 42(14):1465–1476
352. Hogg NG, Stommel HM (1990) How currents in the upper thermocline could advect meddies deeper down. Deep Sea Res 37(4):613–623
353. Holland GJ, Dietachmayer GS (1993) On the interaction of tropical-cyclone-scale vortices: 3. Continuous barotropic vortices. Q J R Meteorol Soc 119(514):1381–1398
354. Holland GJ, Lander M (1993) The meandering nature of tropical cyclone tracks. J Atmos Sci 50(9):1254–1266
355. Holloway G (1986) Eddies, waves, circulation, and mixing: Statistical geofluid mechanics. Ann Rev Fluid Mech 18:91–147
356. Holmboe J (1968) Instability of baroclinic three-layer models of the atmosphere, vol 27. Geofys Publ Universitetsforlaget, Oslo, pp 1–27
357. Hopfinger EJ, van Heijst GJF (1993) Vortices in rotating fluids. Annu Rev Fluid Mech 25:241–289
358. Horton W, Liu J, Meiss JD, Sedlak JE (1986) Solitary vortices in a rotating plasma. Phys Fluids 29(4):1004–1010
359. Houghton RW, Olson DJ, Celone PJ (1986) Observation of an anticyclonic eddy near the continental shelf break south of New England. J Phys Oceanogr 16(1):60–71
360. Huang RX (1987) A three-layer model for wind-driven circulation in a subtropical-subpolar basin. Part I: Model formulation and the subcritical state; Part 2: The supercritical and hypercritical state. J Phys Oceanogr 17(5):664–678; 679–687
361. Huang RX (1988) A three-layer model for wind-driven circulation in a subtropical-subpolar basin. Part 3: Potential vorticity analisys. J Phys Oceanogr 18(5):739–752
362. Huang RX, Bryan K (1987) A multilayer model of the thermohaline and wind-driven ocean circulation. J Phys Oceanogr 17(11):1909–1924
363. Huang RX, Flierl GR (1987) Two-layer models for the thermocline and current structure in subtropical/subpolar gyres. J Phys Oceanogr 17(6):872–884
364. Huppert HE (1975) Some remarks on the initiation of in-ertial Taylor columns. J Fluid Mech 67:397–412
365. Huppert HE, Bryan K (1976) Topographically generated eddies. Deep Sea Res 23(8):655–679
366. Huq P, Britter RE (1995) Turbulence evolution and mixing in a two-layer stably stratified fluid. J Fluid Mech 285:41–67
367. Husain HS, Shtern V, Hussain F (2003) Control of vortex breakdown by addition of near-axis swirl. Phys Fluids 15(2):271–279
368. Hyun KH, Hogan PJ (2008) Topographic effects on the anticyclonic vortex evolution: a modeling study. Cont Shelf Res 28(10–11):1246–1260

369. Hyun KH, Hogan PJ (2008) Topographic effects on the path and evolution of Loop Current Eddies. J Geophys Res 113(C12). doi:10.1029/2007JC004155
370. Ibraev RA, Kuksa VI, Skirta AYu (2000) Modeling of the passive admixture transfer by the eddy currents in the eastern part of the Black Sea. Oceanology 40(1):18–25
371. Ikeda M (1981) Meanders and detached eddies of a strong easrward-flowing jet using a two-layer quasi-geostrophic model. J Phys Oceanogr 11(4):526–540
372. Ikeda M (1981) Instability and splitting of mesoscale rings using a two-layer quasi-geostrophic model on an f-plane. J Phys Oceanogr 11(7):987–998
373. Ikeda M (1983) Linear instability of a current flowing along a bottom sloping using a three-layer model. J Phys Oceanogr 13(2):208–223
374. Ikeda M, Apel JR (1981) Mesoscale eddies detached from spatially growing meanders in a eastward-flowing oceanic jet using a two-layer quasi-geostrophic model. J Phys Oceanogr 11(12):1638–1661
375. Ingersoll AP (1969) Inertial Taylor columns and Jupiter's Great Red Spot. J Atmos Sci 26(7):744–752
376. Inoue R, Smyth WD (2009) Efficiency of mixing forced by unsteady shear flow. J Phys Oceanogr 39(5):1150–1166
377. Ishizu M, Kitade Y, Michida Y (2013) Mixing process on the northeast coast of Hokkaido in summer. J Oceanogr 69(1):1–13
378. Ivanov AYu, Ginzburg AI (2002) Oceanic eddies in synthetic aperture radar images. Earth Planet Sci 111(3):281–295
379. Ivanov YuA, Kort VG, Shapovalov SM, Sherbinin AD (1988) Meso-scale intrusion lenses. In: Kort VG (ed) Proceedings hydrophysical studies at "Mesopolygon" program. Nauka, Moscow, pp 41–46 (in Russian)
380. Izrailsky YuG, Koshel KV, Stepanov DV (2008) Determination of optimal excitation frequency range in background flows. Chaos 18(1):013107. doi:10.1063/1.2835349
381. Izrailsky YuG, Kozlov VF, Koshel KV (2003) Some features of chaotization of a pulsating barotropic flow over a seamount with elliptic cross-section. Russ J Numer Anal Math Model 18(3):243–260
382. Izrailsky YuG, Kozlov VF, Koshel KV (2004) Some specific features of chaotization of the pulsating barotropic flow over elliptic and axisymmetric sea-mounts. Phys Fluids 16(8):3173–3190
383. Jacob JP, Chassignet EP, Dewar WK (2002) Influence of topography on the propagation of isolated eddies. J Phys Oceanogr 32(10):2848–2869
384. Jamaloodeen MI, Newton PK (2007) Two-layer quasigeostrophic potential vorticity model. J Math Phys. doi:10.1063/1.2469221
385. Janowitz GS (1975) The effect of bottom topography on stratified flow in the beta-plane. J Geophys Res 80(30):4163–4168
386. Jeong J, Hussain F (1995) On the identification of a vortex. J Fluid Mech 285:69–94
387. Jiménez J (1975) Stability of a pair of co-rotating vortices. Phys Fluids 18(11):1580–1581
388. Jiménez J, Wray AA (1998) On the characteristics of vortex filaments in isotropic turbulence. J Fluid Mech 371:255–285
389. Johnson ER (1977) Stratified Taylor columns on a beta-plane. Geophys Astrophys Fluid Dyn 9(1):159–177
390. Johnson ER (1978) Trapped vortices in rotating flow. J Fluid Mech 86(2):209–224
391. Johnson ER (1978) Topographically bound vortices. Geophys Astrophys Fluid Dyn 11(1):61–71
392. Johnson ER (1979) Finite depth stratified flow over topography on a beta-plane. Geophys Astrophys Fluid Dyn 12(1):35–43
393. Johnson J, Ambar I, Serra N, Stevens I (2002) Comparative studies of the spreading of Mediterranean water through the Culf Cadiz. Deep-Sea Res II 49:4179–4193
394. Jones C, Winkler S (2002) Invariant manifolds and Lagrangian dynamics in the ocean and atmosphere. In: B Hasselblatt, B Fiedler, AB Katok (eds) Handbook of dynamical systems, vol 2, Chap. 2. Elsevier, Amsterdam/New York/North Holland, pp 55–29

395. Jones H, Marshall J (1993) Convection with rotation in a neutral ocean: a study of deep-ocean convection. J Phys Oceanogr 23(6):1009–1039
396. Josserand Ch, Rossi M (2007) The merging of two co-rotating vortices: a numerical study. Eur J Mech B/Fluids 26(6):779–794
397. Junglaus JH (1999) A three-dimensional simulation of the formation of anticyclonic lenses (meddies) by the instability of an intermediate depth boundary current. J Phys Oceanogr 29(6):1579–1598
398. Juza M, Penduff T, Brankart J-M, Barnier B (2012) Estimating the distortion of mixed layer porperty distributions by the ARGO sampling. J Oper Oceanogr 5(1):45–58
399. Kalashnik MV, Svirkunov PN (1996) Symmetric stability of the cyclostrophic and geostrophic balance states in a stratified medium. (Dokl) Trans RAS/Earth Sci Sec 349(5):829–831
400. Kalashnik MV, Svirkunov PN (1996) Cyclostrophic and geostrophic lalance states in a shallow-water model. Izv Atmos Ocean Phys 32(3):370–377
401. Kalashnik MV, Visheratin KN (2008) Cyclostrophic adjustment in swirling gas flows and the Ranque-Hilsch vortex tube effect. J Exp Theor Phys 106(4):819–829
402. Kalashnik MV, Visheratin KN (2010) Cyclostrophic adjustment and nonlinear oscillations in the core of intense atmospheric vortex. Izv Atmos Ocean Phys 46(5):591–596
403. Kamenkovich VM (1977) Fundamentals of ocean dynamics. Elsevies Sci. Publ. Co., Amsterdam/New York 249 pp
404. Kamenkovich VM, Koshlyakov MN, Monin AS (1986) Synoptic eddies in the ocean. EFM, D. Reidel Publishing Company, Dordrecht, 433 pp
405. Kamenkovich VM, Larichev VD, Khar'kov BV (1982) A numerical barotropic model for analysis of synoptic eddies in the open ocean. Oceanology 21(6):549–558
406. Kapela T, Simó C (2007) Computer assisted proofs for nonsymmetric planar choreographies and for stability of the Eight. Nonlinearity 20(5):1241–1255
407. Kapela T, Zgliczyński P (2003) The existence of simple choreographies for N-body problem—a computer assisted proof. Nonlinearity 16(6):1899–1918
408. Käse RH, Zenk W (1987) Reconstructed Mediterranean salt lens trajectories. J Phys Oceanogr 17(1):158–161
409. Karsten R, Jones H, Marshal J (2002) The role of eddy transfer in setting the stratification and transport of a circumpolar current. J Phys Oceanogr 32(1):39–54
410. Kawakami A, Funakoshi M (1999) Chaos motion of fluid particles around a rotating elliptic vortex in a linear shear flow. Fluid Dyn Res 25(4):167–193
411. Kennelly MA, Evans RH, Joyce TM (1985) Small-scale cyclones on the periphery of Gulf Stream warm-core rings. J Geophys Res 90(C5):8845–8857
412. Kerswell RR (2002) Elliptical instability. Annu Rev Fluid Mech 34:83–113
413. Khvoles R, McWilliams JC, Kizner Z (2007) Non-coincidence of separatrices in two-layer modons. Phys Fluids 19(5):056602. doi:10.1063/1.2731741
414. Kida S (1981) Motion of an elliptic vortex in an uniform shear flow. J Phys Soc Jpn 50(10):3517–3520
415. Killworth PD (1983) On the motion of isolated lenses on a beta-plane. J Phys Oceanogr 13(3):368–376
416. Kirchhoff G (1876) Vorlesungen über mathematische Physik: Mechanik. Taubner, Leipzig
417. Kiya M, Takeo H, Mochizuki O, Kudo D (1999) Simulating vortex pairs interacting with mixing-layer vortices. Fluid Dyn Res 24(1):61–79
418. Kizner ZI (1984) Rossby solitons with axially symmetric baroclinic modes. Dokl (Trans) USSR Acad Sci 275:211–214
419. Kizner ZI (1985) Interpretation of soliton solutions to the equation of quasi-geostrophic vorticity in a baroclinic ocean. Izv Atmos Ocean Phys 21:330–332
420. Kizner ZI (1986) Intensity of synoptic eddies and the quasigeostrophic approximation. Oceanology 26(1):16–20
421. Kizner ZI (1986) Strongly nonlinear solitary Rossby waves. Oceanology 26(3):284–289

422. Kizner ZI (1988) On the theory of intrathermocline eddies. Dokl (Trans) USSR Akad Sci 300:213–216
423. Kizner ZI (1997) Solitary Rossby waves with baroclinic modes. J Mar Res 55(4):671–685
424. Kizner Z (2006) Stability and transitions of hetonic quartets and baroclinic modons. Phys Fluids 18(5):056601. doi:10.1063/1.2196094
425. Kizner Z (2008) Hetonic quartet: exploring the transitions in baroclinic modons. In: Borisov AV, Kozlov VV, Mamaev IS, Sokolovskiy MA (eds) IUTAM symposium on Hamiltonian dynamic, vortex strictures, turbulence (IUTAM Bookseries, vol 6). Springer, New York, pp 125–133
426. Kizner Z (2011) Stability of point-vortex multipoles revisited. Phys Fluids 23(6):064104. doi:10.1063/1.3596270
427. Kizner Z, Berson D, Khvoles R (2002) Baroclinic modon equilibria on the beta-plane: stability and transitions. J Fluid Mech 468:239–270
428. Kizner Z, Berson D, Khvoles R (2003) Non-circular baroclinic modons: constructing stationary solutions. J Fluid Mech 489:199–228
429. Kizner Z, Berson D, Reznik G, Sutyrin G (2003) The theory of the beta-plane baroclinic topographic modons. Geophys Astrophys Fluid Dyn 97(3):175–211
430. Kizner Z, Khvoles R (2004) The tripole vortex: experimental evidence and explicit solutions. Phys Rev E 70(1):016307. doi:10.1103/PhysRevE.70.0163072004
431. Kizner Z, Khvoles R (2004) Two variations on the theme of Lamb-Chaplygin: supersmooth dipole and rotating multipoles. Regul Chaotic Dyn 9(4):509–518
432. Kizner Z, Khvoles R, McWilliams JC (2007) Rotating multipoles on the f- and γ-planes. Phys Fluids 19(1):016603. doi:10.1063/1.2432915
433. Kizner Z, Reznik GM, Fridman B, Khvoles R, McWilliams JC (2008) Shallow-water modons on the f-plane. J Fluid Mech 603:305–329
434. Klocker A, Ferrari R, LaCasce JH (2012) Estimating suppression of eddy mixing by mean flows. J Phys Oceanogr 42(9):1566–1576
435. Kloosterziel RC, van Heijst GJF (1989) On tripolar vortices. In: Nihoul JCJ, Jamart BM (eds) Mesoscale/synoptic coherent structures in geophysical turbulence. Elsevier, Amsterdam/Oxford/New York/Tokyo, pp 609–625
436. Kloosterziel RC, van Heijst GJF (1991) An experimental study of unstable barotropic vortices in a rotating fluid. J Fluid Mech 223:1–24
437. Klyatskin VI (2007) Stochastic equations. Theory and its applications in acoustics, hydrodynamics, and radio physics. Fizmatlit, Moscow. V. 1: Basic regulations, exact results, and asymptotic approximations, 318 pp. V. 2: Coherent phenomena in stochastic dynamic systems, 343 pp (in Russian)
438. Kochin NE, Kibel XA, Roze NV (1965) Theoretical hydrodynamics. Wiley, New York, 577 pp
439. Koiller J, Carvalho SP, Silva RR, Oliveira LCG (1985) On Aref's vortex motions with a symmetry center. Phys D 16(1):27–61
440. Koshel KV, Izrail'ski YuG, Stepanov DV (2006) Determining the optimal frequency of perturbation in the problem of chaotic transport of particles. Dokl Phys 51(4):219–222
441. Koshel KV, Prants SV (2006) Chaotic advection in the ocean. Phys Usp (Adv Phys Sci) 49:1151–1178
442. Koshel KV, Prants SV (2008) Chaotic advection in the ocean. Institute of Computer Science, Moscow–Izhevsk, 364 pp (in Russian)
443. Koshel KV, Sokolovskiy MA, Davies PA (2008) Chaotic advection and nonlinear resonances in an oceanic flow above submerged obstacle. Fluid Dyn Res 20(10):695–736
444. Koshel KV, Sokolovskiy MA, Verron J (2013) Three-vortex quasi-geostrophic dynamics in a two-layer fluid. Part 2. Regular and chaotic advection around the perturbed steady states. J Fluid Mech 717:255–280
445. Koshel KV, Stepanov DV (2005) Boundary effect on the mixing and transport of passive impurities in a nonstationary flow. Tech Phys Lett 31(2):135–137

446. Koshel KV, Stepanov DV (2006) Chaotic advection induced by a topographic vortex in baroclinic ocean. Dokl Earth Sci 407(2):455–459
447. Koshlyakov MN, Panteleev GG (1988) Termohaline characteristic of the Mediterranean water lens at tropical zone North Atlantic. In: Kort VG (ed) Proceedings hydrophysical studies at "Mesopolygon" program, Nauka, Moscow, pp 46–57 (in Russian)
448. Koszalka I, Caballos L, Bracco A (2010) Vertical mixing and coherent anticyclones in the ocean: the role of stratification. Nonlinear Process Geophys 17(1):37–47
449. Kozlov VF (1969) Lectures on the theory of stationary ocean currents (study guide for students in oceanography). Far Eastern State University, Vladivostok, 383 pp (in Russian)
450. Kozlov VF (1975) Mutual adaptation of the mass and current fields to the bottom relief in a baroclinic ocean. Izv Atmos Ocean Phys 11(1):23–28
451. Kozlov VF (1977) Geostrophic motion of a stratified fluid above an uneven bottom. Izv Atmos Ocean Phys 13(9):657–662
452. Kozlov VF (1980) Formation of a Rossby wave under the action of diturbances in a nonstationary barotropic oceanic flow. Izv Atmos Ocean Phys 16(4):275–279
453. Kozlov VF (1981) On a stationary problem of topographical cyclogenesis in a rotating fluid. Izv Atmos Ocean Phys 17(11):878–882
454. Kozlov VF (1982) Quasistationary geostrophic motion of weakly stratified fluid in the ocean with arbitrary bottom relief. Izv Atmos Ocean Phys 18(7):574–578
455. Kozlov VF (1983) The method of contour dynamics in model problems of the ocean topographic cyclogenesis. Izv Atmos Ocean Phys 19(8):635–640
456. Kozlov VF (1984) Models of the topographic vortices in ocean. Nauka, Moscow, 200 pp (in Russian)
457. Kozlov VF (1985) Construction of a numerical model of geostrophic eddies in a baroclinic fluid based on the Contour Dynamics Method. Izv Atmos Ocean Phys 21(2):161–163
458. Kozlov VF (1991) Construction of the stationary states of vortex patches by the method of perturbations. Izv Atmos Ocean Phys 27(2):77–86
459. Kozlov VF (1992) Model of two-dimensional vortex motion with an entrainment mechanism. Fluid Dyn 27(6):793–798
460. Kozlov VF (1992) A nonlinear model for Kirchoff vortex dissipation. Oceanology 32(4):427–430
461. Kozlov VF (1993) Model of the interaction of elliptic vortex patches with entrainment effects. Izv Atmos Ocean Phys 29(1):90–96
462. Kozlov VF (1994) Geophysical hydrodynamics of vortical patches. Phys Oceanogr 6(1):25–34
463. Kozlov VF (1995) Background currents in geophysical hydrodynamics. Izv Atmos Ocean Phys 31(2):229–234
464. Kozlov VF, Gurulev AYu (1996) Barotropic eddy evolution near a rectilinear bottom break. Izv Atmos Ocean Phys 32(2):249–256
465. Kozlov VF, Gurulev AYu (1997) Moment model of the dynamics of barotropic vortex over a marine trench (ridge) of rectangular section. Izv Atmos Ocean Phys 33(6):837–844
466. Kozlov VF, Gurulev AYu (1998) Dynamics of the front of potential vorticity in the field of background currents. Izv Atmos Ocean Phys 34(3):395–403
467. Kozlov VF, Koshel KV (1999) Barotropic model of chaotic advection in background flows. Izv Atmos Ocean Phys 31(1):123–130
468. Kozlov VF, Koshel KV (2000) A model of chaotic transport in the barotropic background flow. Izv Atmos Ocean Phys 36(1):109–118
469. Kozlov VF, Koshel KV (2001) Some features of chaos development in an oscillatory barotropic flow over an axisymmetric submerged obstacle. Izv Atmos Ocean Phys 37(3):351–361

470. Kozlov VF, Koshel KV (2003) Chaotic advection in the models for background flows of geophysical hydrodynamics. In: Borisov AV, Mamaev IS, Sokolovskiy MA (eds) Fundamental and applied problems of the vortex theory. Institute of Computer Science, Moscow–Izhevsk, pp 471–504 (in Russian)
471. Kozlov VF, Koshel KV, Stepanov DV (2005) Influence of the boundary on the chaotic advection in the simplest model of a topographic vortex. Izv Atmos Ocean Phys 41(2):217–227
472. Kozlov VF, Makarov VG (1984) Evolution modeling of unstable geostrophic eddies in a barotropic ocean. Oceanology 24(5):556–560
473. Kozlov VF, Makarov VG (1985) Simulation of the instability of axisymmetric vortices using the contour dynamics method. Fluid Dyn 20(1):28–34
474. Kozlov VF, Makarov VG (1995) Background currents in the Sea of Japan (a barotropic model). Oceanology 35(5):601–604
475. Kozlov VF, Makarov VG (1996) Background currents in the Sea of Japan (a two-layer quasigeostrophicc model). Oceanology 36(4):453–457
476. Kozlov VF, Makarov VG (1996) Background currents in the Sea of Okhotsk. Russ Meteorol Hydrol 1(9):39–44
477. Kozlov VF, Makarov VG, Sokolovskiy MA (1986) Numerical model of the baroclinic instability of axially symmetric eddies in two-layer ocean. Izv Atmos Ocean Phys 22(8):674–678
478. Kozlov VF, Sal'nikov PYu (1989) Mechanism for the formation of mushroom-shaped flows with dense packing of vortices. Izv Atmos Ocean Phys 25(4):324–326
479. Kozlov VF, Sal'nikov PYu (1990) The jet (pulse) model of mushroom-like flow formation. Phys Oceanogr 1(3):171–175
480. Kozlov VF, Shavlyugin AI (1992) Stationary arrays of vortical patches near linear boundaries. Phys Oceanogr 3(4):241–250
481. Kozlov VF, Sokolovskiy MA (1978) Stationary motion of a stratified fluid above a rough bottom (geostrophic approximation on the β–plane). Oceanology 18(4):383–386
482. Kozlov VF, Sokolovskiy MA (1980) Influence of cylindrical topographic disturbanced on a nonstationary zonal flow of a stratified fluid on the beta plane. Izv Atmos Ocean Phys 16(8):596–604
483. Kozlov VF, Sokolovskiy MA (1981) Meander of a barotropic zonal current crossing a bottom ridge (periodic regime). Oceanology 21(6):684–687
484. Kozlov VF, Yaroshchuk EV (1986) Numerical modeling of structural transitions in a plane shear layer. Fluid Dyn 21(5):712–715
485. Kozlov VV (2003) General theory of vortices. Dynamical Systems, X, Encyclopaedia of Mathematical Sciences, vol 67. Springer, Berlin
486. Kozlov VV (1996) Symmetries, topology and resonances in Hamiltonian mechanics. Springer, Berlin
487. Krasnopolskaya TS, Meleshko VV, Peters GWM, Meijer HEH (1999) Mixing in Stokes flow in an annular wedge cavity. Eur J Mech B/Fluids 18(5):793–822
488. Kuksa VI (1983) The intermediate waters of the World Ocean. Gydrometeoizdat, Leningrad, 272 pp (in Russian)
489. Kulik KN, Tur AV, Yanovsky VV (2010) Interaction of point and dipole vortices in an incompressible liquid. Theor Math Phys 162(3):383–400
490. Kundu PK, Cohen IM, Hu HH (2004) Fluid mechanics. Academic, New York 759 pp
491. Kunze E, Sanford T (1993) Submesoscale dynamics near a seamount. Part I: measurements of ertel vorticity. J Phys Oceanogr 23(12):2567–2588
492. Kuo AC, Polvani LM (2000) Nonlinear geostrophic adjustment, cyclone/anticyclone asymmetry, and potential vorticity rearrangement. Phys Fluids 12(5):1087. http://dx.doi.org/10.1063/1.870363
493. Kuo H-C, Chen GT-J, Lin C-H (2000) Merger of tropical cyclones Zeb and Alex. Mon Weather Rev 128(8):2967–2975
494. Kuo H-C, Williams RT, Chen GT-J, Chen Y-L (2001) Topographic effects on barotropic vortex motion: no mean flow. J Atmos Sci 58(10):1310–1327

495. Kurakin LG, Yudovich VI (2002) On nonlinear stability of steady rotation of a regular vortex polygon. Dokl Phys 47(6):465–470
496. Kurakin LG, Yudovich VI (2002) The stability of stationary rotation of a regular vortex polygon. Chaos 12(3):574–595
497. Kurganskiy MV (1990) On the motion of a pair of vortices on the beta-plane. In: Nikiforov EG, Romanov VF (eds) The investigations of vortex dynamics and energetics of the atmosphere, and the problems of climate. Gydrometeoizdat, Leningrad, pp 123–130 (in Russian)
498. Kusch HA, Ottino JM (1992) Experiments on mixing in continuous chaotic flows. J Fluid Mech 236:319–348
499. Kuzmina NP, Zhurbas VM, Rudels B, Stipa T, Paka VT, Muraviev SS (2008) Role of eddies and intrusions in the exchange processes in the Baltic halocline. Oceanology 48(2):149–158
500. Kuznetsov L, Zaslavsky GM (1998) Regular and chaotic advection in the flow field of a three-vortex system. Phys Rev E 58(6):7330–7349
501. Kuznetsov L, Zaslavsky GM (2000) Passive particle transport in three-vortex flow. Phys Rev E 61(4):3777–3792
502. Kvaleberg E, Morey SL, O'Brien JJ (2003) Frontogenesis and subsequent formation of cold filaments and eddies on an idealized shelf. Oceans 5:2831–2834
503. LaCasce JH (1998) A geostrophic vortex over a slope. J Phys Oceanogr 28(12):2362–2381
504. Ladyzhenskaya OA (1969) The mathematical theory of viscous incimpressible flow. 2nd Eng. edition revised and enlarged. Gordon and Breach Science Publishers, New York/London/Paris/Mintreux/Tokyo/Melbourne, 224 pp
505. Lahaye N, Zeitlin V (2011) Collisions of ageostrophic modons and formation of new types of coherent structures in rotating shallow water model. Phys Fluids 23(6):061703 http://dx.doi.org/10.1063/1.3597608
506. Lahaye N, Zeitlin V (2012) Existence and properties of ageostrophic modons and coherent tripoles in the two-layer rotating shallow water model on the f-plane. J Fluid Mech 706:71–107
507. Lahaye N, Zeitlin V (2012) Shock modon: a new type of coherent structure in rotating shallow water. Phys Rev Lett 108(4):044502. doi:10.1103/PhysRevLett.108.044502
508. Lam JS-L, Dritschel DG (2001) On the beta-drift of an initially circular vortex patch. J Fluid Mech 436:107–129
509. Lamb H (1885) Hydrodynamics, 2nd edn. Cambridge University Press, Cambridge
510. Lamb H (1932) Hydrodynamics, 6th edn. Cambridge University Press, Cambridge
511. Landau LD, Lifshits EM (1959) Fluid mechanics. Pergamon Press, New York, 536 pp
512. Lander MA (1995) The merger of two tropical cyclones. Mon. Weather Rev 123(7):2260–2265
513. Lander M, Holland GJ (1993) On the interaction of tropical-cyclone-scale vortices: 1. Observations. Q J R eteorol Soc 119(514):1347–1361
514. Lansky IM, O'Neil TM, Schecter DA (1997) A theory of vortex merger. Phys Rev Lett 79(8):1479–1482
515. Large WG, McWilliams JC, Doney SC (1994) Oceanic vertical mixing: a review and a model with a nonlocal boundary layer parametrization. Rev Geophys 32:363–403
516. Larichev VD, Reznik GM (1976) Two-dimensional solitary Rossby waves. Dokl Akad Nauk SSSR 231(5):1077–1079
517. Larichev VD, Reznik GM (1976) Strongly nonlinear, two-dimensional solitary Rossby waves. Oceanology 16(6):547–550
518. Larichev VD, Reznik GM (1982) Numerical experiments on the study of collision of two-dimensional solitary Rossby waves. Dokl Akad Nauk SSSR 264:229–233
519. Larichev VD, Reznik GM (1983) On collisions between two-dimensional solitary Rossby waves. Oceanology 23(5):545–552
520. Laurent LSt, Naveira Garabato AC, Ledwell JR, Thurnherr AM, Toole JM, Watson AJ (2012) Turbulence and diapycnal mixing in Drake Passage. J Phys Oceanogr 42(12):2143–2152

521. Lavrova OYu, Kostyanoy AG, Lebedev SA, Mityagina VI, Ginzburg AI, Sheremet NA (2011) Complex satellite monitoring of the Russian seas. Space Research Institute of RAS, Moscow, 470 pp
522. Lazutkin VF (1990) Analytic integrals of a semistandard mapping, and separatrix splitting. Leningr Math J 1(2):427–445
523. Le Dizés S, Verga A (2002) Viscous interactions of two co-rotating vortices before merging. J Fluid Mech 467:389–410
524. Legg S (2004) A simple criterion to determine the transition from a localized convection to a distributed convection regime. J Phys Oceanogr 34(12):2843–2846
525. Legg S, Jones H, Visbeck M (1996) A heton perspective of baroclinic eddy transfer in localized open ocean convection. J Phys Oceanogr 26(10):2251–2266
526. Legg S, Marshall J (1993) A heton model of the spreading phase of open-ocean deep convection. J Phys Oceanogr 23(6):1040–1056
527. Legg S, Marshall J (1998) The influence of the ambient flow on the spreading of convected water masses. J Mar Res 56(1):107–139
528. Legg S, McWilliams J, Gao J (1998) Localization of deep ocean convection by a mesoscale eddy. J Phys Oceanogr 28(5):944–970
529. Legras B, Dritschel DG (1991) The elliptical model of two-dimensional vortex dynamics. I. The basic state. Phys Fluids A3(5):845–854
530. Legras B, Dritschel DG (1993) A comparison of the contour surgery and pseudo-spectral method. J Comput Phys 104(2):287–302
531. Leibovich S, Stewartson K (1983) A sufficient condition for the instability of columnar vortices. J Fluid Mech 126:335–356
532. Leith CE (1984) Minimum enstrophy vortices. Phys Fluids 27(6):1388–1395
533. Lenn Y-D, Rippeth TP, Old CP, Bacon S, Polyakov I, Ivanov V, Hölemann J (2011) Intermittent intense turbulent mixing under ice in the Laptev Sea continental shelf. J Phys Oceanogr 41(3):531–547
534. Leoncini X, Kuznetsov L, Zaslavsky GM (2000) Motion of three vortices near collapse. Phys Fluids 12(8):1911–1927
535. Leoncini X, Kuznetsov L, Zaslavsky GM (2001) Chaotic advection near a three-vortex collapse. Phys Rev E 63(3):036224. doi:10.1103/PhysRevE.63.036224
536. Leoncini X, Zaslavsky GM (2002) Jets, stickiness, and anomalous transport. Phys Rev E 65(4):046216. doi:10.1103/PhysRevE.65.046216
537. Levina GV, Montgomery MT (2010) A first examination of the helical nature of tropical cyclogenesis. Dokl Earth Sci 434(part 1):1285–1289
538. Li M, Trowbridge J, Geyer R (2008) Asymmetric tidal mixing due to the horizontal density gradient. J Phys Oceanogr 38(2):418–434
539. Liang X, Thurnherr AM (2012) Eddy-modulated internal waves and mixing on a midocean ridge. J Phys Oceanogr 42(7), 1242–1248
540. Lim CC, Majda AJ (2001) Point vortex dynamics for coupled surface/interior QG and propagating heton clasters in models for ocean convection. Geophys Astrophys Fluid Dyn 94(3–4):177–220
541. Lim CC, Nebus J (2007) Vorticity, statistics, and Monte Carlo simulation. Springer, New York, 290 pp (Springer monographs in mathematics)
542. Lin S-J (1992) Contour dynamics of tornado-like vortices. J Atmos Sci 49(18):1745–1756
543. Liu C, Köhl A, Stammer D (2012) Adjoint-based estimation of eddy-induced tracer mixing parameters in the global ocean. J Phys Oceanogr 42(7):1186–1206
544. Love AEH (1893) On the stability of certain vortex motion. Proc Lond Math Soc s1–25:18–43
545. Love AEH (1893) On the motion of paired vortices with a common axis. Proc Lond Math Soc s1–25:185–194
546. Lovegrove AF, Moroz IM, Read PL (2001) Bifurcation and instabilities in rotating two-layer fluids: 1. f-plane. Nonlinear Process Geophys 8(1):21–36

547. Lovegrove AF, Moroz IM, Read PL (2002) Bifurcation and instabilities in rotating, two-layer fluids: 2. β-plane. Nonlinear Process Geophys 9(3–4):289–309
548. Lugt HJ (1996) Introduction to vortex theory. Vortex Flow Press. Potomac 627 pp
549. Lumpkin R, Flament P, Kloosterziel R, Armi L (2000) Vortex merging in a $1\frac{1}{2}$-layer fluid on an f-plane. J Phys Oceanogr 30(1):233–242
550. Luzzatto-Fegiz P, Williamson CHK (2010) Stability of elliptical vortices from "Imperfect–Velocity–Impulse" diagrams. Theor Comput Fluid Dyn 24(1–4):181–188
551. Luzzatto-Fegiz P, Williamson CHK (2010) Stability of conservative flows and new steady-fluid solutions from bifurcation diagrams exploiting a variational argument. Phys Rev Lett 104:044504. doi:10.1103/PhysRevLett.104.044504
552. Luzzatto-Fegiz P, Williamson CHK (2011) An efficient and general numerical method to compute steady uniform vortices. J Comput Phys 230(17):6495–6511
553. Luzzatto-Fegiz P, Williamson CHK (2012) Determining the stability of steady two-dimensional flows through imperfect velocity-impulse diagrams. J Fluid Mech 706:323–350
554. Luzzatto-Fegiz P, Williamson CHK (2012) Structure and stability of the finite-area von Kármán street. Phys Fluids 24(6):066602. http://dx.doi.org/10.1063/1.4724307
555. Maas LRM, Zahariev K (1996) An exact, stratified model of a meddy. Dyn Atmos Oceans 24:215–225
556. Madelain F (1970) Influence de la topographie du fond sur l'écoulement Méditerranéen entre le Détroit de Gibralter et le cap Saint-Vincent. Cah Oceanogr 22:43–61
557. Makarov VG (1990) A program code for investigation of plane vortex currents in an ideal fluid by the method of contour dynamics. In: Kozlov VF (ed) Method of contour dynamics in oceanological investigations. FED USSR Acad Sci, Vladivostok, pp 28–39 (in Russian)
558. Makarov VG (1991) Computational algorithm of the contour dynamics method with changeable topology of domains under study. Model Mekh 5(4):83–95
559. Makarov VG (1996) Numerical simulation of the formation of tripolar vortices by the method of contour dynamics. Izv Atmos Ocean Phys 32(1):40–49
560. Makarov VG, Bulgakov SN (2008) Regimes of near-wall vortex dynamics in potential flow through gaps. Phys Fluids 20(12):086605. doi:10.1063/1.2969471
561. Makarov VG, Kizner Z (2011) Stability and evolution of uniform-vorticity dipoles. J Fluid Mech 672:307–325
562. Makarov VG, Sokolovskiy MA, Kizner Z (2012) Doubly symmetric finite-core heton equilibria. J Fluid Mech 708:397–417
563. Mallik DD (1979) Influence of bottom topography on the zonal geostrophic flow in a stratified-ocean model. Izv Atmos Ocean Phys 15(10):781–783
564. Malikova NP, Permyakov MS (2010) Effect of the Ekman boundary layer on the evolution of vortex formations. Fluid Dyn 45(6):905–908
565. Malvern LE (1969) Introduction to the mechanics of continous medium. Englewood Cliffs, Prentice-Hall, 713 pp
566. Mancho AM, Small D, Wiggins S (2006) A tutorial on dynamical systems concepts applied to Lagrangian transport in oceanic flows defined as finite time data sets: theoretical and computational issues. Phys Rep 437(3–4):55–124
567. Mao X, Sherwin ST, Blackburn HM (2012) Non-normal dynamics of time-evolving co-rotating vortex pairs. J Fluid Mech 701:430–459
568. Mariotti A, Legras B, Dritschel DG (1994) Vortex stripping and the erosion of coherent structures in two-dimensional flows. Phys Fluids 6(12):3954–3962
569. Marshall JS (1995) Chaotic oscillations and breakup of quasigeostrophic vortices in the N-layer approximation. Phys Fluids 7(5):983–992
570. Marshall JS, Parthasarathy B (1993) Tearing of an aligned vortex by a current difference in two-layer quasi-geostrophic flow. J Fluid Mech 225:157–182
571. Marshall J, Schott F (1999) Open-ocean convection: observation, theory, and models. Rev Geophys 37(1):1–64

572. Martinsen-Burrell N, Julien K, Patersen MR, Weiss JB (2006) Merger and alignment in a reduced model for three-dimensional quasigeostrophic ellipsoidal vortices. Phys Fluids 18(5):057101. doi:10.1063/1.21918872006
573. Masina S, Pinardi N (1993) The halting effect of baroclinicity in vortex merging. J Phys Oceanogr 23(8):1618–1637
574. Matsuura T (1980) On a decay process if isolated, intense vortices in a two-layer ocean. J Oceanogr Soc Jpn 36(1):39–45
575. Matsuura T (1995) The evolution of frontal-geostrophic vortices in two-layer ocean. J Phys Oceanogr 25(10):2298–2318
576. Maxworthy T, Narimousa S (1994) Unsteady, turbulent convection into a homogeneous rotating fluid, with oceanographic applications. J Phys Oceanogr 24(5):865–887
577. Mazé JP, Arhan M, Mercier H (1997) Volume budget of the eastern boundary layer off the Iberian Peninsula. Deep-Sea Res I 44(9–10):1543–1574
578. McDonald NR (1998) The motion of an intense vortex near topography. J Fluid Mech 367:359–377
579. McDonald NR (2000) The interaction of two baroclinic geostrophic vortices on the β-plane. Proc R Lond A 456:1029–1049
580. McDonald NR (2004) A new translating quasigeostrophic V-state. Eur J Mech B/Fluids 23(4):633–644
581. McWilliams JC (1983) Interaction of isolated vortices. II: Modon generation by monopole collision. Geophys Astrophys Fluid Dyn 24(1):1–22
582. McWilliams JC (1985) Submesoscale, coherent vortices in the ocean. Rev Geophys 23(2):165–182
583. McWilliams JC (1996) Modeling the oceanic general circulation. Annu Rev Fluid Mech 28:215–248
584. McWilliams JC (2006) Fundamentals of geophysical fluid dynamics. Cambridge University Press, Cambridge, 266 pp
585. McWilliams JC, Flierl GR, Larichev VD, Reznik GM (1981) Numerical studies of barotropic modons. Dyn Atmos Oceans 5(4):219–238
586. McWilliams JC, Zabusky N (1982) Interaction of isolated vortices. I: Modons colliding with modons. Geophys Astrophys Fluid Dyn 19(3–4):207–227
587. Meacham SP (1991) Meander evolution on piecewise-uniform, quasi-geostrophic jets. J Phys Oceanogr 21(8):1139–1170
588. Meacham SP, Pankratov KK, Shchepetkin AF, Zhmur VV (1994) The interaction of ellipsoidal vortices with background shear flows in a stratified fluid. Dyn Atmos Oceans 21(2–3):167–212
589. Melander MV, McWilliams JC, Zabusky NJ (1987) Asymmetrization and vorticity-gradient intensification of an isolated two-dimensional vortex through filamentation. J Fluid Mech 178:137–159
590. Melander MV, Zabusky NJ, McWilliams JC (1987) Asymmetric vortex merger in two dimensions: which vortex is 'victorious'? Phys Fluids 30(9):2610–2612
591. Melander MV, Zabusky NJ, McWilliams JC (1988) Symmetric vortex merger in two dimensions: causes and conditions. J Fluid Mech 195:303–340
592. Melander MV, Zabusky NJ, Styczek AS (1986) A moment model for vortex interactions of the two-dimensional Euler equations. Part 1. Computational validation of a Hamiltonian elliptical representation. J Fluid Mech 167:95–115
593. Meleshko VV, Aref H (2007) A bibliography of vortex dynamics 1858–1956. Adv Appl Mech 41:197–292
594. Meleshko VV, Galaktionov OS, Peters GWM, Meijer HEH (1999) Three-dimensional mixing in Stokes flow: the partitioned pipe mixer problem revisited. Eur J Mech B/Fluids 18(5):783–792
595. Meleshko VV, Konstantinov MYu (1993) Dynamics of vortex structures. Kiev Naukova Dumka 280 pp (in Russian)
596. Meleshko VV, Konstantinov MYu, Gurzhi AA, Konovaljuk TP (1992) Advection of a vortex pair atmosphere in a velocity field of point vortices. Phys Fluids A 4(12):2779–2797

597. Meleshko VV, van Heijst GJF (1994) Interaction of two-dimensional vortex structures: point vortices, contour kinematics and stirring properties. Chaos Solitons Fract 4(6):977–1010
598. Meleshko VV, van Heijst GJF (1994) On Chaplygin's investigations of two-dimensional vortex structures in an inviscid fluid. J Fluid Mech 272:157–182
599. Mellor GL (1996) Introduction to physical oceanography. Springer, New York, 260 pp
600. Meschanov SL, Shapiro GI (1998) A young lens of Red Sea water in Arabian Sea. Deep-Sea Res Part I 45(1):1–13
601. Mesquita SB, Prahalad YS (1999) Statistical stationary states for a two-layer quasi-geostrophic system. Proc Indian Acad Sci (Math Sci) 109(1):107–115
602. Meunier P, Ehrenstein U, Leweke T, Rossi M (2002) A merger criterion for two-dimensional co-rotating vortices. Phys Fluids 14(8):2757–2766
603. Meunier P, Leweke T (2001) Three-dimensional instability during vortex merging. Phys Fluids 13(10):2747–2751
604. Meunier P, Leweke T (2005) Elliptic instability of a co-rotating vortex pair. J Fluid Mech 533:125–159
605. Middleton JH (2009) Topographic eddies. In: Steele JH, Thorpe SA, Turekian KK (eds) Ocean currents. Elsevier, New York, pp 462–469
606. Mied RP, Lindemann GJ (1982) The birth and evolution of eastward-propagating modons. J Phys Oceanogr 12(3):213–230
607. Mied RP, McWilliams JC, Lindemann GJ (1991) The generation and evolution of mushroom-like vortices. J Phys Oceanogr 21(4):489–510
608. Mikhailovskii AB, Kudashev VR, Lakhin VP, Mikhailovskaya LA, Smolyakov AI, Shishkov SYu (1984) Chains of Rossby solitons and gradient solitons. JETP Lett 40(7):1054–1056
609. Miller PD, Pratt LJ, Helfrich KR, Jones CKRT (2002) Chaotic transport of mass and potential vorticity for an island recirculation. J Phys Oceanogr 32(1):80–102
610. Milne-Thomson LM (1962) Theoretical hydrodynamics, 4th edn. London: MacMillan & co.
611. Mindlin IM (1984) On vorticity-induced waves in a homogeneous incompressible fluid. J Appl Math Mech 48(5):550–555
612. Minobe S, Kanamoto Y, Okada N, Ozawa H, Ikeda M (2000) Plume structures in deep convection of rotating fluid. J Jpn Soc Fluid Mech 19(6):395–396
613. Mirabel AP, Monin AS (1988) Instability of ocean circulation patterns. Trans Dokl USSR Acad Sci Earth Sci Sect 303(6):253–256
614. Mirabel AP, Monin AS (1989) Instability of gyres in a continuously startified ocean. Trans Dokl USSR Acad Sci Earth Sci Sect 309(6):267–270
615. Mitchell TB, Rossi LF (2008) The evolution of Kirchhoff elliptic vortices. Phys Fluids 20:054103. doi:10.1063/1.291299120
616. Miyama T, McCreary JP Jr, Jensen TG, Loschnigg J, Godfrey S, Ishida A (2003) Structure and dynamics of the Indian-Ocean cross-equatorial cell. Deep Sea Res Part II 50(12–13):2023–2047
617. Miyazaki T (1992) Elliptical instability in a stably stratified rotating fluid. Phys Fluids A 5(11):2702–2709
618. Miyazaki T, Fukumoto Y (1992) Three-dimensional instability of strained vortices in a stably stratified fluid. Phys Fluids A 4(11):2515–2522
619. Miyazaki T, Imai T, Fukumoto Y (1995) Three-dimensional instability of Kirchhoff's elliptic vortex. Phys Fluids 7(1):195–202
620. Mizuta R, Yoden S (2001) Chaotic mixing and transport barriers in an idealized stratospheric polar vortex. J Atmos Sci 58(17):2616–2629
621. Mokhov II, Doronina TN, Gryanik VM, Khairullin RR, Korovkina LV, Lagun VE, Mokhov OI, Naumov EP, Petukhov VK, Senatorsky AO, Tevs MV (1994) Extratropical cyclones and anticyclones: tendencies of change. In: Gronas S, Shapiro MA (eds) The life of extratropical cyclones, vol 2. Geophysical Institute, University of Bergen, Bergen, pp 56–60
622. Mokhov II, Gryanik VM, Doronina TN, Lagun DE, Mokhov OI, Naumov EP, Petukhov VK, Tevs MV, Khairullin RR (1993) Vortex activity in the atmosphere: tendencies of changes. Moscow: Institute of Atmospheric Physics of RAS, Preprint N 2, 97 pp

623. Möller JD, Montgomery MT (2000) Tropical cyclone evolution via potential vorticity anomalies in a three-dimensional balance model. J Atmos Sci 57(20):3366–3387
624. Monin AS (1990) Theoretical geophysical fluid dynamics. Kluwer Academic Publishers, Dordrecht, 429 pp
625. Monin AS, Yaglom AM (1971) Statistical fluid mechanics, vols I and II. MIT Press, Cambridge
626. Monin AS, Zhikharev GM (1990) Ocean eddies. Sov Phys Usp 33(5):313–339
627. Moore C (1993) Braids in classical dynamics. Phys Rev Lett 70(24):3675–3679
628. Moore DV, Saffman PG (1975) The density of organized vortices in a turbulent mixing layer. J Fluid Mech 69(3):465–473
629. Moore DW, Saffman PG, Tanveer S (1988) The calculation of some Batchelor flows: the Sadovskii vortex and rotational corner flow. Phys Fluids 31(5):978–990
630. Morel YG (1995) The influence of an upper thermocline current on intrathermocline eddies. J Phys Oceanogr 25(12):3247–3252
631. Morel YG, Carton XJ (1994) Multipolar vortices in two-dimensional incompressible flows. J Fluid Mech 267:23–51
632. Morel YG, McWilliams J (1997) Evolution of isolated interior vortices in the ocean. J Phys Oceanogr 27(5):727–748
633. Morel YG, McWilliams J (2001) Effect of isopycnal and diapycnal mixing on the stability of oceanic currents. J Phys Oceanogr 31(8):2280–2296
634. Morikawa GK (1960) Geostrophic vortex motion. J Meteorol 17(6):148–158
635. Morikawa GK, Swenson EV (1971) Interacting motion of rectilinear geostrophic vortices. Phys Fluids 14(6):1058–1073
636. Nakamura N (2008) Sensitivity of global mixing and fluxes to isolated transport barriers. J Atmos Sci 65(12):3800–3818
637. Narimousa S, Maxworthy T (1985) Two-layer model of shear-driven coastal upwelling in the presence of bottom topography. J Fluid Mech 159:503–531
638. Nauw JJ, Dijkstra HA, Simonnet E (2004) Regimes of low-frequency variability in a three-layer quasi-geostrophic ocean model. J Mar Res 62(5):685–720
639. Negretti ME, Billant P (2013) Stability of a Gaussian pancake vortex in a stratified fluid. J Fluid Mech 718:457–480
640. Newton PK (2001) The N-vortex problem: analytical techniques. Applied Mathematical Sciences, vol 145. Springer, New York/Berlin/Heidelberg, 421 pp
641. Newton PK, Ross SD (2006) Chaotic advection in the restricted four-vortex problem on a sphere. Phys D 223:36–53
642. Nezlin MV (1986) Rossby solitons (Experimental investigations and laboratory model of natural vortices of the Jovian Great Red Spot type). Sov Phys Usp 29(9):807–842
643. Nezlin MV, Rylov AYu, Trubnikov AS, Khutoretski AV (1990) Cyclonic-anticyclonic asymmetry and a new soliton concept for rossby vortices in the laboratory, oceans and the atmospheres of giant planets. Geophys Astrophys Fluid Dyn 52(41):211–247
644. Nezlin MV, Sutyrin GG (1994) Problems of simulation of large, long-lived vortices in the atmospheres of the giant planets (Jupiter, Saturn, Neptune). Surv Geophys 15(1):63–99
645. Ngan K, Shepherd TG (1997) Chaotic mixing and transport in Rossby-wave critical layer. J Fluid Mech 334:315–351
646. Ngan K, Shepherd TG (1999) A closer look at chaotic advection in the stratosphere. Part I: Geometric structure. J Atmos Sci 56(24):4134–4152
647. Ngan K, Shepherd TG (1999) A closer look at chaotic advection in the stratosphere. Part II: Statistical diagnostics. J Atmos Sci 56(24):4153–4166
648. Nikitin OP (1997) Vertical structure of synoptic currents in the northeast tropical Pacific. Oceanology 37(6):737–748
649. Nof D (1983) The translation of isolated cold eddies on a sloping bottom. Deep-Sea Res 30(2A):171–182
650. Nof D (1991) Lenses generated by intermittent currents. Deep-Sea Res 38(3):325–345
651. Nof D (1993) Generation of ringlets. Tellus 45A(4):299–310

652. Nof D, Simon LM (1987) Laboratory experiments on the merging of nonlinear anticyclonic eddies. J Phys Oceanogr 17(3):343–357
653. Norbury J (1975) Steady planar vortex pairs in an ideal fluid. Commun Pure Appl Maths 38:697–700
654. Novikov EA (1975) Dynamics and statistics of a system of vortices. Sov Phys JETP 41(5):937–943
655. Novikov EA, Sedov YuB (1978) Stochastic properties of a four-vortex system. Sov Phys JETP 48:440–444
656. Novikov EA, Sedov YuB (1979) Stochastization of vortices. JETP Lett 29(12):677–679
657. Nycander J (1987) Propagation of discontinuities in the Hasegawa-Mima equation. Phys Fluids 30(6):1585–1587
658. Nycander J (1988) New stationary vortex solutions of the Hasegawa-Mima equation. J Plasma Phys 39(3):413–430
659. Nycander J (1992) Refutation of stability proofs for dipole vortices. Phys Fluids A 4(3):467–476
660. Nycander J (1994) Steady vortices in plasmas and geophysical flows. Chaos 4(2):253–268
661. Nycander J (1995) Existence and stability of stationary vortices in a uniform shear flow. J Fluid Mech 287:193–132
662. Nycander J, Döös K, Coward AC (2002) Chaotic and regular trajectories in the Antarctic Circumpolar Current. Tellus 54A(1):99–106
663. Nycander J, Lacasce JH (2004) Stable and unstable vortices attached to seamounts. J Fluid Mech 507:71–94
664. Nycander J, Pavlenko VP (1987) Global vortex pattern in a rotating plasma. Phys Fluids 30(7):2097–2100; 507:71–94
665. Nycander J, Pavlenko VP (1991) Stationary propagating magnetic electron vortices. Phys Fluids B 3(6):1386–1391
666. Oey L-Y (1988) A model of Gulf Stream frontal instabilities, meanders and eddies along the continental slope. J Phys Oceanogr 18(2):211–229
667. Okamoto A, Hara K, Nagaoka K, Yoshimura S, Vranješ J, Kono M, Tanaka MY (2003) Experimental observation of a tripolar vortex in a plasma. Phys Plasmas 10(6):2211. http://dx.doi.org/10.1063/1.1571059
668. Olbers D, Wolff JO, Volker C (2000) Eddy fluxes and second-order moment balances for non-homogeneous quasigeostrophic turbulence in wind-driven zonal flows. J Phys Oceanogr 30(7):1645–1668
669. Oliver KIC, Eldevik T, Stevens DP, Watson AJ (2008) A Greenland See perspective on the dynamics of postconvective eddies. J Phys Oceanogr 38(12):2755–2771
670. Olson DB (1991) Rings in the Ocean. Ann Rev Earth Planet Sci 19:283–311
671. Olson DB, Brown OB, Emmerson SR (1983) Gulf Stream statistics from Florida Straits to Cape Hatteras derived from satellite and historical data. J Geophys Res 88(C8):4569–4577
672. Olson DB, Schmitt RW, Kennelly MA, Joyce TM (1985) A two-layer diagnostic model of a long-time physical evolution of warm-core ring 82 B. J Geophys Res 90(C5):8813–8822
673. Onsager L (1949) Statistical hydrodynamics. Il Nuovo Cim (1943–1954) 6(2):279–287
674. Orlandi P (1990) Vortex dipole rebound from a wall. Phys Fluids 2(A8):1429–1436
675. Orlandi P, van Heijst GF (1992) Numerical simulation of tripolar vortices in 2D flow. Fluid Dyn Res 9(4):179–206
676. Ottino JM (1989) The kinematics of mixing: Stretching, chaos and transport. Cambridge University Press, Cambridge, 362 pp
677. Ottino JM (1990) Mixing, chaotic advection, and turbulence. Annu Rev Fluid Mech 22:207–253
678. Ottino JM, Khakhar DV (2000) Mixing and segregation of rganular. Annu Rev Fluid Mech 32:55–91
679. Overman EA II (1986) Steady-state solutions of the Euler equations in two dimensions. II. Local analysis of limiting V-states. SIAM J Appl Math 46(5):765–800

680. Overman EA II, Zabusky NJ (1982) Evolution and merger of isolated vortex structures. Phys Fluids 25(8):1297–1305
681. Paillet J, Le Cann B, Carton X, Morel Y, Serpette A (2002) Dynamics and evolution of a northern meddy. J Phys Oceanogr 32(1):55–79
682. Paillet J, Le Cann B, Serpette A, Morel Y, Carton X (1999) Real-time tracking of a Galician meddy. Geophys Res Lett 26(13):1877–1880
683. Paldor N, Boss E (1994) Chaotic trajectories of tidally perturbed internal oscillations. J Atmos Sci 49(23):2306–2318
684. Paldor N, Nof D (1990) Linear instability of an anticyclonic vortex in a two-layer ocean. J Geophys Res 95(C10):18075–18079
685. Pallás-Sauz E, Viúdez Á (2008) Spontaneous generation of inertia-gravity waves during the merging of two baroclinic anticyclones. J Phys Oceanogr 38(1):213–234
686. Pavia EG, Cushman-Roisin B (1990) Merging of frontal eddies. J Phys Oceanogr 20(12):1886–1906
687. Pavlenko VP, Petviashvili VI (1983) Solitary vortex in a flute instability. Sov J Plasma Phys 9:603–604
688. Pedlosky J (1985) The instability of continuous heton clouds. J Atmos Sci 42(14):1477–1486
689. Pedlosky J (1987) Geophysical fluid dynamics. Springer, New York, 710 pp
690. Pedlosky J (1996) Ocean circulation theory. Springer, New York, 453 pp
691. Perepelkin VV, Petrov AG (1983) Dynamics of an elliptic vortex. Fluid Dyn 18(4):539–544
692. Perrot X, Carton X (2009) Point-vortex interaction in an oscillatory deformation field: Hamiltonian dynamics, harmonic resonance and transition to chaos. Discret Contin Dyn Syst Ser B 11(4):971–995
693. Peters D, Vargin P, Körnich H (2007) A study of the zonally asymmetric tropospheric forcing of the austral vortex splitting during September 2002. Tellus 59A(3):384–394
694. Pierce RB, Fairlie TDA (1993) Chaotic advection in the stratosphere: implications for the dispersal of chemically perturbed air from the polar vortex. J Geophys Res 98(D10):18589–18596
695. Pierce RB, Fairlie TD, Grose WL, Swinbank R, O'Neil A (1994) Mixing processes within the polar night jet. J Atmos Sci 51(20):2957–2972
696. Pierrehumbert RT (1980) A family of steady, translating vortex pairs with distributed vorticity. J Fluid Mech 99:129–144
697. Pierrehumbert RT (1991) Large-scale horizontal mixing in planetary atmospheres. Phys Fluids A 3(5):1250–1260
698. Pierrehumbert RT (1991) Chaotic mixing of tracer and vorticity by modulated travelling Rossby waves. Geophys Astrophys Fluid Dyn 58(1–4):285–319
699. Pierrehumbert RT, Yang H (1993) Global chaotic mixing on isentropic surfaces. J Atmos Sci 50(15):2462–2480
700. Płotka H, Dritschel DG (2012) Quasi-geostrophic shallow-water vortex-patch equilibria and their stability. Geophys Asrophys Fluid Dyn 106(6):574–595
701. Płotka H, Dritschel DG (2013) Quasi-geostrophic shallow-water doubly-connected vortex equilibria and their stability. J Fluid Mech 723:40–68
702. Poincaré H (1893) Théorie des tourbillons. Gauthier-Villars, Paris
703. Poje AC, Haller G (1999) Geometry of cross-stream mixing in a double-gyre ocean model. J Phys Oceanogr 29(8):1649–1665
704. Polvani LM (1991) Two-layer geostrophic vortex dynamics. 2. Alignment and two-layer V-states. J Fluid Mech 225:241–270
705. Polvani LM, Carton XJ (1990) The tripole: a new coherent vortex structure of incompressible two-dimensional flows. Geophys Astrophys Fluid Dyn 51(1–4):87–102
706. Polvani LM, Flierl GR (1986) Generalized Kirchhoff vortices. Phys Fluids 29(8):2376–2379
707. Polvani LM, Flierl GR, Zabusky NJ (1989) Filamentation of unstable vortex structures via separatrix crossing: a quantitative estimate of onset time. Phys Fluids A1(2):181–184
708. Polvani LM, Plumb RA (1991) Rossby wave breaking, microbreaking, filamentation, and secondary vortex formation: the dynamics of a perturbed vortex. J Atmos Sci 49(6):462–476

709. Polvani LM, Wisdom J (1990) Chaotic Lagrangian trajectories around an elliptical vortex parch embedded in a constant and uniform background shear flow. Phys Fluids A 2(2):123–126
710. Polvani LM, Zabusky NJ, Flierl GR (1988) Applications of contour dynamics to two-layer quasi-geostrophic flows. Fluid Dyn Res 3(1–4):422–424
711. Polvani LM, Zabusky NJ, Flierl GR (1989) Two-layer geostrophic vortex dynamics: 1. Upper-layer V-states and merger. J Fluid Mech 205:215–242
712. Polzin KL, Toole JM, Ledwell JR, Schmitt RW (1997) Spatial variability of turbulent mixing in the abyssal ocean. Science 276(5309):93–96
713. Pozrikidis C (2008) Numerical computation in science and engineering, 2nd edn. Oxford University Press, New York
714. Pozrikidis C (2009) Fluid dynamics: theory, computation and numerical simulation, 2nd edn. Springer, London
715. Pozrikidis C (2011) Introduction to theoretical and computational fluid dynamics, 2nd edn. Cambridge University Press, Cambridge
716. Prandtl L (1952) Essentials of fluid dynamics. Blackie and Son, London
717. Prants SV (2013) Dynamical systems theory methods to study mixing and transport in the ocean. Phys Scr 87(3):038115. doi:10.1088/0031-8949/87/03/038115
718. Prants SV, Budyansky MV, Uleysky MYu, Zaslavsky GM (2006) Chaotic mixing and transport in a meandering jet flow. Chaos 16(3):033117. http://dx.doi.org/10.1063/1.2229263
719. Prater MD, Sanford TB (1994) A meddy off Cape St. Vincent. Part I: description. J Phys Oceanogr 24(7):1572–1586
720. Pratt LJ (1983) On inertial flow over topography. Part 1. Semigeostrophic adjustment to an obstacle. J Fluid Mech 131:195–218
721. Price JF, O'Neil Baringer M, Lueck RG, Johnson GC, Ambar I, Parrilla G, Cantos A, Kennelly MA, Sanford TB (1993) Mediterranean outflow mixing and dynamics. Science 259(5099):1277–1282
722. Price T (1993) Chaotic scattering of two identical point vortex pairs. Phys Fluids A5(10):2479–2483
723. Prieto R, McNoldy BD, Fulton SR, Schubert WH (2003) A classification of binary tropical cyclone-like vortex interactions. Mon Weather Rev 131(11):2656–2666
724. Prudnikov AP, Brychkov YA, Marichev OI (1986) Integrals and series, vol. 1: elementary functions. Gordon and Breach, New York
725. Prudnikov AP, Brychkov YA, Marichev OI (1990) Integrals and series, vol. 2: special functions. Gordon and Breach, New York
726. Pullin DL (1992) Contour dynamics methods. Annu Rev Fluid Mech 24:89–115
727. Rabinovich AB (1997) Spectral analysis of tsunami waves: separation of source and topography effects. J Geophys Res 102(C6):12663–12676
728. Radko T (2008) Long-range interaction and elastic collisions of isolated vortices. J Fluid Mech 610:285–310
729. Raymond DJ, Jiang H (1990) A theory of long-living mesoscale convective system. J Atmos Sci 47(24):3067–3077
730. Reinaud JN, Carton X (2009) The stability and the nonlinear evolution of quasi-geostrophic hetons. J Fluid Mech 636:109–135
731. Reinaud JN, Dritschel DG (2002) The merger of vertically offset quasi-geostrophic vortices. J Fluid Mech 469:287–315
732. Reinaud JN, Dritschel DG (2005) The critical merger distance between two co-rotating quasi-geostrophic vortices. J Fluid Mech 522:357–381
733. Reinaud JN, Dritschel DG, Koudella CR (2003) The shape of vortices in quasi-geostrophic turbulence. J Fluid Mech 474:175–192
734. Reznik GM (1987) Synoptic movements above a strongly dissected bottom relief. Trans Dokl USSR Acad Sci Earth Sci Sect 296(5):230–233
735. Reznik GM (1999) On the generation of subsurface motions over a sloping bottom in a two-layer ocean. Oceanology 39(3):293–295

736. Reznik GM, Grimshaw RHJ, Sriskanderejan K (1997) On basic mechanisms governing two-layer vortices on a beta-plane. Geophys Astrophys Fluid Dyn 86(1–4):1–42
737. Reznik GM, Kizner Z (2007) Two-layer quasi-geostrophic singular vortices embedded in a regular flow: 1. Invariants of motion and stability of vortex pairs. J Fluid Mech 584:185–202
738. Reznik GM, Kizner Z (2007) Two-layer quasi-geostrophic singular vortices embedded in a regular flow: 2. Steady and unsteady drift of individual vortices on a beta plane. J Fluid Mech 584:203–223
739. Reznik GM, Sutyrin GG (2001) Baroclinic topographic modons. J Fluid Mech 437:121–142
740. Reznik GM, Tsybaneva TB (1994) On the influence of topography and stratification on planetary waves in the ocean (two-layer model). Oceanology 34(1):1–9
741. Reznik GM, Tsybaneva TB (1999) Planetary waves in a stratified ocean of variable depth. Part 1. Two-layer model. J Fluid Mech 388:115–145
742. Reznik GM, Zeitlin V, Ben Jellool M (2001) Nonlinear theory of geostrophic adjustment. Part 1. Rotating shallow-water model. J Fluid Mech 445:93–120
743. Rhines PB (1986) Vorticity dynamics of the oceanic general circulation. Annu Rev Fluid Mech 18:433–497
744. Ricca RL (ed) (2001) An introduction to the geometry and topology of fluid flows. NATO ASI Series II, 47. Kluwer, Dordrecht/Boston/London
745. Ricca RL (ed) (2009) Lectures on topological fluid mechanics. Springer-CIME Lecture Notes in Mathematics 1973. Springer, Heidelberg
746. Riccardi G (2004) Motion of an elliptical uniform vortex outside a circular cylinder. Regul Chaotic Dyn 9(4):399–415
747. Riccardi G, Piva R, Benzi R (1995) A physical model for merging in two-dimensional decaying turbulence. Phys Fluids 7(12):3091–3103
748. Richardson PL, Bower AS, Zenk W (1999) Summary of meddies tracked by floats. Int WOCE Newsl 34:18–20
749. Richardson PL, Bower AS, Zenk W (2000) A census of meddies tracked by floats. Progr Oceanogr 45(2):209–250
750. Richardson PL, Maillard C, Sanford TB (1979) The physical structure and life history of cyclonic Gulf Stream ring Allen. J Geophys Res 84(C12):7727–7741
751. Richardson PL, Tychensky A (1998) Meddy trajectories in the Canary Basin measured during the SEMAPHORE experiment 1993–1995. J Geophys Res 103(C11):25029–25045
752. Richardson PL, Walsh D, Armi L, Schröder M, Price JF (1989) Tracking three meddies with SOFAR floats. J Phys Oceanogr 19(3):371–383
753. Ridderinkhof H, Loder JW (1994) Lagrangian characterization of circulation over submarine banks with application to the Outer Gulf of Marine. J Phys Oceanogr 24(6):1184–1200
754. Ridderinkhof H, Zimmerman JTF (1992) Chaotic stirring in a tidal system. Science 258(5085):1107–1111
755. Riley JJ, Lelong M-P (2000) Fluid motions in the presence of strong stable stratification. Annu Rev Fluid Mech 32:613–657
756. Ripa P (1989) On the stability of ocean vortices. In: Nihoul JCJ, Jamart BM (eds) Mesoscale/synoptic coherent structures in geophysical turbulence. Elsevier, Amsterdam/Oxford/New York/Tokyo, pp 167–179
757. Ritchie EA, Holland GJ (1993) On the interaction of tropical-cyclone-scale vortices: 2. Discrete vortex patches. Q J R Meteorol Soc 119(514):1363–1379
758. Riser SC, Rossby TH (1983) Quasi-Lagrangian structure and variability of the subtropical western North Atlantic circulation. J Mar Res 42(1):127–162
759. Robinson AR (1996) Physical processes, field estimation and an approach to interdisciplinary ocean modeling. Earth-Sci Rev 40(1–2):3–54
760. Rocío R-M, Viúdez Á, Ruiz S (2011) Vortex merger in oceanic tripoles. J Phys Oceanogr 41(6):1239–1251
761. Rodríguez-Santana A, Pelegrí JL, Sangrá P, Marrero-Díaz A (1999) Diapycnal mixing in Gulf Stream meanders. J Geophys Res 104:25891–25912

762. Rodríguez-Santana A, Pelegrí JL, Sangrá P, Marrero-Díaz A (2001) On the relevance of diapycnal mixing for the stability of frontal meanders. Sci Mar 65(Suppl 1):259–267
763. Rogachev KA, Carmack EC, Foreman MGG (2008) Bowhead whales feed on plankton concentrated by estuarine and tidal currents in Academy Bay, Sea of Okhotsk. Cont Shelf Res 28(14):1811–1826
764. Rogberg P, Dritschel DG (2000) Mixing in two-dimensional vortex interactions. Phys Fluids 12(12):3285 http://dx.doi.org/10.1063/1.1320838
765. Rogerson AM, Miller PD, Pratt LJ, Jones CKRT (1999) Lagrangian motion and fluid exchange in a barotropic meandering jet. J Phys Oceanogr 29(10):2635–2655
766. Romanov AS (2007) Dipole approximation in three-vortex dynamics. Theor Math Phys 150(3):347–354
767. Romanovskaya ML, Semenova IP, Slezkin LN (2010) Dynamically equilibrium shapes and directions of motion of ocean current rings. J Appl Math Mech 74(3):365–374
768. Roscoe HK, Shanklin JD, Colwell SR (2005) Has the Antarctic Vortex split before 2002? J Atmos Sci 62(3):581–588
769. Rossby CG (1936) Dynamics of steady ocean currents in the light of experimental fluid mechanics. Pap Phys Oceanogr Meteorol 5(1):1–46
770. Rossby T, Dorson D, Fontaine J (1986) The RAFOS system. J Atmos Ocean Tech 3(4):672–679
771. Rossi LF, Lingevitch JF, Bernoff AJ (1997) Quasi-steady monopole and tripole attractors for relaxing vortices. Phys Fluids 9(8):2329–2338
772. Rott N (1989) Three-vortex motion with zero total circulation. J Appl Math Phys (ZAMP) 40(4):473–494
773. Rott N (1990) Constrained three- and four-vortex problems. Phys Fluids A2(8):1477–1480
774. Roullet G, Klein P (2010) Cyclone-anticyclone asymmetry in geophysical turbulence. Phys Rev Lett 104(21):218501. doi:10.1103/PhysRevLett.104.218501
775. Ruddick B (1987) Anticyclonic lenses in large-scale strain and shear. J Phys Oceanogr 17(6):741–749
776. Ruddick B (1992) Intrusive mixing in a Mediterranean salt lens—intrusion slope and dynamical machanisms. J Phys Oceanogr 22(11):1274–1285
777. Ruddick B (2003) Sounding out ocean fine structure. Science 301(5634):772–773
778. Ruddick B, Hebert D (1988) The mixing of meddy "Sharon" In: Nihoul JCJ, Jamart BM (eds) Small-scale mixing in the ocean. Elsevier, New York, pp 249–262
779. Ryzhov EA, Koshel KV (2010) Chaotic transport and mixing of a passive admixture by vortex flows behind obstacles. Izv Atmos Ocean Phys 46(2):184–191
780. Ryzhov EA, Koshel KV (2011) The effect of chaotic advection in a three-layer ocean model. Izv Atmos Ocean Phys 47(2):241–251
781. Ryzhov EA, Koshel KV (2011) Estimating the size of the regular region of a topographically trapped vortex. Geophys Astrophys Fluid Dyn 105(4–5):536–551
782. Ryzhov EA, Koshel KV (2013) Interaction of a monopole vortex with an isolated topographic feature in a three-layer geophysical flow. Nonlinear Process Geophys 20(1):107–119
783. Ryzhov EA, Koshel KV, Stepanov DV (2008) Evaluating the stochastic layer thickness on a two-layer topographic vortex model. Tech Phys Lett 34(6):531–534
784. Ryzhov E, Koshel K, Stepanov D (2010) Background current concept and chaotic advection in an oceanic vortex flow. Theor Comput Fluid Dyn 24(1–4):59–64
785. Sadovskii VS (1971) Vortex regions in a potential stream with a jump of Bernoulli's constant at the boundary. J Appl Math Mech 35(6):729–735
786. Saenko OA, Zhai X, Merryfield WJ, Lee WG (2012) The combined effect of tidally and eddy-driven diapycnal mixing on the large-scale ocean circulation. J Phys Oceanogr 42(4):526–538
787. Saffman PG (1978) The number of waves on unstable vortex rings. J Fluid Mech 84(4):625–639
788. Saffman PG (1992) Vortex dynamics (Cambridge monographs on mechanics and applied mathematics). Cambridge University Press, Cambridge, 311 pp

789. Saffman PG, Baker GK (1979) Vortex interactions. Annu Rev Fluid Mech 11:95–122
790. Saffman P, Schatzman J (1981) Properties of a vortex street of finite vortices. SIAM J Sci Stat Comp 2(3):285–295
791. Saffman P, Schatzman J (1982) An inviscid model for the vortex-street wake. J Fluid Mech 122:467–486
792. Saffman PG, Szeto R (1980) Equilibrium shape of a pair of equal vorteces. Phys Fluids 23(12):2339–2342
793. Saffman PG, Szeto R (1981) Structure of a linear array of uniform vortices. Stud Appl Maths 65:223–248
794. Saffman PG, Tanveer S (1982) The touching pair of equal and opposite uniform vortices. Phys Fluids 25(11):1929–1930
795. Sakamoto K, Akitomo K (2006) Instabilities of the tidally induced bottom boundary layer in the rotating frame and their mixing effect. Dyn Atmos Oceans 41(3–4):191–219
796. Sakamoto T, Yamagata T (1997) Evolution of baroclinic planetary eddies over localized bottom topography in terms of JEBAR. Geophys Asrophys Fluid Dyn 84(1–2):1–27
797. Sakuma H, Ghil M (1991) Stability of propagating modons for small-amplitude perturbations. Phys Fluids A 3(3):408–414
798. Salmon R (1998) Lectures on geophysical fluid dynamics. Oxford University Press, New York, 378 pp
799. Samelson RM (1992) Fluid exchange across a meander jet. J Phys Oceanogr 22(4):431–440
800. Samelson RM, Wiggins S (2006) Lagrangian transport in geophysical jets and waves: the dynamical systems approach. Springer Science+Business Media, LLC, New York, 149 pp
801. Sangrá P, Pelegrí JL, Hernández-Guerra A, Arregui I, Martín JM, Marrero-Díaz A, Martínez A, Ratsimandresy AW, Rodríguez-Santana A (2005) Life history of an anticyclonic eddy. J Geophys Res 110:C03021. doi:10.1029/2004JC002526
802. Saunders PM (1973) The instability of a baroclinic vortex. J Phys Oceanogr 3(1):61–65
803. Savchenko VG, Emery WJ, Vladimirov OA (1978) A cyclonic eddy in the Antarctic Circumpolar Current south of Australia: results of Soviet-American observations aboard the R/V Professor Zubov. J Phys Oceanogr 8(9):825–837
804. Schär C, Durran DR (1997) Vortex formation and vortex shedding in continuously stratified flows past isolated topography. J Atmos Sci 54(4):534–554
805. Schär C, Smith RB (1993) Shallow-water flow past isolated topography. Part 1. Vorticity production and wake formation. J Atmos Sci 50(10):1373–1400
806. Schultz Tokos KL, Hinrichsen H-H, Zenk W (1994) Merging and migration of two meddies. J Phys Oceanogr 24(10):2129–2141
807. Schultz Tokos K, Rossby T (1991) Kinematics and dynamics of Mediterranean salt lens. J Phys Oceanogr 21(6):879–892
808. Sedov LI (1997) Mechanics of continuous media, vol 2. Word Sci Pub. Co. Pte. Ltd., Singapore, pp 615–1308
809. Selivanova EN (1994) The topology of the problem of three-point vortices. Proc Steklov Inst Math 205:129–137
810. Semenova IP, Slezkin LN (2003) Dynamically equilibrium shape of intrusive vortex formations in the ocean. Fluid Dyn 38(5):663–669
811. Sengupta D, Piterbarg LI, Reznik GM (1992) Localization of topographic Rossby waves over random relief. Dyn Atmos Oceans 17(1):1–21
812. Serra N, Ambar I (2001) Eddy generation in the Mediterranean undercurrent. Deep-Sea Res Part II 49(19):4225–4243
813. Serra N, Sadux S, Ambar I (2002) Observations and laboratory modeling of meddy generation of cape St. Vincent. J Phys Oceanogr 32(1):3–25
814. Shaffer G, Salinas S, Pizarro O, Vega A, Hormazabal S (1995) Currents in the deep ocean off Chile (30° S). Deep Sea Res Part I 42(4):425–436
815. Shapiro GI, Meschanov SL (1996) Spreading pattern and mesoscale structure of Mediterranean outflow in the Iberian Basin estimated from historical data. J Mar Syst 7(2–4):337–348

816. Shapiro GI, Meschanov SL, Emelianov MV (1992) Mediterranean lens after collision with seamounts. Oceanology 32:420–427
817. Shapiro GI, Meschanov SL, Emelianov MV (1995) Mediterranean lens "Irving" after its collision with seamounts. Oceanol Acta 18(3):309–318
818. Shapiro LJ (1992) Hurricane vortex motion and evolution in a three-layer model. J Atmos Sci 49(2):140–154
819. Shapiro LJ (2000) Potential vorticity assymmetry and tropical cyclone evolution in a moist three-layer model. J Atmos Sci 57(21):3645–3662
820. Shavlyugin AI (2011) Two-layer quasi-geostrophic model of contour dynamics for a round basin. Izv Atmos Ocean Phys 47(5):619–627
821. Shaw P-T, Chao S-Y (2003) Effects of a baroclinic current on a sinking dense water plume from a submarine canyon and heton shedding. Deep Sea Res Part I 50(3):357–370
822. Shepherd TG (1990) Symmetries, conservation laws, and Hamiltonian structure in geophysical fluid dynamics. Adv Geophys 32:287–339
823. Shimada K, Kubokawa A (1997) Nonlinear evolution of linearly unstable barotropic boundary currents. J Phys Oceanogr 27(7):1326–1348
824. Shirshov PP (1938) Oceanological observations. Dokl AN USSR 8:569–580 (in Russian)
825. Siedler G, Armi L, Müller TJ (2005) Meddies and decadal changes at the Azores Front from 1980 to 2000. Deep-Sea Res II 52(3–4):583–604
826. Simiu E (1969) Melnikov process for stochastically perturbed, slowly varying oscillators: application to a model of wind-driven coastal currents. J Appl Mech 63(2):429–435
827. Simó C (2001) New families of solutions to the N-body problems. In: Casacuberta C, Miró-Roig RM, Verdera J, Xambó, S (eds) Proceedings of the European 3rd congress of mathematics, vol I (Barcelona, 2000). Progress in mathematics series, vol 201. Birkhäuser, Basel, pp 101–115
828. Simó C (2002) Dynamical properties of the figure eight solution of three-body problem. In: Chenciner A, Cushman R, Robinson C, Xia ZJ (eds) Celestial mechanics, dedicated to Donald Saari for his 60th birthday. Proceedings of the international conference on celestial mechanics, Northwestern University, Evanston, Illinois, 15–19 December 1999. Contemporary Mathematics, vol 292, American Mathematical Society, Providence, pp 209–228
829. Sitnikov IG, Pokhil AE (1998) Interaction of tropical cyclones with each other and with other weather sistems (Part I). Russ Meteorol Hydrol 5:22–28
830. Sitnikov IG, Pokhil AE (1999) Interaction of tropical cyclones with each other and with other weather sistems (Part II). Russ Meteorol Hydrol 7:26–37
831. Skyllingstad ED, Denbo DW (1995) An ocean large-eddy simulation of Langmuir circulations and convection in the surface mixed layer. J Geophys Res 100(C5):8501–8522
832. Smeed DA (1988) Baroclinic instability of three-layer flows: 1. Linear stability. J Fluid Mech 194:217–231
833. Smeed DA (1988) Baroclinic instability of three-layer flows: 2. Experiments with eddies. J Fluid Mech 194:233–259
834. Smith GB, Wei T (1994) Small-scale structure in colliding off-axis vortex rings. J Fluid Mech 259:281–290
835. Smith KS, Marshall J (2009) Evidence for enhanced eddy mixing at middepth in the Southern Ocean. J Phys Oceanogr 39(1):50–69
836. Smith RK (1981) The cyclostrophic adjustment of vortices with application to tropical cyclone modification. J Atmos Sci 38(9):2021–2030
837. Smyth WD, Carpenter JR, Lawrence GA (2007) Mixing in symmetric Holmboe waves. J Phys Oceanogr 37(6):1566–1583
838. Smyth WD, Kimura S (2011) Mixing in a moderately sheared salt-fingering layer. J Phys Oceanogr 41(7):1364–1384
839. Sokolovskiy MA (1986) Numerical modelling of the vortex structure evolution by contour dynamics method. Preprint. Pacific Oceanological Institute of Far East Division of Academy Sciences of USSR, Vladivostok, 19 pp (in Russian)

840. Sokolovskiy MA (1988) Numerical modelling of nonlinear instability for axisymmetric two-layer vortices. Izv Atmos Ocean Phys 24(7):536–542
841. Sokolovskiy MA (1989) Head-on collisions of distributed hetons. Trans Dokl USSR Acad Sci Earth Sci Sect 306(3):215–217
842. Sokolovskiy MA (1990) Numerical modelling of the interaction of distributed hetons during their head-on collisions. In: Kozlov VF (ed) Method of contour dynamics in oceanological investigations. FED USSR Acad Sci, Vladivostok, pp 40–57 (in Russian)
843. Sokolovskiy MA (1990) On the interaction of distributed hetons. Preprint. Pacific Oceanological Institute of Far East Division of Academy Sciences of USSR, Vladivostok, 20 pp (in Russian)
844. Sokolovskiy MA (1991) Modeling triple-layer vortical motions in the ocean by the Contour Dynamics Method. Izv Atmos Ocean Phys 27(5):380–388
845. Sokolovskiy MA (1997) Stability of an axisymmetric three-layer vortex. Izv Atmos Ocean Phys 33(1):16–26
846. Sokolovskiy MA (1997) Stability analysis of the axisymmetric three-layered vortex using contour dynamics method. Comput Fluid Dyn J 6(2):133–156
847. Sokolovskiy MA, Carton X (2010) Baroclinic multipole formation from heton interaction. Fluid Dyn Res 42:045501. doi:10.1088/0169-5983/42/4/045501
848. Sokolovskiy MA, Filyushkin BN, Carton X (2013) Dynamics of intrathermocline vortices in a gyre flow over a seamount chain. Ocean Dyn 63. doi:10.1007/s10236-013-0628-y
849. Sokolovskiy MA, Koshel KV, Carton X (2011) Baroclinic multipole evolution in shear and strain. Geophys Asrophys Fluid Dyn 105(4–5):506–535
850. Sokolovskiy MA, Koshel KV, Verron J (2013) Three-vortex quasi-geostrophic dynamics in a two-layer fluid. Part 1. Analysis of relative and absolute motions. J Fluid Mech 717:232–254
851. Sokolovskiy MA, Verron J (2000) Finite-core hetons: stability and interactions. J Fluid Mech 423:127–154
852. Sokolovskiy MA, Verron J (2000) Four-vortex motion in the two layer approximation: integrable case. Regul Chaotic Dyn 5(4):413–436
853. Sokolovskiy MA, Verron J (2002) Dynamics of the triangular two-layer vortex structures with zero total intensity. Regul Chaotic Dyn 7(4):435–472
854. Sokolovskiy MA, Verron J (2002) New stationary solutions to the problem of three vortices in a two-layer fluid. Dokl Phys 47(3):233–237
855. Sokolovskiy MA, Verron J (2004) Dynamics of the three vortices in two-layer rotating fluid. Regul Chaotic Dyn 9(4):417–438
856. Sokolovskiy MA, Verron J (2006) Some properties of motion of A + 1 vortices in a two-layer rotating fluid. Russ Sci J Nonlinear Dyn 2(1):27–54 (in Russian)
857. Sokolovskiy MA, Verron J (2008) Motion of $A+1$ vortices in a two-layer rotating fluid. In: Borisov AV, Kozlov VV, Mamaev IS, Sokolovskiy MA (eds) IUTAM symposium on hamiltonian dynamic, vortex strictures, turbulence (IUTAM Bookseries, vol 6). Springer, New York, pp 481–490
858. Sokolovskiy MA, Verron J (2011) Dynamics of vortex structures in a stratified rotating fluid. Moscow-Izhevsk, Izhevsk Institution of Computer Science, 372 pp (in Russian)
859. Sokolovskiy M, Verron J, Carton X, Gryanik V (2010) On instability of elliptical hetons. Theor Comput Fluid Dyn 24(1–4):117–123
860. Sokolovskiy MA, Verron J, Vagina IM (2001) Effect of a submerged small-height obstacle on the dynamics of a distributed heton. Izv Atmos Ocean Phys 37(1):122–133
861. Sokolovskiy MA, Zyryanov VN, Davies PA (1998) On the influence of an isolated submerged obstacle on a barotropic tidal flow. Geophys Astrophys Fluid Dyn 88(1):1–30
862. Solomon TH, Weeks ER, Swinney HL (1994) Chaotic advection in a two-dimensional flow: Lévy flights and anomalous diffusion. Phys D 76(1–3):70–84
863. Spall MA (1994) Mechanism for low-frequency variability and salt flux in Mediterranean salt tongue. J Geophys Res 99(C5):10121–10129
864. Spall MA, Chapman DC (1998) On the efficiency of baroclinic eddy heat transport across narrow fronts. J Phys Oceanogr 28(11):2275–2287

865. Spall MA, Richardson PL, Price J (1993) Advection and eddy mixing in the Mediterranean salt tongue. J Mar Res 51(4):797–818
866. Spohn A, Mory M, Hopfinger EJ (1993) Observations of vortex breakdown in an open cylindrical container with a rotating bottom. Exp Fluids 14(1–2):70–77
867. Spohn A, Mory M, Hopfinger EJ (1998) Experiments on vortex breakdown in a confined flow generated by a rotating disc. J Fluid Mech 370:73–99
868. Stammer D (1998) On eddy characteristics, eddy transports, and mean flow properties. J Phys Oceanogr 28(4):727–739
869. Stammer D, Hinrichsen H-H, Käse RH (1991) Can meddies be detected by satellite altimetry? J Geophys Res 96(C4):7005–7014
870. Stammer D, Wunsch C, Neyoshi K (2006) Temporal changes in ocean eddy transport. J Phys Oceanogr 36(3):543–550
871. Stegner A, Zeitlin V (1996) Asymptotic expansions and monopolar solitary Rossby vortices in barotropic and two-layer models. Geophys Astrophys Fluid Dyn 83(3–4):159–194
872. Stern ME (1975) Minimal properties of planetery eddies. J Mar Res 33(1):1–13
873. Stern ME (2000) Scattering of an eddy advected by a current towards a topographic obstacle. J Fluid Mech 402:211–223
874. Stewart KD, Hughes GO, Griffiths RW (2012) The role of turbulent mixing in an overturning circulation maintained by surface buoyancy forcing. J Phys Oceanogr 42(11):1907–1922
875. Stirling JR (2003) Chaotic advection, transport and patchiness in clouds of pollution in an estuarine flow. Discret Contin Dyn Syst Ser B 3(2):263–284
876. Stremler MA, Aref H (1999) Motion of three point vortices in a periodic parallelogram. J Fluid Mech 392:101–128
877. Su CH (1979) Motion of fluid with constant vorticity in a singly-connected region. Phys Fluids 22(10):2032–2033
878. Sutyrin GG, Herbette S, Carton X (2011) Deformation and splitting of baroclinic eddies encountering a tall seamount. Geophys Astrophys Fluid Dyn 105(4–5):478–505
879. Sutyrin GG, Hesthaven JS, Lynov JP, Rasmussen JJ (1994) Dynamical properties of vortical structures on the beta-plane. J Fluid Mech 268:103–131
880. Sutyrin GG, McWilliams JC, Saravanan R (1998) Co-rotating stationary states and vertical alignment of geostrophic vortices with thin cores. J Fluid Mech 357:321–349
881. Sutyrin GG, Perrot X, Carton X (2008) Integrable motion of a vortex dipole in an axisymmetric flow. Phys Lett 372A:5452–5457
882. Sutyrin GG, Stegner A, Taupier-Letage I, Teinturier S (2009) Amplification of a surface-intensified eddy drift along a steep shelf in the Eastern Mediterranean Sea. J Phys Oceanogr 39(7):1729–1741
883. Swallow JG (1969) A deep eddy off Cape St. Vincent. Deep-Sea Res 16(Suppl):285–295
884. Swaters GE (1986) Stability conditions and apriori estimates for equivalent barotropic modons. Phys Fluids 29(5):1419–1422
885. Swenson M (1987) Instability of equivalent-barotropic riders. J Phys Oceanogr 17(4):492–506
886. Synge J (1949) On the motion of three vortices. Can J Math 1:257–270
887. Tailleux R (2012) On the generalized eigenvalue problem for the Rossby wave vertical velocity in the presence of mean flow and topography. J Phys Oceanogr 42(6):1045–1050
888. Takahashi J, Masuda A (1998) Mechanisms of the southward translation of meddies. J Oceanogr 54(6):669–680
889. Tan B, Boyd JP (1977) Dynamics of the Flierl-Petviashvili monopoles in a barotropic model with topographic forcing. Wave Motion 26(3):239–251
890. Tang B, Cushman-Roisin B (1992) Two-layer geostrophic dynamics: 2. Geostrophic turbulence. J Phys Oceanogr 22(2):128–138
891. Tansley CE, Marshall DP (2000) On the influence of bottom topography and the Deep Western Boundary Current on Gulf Stream separation. J Mar Res 58(2):297–325
892. Tavantzis J, Ting L (1988) The dynamics of three vortices revisited. Phys Fluids 31(6):1392–1409

893. Taylor GI (1921) Experiments with rotating fluids. Proc R Soc Ser A 100:114–121
894. Taylor GI (1923) Experiments on the motion of solid bodies in rotating fluids. Proc R Soc Ser A 104:213–218
895. Tél T, Gruiz M (2006) Chaotic dynamics. An introduction based on classical mechanics. Cambridge University Press, Cambridge, 412 pp
896. Temam R, Ziane M (2005) Some mathematical problems in geophysical fluid dynamics. In: Friedlander S, Serre D (eds) Handbook of mathematical fluid dynamics, vol 3. Elsevier, New York, pp 535–658
897. Tevs MV (1999) Kinematic study of the vertical structure of tropical cyclones on the basis of an *N*-level quasi-geostrophic atmospheric model. Izv Atmos Ocean Phys 35(4):435–439
898. Thierry V, Morel Y (1999) Influence of a strong bottom slope on the evolution of a surface-intensified vortex. J Phys Oceanogr 29(5):911–924
899. Thivolle-Cazat E, Sommeria J, Galmiche M, Verron J (2001) An experimental investigation of heton instability in a rotating, two-layer fluid. In: Chashechkin YuD (ed) Proceedings of the Moscow international conference on fluxes and structures in fluids, Institute for Problems in Mechanics of RAS, Moscow, pp 205–206
900. Thivolle-Cazat E, Sommeria J, Galmiche M (2005) Baroclinic instability of two-layer vortices in laboratory experiments. J Fluid Mech 544:69–97
901. Thompson AF, Sallée J-B (2012) Jets and topography: jet transitions and the impact on transport in the Antarctic Circumpolar Current. J Phys Oceanogr 42(6):956–972
902. Thompson L (1993) Two-layer quasigeostrophic flow over finite isolated topography. J Phys Oceanogr 23(7):1297–1314
903. Thompson L, Flierl GR (1993) Barotropic flow over finite isolated topography: steady solutions on the beta-plane and the initial value problem. J Fluid Mech 250:553–586
904. Thomson W (Lord Kelvin) (1875) Vortex statics. Math Phys Pap IV:115–128
905. Thomson W (Lord Kelvin) (1887) On the stability of steady and of periodic fluid motion. Philos Mag ser 5. 23(144):459–469
906. Tilburg CE, Hurlburt HE, O'Brien JJ, Shriver JF (2002) Remote topographic forcing of a baroclinic western boundary current: an explanation for the Southland Current and the pathway of the subtropical front east of New Zealand. J Phys Oceanogr 32(11):3216–3232
907. Tkachenko VK (1966) On vortex lattices. Sov Phys JETP 22(6):1282–1286
908. Tkachenko VK (1966) Stability of vortex lattices. Sov Phys JETP 23(6):1049–1056
909. Trieling RR, van Heijst GJF, Kizner Z (2010) Laboratory experiments on multipolar vortices in a rotating fluid. Phys Fluids 22(9):094104. doi:10.1063/1.3481797
910. Trieling RR, Velasco Fuentes OU, van Heijst GJFR (2005) Interaction of two unequal corotating vortices. Phys Fluids 17(8):087103. doi:10.1063/1.1993887
911. Tritton DJ (1988) Physical fluid dynamics. Clarendon Press, Oxford, 536 pp
912. Tronin KG (2006) Absolute choreographies of point vortices on a sphere. Regul Chaotic Dyn 11(1):123–130
913. Turkington B (1985) Corotating steady vortex flows with N-fold symmetry. Nonlinear Anal Theor Methods Appl 9(4):351–369
914. Tychensky A, Carton X (1998) Hydrological and dynamical characterization of meddies in the Azores region: a paradigm for baroclinic vortex dynamics. J Geophys Res 103(C11):25061–25079
915. Tychensky A, Le Traon P-Y, Hernandez F, Jourdan D (1998) Large structures and temporal change in the Azores Front during the SEMAPHORE experiment. J Geophys Res 103(C11):25009–25027
916. Uleysky MYu, Budyansky MV, Prants SV (2007) Effect of dynamical traps on chaotic transport in a meandering jet flow. Chaos 17(4):043105. doi:10.1063/1.2783258
917. Uleysky MYu, Budyansky MV, Prants SV (2008) Genesis and bifurcations of unstable periodic orbits in a jet flow. J Phys A Math Theor 41(2):215102. doi:10.1088/1751-8113/41/21/215102

918. Uleysky MYu, Budyansky MV, Prants SV (2010) Mechanism of destruction of the transport barriers in geophysical jets with Rossby waves. Phys Rev E 81(1):017202. doi:10.1103/PhysRevE.81.017202
919. Valcke S, Verron J (1993) On interactions between two finite-core hetons. Phys Fluids A5(8):2058–2060
920. Valcke S, Verron J (1996) Cyclone-anticyclone asymmetry in the merging process. Dyn Atmos Oceans 24(1–4):227–236
921. Valcke S, Verron J (1997) Interactions of baroclinic isolated vortices: the dominant effect of shielding. J Phys Oceanogr 27(4):524–541
922. Vallis GK (2006) Atmospheric and oceanic fluid dynamics: fundamentals and large-scale circulation. Cambridge University Press, Cambridge, 745 pp
923. Vandermeirsch FO, Carton XJ, Morel YG (2003) Interaction between an eddy and a zonal jet. Part I. One-and-a-half-layer model. Dyn Atmos Oceans 36(4):247–270
924. Vandermeirsch FO, Carton XJ, Morel YG (2003) Interaction between an eddy and a zonal jet. Part II. Two-and-a-half-layer model. Dyn Atmos Oceans 36(4):271–296
925. Vandermeirsch F, Morel Y, Sutyrin G (2001) The net convective effect of a vertically sheared current on a coherent vortex. J Phys Oceanogr 31(8):2210–2225
926. Vandermeirsch F, Morel Y, Sutyrin G (2002) Resistance of a coherent vortex to a vertical shear. J Phys Oceanogr 32(11):3089–3100
927. van der Toorn R, Zimmerman JTF (2010) Angular momentum dynamics and the intrinsic drift of monopolar vortices on a rotating sphere. J Math Phys 51(8):083102. doi:10.1063/1.3455315
928. van Geffen JHGM, Davies PA (1999) Interaction of a monopolar vortex with a topographic ridge. Geophys Astrophys Fluid Dyn 90(1–2):1–41
929. van Geffen JHGM, Davies PA (2000) A monopolar vortex encounters a north-south ridge or trough. Fluid Dyn Res 26(3):157–179
930. van Geffen JHGM, Davies PA (2000) A monopolar vortex encounters an isolated topographic feature on a β-plane. Dyn Atmos Oceans 32(1):1–26
931. Van-Dyke M (1982) An album of fluid motion. Parabolic Press, Stanford.
932. van Heijst GJF (1994) Topography effects on vortices in a rotating fluid. Meccanica 29(4):431–451
933. van Heijst GJF, Clercx HJH (2009) Laboratory modeling of geophysical vortices. Annu Rev Fluid Mech 41:143–164
934. van Heijst GJF, Flór JB (1989) Laboratory experiments on dipole structures in a stratified fluids. In: Nihoul JCJ, Jamart BM (eds) Mesoscale/synoptic coherent structures in geophysical turbulence. Elsevier, Amsterdam/Oxford/New York/Tokyo, pp 591–608
935. van Heijst GJF, Kloosterziel RC (1989) Tripolar vortices in a rotating fluid. Nature 338:569–571
936. van Heijst GJF, Kloosterziel RC, Williams CWM (1991) Laboratory experiments on the tripolar vortex in a rotating fluid. J Fluid Mech 225:301–331
937. Velasco Fuentes OU (1994) Propagation and transport properties of dipolar vortices on a γ plane. Phys Fluids 6(10):3341–3352
938. Velasco Fuentes OU (2001) Chaotic advection by two interacting finite-area vortices. Phys Fluids 13(4):901–912
939. Velasco Fuentes OU, van Heijst GJF (1994) Experimental study of dipolar vortices on a topographic β-plane. J Fluid Mech 259:79–106
940. Velasco Fuentes OU, van Heijst GJF, Cremers BE (1995) Chaotic transport by dipolar vortices on the β-plane. J Fluid Mech 291:139–161
941. Velasco Fuentes OU, van Heijst GJF, van Lipzing NPM (1996) Unsteady behaviour of a topography-modulated tripole. J Fluid Mech 307:11–41
942. Venaille A (2012) Bottom-trapped currents as statistical equilibrium states above topographic anomalies. J Fluid Mech 699:500–510
943. Verron J (1986) Topographic eddies in temporally varying oceanic flows. Geophys Astophys Fluid Dyn 35(1–4):257–276

944. Verron J, Hopfinger E (1991) The enigmatic merging conditions of two-layer baroclinic vortices. C R Acad Sci Paris Ser II 313(7):737–742
945. Verron J, Hopfinger E, McWilliams JC (1990) Sensitivity to initial conditions in the merging of two-layer baroclinic vortices. Phys Fluids A2(6):886–889
946. Verron J, Le Provost C (1985) A numerical study of quasigeostrophic flow over isolated topography. J Fluid Mech 154:231–252
947. Verron J, Le Provost C, Holland WR (1987) On the effects of a midocean ridge on the general circulation: numerical simulations with an eddy-resolved ocean model. J Phys Oceanogr 17(3):301–312
948. Verron J, Valcke S (1994) Scale-dependent merging of baroclinic vortices. J Fluid Mech 264:81–106
949. Vilela RD, de Moura APS, Grebory C (2006) Finite-size effects on open chaotic advection. Phys Rev E 73:026302. doi:10.1103/PhysRevE.73.026302
950. Villat H (1930) Leçons sur la théorie des tourbillons. Gauthier-Villars et cie, Paris, 300 pp
951. Visbeck M, Marshall J, Haine T, Spall M (1997) On the specification of eddy transfer coefficients in coarse-resolution ocean circulation models. J Phys Oceanogr 27(3):381–402
952. Visbeck M, Marshall J, Jones H (1996) Dynamics of isolated convective regions in the ocean. J Phys Oceanogr 26(9):1721–1734
953. Visheratin KN, Kalashnik MV (2007) Cyclostrophic adjustment and cooling processes in Ranque-vortex tube. Int J Low-Carbon Tech 2(1):10–19
954. Viúdez A (2010) Vertical splitting of vortices in geophysical dipoles. J Phys Oceanogr 40(9):2170–2179
955. von Hardenberg J, McWilliams JC, Provenzale A, Shchepetkin A, Weiss JB (2000) Vortex merging in quasi-geostrophic flows. J Fluid Mech 412:331–353
956. Voropayev SI (1992) Mushroom-like currents: the laboratory experiments, theory, numerical calculations. In: Barenblatt GI, Seidov DG, Sutyrin GG (eds) Coherent structures and self-organisation of currents in the ocean. Nauka, Moscow, pp 177–189 (in Russian)
957. Voropayev SI, Afanasyev YaD (1992) Two-dimensional vortex-dipole interactions in a stratified fluid. J Fluid Mech 236:665–689
958. Vosbeek PWC, van Heijst GJF, Mogendorff VP (2001) The strain rate in evolution of elliptical vortices in inviscid two-dimensional flows. Phys Fluids 13(12):3699–3708
959. Vranješ J, Marić G, Shukla PK 1999) Tripolar vortices and vortex chains in dusty plasma. Phys Lett A 258(4–6):317–322
960. Vranješ J, Okamoto A, Yoshimura S, Poedts S, Kono M, Tanaka MY (2002) Analytical description of a neutral-induced tripole vortex in a plasma. Phys Rev Lett 89(26). doi:10.1103/PhysRevLett.89.265002
961. Vranješ J, Stenflo L, Shukla PK (2000) Tripolar vortices and vortex chains in a shallow atmosphere. Phys Lett A 267(2–3):184–187
962. Vukovich FM, Waddell E (1991) Interaction of a warm core ring with the western slope in the Gulf of México. J Phys Oceanogr 21(7):1062–1074
963. Waite ML, Smolarkiewicz PK (2008) Instability and breakdown of a vertical vortex pair in a strongly stratified fluid. J Fluid Mech 606:239–273
964. Wåhlin AK (2004) Topographic advection of dense bottom water. J Fluid Mech 510:95–104
965. Walsh D, Pratt LJ (1995) The interaction of a pair of point potential vortices in uniform shear. Dyn Atmos Oceans 22(3):135–160
966. Wan Y-H (1986) The stability of rotating vortex patches. Commun Math Phys 107(1):1–20
967. Wan Y-H, Pulvirenti M (1985) Nonlinear stability of circular vortex patches. Commun Math Phys 99(3):435–450
968. Wang GH, Dewar WK (2003) Meddy-seamount interactions: Implications for the Mediterranean salt tongue. J Phys Oceanogr 33(11):2446–2461
969. Warren BA, Wunsch C (1981) Evolution of physical oceanography. MIT Press, Cambridge, 623 pp
970. Waseda T, Mitsudera H (2002) Chaotic advection of the shallow Kuroshio coastal waters. J Oceanogr 58(5):627–638

971. Waseda T, Mitsudera H, Taguchi B, Yoshikawa Y (2002) On the eddy-Kuroshio interaction: evolution of the mesoscale eddy. J Geophys Res. doi:10.1029/2000JC000756
972. Waugh DW (1992) The efficiency of symmetric vortex merger. Phys Fluids A 4(8):1745–1758
973. Waugh DW (1993) Subtropical stratospheric mixing linked to distubances in the polar vortices. Nature 365(6446):535–537
974. Waugh DW, Keating SR, Chen M-L (2012) Diagnosing ocean stirring: comparison of relative dispersion and finite-time Lyapunov exponents. J Phys Oceanogr 42(35):1173–1185
975. Welander P (1955) Studies of the general development of motion in a two-dimensional, ideal fluid. Tellus 7(2):141–156
976. White AJ, McDonald NR (2004) The motion of a point vortex near large- amplitude topography in a two-layer fluid. J Phys Oceanogr 34(12):2808–2824
977. Whitehead JA (1995) Thermohaline ocean processes and models. Annu Rev Fluid Mech 27:89–113
978. Whitehead JA, Marshall J, Hufford GE (1996) Localized convection in a rotating stratified fluid. J Geophys Res 101(C10):25705–25721
979. Whitehead JA, Wang W (2008) A laboratory model of vertical ocean circulation driven by mixing. J Phys Oceanogr 38(5):1091–1106
980. Widnall SE, Sullivan JP (1973) On the stability of vortex rings. Proc R Soc Lond A 332:335–353
981. Wiggins S (2005) The dynamical systems approach to Lagrangian transport in oceanic flows. Annu Rev Fluid Mech 37:295–328
982. Winant CD, Browand FK (1974) Vortex pairing: the mechanism of turbulent mixing-layer growth at moderate Reynolds number. J Fluid Mech 63(2):237–255
983. Wirth A (2000) The parametrization of baroclinic instability in a simple model. J Mar Res 58(4):571–583
984. Wirth A, Barnier B (2008) Mean circulation and structures of tilted ocean deep convection. J Phys Oceanogr 38(4):803–816
985. Woodgate RA, Aagaard K, Muench RD, Gunn J, Björk G, Rudels B, Roach AT, Schauer U (2001) The Arctic ocean boundary current along the Eurasian slope and the adjacent Lomonosov Ridge: water mass properties, transports and transformations from moored instruments. Deep Sea Res I 48(8):1757–1792
986. Wu HM, Overman EA II, Zabusky NJ (1984) Steady-state solutions of the Euler equations: rotating and translating V-states with limiting cases. I. Numerical algorithms and results. J Comput Phys 53(1):42–71
987. Wu J-Z, Ma H-Y, Zhou M-D (2006) Vorticity and vortex dynamics. Springer, Berlin/Heidelberg 776 pp
988. Wunsch C, Ferrari R (2004) Vertical mixing, energy, and the general circulation of the oceans. Annu Rev Fluid Mech 36:281–314
989. Wüst G (1935) Schichtung und zirculation des Atlantischen Ozean. Die Stratosphäre. Exped. "Meteor" 1925–27. Wiss Ergeb Bd 6(1):106 pp
990. Yang H (1993) Chaotic mixing and transport in wave systems and atmosphere. Int J Bifurc Chaos 3(6):1423–1445
991. Yang H (1996) Chaotic transport and mixing by ocean gyre circulation. In: Adler J, Muller P, Rozovskii B (eds) Stochastic modeling in physical oceanography. Birkhauser, Boston, pp 439–466
992. Yang H (1996) The subtropical/subpolar gyre exchange in the presence of annually migrating wind and meandering jet: water mass exchange. J Phys Oceanogr 26(1):115–139
993. Yang H (1996) Lagrangian modeling of potential vorticity homogenization and the associated front in the Gulf Stream. J Phys Oceanogr 26(11):2480–2496
994. Yang H (1998) The central barrier, asymmetry and random phase in chaotic transport and mixing by Rossby waves in a jet. Int J Bifurc Chaos 8(6):1131–1152
995. Yang H, Liu Z (1994) Chaotic transport in a double gyre ocean. Geophys Res Lett 21(7):545–548

996. Yang H, Liu Z (1997) The three-dimensional chaotic transport and the great ocean barrier. J Phys Oceanogr 27(7):1258–1273
997. Yasuda I (1995) Geostrophic vortex merger and streamer development in the ocean with special reference to the merger of Kuroshio warm core ring. J Phys Oceanogr 25(5):979–996
998. Yasuda I, Flierl GR (1995) Two-dimensional asymmetric vortex merger: contour dynamics experiment. J Oceanogr 51(2):145–170
999. Yasuda I, Flierl GR (1997) Two-dimensional asymmetric vortex merger: merger dynamics and critical merger distance. Dyn Atmos Oceans 26(3):159–181
1000. Yatsuyanagi Y, Hatori T, Kato T (2001) Chaotic reconnection due to fast mixing of vortex-current filaments. Earth Planets Space 53:615–618
1001. Yegorikhin VD, Ivanov YA, Kort VG, Koshlyakov MN, Lukashev YF, Morozov EG, Ovchinnikov IM, Paka VT, Tsybaneva TB, Shadrin IF, Shapovalov SM (1987) An intrathermocline lens of Mediterranean water in the tropical North Atlantic. Oceanology 27(2):121–127
1002. Yemel'yanov MN, Fedorov KN (1985) Structure and transformation of intermediate waters of the Mediterranean Sea and Atlantic Ocean. Oceanology 25(2):155–161
1003. Young WR (1985) Some interactions between small numbers of baroclinic, geostrophic vortices. Geophys Astrophys Fluid Dyn 33(1–4):35–61
1004. Youssef A, Marcus PS (2003) The dynamics of Jovian white ovals from formation to merger. Icarus 162(1):74–93
1005. Yuan G-C, Pratt LJ, Jones CKRT (2002) Barrier destruction and Lagrangian predictability at depth in a meandering jet. Dyn Atmos Oceans 35(1):41–61
1006. Zabusky NJ, Hughes MH, Roberts KV (1979) Contour dynamics for the Euler equations in two dimensions. J Comput Phys 30(1):96–106
1007. Zaslavsky GM (2007) The physics of chaos in Hamiltonian systems, 2nd edn. Imperial College Press, London, 315 pp
1008. Zatsepin AG, Ginzburg AI, Kostyanoy AG, Kremenetskiy VV, Krivosheya VG, Stanichny SV, Poulain P-M (2003) Observation of Black Sea mesoscale eddies and associated horizontal mixing. J Geophys Res 108(C8):3246. doi:10.1029/2002JC001390
1009. Zatsepin AG, Kostyanoy AG (1992) Laboratory study of the instability of baroclinic eddies and fronts. In: Barenblatt GI, Seidov DG, Sutyrin GG (eds) Coherent structures and self-organisation of currents in the ocean. Nauka, Moscow, pp 163–177 (in Russian)
1010. Zavala Sansón L (2000) The effects of topography on rotating barotropic flows. Proefschrift. Technische Universiteit Eindhoven, Eindhoven, 152 pp
1011. Zavala Sansoón L (2002) Vortex-ridge interaction in a rotating fluid. Dyn Atmos Oceans 35(4):299–325
1012. Zavala Sansón L (2010) Solutions of barotropic trapped waves around seamounts. J Fluid Mech 661:32–44
1013. Zavala Sansón L, Aguiar ACB, van Heijst GJF (2012) Horizontal and vertical motions of barotropic vortices over a submarine mountain. J Fluid Mech 695:173–198
1014. Zavala Sansón L, van Heijst GJF (2000) Interaction of barotropic vortices with coastal topographies: laboratory experiments and numerical simulation. J Phys Oceanogr 30(9):2141–2162
1015. Zavala Sansón L, van Heijst GJF (2002) Ekman effects in a rotating flow over bottom topography. J Fluid Mech 471:239–255
1016. Zehnder JA (1993) The influence of large-scale topography on barotropic vortex motion. J Atmos Sci 50(15):2519–2532
1017. Zelenko AA, Resnyansky YuD (2007) Deep convection in the ocean general circulation model: variability on the diurnal, seasonal, and interannual time scales. Oceanology 47(2):191–204
1018. Zeng X, Pielke RA, Eykholt R (1993) Chaos theory and its applications to the atmosphere. Bull Am Meteorol Soc 74(4):631–644
1019. Zenk W, Schultz Tokos K, Boebel O (1992) New observations of meddy movement south of the Tejo Plateau. Geophys Res Lett 12(24):2389–2392

1020. Zhikharev G (1989) Stability of steady circulation regims with a nonzonal mean flow over wary topography in a barotropic model of the open ocean. J Phys Oceanogr 19(3):392–395
1021. Zhikharev GM (1990) On steady and travelling waves over bottom topography in the model of homogeneous flow in a beta-plane channel. Geophys Astrophys Fluid Dyn 54(3–4):159–279
1022. Zhikharev GM (1994) On steady quasi-geostrophic flow component formation over undulating bottom topography: Part I. Low-order and direct numerical simulations of homogeneous flows. Geophys Astrophys Fluid Dyn 74(1–4):99–122
1023. Zhikharev GM (1995) On steady quasi-geostrophic flow component formation over undulating bottom topography: Part II. A two-layer low-order model of the open ocean. Geophys Astrophys Fluid Dyn 80(3–4):145–166
1024. Zhmur VV (2011) Mesoscale ocean eddies. GEOS, Moscow, 290 pp (in Russian)
1025. Zhmur VV, Pankratov KK (1990) Distant interaction for an ensemble of quasigeostrophic eddies. The hamiltonian formulation. Izv Atmos Ocean Phys 26(9):714–720
1026. Zhmur VV, Ryzhov EA, Koshel KV (2011) Ellipsoidal vortex in a nonuniform flow: dynamics and chaotic advection. J Mar Res 69(2–3):435–461
1027. Zhmur VV, Shchepetkin AF (1991) Evolution of an ellipsoidal vortex in a stratified ocean in the f-plane approximation. Izv Atmos Ocean Phys 27(5):337–345
1028. Zhmur VV, Shchepetkin AF (1992) Interaction between two quasigeostrophic baroclinic vortices: tendency to come together and merge. Izv Atmos Ocean Phys 28(5):407–417
1029. Zhurbas VM, Oh IS, Pyzhevich ML (2003) Maps of horizontal diffusivity and Lagrangian scales in the Pacific Ocean obtained from drifter data. Oceanology 43(5):622–631
1030. Ziglin SL (1980) Nonintegrability of the problem on the motion of four point vortices. Sov Math Dokl 21:296–299
1031. Zyryanov VN (1981) A contribution to the theory of Taylor columns in a stratified ocean. Izv Atmos Ocean Phys 17(10):793–800
1032. Zyryanov VN (1985) Theory of steady ocean currents. Gydrometeoizdat, Lenigrad, 248 pp (in Russian)
1033. Zyryanov VN (1986) Meandering flow past bottom relief. Izv Atmos Ocean Phys 22(12):1009–1014
1034. Zyryanov VN (1995) Topographic eddies in sea currents dynamics. WPI RAS, Moscow, 240 pp (in Russian)
1035. Zyryanov VN (2003) Topographic vortices in a stratified ocean. In: Borisov AV, Mamaev IS, Sokolovskiy MA (eds) Fundamental and applied problems of the vortex theory. Institute of Computer Science, Moscow–Izhevsk, pp 623–673 (in Russian)
1036. Zyryanov VN (2006) Topographic eddies in a stratified ocean. Regul Chaotic Dyn 11(4):491–521
1037. Zyryanov VN (2009) Secondary toroidal Taylor vortices above bottom perturbations in a rotating fluid. Dokl Phys 54(7):338–344
1038. Zyryanov VN (2011) Secondary toroidal vortices above seamounts. J Mar Res 69(2–3):463–481

Index

M.A. Sokolovskiy and J. Verron, *Dynamics of Vortex Structures in a Stratified Rotating Fluid*, Atmospheric and Oceanographic Sciences Library 47,
DOI 10.1007/978-3-319-00789-2,

O

P

R

S

T

U

V

MIX
Papier aus verantwortungsvollen Quellen
Paper from responsible sources
FSC® C105338

If you have any concerns about our products,
you can contact us on
ProductSafety@springernature.com

In case Publisher is established outside the EU,
the EU authorized representative is:
Springer Nature Customer Service Center GmbH
Europaplatz 3, 69115 Heidelberg, Germany

Printed by Libri Plureos GmbH
in Hamburg, Germany